COLLOQUIA MATHEMATICA
SOCIETATIS JÁNOS BOLYAI, 10.

INFINITE AND FINITE SETS

to **PAUL ERDŐS**

on his 60th birthday

Vol. III.

Edited by: A. HAJNAL
R. RADO
VERA T. SÓS

NORTH–HOLLAND PUBLISHING COMPANY
AMSTERDAM-LONDON

Budapest, Hungary, 1975

ISBN *North-Holland:* 0 7204 2814 9

Joint edition published by

JÁNOS BOLYAI MATEMATICAL SOCIETY

and

NORTH-HOLLAND PUBLISHING COMPANY

Amsterdam-London

Printed in Hungary

ÁFÉSZ, VÁC

Sokszorosító üzeme

CONTENTS

Volume I

Volume II

Volume III

COLLOQUIA MATHEMATICA SOCIETATIS JÁNOS BOLYAI
10. INFINITE AND FINITE SETS, KESZTHELY (HUNGARY), 1973.

EXTREMAL CONNECTIVITY PROBLEMS

W. MADER

All graphs considered are undirected and finite, neither multiple edges nor loops are admitted. We deal with the following problems: How many edges, dependent on the number of vertices, guarantee the existence of two vertices (or two adjacent vertices) joint by n openly disjoint [resp. edge-disjoint] paths? How many edges guarantee the existence of an n-connected [n-edge-connected] subgraph? Characterize the graphs with maximal number of edges containing no configuration as just mentioned.

In the first place some notations. Let $V(G)$ [resp. $E(G)$] be the set of vertices [resp. edges] of the graph G and let $|G|$ [resp. $e(G)$] denote the number of vertices [resp. edges] of G. For the edge between the vertices x and y we write $[x, y]$. Furthermore, $d(x, G) := |\{y \in V(G) \mid [x, y] \in E(G)\}|$ and $V_n(G) := \{x \in V(G) \mid d(x, G) < n\}$. We define $\mu(x, y;\ G)$ [resp. $\lambda(x, y;\ G)$] to be the maximal number of openly disjoint [edge-disjoint] paths between the vertices x and y in the graph G.

Let G be a graph, all vertices of which have degree $n-1$ except just one vertex, which is adjacent to all other vertices. Then we have

$e(G) = \frac{n}{2}(|G| - 1)$ and obviously there are no two vertices x and y with $\lambda(x, y; G) \geqslant n$. At the Conference on Graph Theory at Oberwolfach (Germany) in 1972 B. Bollobás posed the question, if every graph G with $e(G) > \frac{n}{2}(|G| - 1)$ contains two vertices x and y with $\lambda(x, y; G) \geqslant n$. For $n = 4$ this is a consequence of the results in [1], for $n = 5$ it was proved in [2].* I could settle this conjecture for all n by proving the following generalization.

Theorem [3]. *Every graph G with $e(G) > \frac{n}{2}(|G| - 1) - \frac{1}{2} \sum_{x \in V_n(G)} (n - 1 - d(x, G))$ and $|G| \geqslant n$ contains two vertices x and y with $\lambda(x, y; G) \geqslant n$.*

Whereas after [1] for $n = 4$ every 2-connected graph G with $e(G) = \frac{n}{2}(|G| - 1)$ and $\lambda(x, y; G) < n$ for all $x \neq y$ is of the type given above (namely a wheel), we have not succeeded in characterizing the extremal graphs for $n \geqslant 5$. But in an extremal graph G (with $e(G) = \frac{n}{2}(|G| - 1)$) every least separating set of edges T has a simple structure: It is $|T| = n - 1$ and there is an edge $[t, s] \in T$, which is adjacent to each edge of T. By splitting G "along this edge $[t, s]$" we get two extremal graphs again.

The corresponding problem for openly disjoint paths is solved only for $n \leqslant 4$ and for these n the solution is the same as for edge-disjoint paths [1]. For every $n \geqslant 5$ and every c, however, it is possible to construct an infinite set of graphs G with $e(G) \geqslant \frac{n}{2}|G| + c$ and $\mu(x, y; G) < n$ for all $x \neq y$. But by prohibiting "small" circuits we get the following

Theorem [3]. *Every graph G of girth greater than $n + 1$ with $e(G) > \frac{n}{2}(|G| - (n - 1))$ and $|G| \geqslant n + 1$ contains two vertices x and y with $\mu(x, y; G) \geqslant n$ (for $n \geqslant 4$).*

*For $n = 6$ it was also proved by J.L. Leonard in "Graphs with 6-ways", *Can. J. Math.*, 25 (1973), 687-692.

If we ask for the maximal number of edges in a graph G containing no *adjacent* vertices x and y with $\mu(x, y;\ G) > n$ [resp. $\lambda(x, y;\ G) > n$], both the problems (for openly disjoint and for edge-disjoint paths) concide and surprisingly have a solution simple to state. From the

Theorem [7] *Every finite graph G with $E(G) \neq \phi$ contains two adjacent vertices x and y with* $\mu(x, y;\ G) = \min\{d(x, G), d(y, G)\}$;

we easily deduce

Theorem [5]. *Every graph G with $e(G) > n|G| - \binom{n+1}{2}$ and $|G| \geqslant n$ contains two adjacent vertices x and y with $\mu(x, y;\ G) > n$.*

This result is not sharp for graphs with a number of vertices greater than $n + 1 \geqslant 3$. But we can show that every graph G with sufficiently large number of vertices and $e(G) > n|G| - n^2 = e(K_{n,|G|-n})$ contains two adjacent vertices x and y with $\mu(x, y;\ G) > n$ and that the extremal graphs of this problem are the complete bipartite graphs $K_{n,n'}$.

Theorem [5]. *For each $n \geqslant 2$ there is an integer $m(n)$ with the property that every graph $G \neq K_{n,|G|-n}$ with $e(G) \geqslant n|G| - n^2$ and $|G| \geqslant m(n)$ contains two adjacent vertices x and y with $\mu(x, y;\ G) > n$.*

Let $f(n)$ denote the least possible integer $m(n)$. We are able to show $3n \leqslant f(n) \leqslant \binom{n}{2} + 2n + 1$. I should suppose $f(n) = 3n$, but I can only prove that every graph $G \neq K_{n,|G|-n}$ with $e(G) \geqslant n|G| - n^2$ and $|G| \geqslant 3n$ contains two adjacent vertices x and y with $\lambda(x, y;\ G) > n$.

Let us now consider the question, how many edges a graph without an n-connected (n-edge-connected) subgraph can contain. It is easy to give an answer in case of edge-connectivity.

Theorem [4]. *Every graph G with $e(G) > (n-1)|G| - \binom{n}{2}$ and $|G| \geqslant n$ has an n-edge-connected subgraph.*

This result is best possible for all n and every number of vertices. For (vertex-) connectivity we can give only a complete solution for small

n and inequalities in general case. The following theorem is immediately proved by induction.

Theorem [6]. *Every graph* G *with* $e(G) > (2n-3)(|G|-(n-1))$ *and* $|G| \geqslant 2n-1$ *contains an* n*-connected subgraph.*

From this we get the existence of a number $g(n)$ with the following property: There is an infinite set of graphs G without an n-connected subgraph, for which the equality $e(G) = g(n)(|G|-(n-1))$ holds, but every graph H with a sufficiently large number of vertices and $e(H) > g(n)(|H|-(n-1))$ has an n-connected subgraph. We have obtained the following inequalities.

Theorem [6]. $\frac{3}{2}n - 2 \leqslant g(n) < (n-1)\left(1 + \frac{1}{\sqrt{2}}\right)$ *for all* $n \geqslant 2$.

For all n with $2 \leqslant n \leqslant 7$ we have proved $g(n) = \frac{3}{2}n - 2$ and for $n \leqslant 6$ we can characterize the graphs G with $e(G) = g(n)(|G|-(n-1))$ containing no n-connected subgraph. But I don't know, if $g(n) > \frac{3}{2}n - 2$ for any $n \geqslant 2$.

REFERENCES

[1] B. Bollobás, On graphs with at most three independent paths connecting any two vertices. *Studia Sci. Math. Hungar.*, 1 (1966), 137-140.

[2] J.L. Leonard, On graphs with at most four line-disjoint paths connecting any two vertices. *J. Combinat. Theory, Ser. B,* 13 (1972), 242-250.

[3] W. Mader, Ein Extremalproblem des Zusammenhangs von Graphen. *Math. Z.*, 131 (1973), 223-231.

[4] W. Mader, Minimale n-fach kantenzusammenhängende Graphen. *Math. Ann.*, 191 (1971), 21-28.

[5] W. Mader, Existenz gewisser Konfigurationen in *n*-gesättigten Graphen und in Graphen genügend grosser Kantendichte. *Math. Ann.*, 194 (1971, 295-312.

[6] W. Mader, Existenz *n*-fach zusammenhängender Teilgraphen in Graphen genügend grosser Kantendichte. *Abh. Math. Sem. Univ. Hamburg,* 37 (1972), 86-97.

[7] W. Mader, Grad und lokaler Zusammenhang in endlichen Graphen. *Math. Ann.*, 205 (1973), 9-11.

COLLOQUIA MATHEMATICA SOCIETATIS JÁNOS BOLYAI
10. INFINITE AND FINITE SETS, KESZTHELY (HUNGARY), 1973.

A REMARK ON RARE FILTERS

A.R.D. MATHIAS

Definition. A filter F on ω is *rare* if it contains all cofinite subsets of ω and for any function π from ω onto ω such that $\pi^{-1}\{i\}$ is finite for each $i \in \omega$, F contains a subset A of ω on which π is $1-1$.

Theorem. *No* Σ_1^1 *subset of* 2^ω *can be a rare filter.*

First, notation. A, B, C will denote infinite subsets of ω, and $\langle a_n \mid n < \omega \rangle$, $\langle b_n \mid n < \omega \rangle$, $\langle c_n \mid n < \omega \rangle$ the enumerations of their elements in increasing order. F will always denote a filter on ω containing the Fréchet filter of all cofinite sets.

A family P of infinite subsets of ω is called a *Scott family* if $\forall A\, \exists B \subseteq A (A \in P \Leftrightarrow B \notin P)$; if, in other words, every infinite subset of ω has an infinite subset in P and an infinite subset not in P. It is a theorem of Silver [4] that no Σ_1^1 subset of 2^ω can be a Scott family; and in [2] it was shown that in Solovay's model in which all sets of reals are Lebesgue measurable, there is no Scott family; or, equivalently, that the partition relation $\omega \to (\omega)^\omega$ holds. Proofs of Silver's theorem that eschew forcing are known: one due to the author, uses Ramsey ultrafilters and another, recently given by Ellentuck [5], proceeds by reduction to a classical result of analytic topology.

The proof presented at Keszthely of the theorem, which was announced in [3, page 209], used forcing and, though not without its charms, was long. The argument below is inspired by a recent letter of Baumgartner for which the author here records his gratitude.

Given $A = \{a_n \mid n < \omega\}$, define $O(A) = \{m \mid m \leqslant a_0\} \cup \cup \{m \mid \exists n(a_{2n+1} < m \leqslant a_{2n+2})\}$ and $E(A) = \{m \mid \exists n(a_{2n} < m \leqslant a_{2n+1})\}$. Then $E(A)$ is the complement of $O(A)$ in ω. Now given F define $P^{(F)} = \{A \mid O(A) \in F\}$ and $Q^{(F)} = \{A \mid E(A) \in F\}$. Then $P^{(F)}$ and $Q^{(F)}$ are disjoint; further if $A \in P^{(F)}$ and $n \in A$, then $A - \{n\} \in Q^{(F)}$, so that every infinite A has an infinite subset not in $P^{(F)}$.

Lemma. *If F is rare then $P^{(F)}$ is a Scott family.*

The theorem is an immediate consequence of the lemma and Silver's theorem, for if F is Σ^1_1 so is $P^{(F)}$; similarly the lemma yields a new proof of the author's result that there is no rare filter in Solovay's model, which was originally established using forcing and some absoluteness arguments from [1]. More generally, an adequate class in the sense of [1] which contains no Scott family contains no rare filter.

To prove the lemma we have to show that if F is rare, then every infinite subset A of ω has an infinite subset in $P^{(F)}$. Define $\pi\colon \omega \to \omega$ by $\pi(i) =$ the least n with $i \leqslant a_{2n}$. As F is rare it contains a B on which F is $1-1$. Then $\{n \mid \neg\exists m(m \in B$ and $a_n < m \leqslant a_{n+1})\}$ is infinite, so there is an infinite subset C of A such that $E(C) \cap B = O$, whence $O(C) \in F$ and $C \in P^{(F)}$ as required.

In fact F's having a much weaker property than rarity is sufficient for $P^{(F)}$ to be a Scott family, as examination of the argument will show: call F *feeble* if there is a weakly monotonic function $\psi\colon \omega \to \omega$ such that $\{B \mid \psi^{-1}``B \in F\}$ is the Fréchet filter. Then $P^{(F)}$ is a Scott family when and only when F is not feeble; no ultrafilter is feeble; and hence the

Proposition. *If* $\omega \to (\omega)^\omega$ *then every filter containing the cofinite sets is feeble and every ultrafilter principal.*

REFERENCES

[1] A.S. Kechris – Y.N. Moschovakis, Notes on the theory of scales.

[2] A.R.D. Mathias, On a generalization of Ramsey's theorem, *Fellowship dissertation,* Peterhouse, Cambridge, 1969.

[3] A.R.D. Mathias, Solution of problems of Choquet and Puritz, Conference in Mathematical Logic, London 1970, *Lecture Notes in Mathematics*, No. 255, Springer-Verlag.

[4] J. Silver, Every analytic set is Ramsey, *The Journal of Symbolic Logic,* 35 (1970), 60-64.

[5] E. Ellentuck, A new proof that analytic sets are Ramsey, *The Journal of Symbolic Logic,* 39 (1974), 163-165.

COLLOQUIA MATHEMATICA SOCIETATIS JÁNOS BOLYAI
10. INFINITE AND FINITE SETS, KESZTHELY (HUNGARY), 1973.

ON THE GOOD COLORINGS, II: CASE OF THE COMPLETE h-PARTITE HYPERGRAPHS

J.-C. MEYER

1. DEFINITIONS AND NOTATIONS

Let $h, n_1, n_2, \ldots, n_h$ be positive integers. We shall denote by $K^h_{n_1, n_2, \ldots, n_h}$ a hypergraph whose vertex set is the union of h disjoint sets $X_1, X_2, \ldots, X_h$, with $|X_i| = n_i > 0$, $1 \leqslant i \leqslant h$, and whose edges are all subsets $\{x_1, x_2, \ldots, x_h\}$ such that $(x_1, x_2, \ldots, x_h) \in$ $\in X_1 \times X_2 \times \ldots \times X_h$. This hypergraph is also called the complete h-partite hypergraph on $\{X_1, X_2, \ldots, X_h\}$. For $h = 2$, it reduces to the complete bipartite graph K_{n_1, n_2}.

In the following, we shall put $X_i = \{x_i^0, x_i^1, \ldots, x_i^{n_i - 1}\}$ for $1 \leqslant i \leqslant h$; we shall assume that $n_1 \leqslant n_2 \leqslant \ldots \leqslant n_h$ and we put $q_i =$ $= \frac{1}{n_i} \prod_{j=1}^{h} n_j$ for $1 \leqslant i \leqslant h$. Thus we have $q_1 \geqslant q_2 \geqslant \ldots \geqslant q_h$.

Some coloring numbers for the complete h-partite hypergraphs have been obtained in [2] and some results and unsolved problems concerning the cliques of these hypergraphs can be found in [3].

Let $H = (E_i \mid i \in I)$ be a hypergraph with $X = \bigcup_{i \in I} E_i$ as set of vertices. Let $S_1, S_2, \ldots, S_p$ be a partition of X with cardinality p. Let us denote by $p(i)$ the number of indices j such that $E_i \cap S_j \neq \phi$. Then $p(i) \leqslant \min\{|E_i|, p\}$. We say that $(S_1, S_2, \ldots, S_p)$ is *a good vertex p-coloring in H*, if $p(i) = \min\{|E_i|, p\}$ for all $i \in I$.

A good edge p-coloring of H is a good vertex p-coloring of the dual H^* of H.

If $H = K^h_{n_1, n_2, \ldots, n_h}$, we denote by $x_1^{i_1} x_2^{i_2} \ldots x_h^{i_h}$ the vertex of H^* associated with the edge $\{x_1^{i_1}, x_2^{i_2}, \ldots, x_h^{i_h}\}$ of H. Thus, the edges of H^* are the subsets

$$E_l^\alpha = \{x_1^{i_1} x_2^{i_2} \ldots x_l^\alpha \ldots x_h^{i_h} \mid 0 \leqslant i_j < n_j,\ 1 \leqslant j \leqslant h,\ j \neq l\} .$$

Clearly, this edge E_l^α has cardinality q_l.

If a and b are integers, $b > 0$, we shall denote by $(a)_b$ the integer that is congruent with a modulo b and satisfies $0 \leqslant (a)_b < b$.

We shall use the lexicographic ordering on the k-tuples of non-negative integers: given $(\alpha_1, \alpha_2, \ldots, \alpha_k)$ and $(\beta_1, \beta_2, \ldots, \beta_k)$, two k-tuples of non-negative integers, we shall write $(\alpha_1, \alpha_2, \ldots, \alpha_k) < (\beta_1, \beta_2, \ldots, \beta_k)$, if there is an index j, $1 \leqslant j \leqslant k$, such that $\alpha_i = \beta_i$ for $1 \leqslant i < j$ and $\alpha_j < \beta_j$.

The main result of this paper is the following theorem: For each integer p such that $1 \leqslant p \leqslant \prod_{j=1}^{h} n_j$, the hypergraph $K^h_{n_1, n_2, \ldots, n_h}$ has a good edge p-coloring.

2. THE MAIN RESULT

Lemma 1. *Let k and l be two integers such that $1 \leqslant k < l \leqslant h$. For $1 \leqslant j \leqslant h$, let i_j and i'_j be integers such that $0 \leqslant i_j < n_j$ and $0 \leqslant i'_j < n_j$ and such that $i_l = i'_l$. For $1 \leqslant i \leqslant h$ and $i \neq k$, let m_i be integers such that $m_i \geqslant n_i$ for $i \neq k + 1$ and $n_k \leqslant m_{k+1} \leqslant n_{k+1}$. Then the equation system*

$$\begin{cases} i_{j-1} + i_j \equiv i'_{j-1} + i'_j \pmod{m_{j-1}} & \text{for} \quad 2 \leqslant j \leqslant k & (*) \\ i_{j-1} + i_j \equiv i'_{j-1} + i'_j \pmod{m_j} & \text{for} \quad k < j \leqslant h & (**) \end{cases}$$

yields that $i_j = i'_j$ *for* $1 \leqslant j \leqslant h$.

Proof. It suffices to remark that if r and n are two integers with $r \geqslant n > 0$, the conditions $x \equiv y \pmod r$, $0 \leqslant x < n$, $0 \leqslant y < n$ imply $x = y$. Starting with $i_l = i'_l$ we successively see that $i_j = i'_j$ for all $j < l$, and the same applies for all $j > l$.

Lemma 2. *Let* k *and* l *be two integers with* $1 \leqslant l \leqslant k \leqslant h$. *For* $1 \leqslant i \leqslant h$, $i \neq k$, *let* m_i *be integers such that* $0 < m_i \leqslant n_i$. *For* $2 \leqslant j \leqslant h$, *let* α_j *be arbitrary integers. Let* α *be an integer with* $0 \leqslant \alpha < n_l$. *then the equation system*

$$\begin{cases} i_{j-1} + i_j \equiv \alpha_j \pmod{m_{j-1}} & \text{for} \quad 2 \leqslant j \leqslant k & (***) \\ i_{j-1} + i_j \equiv \alpha_j \pmod{m_j} & \text{for} \quad k < j \leqslant h & (****) \\ i_l = \alpha\,, \end{cases}$$

has always a solution $(i_1, i_2, \ldots, i_h)$ *satisfying* $0 \leqslant i_j < n_j$ *for* $1 \leqslant j \leqslant h$.

Proof. It suffices to remark that if β, γ, r, n are integers with $n \geqslant r > 0$, the equation

$$x + \beta \equiv \gamma \pmod r$$

has always a solution x_0 such that $0 \leqslant x_0 < r$. This solution satisfies also $0 \leqslant x_0 < n$, since $n \geqslant r$. This fact permits to successively determine the integers i_j for $j < l$, and similarly for $j > l$.

Proposition 1. *Let* k *and* m *be integers such that* $1 \leqslant k < h$ *and*

*These equations take place in the system only if $k \geqslant 2$.

**These equations take place in the system only if $k < h$.

***These equations take place in the system only if $k \geqslant 2$.

****These equations take place in the system only if $k < h$.

$n_k \leqslant m \leqslant n_{k+1}$. *For* $p = \dfrac{mq_k}{n_{k+1}} = \dfrac{mq_{k+1}}{n_k}$, *the hypergraph* $K^h_{n_1, n_2, \ldots, n_h}$ *has a good edge* p*-coloring, the color of edge* $x_1^{i_1} x_2^{i_2} \ldots x_h^{i_h}$ *being given by the* $(h-1)$*-tuple* $(\alpha_2, \alpha_3, \ldots, \alpha_h)$, *where*

$$\alpha_j = \begin{cases} (i_{j-1} + i_j)_{n_{j-1}} & \text{if} \quad 2 \leqslant j \leqslant k \\ (i_k + i_{k+1})_m & \text{if} \ j = k+1 \\ (i_{j-1} + i_j)_{n_j} & \text{if} \quad k+1 < j \leqslant h \,. \end{cases}$$

(The case $2 \leqslant j \leqslant k$ occurs only if $k \geqslant 2$; the case $k+1 < j \leqslant h$ occurs only if $k < h-1$).

Proof. In the above coloring, there are at most $p = \dfrac{mq_k}{n_{k+1}}$ different colors. It suffices to show that in $(K^h_{n_1, n_2, \ldots, n_h})^*$ the number of colors which occur in an edge $E_l^\alpha = \{x_1^{i_1} x_2^{i_2} \ldots x_l^\alpha \ldots x_h^{i_h} \mid 0 \leqslant i_j < n_j,\ 1 \leqslant j \leqslant h,\ j \neq l\}$ is equal to

$$\begin{aligned} &q_l \quad \text{if} \quad l > k\,, \\ &p = \frac{mq_k}{n_{k+1}} \quad \text{if} \quad l \leqslant k\,. \end{aligned}$$

For the case $l > k$, since the edge E_l^α has cardinality $q_l \leqslant p$, it suffices to show that two distinct vertices $x_1^{i_1} x_2^{i_2} \ldots x_l^\alpha \ldots x_h^{i_h}$ and $x_1^{i'_1} x_2^{i'_2} \ldots \ldots x_l^\alpha \ldots x_h^{i'_h}$ are of different colors; or, equivalently, that the equation system:

$$\begin{cases} i_{j-1} + i_j \equiv i'_{j-1} + i'_j \pmod{m_{j-1}} & \text{for} \quad 2 \leqslant j \leqslant k \\ i_{j-1} + i_j \equiv i'_{j-1} + i'_j \pmod{m_j} & \text{for} \quad k < j \leqslant h\,, \end{cases}$$

where $m_j = n_j$ for $1 \leqslant j \leqslant h$, $j \neq k$, $j \neq k+1$, and $m_{k+1} = m$, implies that $i_j = i'_j$ for any $1 \leqslant j \leqslant h$. This follows immediately from Lemma 1.

For the case $l \leqslant k$, we have to show that if the α_j, for $2 \leqslant j \leqslant h$, are integers such that $0 \leqslant \alpha_j < n_{j-1}$ if $2 \leqslant j \leqslant k$, $0 \leqslant \alpha_j \leqslant n_j$ if $k+1 < j \leqslant h$, and $0 \leqslant \alpha_{k+1} < m$, then there is a vertex $x_1^{i_1} x_2^{i_2} \ldots \ldots x_l^{\alpha} \ldots x_h^{i_h}$ in E_l^{α} with color $(\alpha_2, \alpha_3, \ldots, \alpha_h)$; or, in other words, that the equation system

$$\begin{cases} i_{j-1} + i_j \equiv \alpha_j \pmod{m_{j-1}} & \text{for} \quad 2 \leqslant j \leqslant k \\ i_{j-1} + i_j \equiv \alpha_j \pmod{m_j} & \text{for} \quad k < j \leqslant h \\ i_l = \alpha\,, \end{cases}$$

where $m_j = n_j$ for $1 \leqslant j \leqslant h$, $j \neq k$, $j \neq k+1$, and $m_{k+1} = m$, has at least one solution $i_1, i_2, \ldots, i_h$ such that $0 \leqslant i_j < n_j$ for all j, $1 \leqslant j \leqslant h$. This follows immediately from Lemma 2.

Corollary. *For every* $1 \leqslant j \leqslant h$, *the* $K^h_{n_1, n_2, \ldots, n_h}$ *hypergraph has a good edge* q_j*-coloring.*

Proof. If $1 \leqslant j < h$, it suffices to set $k = j$ and $m = n_{k+1} = n_{j+1}$ in Proposition 1, since, then, we have

$$\frac{mq_k}{n_{k+1}} = q_k = q_j\,.$$

If $j = h$, it suffices to set $k = h-1$ and $m = n_k = n_{h-1}$ in Proposition 1, since, then, we have

$$\frac{mq_{k+1}}{n_k} = q_{k+1} = q_h\,.$$

Remark. In [2], C. Berge has proved that the chromatic index of $K^h_{n_1, n_2, \ldots, n_h}$ is equal to q_1, and therefore that this hypergraph has a good edge q_1-coloring.

Proposition 2. *For every integer* p *such that* $q_1 < p \leqslant \prod_{j=1}^{h} n_j$, *the hypergraph* $K^h_{n_1, n_2, \ldots, n_h}$ *has a good edge* p*-coloring.*

Proof. By the corollary to Proposition 1, $K^h_{n_1,n_2,\ldots,n_h}$ possesses a good edge q_1-coloring and every edge E in $(K^h_{n_1,n_2,\ldots,n_h})^*$ satisfies $|E| \leqslant q_1$. Thus, Proposition 2 follows from the more general result:

If $H = (X, \mathscr{E})$ is a hypergraph having a good vertex p-coloring, with $p \geqslant \max_{E\in\mathscr{E}} |E|$, $p < |X|$, then H possesses a good vertex $(p+1)$-coloring.

Let the good vertex p-coloring be given by a partition $(S_1, S_2, \ldots, S_p)$ of X. Since $p < |X|$, at least one of the sets S_i has at least two elements, and this set can therefore be divided in two non empty subsets. This yields a new partition of X, with cardinality $p+1$, which is a good vertex $(p+1)$-coloring of H.

Proposition 3. *For every integer* p *such that* $1 \leqslant p < q_h = \prod_{j=1}^{h-1} n_j$, *the hypergraph* $K^h_{n_1,n_2,\ldots,n_h}$ *has a good edge* p*-coloring.*

Proof. According to the corollary to Proposition 1, $K^h_{n_1,n_2,\ldots,n_h}$ possesses a good edge q_h-coloring and every edge E in $(K^h_{n_1,n_2,\ldots,n_h})^*$ satisfies $|E| \geqslant q_h$. Thus, Proposition 3 follows from the more general result:

If $H = (X, \mathscr{E})$ is a hypergraph having a good vertex p-coloring with $1 < p \leqslant \min_{E\in\mathscr{E}} |E|$, then H possesses also a good vertex q-coloring for $1 \leqslant q < p$. Let the good vertex p-coloring in H be given by the partition $(S_1, S_2, \ldots, S_p)$ of X. Since $p \leqslant \min_{E\in\mathscr{E}} |E|$, we have $E \cap S_j \neq \phi$ for every edge $E \in \mathscr{E}$ and every j, $1 \leqslant j \leqslant p$. Let us define $S'_j = S_j$ for $1 \leqslant j \leqslant q-1$ and $S'_q = \bigcup_{j=q}^{p} S_j$. Then $(S'_1, S'_2, \ldots, S'_q)$ is a partition of X and we have $E \cap S'_j \neq \phi$ for every edge $E \in \mathscr{E}$ and every j, $1 \leqslant j \leqslant q$. Since $1 \leqslant q \leqslant \min_{E\in\mathscr{E}} |E|$, the partition $(S'_1, S'_2, \ldots, S'_q)$ is a good vertex q-coloring of H.

Lemma 3. *Let* $m_1, m_2, \ldots, m_l$ *be positive integers. For* $1 \leqslant j \leqslant l$, *let us put* $d_j = \prod_{i=j}^{l} m_i$ *and* $d_{l+1} = 1$.

Every integer $p \geqslant 0$ *can be written as*

$$p = \sum_{j=1}^{l+1} a_j d_j ,$$

where the a_j *are integers,* $a_1 \geqslant 0$, *and* $0 \leqslant a_j < m_{j-1}$ *for* $2 \leqslant j \leqslant l+1$. *Furthermore, this decomposition is unique.*

Proof. We shall define the integers $a_1, a_2, \ldots, a_{l+1}$ as follows: a_1 is the quotient of p by d_1; for $1 \leqslant k \leqslant l$, a_{k+1} is the quotient of $p - \sum_{j=1}^{k} a_j d_j$ by d_{k+1}.

Since $d_{l+1} = 1$, the last division is exact, and we have

$$p - \sum_{j=1}^{l} a_j d_j = a_{l+1} d_{l+1}, \quad \text{i.e.} \quad p = \sum_{j=1}^{l+1} a_j d_j .$$

Since $d_j = m_j d_{j+1}$ for $1 \leqslant j \leqslant l$, we can successively check that $0 \leqslant a_k < m_{k-1}$ for $2 \leqslant k \leqslant l+1$.

Furthermore, we see that if $\alpha_1, \alpha_2, \ldots, \alpha_{l+1}$ are integers satisfying $\alpha_1 \geqslant 0$ and $0 \leqslant \alpha_j < m_{j-1}$ for $2 \leqslant j \leqslant l+1$, we have, for $2 \leqslant k \leqslant l+1$,

$$\sum_{j=k}^{l+1} \alpha_j d_j < d_{k-1} .$$

Thus, if the $(l+1)$-tuples $(\alpha_1, \alpha_2, \ldots, \alpha_{l+1})$ of integers satisfying $\alpha_1 \geqslant 0$ and $0 \leqslant \alpha_j < m_{j-1}$ for $2 \leqslant j \leqslant l+1$, are lexicographically ordered, the function

$$(\alpha_1, \alpha_2, \ldots, \alpha_{l+1}) \rightarrow \sum_{j=1}^{l+1} \alpha_j d_j$$

is strictly increasing. This shows the unicity of the decomposition.

Remark. We can see that, if $p = \sum_{j=1}^{l+1} a_j d_j$, the above defined function is a bijection from the set of $(l+1)$-tuples $(\alpha_1, \alpha_2, \ldots, \alpha_{l+1})$, such that $(\alpha_1, \alpha_2, \ldots, \alpha_{i+1}) < (a_1, a_2, \ldots, a_{i+1})$ on the set $\{0, 1, \ldots$ $\ldots, p-1\}$.

Theorem. *The hypergraph* $K^h_{n_1,n_2,\dots,n_h}$ *has a good edge p-coloring for every integer p such that* $1 \leqslant p \leqslant \prod_{j=1}^{h} n_j$.

Proof. By Proposition 2, Proposition 3 and the corollary to Proposition 1, the theorem is proved when $q_1 < p \leqslant \prod_{j=1}^{h} n_j$, or when $1 \leqslant p < q_n$, or when $p = q_k$ for some k. The only case which remains to be proved is when there is an integer k with $1 \leqslant k \leqslant h-1$ such that $q_{k+1} < < p < q_k$.

For $1 \leqslant j \leqslant h+1$, $j \neq k$, $j \neq k+1$, put

$$r_j = \begin{cases} \dfrac{1}{n_k n_{k+1}} \prod_{i=j}^{h} n_i & \text{if} \quad 1 \leqslant j < k \\ \prod_{i=j}^{h} n_i & \text{if} \quad k+1 < j \leqslant h \\ 1 & \text{if} \quad j = h+1 \,. \end{cases}$$

We shall consider separately the case $k > 1$ and the case $k = 1$.

Case $k > 1$. Thus, we have $h \geqslant k+1 \geqslant 3$.

By Lemma 3 with $l = h-2$, $m_i = n_i$ for $1 \leqslant i \leqslant k-1$, and $m_i = n_{i+2}$ for $k \leqslant i \leqslant h-2$, there exist integers b_j, $2 \leqslant j \leqslant h+1$, $j \neq k$, $j \neq k+1$, satisfying $0 \leqslant b_j < n_{j-1}$ for $j \neq k+2$ and $0 \leqslant b_{k+2} < n_{k-1}$, and a non-negative integer b, such that

$$p = br_1 + \sum_{\substack{2 \leqslant j \leqslant h+1 \\ j \neq k \\ j \neq k+1}} b_j r_j \,.$$

Since $q_{k+1} < p < q_k$, and b is the quotient of p by r_1, we have $n_k \leqslant b < n_{k+1}$.

Let us color the vertices $x_1^{i_1}, x_2^{i_2} \dots x_h^{i_h}$ of $(K^h_{n_1,n_2,\dots,n_h})^*$ as follows:

If

$$(1) \qquad \begin{aligned} &((i_1 + i_2)_{n_1}, (i_2 + i_3)_{n_2}, \ldots \\ &\ldots, (i_{k-1} + i_k)_{n_{k-1}}, (i_{k+1} + i_{k+2})_{n_{k+2}}, \ldots \\ &\ldots, (i_{h-1} + i_h)_{n_h}) < (b_2, b_3, \ldots, b_{k+2}, b_{k+3}, \ldots, b_{h+1}), \end{aligned}$$

we shall color vertex $x_1^{i_1} x_2^{i_2} \ldots x_h^{i_h}$ with color

$$\begin{aligned} &((i_1 + i_2)_{n_1}, (i_2 + i_3)_{n_2}, \ldots \\ &\ldots, (i_{k-1} + i_k)_{n_{k-1}}, (i_k + i_{k+1})_{b+1}, (i_{k+1} + i_{k+2})_{n_{k+2}}, \ldots \\ &\ldots, (i_{h-1} + i_h)_{n_h}). \end{aligned}$$

If $(i_1, i_2, \ldots, i_h)$ does not satisfy (1) we shall color vertex $x_1^{i_1} x_2^{i_2} \ldots x_h^{i_h}$ with color

$$\begin{aligned} &((i_1 + i_2)_{n_1}, (i_2 + i_3)_{n_2}, \ldots \\ &\ldots, (i_{k-1} + i_k)_{n_{k-1}}, (i_k + i_{k+1})_b, (i_{k+1} + i_{k+2})_{n_{k+2}}, \ldots \\ &\ldots, (i_{h-1} + i_h)_{n_h}). \end{aligned}$$

To show that this is a good edge p-coloring of $K^h_{n_1, n_2, \ldots, n_h}$, we shall first show that there is at most p different colors. Clearly, color $(\alpha_2, \alpha_3, \ldots, \alpha_h)$ satisfies

$$(2) \qquad \begin{cases} 0 \leqslant \alpha_i < n_{i-1} \quad \text{for} \quad 2 \leqslant i \leqslant k, \\ 0 \leqslant \alpha_i < n_i \quad \text{for} \quad k + 1 < i \leqslant h, \ \alpha_k \geqslant 0, \quad \text{and} \\ (\alpha_{k+1}, \alpha_2, \ldots, \alpha_{k-1}, \alpha_k, \alpha_{k+2}, \ldots, \alpha_h) < \\ < (b, b_2, \ldots, b_{k-1}, b_{k+2}, b_{k+3}, \ldots, b_{h+1}). \end{cases}$$

By the remark following the proof of Lemma 3, the number of $(h-1)$-tuples satisfying (2) is equal to

$$br_1 + \sum_{\substack{2 \leq j \leq h+1 \\ j \neq k \\ j \neq k+1}} b_j r_j = p \, .$$

It suffices now to show that an edge

$$E_l^\alpha = \{x_1^{i_1} x_2^{i_2} \ldots x_l^\alpha \ldots x_h^{i_h} \mid 0 \leq i_j < n_j, \ 1 \leq j \leq h, \ j \neq l\} \, ,$$

meets

(i) q_l colors if $l > k$, or

(ii) p colors if $l \leq k$.

(i) Assume $l > k$. Since edge E_l^α has cardinality q_l, it suffices to show that in E_l^α two distinct vertices, viz. $x_1^{i_1} x_2^{i_2} \ldots x_l^\alpha \ldots x_h^{i_h}$ and $x_1^{i'_1} x_2^{i'_2} \ldots x_l^\alpha \ldots x_h^{i'_h}$ have different colors. Put $i_l = i'_l = \alpha$.

If $(i_1, i_2, \ldots, i_h)$ satisfies (1), and $(i'_1, i'_2, \ldots, i'_h)$ does not satisfy (1), the two corresponding vertices have different colors, since we have

$$((i_1 + i_2)_{n_1}, \ldots, (i_{k-1} + i_k)_{n_{k-1}}, (i_{k+1} + i_{k+2})_{n_{k+2}}, \ldots$$
$$\ldots, (i_{n-1} + i_n)_{n_k}) \neq$$
$$\neq ((i'_1 + i'_2)_{n_1}, \ldots, (i'_{k-1} + i'_k)_{n_{k-1}}, (i'_{k+1} + i'_{k+2})_{n_{k+2}}, \ldots$$
$$\ldots, (i'_{h-1} + i'_h)_{n_h}) \, .$$

If either both $(i_1, i_2, \ldots, i_h)$ and $(i'_1, i'_2, \ldots, i'_h)$ satisfy condition (1), or none of them fulfills (1), we have to show that the equation system

$$\begin{cases} i_{j-1} + i_j \equiv i'_{j-1} + i'_j \pmod{m_{j-1}} & \text{for} \quad 2 \leq j \leq k \\ i_{j-1} + i_j \equiv i'_{j-1} + i'_j \pmod{m_j} & \text{for} \quad k < j \leq h \, , \end{cases}$$

where $m_i = n_i$ for $1 \leq i \leq h$, $i \neq k$, $i \neq k+1$, and $m_{k+1} = b + 1$ if $(i_1, i_2, \ldots, i_h)$ satisfies (1), and $m_{k+1} = b$ if $(i_1, i_2, \ldots, i_h)$ does not satisfy (1), yields $i_j = i'_j$. This follows from Lemma 1.

(ii) Assume $l \leq k$.

Let $(\alpha_2, \alpha_3, \ldots, \alpha_h)$ be a color satisfying (2).

If
$$(\alpha_2, \alpha_3, \ldots, \alpha_{k-1}, \alpha_k, \alpha_{k+2}, \ldots, \alpha_h) <$$
$$< (b_2, b_3, \ldots, b_{k-1}, b_{k+2}, b_{k+3}, \ldots, b_{h+1}),$$

the equation system

$$\begin{cases} i_{j-1} + i_j \equiv \alpha_j \pmod{m_{j-1}} & \text{if} \quad 2 \leqslant j \leqslant k \\ i_{j-1} + i_j \equiv \alpha_j \pmod{m_j} & \text{if} \quad k < j \leqslant h \\ i_l = \alpha, \end{cases}$$

where $m_i = n_i$ for $1 \leqslant i \leqslant k$, $i \neq k$, $i \neq k+1$, and $m_{k+1} = b+1$, has always a solution $(i_1, i_2, \ldots, i_h)$ satisfying $0 \leqslant i_j < n_j$ for all j, (by Lemma 2). Furthermore, the vertex $x_1^{i_1} x_2^{i_2} \ldots x_h^{i_h}$ will be colored with color

$$((i_1 + i_2)_{n_1}, (i_2 + i_3)_{n_2}, \ldots$$
$$\ldots, (i_{k-1} + i_k)_{n_{k-1}}, (i_k + i_{k+1})_{b+1}, \ldots, (i_{h-1} + i_h)_{n_h}) =$$
$$= (\alpha_2, \alpha_3, \ldots, \alpha_h).$$

If
$$(\alpha_2, \alpha_3, \ldots, \alpha_{k-1}, \alpha_k, \alpha_{k+2}, \ldots, \alpha_h) \geqslant$$
$$\geqslant (b_2, b_3, \ldots, b_{k-1}, b_{k+2}, b_{k+3}, \ldots, b_{h+1}),$$

we have $\alpha_{k+1} < b$, since otherwise $(\alpha_2, \alpha_3, \ldots, \alpha_h)$ would not satisfy (2). Then the equation system

$$\begin{cases} i_{j-1} + i_j \equiv \alpha_j \pmod{m_{j-1}} & \text{if} \quad 2 \leqslant j \leqslant k \\ i_{j-1} + i_j \equiv \alpha_j \pmod{m_j} & \text{if} \quad k < j \leqslant h \end{cases}$$

where $m_i = n_i$ for $1 \leqslant i \leqslant h$, $i \neq k$, $i \neq k+1$, and $m_{k+1} = b$, has always a solution $i_1, i_2, \ldots, i_h$ satisfying $0 \leqslant i_j < n_j$ for all j (by Lemma 2). Then the vertex $x_1^{i_1} x_2^{i_2} \ldots x_h^{i_h}$ will be colored with color

$$((i_1 + i_2)_{n_1}, (i_2 + i_3)_{n_2}, \ldots, (i_{k-1} + i_k)_{n_{k-1}}, (i_k + i_{k+1})_b, \ldots$$

$$\ldots, (i_{h-1} + i_h)_{n_h}) = (\alpha_2, \alpha_3, \ldots, \alpha_h) .$$

Case $k = 1$.

Note that if $h = 2$, the complete bipartite graph has a good edge p-coloring for every integer p. This follows from Proposition 1, with $k = 1$ and $m = p$ (it follows also from Theorem 1 in [1]).

If $h \geq 3$, by Lemma 3, with $l = h - 2$ and with $m_i = n_{i+2}$ for $1 \leq i \leq h - 2$, there exist integers b_j, $4 \leq j \leq h + 1$, satisfying $0 \leq b_j < n_{j-1}$ and a non-negative integer b so that

$$p = br_3 + \sum_{4 \leq j \leq h+1} b_j r_j .$$

Since $q_2 < p < q_1$, and b is the quotient of p by r_3, we have $n_1 \leq b < n_2$.

We shall color the vertices $x_1^{i_1} x_2^{i_2} \ldots x_h^{i_h}$ of $(K^h_{n_1, n_2, \ldots, n_h})^*$ as follows:

If

(3) $$((i_2 + i_3)_{n_3}, \ldots, (i_{h-1} + i_h)_{n_h}) < (b_4, \ldots, b_{h+1}) ,$$

then we color vertex $x_1^{i_1} x_2^{i_2} \ldots x_h^{i_h}$ with color

$$((i_1 + i_2)_{b+1}, (i_2 + i_3)_{n_3}, \ldots, (i_{h-1} + i_h)_{n_h}) .$$

If $(i_2, i_3, \ldots, i_h)$ does not satisfy (3), we color vertex $x_1^{i_1} x_2^{i_2} \ldots x_h^{i_h}$ with color

$$((i_1 + i_2)_b, (i_2 + i_3)_{n_3}, \ldots, (i_{h-1} + i_h)_{n_h}) .$$

Let us show that this is a good edge p-coloring of $K^h_{n_1, n_2, \ldots, n_h}$.

The above coloring has at most p different colors, because a color $(\alpha_2, \alpha_3, \ldots, \alpha_h)$ satisfies

$$(4) \qquad \begin{cases} 0 \leqslant \alpha_i < n_i \quad \text{for} \quad 2 < i \leqslant h, \ \alpha_2 \geqslant 0, \quad \text{and} \\ (\alpha_2, \alpha_3, \ldots, \alpha_h) < (b, b_4, \ldots, b_{h+1}), \end{cases}$$

and, by the remark following the proof of Lemma 3, the number of $(h-1)$-tuples satisfying (4) is equal to

$$br_3 + \sum_{4 \leqslant j \leqslant h+1} b_j r_j = p .$$

Thus, it suffices to show that an edge

$$E_l^\alpha = \{x_1^{i_1} x_2^{i_2} \ldots x_l^\alpha \ldots x_h^{i_h} \mid 0 \leqslant i_j < n_j, \ 1 \leqslant j \leqslant h, \ j \neq l\}$$

of $(K^h_{n_1, n_2, \ldots, n_h})^*$ meets

(i) q_l colors if $l > 1$.

(ii) p colors if $l = 1$.

(i) Assume $l > 1$. Since edge E_l^α has cardinality q_l, it suffices to show that in E_l^α two distinct vertices, viz. $x_1^{i_1} x_2^{i_2} \ldots x_l^\alpha \ldots x_h^{i_h}$ and $x_1^{i'_1} x_2^{i'_2} \ldots x_l^\alpha \ldots x_h^{i'_h}$, have different colors. Put $i_l = i'_l = \alpha$.

If $(i_2, i_3, \ldots, i_h)$ satisfies (3), and $(i'_2, i'_3, \ldots, i'_h)$ does not satisfy (3), the two corresponding vertices have different colors, because

$$((i_2 + i_3)_{n_3}, \ldots, (i_{h-1} + i_h)_{n_h}) \neq$$
$$\neq ((i'_2 + i'_3)_{n_3}, \ldots, (i'_{h-1} + i'_h)_{n_h}) .$$

If either both $(i_2, i_3, \ldots, i_h)$ and $(i'_2, i'_3, \ldots, i'_h)$ satisfy (3), or none of them fulfills (3), we have to show that the equation system

$$i_{j-1} + i_j \equiv i'_{j-1} + i'_j \ (\mathrm{mod}\ m_j), \qquad (1 < j \leqslant h),$$

where $m_i = n_i$ for $3 \leqslant i \leqslant h$, and $m_2 = b + 1$ if $(i_2, i_3, \ldots, i_h)$ satisfies (3), and $m_2 = b$ if $(i_2, i_3, \ldots, i_h)$ does not satisfy (3), yields $i_j = i'_j$. This follows from Lemma 1.

(ii) Assume $l = 1$. Let $(\alpha_2, \alpha_3, \ldots, \alpha_h)$ be a color satisfying (4). If $(\alpha_3, \ldots, \alpha_h) < (b_4, \ldots, b_{h+1})$, the equation system

$$\begin{cases} i_{j-1} + i_j \equiv \alpha_j \pmod{m_j} & \text{if} \quad 1 < j \leqslant h \\ i_1 = \alpha , \end{cases}$$

where $m_i = n_i$ for $2 < i \leqslant h$, and $m_2 = b + 1$ always has a solution $i_1, i_2, \ldots, i_h$ such that $0 \leqslant i_j < n_j$ for any $1 \leqslant j \leqslant h$ (by Lemma 2). Furthermore, vertex $x_1^{i_1} x_2^{i_2} \ldots x_h^{i_h}$ will be colored with $(\alpha_2, \ldots, \alpha_h)$, because

$$((i_1 + i_2)_{b+1}, (i_2 + i_3)_{n_3}, \ldots, (i_{h-1} + i_h)_{n_h}) =$$
$$= (\alpha_2, \alpha_3, \ldots, \alpha_h) .$$

If $(\alpha_3, \ldots, \alpha_h) \geqslant (b_4, \ldots, b_{h+1})$, we have $\alpha_2 < b$, since otherwise $(\alpha_2, \ldots, \alpha_h)$ would not satisfy (4). Then, the equation system

$$\begin{cases} i_{j-1} + i_j \equiv \alpha_j \pmod{m_j} & \text{for} \quad 1 < j \leqslant h \\ i_1 = \alpha , \end{cases}$$

where $m_i = n_i$ for $2 < i \leqslant h$, and $m_2 = b$ has always a solution $i_1, i_2, \ldots, i_h$ such that $0 \leqslant i_j < n_j$ for all j (by Lemma 2). Furthermore, the vertex $x_1^{i_1} x_2^{i_2} \ldots x_h^{i_h}$ will be colored with color

$$((i_1 + i_2)_b, (i_2 + i_3)_{n_3}, \ldots, (i_{h-1} + i_h)_{n_h}) =$$
$$= (\alpha_2, \alpha_3, \ldots, \alpha_h) .$$

Q.E.D.

Remark. The result for $h = 3$ has also been proved by J.C. Bermond (not published) by another method, with which the extension to the case $h \geqslant 4$ does not seem to be possible.

Acknowledgements. The author wishes to thank very sincerely Professor H. Delange for his helpful comments.

REFERENCES

[1] C. Berge, On the good k-colorings, I. *(These Proceedings).*

[2] C. Berge, Classe chromatique des hypergraphes h-parti complets, *to appear in Proc. Hypergraph Seminar,* Columbus, Ohio, 1972, eds C. Berge et D.K. Ray-Chaudhuri, Springer-Verlag, *Lecture Notes series.*

[3] J.C. Meyer, Quelques problèmes concernant les cliques des hypergraphes h-complets et q-parti h-complets, *to appear in Proc. Hypergraph Seminar,* Columbus, Ohio, 1972. eds C. Berge et D.K. Ray-Chaudhuri, Springer-Verlag, *Lecture Notes series.*

COLLOQUIA MATHEMATICA SOCIETATIS JÁNOS BOLYAI
10. INFINITE AND FINITE SETS, KESZTHELY (HUNGARY), 1973.

SOME THEOREMS ON TRANSVERSALS

E.C. MILNER* – S. SHELAH

1. INTRODUCTION

Let $\mathcal{F} = \langle F_i \mid i \in I \rangle$ be a set system with index set I. We write $|\mathcal{F}|$ to denote the cardinality of the system, i.e. $|\mathcal{F}| = |I|$. We call $\mathcal{F}$ a (κ, λ)-system, and write $\mathcal{F} \in S(\kappa, \lambda)$ if $|\mathcal{F}| = \kappa$ and $|F_i| = \lambda$ for every index $i \in I$. $S(\kappa, < \lambda)$ is defined is an analogous way. $\mathcal{F}_0$ is a *sub-system* of $\mathcal{F}$, $\mathcal{F}_0 \subset \mathcal{F}$, if $\mathcal{F}_0 = \langle F_i \mid i \in I_0 \rangle$ and $I_0 \subset I$. A transversal of $\mathcal{F}$ is a function with domain I such that $f(i) \in F_i$, $(i \in I)$ and $f(i) \neq f(j)$ if $i \neq j$. We denote by Trans $(\mathcal{F})$ the set of all transversals of $\mathcal{F}$. If $f \in$ Trans $(\mathcal{F})$ we call the system of elements $\langle f(i) \mid i \in I \rangle$ a *system of distinct representatives* of $\mathcal{F}$, and we call the set of elements $\{f(i) \mid i \in I\}$ a *transversal set* of $\mathcal{F}$.

A fundamental theorem of transversal theory asserts that, if either of the finiteness conditions

(1.1) $\quad |\mathcal{F}| < \aleph_0$

*Research supported by Canadian National Research Council grant #A. 5198.

or

(1.2) $\quad |F_i| < \aleph_0, \qquad (i \in I)$

holds, then a necessary and sufficient condition for the existence of a transversal of the system $\mathscr{F} = \langle F_i \mid i \in I \rangle$ is that

(1.3) $\quad |\mathscr{F}(K)| \geqslant |K|$ *for every finite set* $K \subset I$,

where

$$\mathscr{F}(K) = \bigcup_{i \in K} F_i .$$

This result was proved by P. Hall [1] in the case that (1.1) holds (and also by D. König [2] in a different form), and by Marshall Hall [3] in the case that (1.2) holds. Combining these, an equivalent formulation of Marshall Hall's theorem is that, if (1.2) holds then

(1.4) $\quad \mathrm{Trans}\,(\mathscr{F}) \neq \phi \Leftrightarrow (\forall \mathscr{F}_0 \subset \mathscr{F})(|\mathscr{F}_0| < \aleph_0 \Rightarrow \mathrm{Trans}\,(\mathscr{F}_0) \neq \phi)$.

We are interested in possible extensions of this result to systems having infinite members.

For an infinite cardinal number μ we write $\mathscr{F} \in T(\mu)$ if and only if

$$\mathrm{Trans}\,(\mathscr{F}') \neq \phi \textit{ whenever } \mathscr{F}' \subset \mathscr{F} \textit{ and } |\mathscr{F}'| < \mu .$$

Using this notation, Marshall Hall's theorem (1.4) asserts that, for any cardinal κ,

$$\mathscr{F} \in S(\kappa, < \aleph_0) \wedge \mathscr{F} \in T(\aleph_0) \Rightarrow \mathrm{Trans}\,(\mathscr{F}) \neq \phi .$$

Our question then is whether there are other triples (κ, λ, μ) such that

(1.5) $\quad \mathscr{F} \in S(\kappa, < \lambda) \wedge \mathscr{F} \in T(\mu) \Rightarrow \mathrm{Trans}\,(\mathscr{F}) \neq \phi$.

A special case of (1.5) $(\kappa = \mu = \aleph_2,\ \lambda = \aleph_1)$ is stated as an unsolved problem in [4], and it appears again in this form (Problem 42C) in the collection of problems [5]. Erdős and Hajnal attribute this formulation of the question to W. Gustin. In a more recent paper [6] (which is a progress report on the collection of unsolved problems [5]. It is mentioned that it follows from a theorem of Jensen [10] that Gustin's problem has a solution in L; more precisely, if $V = L$, then there is a system $\mathcal{F}$ such that

$$\mathcal{F} \in S(\aleph_2, \aleph_0) \wedge \mathcal{F} \in T(\aleph_2) \wedge \mathrm{Trans}(\mathcal{F}) = \phi . \tag{1.6}$$

The hypothesis $V = L$ is not in fact needed here. It can be verified (essentially as in the proof of Theorem 1) that the system* $\mathcal{F} = \langle F_{\alpha\beta} \mid \omega \leqslant \alpha < \omega_1 \leqslant \beta < \omega_2 \rangle$ satisfies (1.6), where

$$F_{\alpha\beta} = \alpha \times \{\alpha, \beta\} = \bigcup_{\nu < \alpha} \{\langle \nu, \alpha \rangle, \langle \nu, \beta \rangle\} .$$

We will prove the following theorem and corollary which is more general. κ^+ denotes the successor cardinal of κ.

Theorem 1. *Let κ, λ be infinite cardinal numbers, κ regular and $\lambda > \aleph_0$. If there is a system $\mathcal{F}$ such that*

$$\mathcal{F} \in S(\kappa, < \lambda) \wedge \mathcal{F} \in T(\kappa) \wedge \mathrm{Trans}(\mathcal{F}) = \phi , \tag{1.7}$$

then there is a system $\mathcal{F}_1$ such that

$$\mathcal{F}_1 \in S(\kappa^+, < \lambda) \wedge \mathcal{F}_1 \in T(\kappa^+) \wedge \mathrm{Trans}(\mathcal{F}_1) = \phi . \tag{1.8}$$

If, in addition, $\mathcal{F} \in S(\kappa, \lambda_1)$ where $\lambda_1 \geqslant \aleph_0$, then $\mathcal{F}_1 \in S(\kappa^+, \lambda_1)$.

Corollary. *For $\alpha \geqslant 0$ and $1 \leqslant n < \omega$, there is a system $\mathcal{F}$ such that*

$$\mathcal{F} \in S(\aleph_{\alpha+n}, \aleph_\alpha) \wedge \mathcal{F} \in T(\aleph_{\alpha+n}) \wedge \mathrm{Trans}(\mathcal{F}) = \phi .$$

The corollary is an immediate deduction from the theorem. Consider the system $\mathcal{F} = \langle \xi \mid \omega_\alpha \leqslant \xi < \omega_{\alpha+1} \rangle$ where, as usual, the ordinal number ξ is the set $\{\eta \mid \eta < \xi\}$ of all smaller ordinals. Clearly, $\mathcal{F} \in S(\aleph_{\alpha+1}, \aleph_\alpha)$ and $\mathcal{F} \in T(\aleph_{\alpha+1})$. Also, by a theorem of Alexandroff and Urysohn [7] on regressive functions, we have $\mathrm{Trans}(\mathcal{F}) = \phi$. The corollary now follows from the theorem by induction on n.

*This is a modification of an example communicated to us by J. Truss, Leeds University, England.

We do not know if the assumed regularity of κ is necessary for the validity of Theorem 1. The simplest open question is whether there is a system $\mathscr{F}$ which satisfies

$$\mathscr{F} \in S(\kappa, \aleph_0) \wedge \mathscr{F} \in T(\kappa) \wedge \text{Trans}(\mathscr{F}) = \phi$$

when $\kappa = \aleph_\omega$ or $\aleph_{\omega+1}$.* Hajnal pointed out to us that the remark in [6] regarding Problem 42C applies more generally, and that Jensen's result actually leads to the following theorem.

Theorem 2. *If* V = L *and* κ *is a regular cardinal which is not weakly compact and* $\kappa > \lambda \geq \aleph_0$, *then there is an* $\mathscr{F}$ *which satisfies*

(1.9) $$\mathscr{F} \in S(\kappa, \lambda) \wedge \mathscr{F} \in T(\kappa) \wedge \text{Trans}(\mathscr{F}) = \phi ,$$

The condition that κ not be weakly compact in Theorem 2 is essential. It is easy to prove the following.

Theorem 3. *If* κ *is weakly compact, then* $\mathscr{F} \in S(\kappa, < \kappa) \wedge \mathscr{F} \in T(\kappa) \Rightarrow \text{Trans}(\mathscr{F}) \neq \phi$.

By way of contrast with the negative results in Theorems 1 and 2 we will establish the following positive Hall-type theorem. A special case of this has been used in [8] to settle a conjecture of Nash-Williams.

Theorem 4. *Let* λ *be an infinite cardinal number and suppose that* $\mathscr{F} = \langle F_i \mid i \in I_0 \cup I_1 \rangle$ *is a set system with*

(i) $I_0 \cap I_1 = \phi, \ |I_1| \leq \lambda,$

(ii) $|F_i| < \aleph_0, \ (i \in I_0),$

(iii) $|F_i| \leq \lambda, \ (i \in I_1).$

Then a necessary and sufficient condition for the existence of a transversal of $\mathscr{F}$ *is that*

(1.10) $\text{Trans}(\mathscr{F}') \neq \phi$ *whenever* $\mathscr{F}' \subset \mathscr{F}$ *and* $|\mathscr{F}| \leq \lambda$.

*Shelah has since proved this is false for $\kappa = \aleph_\omega$ (see his paper in Volume 3 of these proceedings). More generally, he has now proved that if $\text{cf}\,\kappa < \kappa$ and $\lambda < \kappa$, then $\mathscr{F} \in S(\kappa, \lambda) \wedge \mathscr{F} \in T(\kappa) \Rightarrow \text{Trans}(\mathscr{F}) \neq \phi$.

2. NOTATION

We write $K \subset\subset S$ to indicate that K is a finite subset of S.

Greek letters denote ordinal numbers. A cardinal is an initial ordinal. If A is a set of ordinals, then $\sup A$ denotes that the least ξ such that $\alpha \leqslant \xi$ for all $\alpha \in A$. B is a *cofinal* subset of A if $\sup B = \sup A$. A is *closed* if $\sup B \in A$ whenever $B \subset A$ and $\sup B < \sup A$. S is a *stationary* subset of A, $S \in \mathrm{Stat}\,(A)$, if and only if $S \cap B \neq \phi$ for every closed, cofinal subset B of A. The function f on A is *regressive* if $f(\xi) < \xi$ for all $\xi \in A - \{0\}$. The cofinality of ξ, $\mathrm{cf}\,(\xi)$ is the least ordinal α for which there is a function $g\colon \alpha \to \xi$ such that $\sup \{g(\sigma) \mid \sigma < \alpha\} = \xi$.

We use the following well-known facts. Let κ be a regular cardinal, $\kappa > \mu \geqslant \omega$.

1. If $S \in \mathrm{Stat}\,(\kappa)$ and f is regressive on S, then f is not $1-1$; in fact there is $\theta < \kappa$ such that $|f^{-1}(\theta)| = \kappa$;

2. $\{\xi \in \kappa \mid \mathrm{cf}\,(\xi) = \mu\}$ is a stationary subset of κ (see [9]).

3. PROOF OF THEOREM 1

We may assume that the system $\mathscr{F}$ which satisfies the hypothesis (1.7) is indexed by κ, i.e. $\mathscr{F} = \langle F_\nu \mid \nu < \kappa\rangle$. Let $C = \{\rho \mid \kappa \leqslant \rho < \kappa^+, \mathrm{cf}\,(\rho) = \kappa\}$. For each $\rho \in C$ there is an increasing sequence of ordinal numbers $\beta(\rho, \sigma)$, $(\sigma < \kappa)$ such that

$$\rho = \lim_{\sigma < \kappa} \beta(\rho, \sigma)\,.$$

Put

$$G(\rho, \sigma) = (\{\rho\} \times F_\sigma) \cup \{\beta(\rho, \sigma)\} \qquad (\rho \in C _\wedge \sigma < \kappa)\,.$$

We will prove that (1.8) holds with

$$\mathscr{F}_1 = \langle G(\rho, \sigma) \mid \rho \in C _\wedge \sigma < \kappa\rangle\,.$$

Clearly, $|\mathscr{F}_1| = \kappa |C| = \kappa^+$ (here we use the fact that κ is regular; if κ is singular we would have $C = \phi$). Also

$$|G(\rho, \sigma)| = |F_\sigma| + 1 < \lambda \qquad (\rho \in C, \ \sigma < \kappa),$$

and

$$|G(\rho, \sigma)| = |F_\sigma| \ \text{ if } \ F_\sigma \ \text{ is infinite.}$$

It remains to show that

(3.1) $\qquad \in T(\kappa)$,

and

(3.2) $\qquad \text{Trans}\,(\mathscr{F}_1) = \phi$.

In order to prove (3.1) it will be enough to prove that

(3.3) $\qquad \text{Trans}\,(\mathscr{F}_1(\alpha)) \neq \phi$,

where $\mathscr{F}_1(\alpha) = \langle G(\rho, \sigma) \,|\, \rho \in C \wedge \rho < \alpha \wedge \sigma < \kappa \rangle$ and $\kappa^2 \leqslant \alpha < \kappa^+$. For, if $\mathscr{F}' \subset \mathscr{F}_1$ and $|\mathscr{F}'| \leqslant \kappa$, then $\mathscr{F}' \subset \mathscr{F}_1(\alpha)$ for some α with $\kappa^2 \leqslant \alpha < \kappa^+$.

Let α be fixed, $\kappa^2 \leqslant \alpha < \kappa^+$. Then

$$C(\alpha) = \{\rho \in C \,|\, \rho < \alpha\} = \{\rho_\tau \,|\, \tau < \kappa\}_{\neq},$$

i.e. $\rho_\sigma \neq \rho_\tau$ if $\sigma < \tau < \kappa$. We shall define ordinals $\sigma_\tau < \kappa$ for $\tau < \kappa$ so that the κ sets

$$B_\tau = \{\beta(\rho_\tau, \sigma) \,|\, \sigma_\tau \leqslant \sigma < \kappa\} \qquad (\tau < \kappa)$$

are pairwise disjoint. Let $\tau_0 < \kappa$ and suppose that σ_τ has been defined for $\tau < \tau_0$. For each $\tau < \tau_0$ there is $\xi_\tau < \kappa$ such that

(3.4) $\qquad B_\tau \cap \{\beta(\rho_{\tau_0}, \sigma) \,|\, \xi_\tau \leqslant \sigma < \kappa\} = \phi$.

If $\rho_\tau < \rho_{\tau_0}$, then (3.4) holds with any choice for $\xi_\tau < \kappa$ such that $\beta(\rho_{\tau_0}, \xi_\tau) > \rho_\tau$. If, on the other hand, $\rho_\tau > \rho_{\tau_0}$, then the existence of ξ_τ such that (3.4) holds follows from the fact that $\text{cf}\,(\rho_{\tau_0}) = \kappa$ and $|\{\beta \in B_\tau \,|\, \beta < \rho_{\tau_0}\}| < \kappa$. Hence, there are ordinals $\xi_\tau < \kappa$, $(\tau < \tau_0)$ such that (3.4) holds. Now put

$$\sigma_{\tau_0} = \sup_{\tau < \tau_0} \xi_\tau .$$

This defines the $\sigma_\tau < \kappa$, $(\tau < \kappa)$ so that the sets B_τ are pairwise disjoint.

For each $\tau < \kappa$ the sub family $\langle F_\nu \mid \nu < \tau \rangle$ of $\mathscr{F}$ has a transversal, i.e. there is a $1-1$ function f_τ on τ such that

$$f_\tau(\nu) \in F_\nu \qquad (\nu < \tau < \kappa) .$$

Now define a function g on $C(\alpha) \times \kappa$ by putting

$$g(\rho_\tau, \sigma) = \begin{cases} \langle \rho_\tau, f_\tau(\sigma) \rangle & \text{if } \sigma < \sigma_\tau , \\ \beta(\rho_\tau, \sigma) & \text{if } \sigma_\tau \leqslant \sigma < \kappa . \end{cases}$$

Clearly, $g(\rho_\tau, \sigma) \in G(\rho_\tau, \sigma)$, $(\sigma, \tau < \kappa)$ and g is $1-1$ since f is and the sets B_τ $(\tau < \kappa)$ are pairwise disjoint. Therefore, $g \in \text{Trans}\,(\mathscr{F}_1(\alpha))$. This proves (3.3) and hence (3.1).

We now prove (3.2). Suppose, on the contrary, that $\mathscr{F}_1$ has a transversal. Then there is a $1-1$ function h on $C \times \kappa$ such that $h(\rho, \sigma) \in$ $\in G(\rho, \sigma)$. Suppose that for some $\rho \in C$ we have

$$h(\rho, \sigma) \neq \beta(\rho, \sigma) \qquad (\forall \sigma < \kappa) .$$

Then

$$h(\rho, \sigma) = \langle \rho, g(\sigma) \rangle \qquad (\sigma < \kappa) ,$$

where g is a $1-1$ function on κ such that $g(\sigma) \in F_\sigma$. This contradicts the hypothesis that $\text{Trans}\,(\mathscr{F}) = \phi$. Hence, for each $\rho < \kappa$ there is $\sigma(\rho) < \kappa$ such that

$$h(\rho, \sigma(\rho)) = \beta(\rho, \sigma(\rho)) = \theta(\rho) .$$

Then $\theta(\rho) < \rho$ for $\rho \in C$ and, since C is a stationary subset of κ^+ (see 2), it follows that there are $\rho_1, \rho_2 \in C$ such that $\rho_1 \neq \rho_2$ and $\theta(\rho_1) = \theta(\rho_2)$. This contradicts our assumption that h is $1-1$. Therefore, (3.2) holds.

4. PROOF OF THEOREM 2

It follows from a theorem of Jensen [10] that, if $V = L$ and κ is a regular cardinal which is not weakly compact, then there is a set $A \subset \kappa$ such that

(i) $A \in \mathrm{Stat}(\kappa)$,

(ii) $A \cap \xi \notin \mathrm{Stat}(\xi)$, $(\xi < \kappa)$,

(iii) $\alpha \in A \Rightarrow \mathrm{cf}(\alpha) = \omega$.

For $\alpha \in A$, let B_α be a set of ordinals of order type ω such that $\sup(B_\alpha) = \alpha$. Let B be any set of power λ disjoint from $\bigcup_{\alpha \in A} B_\alpha$. We will show that the (κ, λ)-system $\mathscr{F} = \langle B_\alpha \cup B \mid \alpha \in A \rangle$ satisfies (1.9).

Suppose that $\mathscr{F}$ has a transversal f. Let $A' = \{\alpha \in A \mid f(\alpha) \notin B\}$ then $A' \in \mathrm{Stat}(\kappa)$ and f is regressive and $1-1$ on A'. This is impossible and hence $\mathrm{Trans}(\mathscr{F}) = \phi$. To show that $\mathscr{F} \in T(\kappa)$ it will be enough to show that the system $\langle B_\alpha \mid \alpha \in A \cap \xi \rangle$ has a transversal for $\xi < \kappa$. We will actually, by transfinite induction on $\xi < \kappa$, prove the following slightly stronger statement R_ξ: *If* D_α *is a set of ordinals of type* ω *such that* $\sup(D_\alpha) = \alpha$, $(\alpha \in A \cap \xi)$, *then* $\mathrm{Trans}(\langle D_\alpha \mid \alpha \in A \cap \xi \rangle) \neq \phi$.

Let $\xi_0 < \kappa$ and assume that R_ξ holds for $\xi < \xi_0$. If $\xi_0 = \eta + 1$, then $A \cap \xi_0 = A \cap \eta$ and so R_{ξ_0} holds. Now assume that ξ_0 is a limit ordinal. By (ii) there is a closed cofinal subset C of ξ_0 such that $C \cap A = \phi$. Let $C = \{\nu_\sigma \mid \sigma < \rho\}$, where $\nu_0 < \nu_1 < \ldots < \xi_0$. We can assume that $\nu_0 = 0$ since $0 \notin A$. For $\alpha \in A \cap \xi_0$ there is $\sigma = \sigma(\alpha) < \rho$ such that $\nu_\sigma < \alpha < \nu_{\sigma+1}$. Put $E_\alpha = D_\alpha \cap [\nu_\sigma, \nu_{\sigma+1})$. By the induction hypothesis, the system $G_\sigma = \langle E_\alpha \mid \alpha \in A \cap \xi_0 \wedge \sigma(\alpha) = \sigma \rangle$ has a transversal $(\sigma < \rho)$. Moreover, the systems G_σ, $(\sigma < \rho)$ are pairwise strongly disjoint and hence $\langle D_\alpha \mid \alpha \in A \cap \xi_0 \rangle$ also has a transversal. This shows that R_{ξ_0} holds and the proof is complete.

5. PROOF OF THEOREM 3

A partially ordered set $(A, \leqslant)$ is a *tree* if (i) it has a minimal element and (ii), $A(z) = \{x \in A \mid x \leqslant z\}$ is well-ordered by $\leqslant$ for all $z \in A$. The *order,* $O(z)$, *of* $z \in A$ is the ordinal number which is the type of $(A(z), \leqslant)$. The *order* of the tree is $\bigcup_{z \in A} O(z)$. A *branch* is a set $B \subset A$ which is well-ordered by $\leqslant$ and is such that $x \leqslant y \in B \Rightarrow x \in B$. The cardinal κ is weakly compact if it has the *tree property,* i.e. whenever $(A, \leqslant)$ is a tree of order κ having fewer than κ elements of order ξ for all $\xi < \kappa$, then there is a branch of order κ. (Erdős and Tarski [11] proved that if κ has the tree property then $\kappa \to (\kappa, \kappa)^2$, i.e. any graph on κ either contains a complete subgraph of order κ or an edge-free set of order κ. Hanf proved the converse (see [12]). This fact easily implies the following lemma which is stronger than Theorem 3. We cannot find precisely this statement in the literature although equivalents are known; it is expressed in the style of Rado's selection lemma [13] and we give the simple proof.

Lemma. *Let* κ *be weakly compact and let* $\langle F_\nu \mid \nu < \kappa \rangle$ *be a* $(\kappa, < \kappa)$*-system. Suppose that, for each* $\xi < \kappa$; f_ξ *is a function with domain* ξ *such that* $f_\xi(\nu) \in F_\nu$, $(\nu < \xi)$. *Then there is a function* f *defined on* κ *such that*

$$(\forall \xi < \kappa)(\exists \eta < \kappa)(f \upharpoonright \xi = f_\eta \upharpoonright \xi) .$$

Remark. If $f_\xi \in \text{Trans}\,(\langle F_\nu \mid \nu < \xi \rangle)$, $(\xi < \kappa)$, then clearly $f \in \text{Trans}\,(\langle F_\nu \mid \nu < \xi \rangle)$.

Proof of Lemma. Let $A = \{f_\xi \upharpoonright \mu \mid \mu \leqslant \xi < \kappa\}$. Then the partially ordered set $(A, \subseteq)$ is a tree of order κ. Since κ is strongly inaccessible, there are fewer than κ choice functions of $F \upharpoonright \xi$, $(\xi < \kappa)$ and so the tree has fewer than κ elements of order ξ, $(\xi < \kappa)$. Hence there is a branch B of order κ. Let $f = \bigcup B$. For each $\xi < \kappa$ we have $f \upharpoonright \xi \in A$ and hence $f \upharpoonright \xi = f_\eta \upharpoonright \xi$ for some $\eta < \kappa$.

6. PROOF OF THEOREM 4

The necessity of (1.10) is obvious, we have to prove the sufficiency.

Let S be any set. We shall define a set $S^* \supset S$ in the following way. For $B \subset\subset S$, let

$$G_S(B) = \{K \mid K \subset\subset I_0 \wedge S \cap \mathscr{F}(K) = B \wedge$$
$$\wedge \mid \mathscr{F}(K) \setminus B \mid < \mid K \mid \wedge (\forall i \in K)(F_i \not\subset S)\} .$$

If $G_S(B) = \phi$, put $H_S(B) = B$; if $G_S(B) \neq \phi$, select $K \in G_S(B)$ and put $H_S(B) = \mathscr{F}(K)$. Now define

$$S^* = \bigcup_{B \subset\subset S} H_S(B) .$$

Since $H_S(B) \supset B$, we have that $S^* \supset S$. Also, if S is an infinite set, then $|S^*| = |S|$.

Now put $A_0 = \mathscr{F}(I_1)$, $A_{n+1} = A_n^*$, $(n < \omega)$, $\bar{A} = \bigcup_{n<\omega} A_n$. Then $|\bar{A}| \leqslant \lambda$. Put

$$I_3 = \{i \in I \mid F_i \subset \bar{A}\} , \qquad I_4 = I \setminus I_3 .$$

Then $I_1 \subset I_3$ and $I_4 \subset I_0$. The hypothesis implies that any finite subfamily of $\mathscr{F}$ has a transversal and therefore

$$|\{i \in I \mid F_i = F_{i_0}\}| \leqslant |F_{i_0}| \qquad (i_0 \in I_0) .$$

It follows from this that $|I_0 \cap I_3| \leqslant \lambda$ and hence $|I_3| \leqslant \lambda$. Therefore, by assumption, there is a transversal f of $\mathscr{F}_3 = \langle F_i \mid i \in I_3 \rangle$. We will show that f can be extended to a transversal of $\mathscr{F}$, i.e. there is a transversal of $\mathscr{F}_4 = \langle F_i \mid i \in I_4 \rangle$ whose range is disjoint from the set $T = \{f(i) \mid i \in I_3\}$.

Suppose this is false. Then, since the members of $\mathscr{F}_4$ are finite sets, it follows from (1.3) that there is a finite set $K \subset I_4$ such that

$$| \mathscr{F}(K) \setminus T | < |K| .$$

Let $B = \mathscr{F}(K) \cap \bar{A}$. Then

(6.1) $$|\mathscr{F}(K) \setminus B| < |K| .$$

Also, since B is a finite set, there is an integer n_0 such that $B \subset\subset A_n$ $(n_0 \leqslant n < \omega)$. Let $n_0 \leqslant n < \omega$. By (6.1) and the fact that K is a finite subset of I_4 it follows that

$$K \in G_{A_n}(B) \neq \phi .$$

Therefore, there is $K_n \subset\subset I_0$ such that

$$A_n \cap \mathscr{F}(K_n) = B ,$$

(6.2) $$|\mathscr{F}(K_n) \setminus B| < |K_n| ,$$

(6.3) $$(\forall i \in K_n)(F_i \not\subset A_n) ,$$

(6.4) $$(\forall i \in K_n)(F_i \subset A_{n+1}) .$$

By (6.4), $K_n \subset I_3$ and therefore, by (6.2), there is $i_n \in K_n$ such that $f(i_n) \in B$. This defines i_n for $n_0 \leqslant n < \omega$. By (6.3) and (6.4) we see that $i_n \neq i_p$, $(n_0 \leqslant n < p < \omega)$. Therefore, since f is $1-1$,

$$|B| \geqslant |\{f(i_n) | n_0 \leqslant n < \omega\}| \geqslant \aleph_0 .$$

This contradiction proves the theorem.

REFERENCES

[1] P. Hall, On Representatives of Subsets, *J. London Math. Soc.*, 10 (1935), 26-30.

[2] D. König, Gráphok és mátrixok, *Mat. Fiz. Lapok*, 38 (1931), 116-119. [Hungarian with German summary.]

[3] M. Hall, Jr., Distinct Representatives of Subsets, *Bull. Amer. Math. Soc.*, 54 (1948), 922-926.

[4] P. Erdős – A. Hajnal, On a property of families of sets, *Acta Math. Acad. Sci. Hungar.*, 12 (1961), 87-123.

[5] P. Erdős – A. Hajnal, Unsolved problems in set theory, *Proceedings of Symposia in Pure Mathematics,* XIII, Part I, A.M.S. Providence, R.I. (1971), 17-48.

[6] P. Erdős – A. Hajnal, Unsolved and solved problems in set theory, Tarski Symposium (*to appear*).

[7] Alexandroff – Urysohn, Memoire sur les espaces topologiques compacts, *Verh. Nederl. Akad. Wentensch. Sect.* I, 14, Nr. 1, S1 (1929).

[8] R.M. Damerell – E.C. Milner, Necessary and sufficient conditions for transversals of countable set systems, *Journal of Combinatorial Theory* (*to appear*).

[9] W. Neumer, Verallgemeinerung eines Satzes von Alexandroff and Urysohn, *Math. Zeit.,* 54 (1951), 254-261.

[10] R.B. Jensen, The fine structure of the constructible hierarchy, *Annals Math. Logic,* 4 (1972), 229-308. (Theorem 6.1).

[11] P. Erdős – A. Tarski, On some problems involving inaccessible cardinals, in *Essays on the foundations of mathematics,* Jerusalem, (1961), 50-82.

[12] A. Hajnal, Remarks on a theorem of W.P. Hanf, *Fund. Math.,* 54 (1964), 109-113.

[13] R. Rado, Axiomatic treatment of rank in infinite sets, *Canad. J. Math.,* 1 (1949), 337-343.

COLLOQUIA MATHEMATICA SOCIETATIS JÁNOS BOLYAI
10. INFINITE AND FINITE SETS, KESZTHELY (HUNGARY), 1973.

A RAMSEY GRAPH WITHOUT TRIANGLES EXISTS FOR ANY GRAPH WITHOUT TRIANGLES

J. NEŠETŘIL – V. RÖDL

We are going to prove a theorem of Ramsey's type which was proposed by F. Galvin (see e.g. [1] where a partial answer was given).

Theorem. *For every triangle free graph* G *and for all integers* k *there exists a triangle free graph* H *satisfying the following condition.*

For every partition $E(H) = \bigcup_{i=1}^{k} E_i$ *there is an* i *such that* G *is an induced subgraph of* $(V(H), E_i)$.

§1.

The proof uses a representation of a graph as a special intersection graph of a family of sets.

Let $2 \leqslant m \leqslant n$ be naturals. Put $G_{m,n} = (P_m([1, n]), E_{m,n})$ where $[1, n] = \{1, \ldots, n\}$, $P_m(X) = \{M \subseteq X \mid |M| = m\}$ for every set X, and

$[A, B] \in E_{m,n}$ iff $A = \{a_1, \ldots, a_m\}_<$, $B = \{b_1, \ldots, b_m\}_<$ and $A \cap B = \{a_1\} = \{b_j\}$ for a $1 < j \leqslant m$. (We write $\{a_1, \ldots, a_m\}_<$ iff $a_1 < a_2 < \ldots < a_m$).

Denote by $\leqslant$ the partial order of $P_m([1, n])$ induced by the ordering of the first elements of m-subsets of $[1, n]$ (thus $\{a_1, \ldots, a_m\}_< \leqslant \{b_1, \ldots, b_m\}_<$ iff $a_1 \leqslant b_1$).

Claim. The graph $G_{m,n}$ does not contain a triangle for every $2 \leqslant m \leqslant n$.

Proof. Let $A^i = \{a_1^i, \ldots, a_m^i\}_<$, $i = 1, 2, 3$ indices a triangle in $G_{m,n}$. We may assume $a_1^1 < a_1^2 < a_1^3$ as these points are pairwise different. But then $A^1 \cap A^2 \supseteq \{a_1^2, a_1^3\}$, a contradiction.

On the other hand we show that every graph without triangles is a full subgraph of some $G_{m,n}$. In fact we will prove a stronger statement. First we need some preliminaries to do so. We define the type $t(A, B)$ of an edge $[A, B] \in E_{m,n}$, $A \leqslant B$.

$$t(A, B) = (\epsilon_a \mid a \in A \cup B) \text{ where } \epsilon_a = 1 \text{ if } a \in B,\ \epsilon_a = 0 \text{ otherwise.}$$

The types $t(A, B)$ and $t(A', B')$ are said to be equivalent if there exists a bijection $\varphi\colon A \cup B \to A' \cup B'$ such that

(i) φ is a monotonic mapping i.e. $\varphi(a) \leqslant \varphi(b) \Leftrightarrow a \leqslant b$

(ii) $\epsilon_{\varphi(a)} = \epsilon_a$ for every $a \in A \cup B$.

Further, we shall need the notion of an embedding which is sensitive with respect to the types.

Let X be a graph. If there is a one-to-one mapping $f\colon V(X) \to P_m([1, n])$ such that $[x, y] \in E(X) \Leftrightarrow [f(x), f(y)] \in E_{m,n}$ then we say that f is a full embedding of X into $G_{m,n}$ and write $X \leqslant G_{m,n}$. Clearly $G_{m,n} \leqslant G_{m,n'}$ for $n \leqslant n'$. Note that $G_{m,n} \leqslant G_{m',n}$ for $m \leqslant m'$ is not generally true.

Let $(X, \leqslant)$ be a graph with a given linear order of its vertices. A full embedding $f\colon X \to G_{m,n}$ is said to be a good embedding iff g1. $x \leqslant y$ implies $f(x) \leqslant f(y)$ and

g2. $t(f(x), f(y)) = t(f(z), f(y))$ for every $\{[x, y], [z, y]\} \subseteq E(X)$, $y \geqslant \max\{x, z\}$.

Lemma. *For every graph* X *without triangles there exists a good embedding* $f: X \to G_{m,n}$ *for convenient* m, n.

Proof. By induction on $p = |X|$. The statement is clearly true for $|X| \leqslant 2$.

Thus let X be a graph with $p + 1 > 2$ vertices without triangles, assume without loss of generality $V(X) = [1, p + 1]$.

Put

$$\bar{X} = X - \{p + 1\} = ([1, p], E(X) - \{e \mid p + 1 \in e\}) .$$

By the induction hypothesis there exists a good embedding $F: \bar{X} \to G_{m,n}$. Define $f: [1, p + 1] \to P_{m+1}([1, n'])$ where $n' = \max\{n + m + 1, n + p + 1\}$ by

$$f(p + 1) = [n + 1, n + m + 1]$$

$$f(i) = F(i) \cup \{n + 1\} \qquad \text{if} \quad [i, p + 1] \in E(X)$$

$$f(i) = F(i) \cup \{n + i + 1\} \quad \text{if} \quad [i, p + 1] \notin E(X) .$$

We prove that f is a good embedding of X into $G_{m+1,n'}$.

1. f is a monomorphism: $f(i) \cap f(j) = F(i) \cap F(j)$ if $[i, j[\in E(\bar{X})$ as in this case $f(i) \cap f(j) \cap [n + 1, n'] = \phi$ (X does not have triangles); $f(i) \cap f(p + 1) = \{n + 1\}$ if $[i, p + 1] \in E(X)$.

2. f is a full embedding since F is a full embedding and $[f(i), f(p + 1)] \in E_{m+1,n'} \Leftrightarrow [i, p + 1] \in E(X)$.

3. Clearly f is a monotonic mapping $([1, p + 1], \leqslant) \to$ $\to (P_{m+1}([1, n']), \leqslant)$.

4. We prove g2. First, let be $i, j < k \leqslant p$. Then $t(F(i), F(k)) =$ $= t(F(j), F(k))$ by the induction hypothesis. We have $f(l) = f(l) \cup \{\lambda_l\}$,

$l = i, j, k$, $\lambda_l \geqslant n + 1$. As X does not have triangles, $\lambda_i \neq \lambda_k$ and $\lambda_j \neq \lambda_k$. Further if $\lambda_k = n + 1$ then $\lambda_i > \lambda_k$ and $\lambda_j > \lambda_k$. Hence, subsymbolically, $t(f(i), f(k)) = t(f(j), f(k)) = t(F(i), F(k)), 1, 0) = (t(F(j), F(k)), 1, 0)$; and if $\lambda_k \neq n + 1$ then $\lambda_i < \lambda_k$ and $\lambda_j < \lambda_k$. Hence $t(f(i), f(k)) = t(f(j), f(k)) =$ $= (t(F(i), F(k)), 0, 1) = (t(F(j), F(k)), 0, 1)$.

On the other hand, $t(f(i), f(p + 1)) = (\underbrace{0, 0, \ldots, 0}_{m}, \underbrace{1, \ldots, 1}_{m+1})$ holds for all $i < p + 1$. This proves the lemma.

Remark. Actually, it can be deduced from the proof that a graph without triangles with p vertices has a good embedding into $G_{p, (p/2)(p+1)}$.

Before proving the theorem we state two lemmas.

Lemma Folkman [2]. *For every graph X without triangles there exists a graph Y without triangles such that $X \to Y|_{V_i}$ for some $i \in [1, k]$ for every partition $V(Y) = \bigcup_{i=1}^{k} V_i$. In this case we write $(\cdot, X) \underset{k}{\to} Y$. We call this symbol Rado's reversed arrow.* (For the reason see our forthcoming paper.)

Proof. By the first lemma X can be fully embedded into some $G_{m,n}$. Hence it is sufficient to prove the statement for all $G_{m,n}$. Let $N \geqslant R_m(k; n, \ldots, n)$ be the Ramsey number for partitions of m-sets into k parts which ensures the existence of monochromatic n-set. Then $G_{m,n}$ has the desirable property as by the Ramsey theorem for every partition $\bigcup_{i=1}^{k} V_i = P_m([1, N])$ there exists an $i \in [1, k]$ and an $A \subseteq [1, N]$, $|A| = n$ such that $P_m(A) \subset A_i$ and consequently even $G_{m,n} \leqslant$ $\leqslant (A, E_{m,N} \cap P_2(P_m(A)))$.

Lemma. Assume $(\cdot, X) \underset{k}{\to} Y$. Let $<$ be an ordering of $V(Y)$ and let $E(Y) = \bigcup_{i=1}^{k} E_i$ be a partition with the following property:

$\{[i, j] \mid i < j,\ [i, j] \in E(Y)\} \subset E_l$ for an $l \in [1, k]$ and for every $j \in$ $\in V(Y)$.

Then $X \leqslant (V(X), E_l)$ for an $l \in [1, k]$.

Proof. Put $\bar{V}_l = \{i \mid [j, i] \in E_l,\ j < i\}$. Put $V_1 = \left(V(Y) - \bigcup_{l=1}^{k} \tilde{V}_l\right) \cup \tilde{V}_l = V_l$ for $k \geqslant l > 1$. Noe $X \leqslant Y|_{V_l}$ for an $l \in [1, k]$ and than it easily that $X \leqslant (V(X), E_l)$ holds.

Proof of the theorem. Let X be a triangle free graph. By the second lemma there is a triangle free Y such that $(\cdot, X) \underset{k}{\rightarrow} Y$. By the first lemma we can choose m, n and a good embedding f of Y into $G_{m,n}$.

Define recursively the numbers

$$R_m^1(k;\ n, \ldots, n) = R_m(k;\ n, \ldots, n)$$

$$R_m^{a+1}(k;\ n, \ldots, n) =$$

$$= R_m(k;\ R_m^a(k;\ n, \ldots, n), \ldots, R_m^a(k; n, \ldots, n))\,.$$

Put $M = m^m$ and let $N \geqslant R_m^M(k;\ n, \ldots, n)$. We claim that $G_{m,N}$ satisfies the requirements of the theorem.

Every edge $[A, B] \in E_{m,N}$ satisfies $|A \cup B| = 2m - 1$ and for every set $C \in P_{2m-1}([1, N])$ there are at most M different edges $[A, B] \in E_{m,N}$ satisfying $A \cup B = C$. Moreover $t(A, B) = t(A', B')$ and $A \cup B = A' \cup B'$ implies $[A, B] = [A', B']$.

Hence given a partition $E_{m,N} = \bigcup_{i=1}^{k} E_i$ we have induced partitions $\bigcup_{i=1}^{k} E_i^t = E_{m,N}^t$ where $E_{m,N}^t$ is the set of all edges $G_{m,N}$ with the type t and E_i^t is the set of all edges E_i with the type t. But as we have seen $E_{m,N}^t = P_{2m-1}([1, N])$. Hence we may use repeatedly the Ramsey theorem to the set systems $(P_m([1, N]), E_i^t)$, t a type, $i \in [1, k]$, and find a set $R \subset [\ , N]$ such that

1. $|R| = n$

2. For every type t there exist an $i \in [1, k]$ such that $E_i^t \supseteq P_2(P_m(R))$.

Since $G_{m,n}$ and $G_{m,N}|_R = (P_m(R), E_{m,N} \cap P_2(P_m(R)))$ are isomorphic graphs under a monotonic mapping there exists a good embedding $g\colon Y \to G_{m,N}|_R$. But now using that g is good and the third lemma there exists a full embedding of X into one of the graphs $(P_m(R), E_i \cap P_2(P_m(R))) \leqslant P_m([1, N]), E_i)$ for an $i \in [1, k]$. This finishes the proof.

REFERENCES

[1] P. Erdős,– A. Hajnal, Problems and results in finite and infinite combinatorial analysis, *Ann. of N. Y. Acad. of Sci.*, 175 (1970), 115-124.

[2] J. Folkman, Graphs with monochromatic complete subgraphs in every edge colouring, *SIAM J. Applied Math.*, 18 (1970), 19-29.

COLLOQUIA MATHEMATICA SOCIETATIS JÁNOS BOLYAI
10. INFINITE AND FINITE SETS, KESZTHELY (HUNGARY), 1973.

k-HAMILTONIAN GRAPHS WITH GIVEN GIRTH

V. NEUMANN-LARA

1. INTRODUCTION

In 1963 H. Sachs proved that for every $r \geqslant 2$ and $g \geqslant 3$ there exist Hamiltonian regular graphs of degree r and girth g ([1] and [2]).

We prove here that for each pair (k, g) such that $k \geqslant 2$, $g \geqslant 3$ there exist an infinite number of graphs G with girth g which can be factorized into k edge-disjoint Hamiltonian cycles. G can be chosen bipartite provided g is even. (Theorems 5 and 9).

The techniques introduced by Sachs in [1] are used widely in this paper. In special Constuction 1 and the content of Theorem 1 are included in [1].

2. TERMINOLOGY

We shall employ Harary's terminology [3]. $V(G)$ and $E(G)$ are the set of points and edges of G respectively. G is a H_k-graph if it is factorizable in k pairwise edge-disjoint Hamiltonian cycles $C_1, \ldots, C_k$.

If moreover G contains a set of independent edges $a_1, \ldots, a_k$; $a_i \in E(C_i)$ $i = 1, \ldots, k$ we call G a *RH_k-graph* and the edges $a_1, \ldots, a_k$ are called *independent representative edges of* $C_1, \ldots, C_k$. Let G and G' be two graphs. A function $\pi: V(G) \to V(G')$ is an *homomorphism* from G into G' if it preserves adjacency. π is a *covering-homomorphism* if it is surjective and for each $(u', v') \in E(G')$ there exists an edge $(u, v) \in E(G)$ such that $u \in \pi^{-1}(u')$ and $v \in \pi^{-1}(v')$.

The girth of G will be denoted by $g(G)$ and for any walk W in G, $l(W)$ will denote the length of W. Finally Z_r represents the ring of integers modulo r.

3. EXISTENCE THEOREMS

CONSTRUCTION 1. THE CLASS $Q[B, X]$

Let B be any multigraph without isolated points. For every $a \in E(B)$ with endpoints u and v define $a^* = ((u, a), (v, a))$ and for every $u \in V(B)$ put $S(u) = \{(u, a) \mid a \in E(B)$ and u is an endpoint of $a\}$.

The graph B^* is defined by

$$V(B^*) = \bigcup_{u \in V(B)} S(u); \qquad E(B^*) = \{a^* \mid a \in E(B)\}.$$

Let X be a graph and suppose that B is regular of degree $|V(X)|$. Consider a family $(X_u)_{u \in V(B)}$ of pairwise disjoint isomorphic copies of X, each of them disjoint to B^* and bijections $\theta_u: S(u) \to V(X_u)$. (Such bijections exist because $|V(X_u)| = \text{degree } B = |S(u)|$). Identifying each point $w \in S(u)$ with $\theta_u(w)$ in $B^* \cup \bigcup_{u \in V(B)} X_u$ for every $u \in V(B)$ a graph G is obtained. After this identification $E(B^*)$ remains an independent set of edges in G and each X_u becomes an induced subgraph of G. The class of all graphs obtained in this way is denoted by $Q[B, X]$.

Theorem 1. *If X is a regular graph of degree r and girth g and B is a regular multigraph of degree $|V(X)|$ and girth at least $\left\{\frac{1}{2} g\right\}$ then every member $G \in Q[B, X]$ is a regular graph of degree $r + 1$ and girth g.*

Theorem 2. *Let* X *be a graph containing a spanning* RH_k*-subgraph* X_0 *and* B *a regular multigraph of degree* $|V(X)|$ *contaning a spanning* H_k*-submultigraph* B_0. *Then there exists a graph* $G \in Q[B, X]$ *which contains a spanning* RH_k*-subgraph.*

Proof. Let $\gamma_1, \ldots, \gamma_k$ be pairwise edge-disjoint Hamiltonian cycles of B_0; $C_1, \ldots, C_k$ pairwise edge-disjoint Hamiltonian cycles of X_0 and $a_1, \ldots, a_k$ independent representative edges of $C_1, \ldots, C_k$ respectively. Denote by C_{iu} the cycle which corresponds to C_i in X_u in Construction 1 and by $a_{iu} = (x_{iu}, y_{iu})$ the edge which corresponds to a_i in X_u. Let P_{iu} be the path obtained from C_{iu} by deletion of the edge a_{iu}.

Put $\gamma_i = (v_i^0, e_i^0, v_i^1, \ldots, v_i^{m-1}, e_i^{m-1}, v_i^0)$ where the upper indices are taken modulo $m = |V(B)|$ and suppose $u = v_i^j$. If Construction 1 is made in such a way that θ_u sends $\{(u, e_i^j), (u, e_i^{j+1})\}$ onto $\{x_{iu}, y_{iu}\}$, the set $E^*(\gamma_i) = \{a^* \mid a \in E(\gamma_i)\}$ together with the paths P_{iu} form a Hamiltonian cycle Γ_i in G. Clearly $\Gamma_1, \ldots, \Gamma_k$ are pairwise edge-disjoint. Choose $\alpha_i \in E(\gamma_i)$. Then $\alpha_1^*, \ldots, \alpha_k^*$ are independent representative edges of $\Gamma_1, \ldots, \Gamma_k$ respectively.

Lemma 1. *Let* $k_0 \geqslant 1$ *be an integer. Suppose that for every* $g \geqslant 3$ *there exists an* RH_{k_0}*-graph with girth* g. *Then for every triplet* (k_0, g, r) *such that* $g \geqslant 3$ *and* $r \geqslant 2k_0$ *there exists a regular graph of degree* r *and girth* g *which contains a spanning* RH_{k_0}*-subgraph.*

Proof. Suppose the lemma is not true. Let g_1 be the minimum value of g for which the Lemma fails to be true and r_1 the minimum value of r for which this occurs when $g = g_1$. By hypothesis $r_1 > 2k_0$. Let X be a regular graph of degree $r_1 - 1$ and girth g_1 containing a spanning RH_{k_0}-subgraph. Clearly $|V(X)| > 2k_0$.

We consider two cases:

Case 1. $g_1 = 3$. Take as B a multigraph with two points and $|V(X)|$ edges and let B_0 be a submultigraph of B consisting of these two points and any $2k_0$ edges.

Case 2. $g_1 > 3$. Let B be a regular graph of degree $|V(X)|$ and girth $\{\frac{1}{2}(g_1+1)\}$ containing a spanning RH_{k_0}-subgraph B_0. (Notice that $3 \leqslant \{\frac{1}{2}(g_1+1)\} < g_1$.)

By Theorem 2 there exists a graph $G \in Q[B, X]$ containing a spanning RH_{k_0}-subgraph. By Theorem 1, G is regular of degree r_1 and girth g_1. This contradicts the definition of g_1 and r_1.

Theorem 3. *Let G be a regular graph of degree $2km$ and girth $g \geqslant 3$ containing a spanning RH_k-subgraph. Suppose $m \geqslant g$. Then there exists a RH_{k+1}-graph $\widetilde{G}$ of girth g which can be contracted to G in such a way that each point of G is covered by exactly m points of $\widetilde{G}$.*

Proof.

(a) Let G_0 be a spanning RH_k-subgraph of G and $C_1, \ldots, C_k$ pairwise edge-disjoint Hamiltonian cycles of G_0. Delete in G all the edges belonging to $\bigcup_{i=1}^{k} C_i$. The resulting graph is regular of degree $2(m-1)k$ and therefore is 2-factorizable (Petersen's Theorem). Grouping 2-factors in a suitable way a $2(m-1)$-factorization is obtained. Let $E'_1, \ldots, E'_k$ be these $2(m-1)$-factors. Taking $E_i = E'_i \cup C_i$ we obtain a $2m$-factorization of G in which each factor E_i is Hamiltonian and therefore connected. Let $P_i = (u_i^0, e_i^0, u_i^1, \ldots, u_i^{s-1}, e_i^{s-1}, u_i^s)$, $u_i^0 = u_i^s$ be an Eulerian trail in E_i. Here $s = pm$ where $p = |V(G)|$. Put $P_{i\sigma} = (u_i^0, e_i^0, u_i^1, \ldots, u_i^\sigma)$ for $0 \leqslant \sigma \leqslant s$ and let S_i be the set obtained from $\{P_{i\sigma} \mid 0 \leqslant \sigma \leqslant s\}$ by identification of P_{i0} and P_{is}. Thus we can put $S_i = \{P_{i\sigma} \mid \sigma \in Z_s\}$. Define the cycle $\widetilde{E}_i$ by $V(\widetilde{E}_i) = S_i$; $E(\widetilde{E}_i) = \{(P_{i\sigma}, P_{i(\sigma+1)}) \mid \sigma \in Z_s\}$. The function $\pi_i \colon V(\widetilde{E}_i) \to V(G)$ defined by $\pi_i(P_{i\sigma}) = u_{i\sigma}$ is an homomorphism from $\widetilde{E}_i$ into G in which each point of G is covered by m points of $\widetilde{E}_i$.

Consider now the Hamiltonian cycle $C_1 = (x_0, a_0, x_1, \ldots, a_{p-1}, x_0)$ in G and suppose π_1 carry $\widetilde{a}_j = (y_j, w_{j+1})$ to a_j. Thus for each $j \in Z_p$, $\pi_1(y_j) = \pi_1(w_j) = x_j$. We can suppose that P_1 was chosen so that $y_0 \neq w_0$.

(b) Define the cycle Γ by $V(\Gamma) = Z_{pm}$, $E(\Gamma) = \{(i, i+1) \mid i \in Z_{pm}\}$ and denote by Γ_j the arc $(jm, jm+1, \ldots, jm+m-1)$ of Γ. Let $\theta_{ij}\colon V(\Gamma_j) \to \pi_i^{-1}(x_j)$; $j = 0, \ldots, p-1$; $i = 1, \ldots, k$ be bijections such that

$$\theta_{1j}^{-1}(w_j) = jm + \left\{\tfrac{1}{2} g\right\} - 1$$

$$\theta_{1j}^{-1}(y_j) = \begin{cases} \theta_{1j}^{-1}(w_j) & \text{if } \quad y_j = w_j \\ (j+1)m - \left[\tfrac{1}{2} g\right] & \text{if } \quad y_j \neq w_j \end{cases}$$

for $j = 0, \ldots, p-1$.

For each j and each $u \in V(\Gamma_j)$ identify in $\Gamma \cup \bigcup_{i=1}^{k} \widetilde{E}_i$ the elements $u, \theta_{1j}(u), \ldots, \theta_{kj}(u)$. In this way a graph $\widetilde{G}$ is obtained. After the identification Γ and the $\widetilde{E}_i$'s become edge-disjoint Hamiltonian cycles of $\widetilde{G}$ and the homomorphisms π_i induce a contraction $\pi\colon \widetilde{G} \to G$. (Notice that Γ_j is the subgraph of $\widetilde{G}$ induced by $\pi^{-1}(x_j)$). It is easy to see that each point of G is covered by m points of $\widetilde{G}$, each edge of G not in C_1 is covered by exactly one edge of $\widetilde{G}$ and each edge a_j is covered by two edges of $\widetilde{G}$ namely $\widetilde{a}_j$ and the edge $\widetilde{a}_j' = (jm+m-1, (j+1)m)$ of Γ.

(c) Let λ_j be the cycle of $\widetilde{G}$ formed by $\widetilde{a}_j$; and the arc $\left(y_j, \ldots, (j+1)m + \left\{\tfrac{1}{2} g\right\} - 2, (j+1)m + \left\{\tfrac{1}{2} g\right\} - 1\right)$ of Γ. This arc contains the edge $\widetilde{a}_j'$ and its number of points is

$$(j+1)m + \left\{\tfrac{1}{2} g\right\} - 1 - \left(jm + \left\{\tfrac{1}{2} g\right\} - 2\right) = m + 1 \qquad \text{or}$$

$$(j+1)m + \left\{\tfrac{1}{2} g\right\} - 1 - \left((j+1)m - \left[\tfrac{1}{2} g\right] - 1\right) = g$$

according to whether $w_j = y_j$ or $w_j \neq y_j$. Therefore $l(\lambda_j) \geqslant g$ for every j and $l(\lambda_0) = g$. Notice that $\lambda_0, \ldots, \lambda_{p-1}$ are edge-disjoint. With help of the simple following Lemma we shall prove that $g(\widetilde{G}) = g$.

Lemma 2. *Let G_1 be a graph consisting of a Hamiltonian cycle H and chords $c_1, \ldots, c_r$ of H. Let $A_1, \ldots, A_r$ be arcs of H determined* *by $c_1, \ldots, c_r$ respectively and suppose they are pairwise edge-disjoint. Then the cycles of G_1 are those formed by one chord c_i and its correspondent arc A_i and those obtained from H by substitution of some of the arcs A_i's by their correspondent c_i's.*

Let now C be a cycle of $\widetilde{G}$. We need only to prove that $l(C) \geqslant g$. The proof is divided into two cases.

Case 1. C contains an edge $\xi \notin \{\widetilde{a}_0, \ldots, \widetilde{a}_{p-1}\} \cup E(\Gamma)$. Then $\pi(C)$ contains a closed walk which uses $\pi(\xi)$ once and therefore contains a cycle in G. It follows that $l(C) \geqslant g$.

Case 2. $E(C) \subseteq \{\widetilde{a}_0, \ldots, \widetilde{a}_{p-1}\} \cup E(\Gamma)$. By Lemma 2 either $C = \lambda_j$ for some j and $l(C) \geqslant g$ or $E(\pi(C)) = \{a_0, \ldots, a_{p-1}\}$ and then $l(C) \geqslant p > g$.

(d) Let now $a_{j_1}, \alpha_2, \ldots, \alpha_k$ be a set of independent representative edges of $C_1, \ldots, C_k$ and let $\widetilde{\alpha}_2, \ldots, \widetilde{\alpha}_k$ be edges of $\widetilde{G}$ such that $\pi(\widetilde{\alpha}_i) = \alpha_i$; $i = 2, \ldots, k$. Since $|V(\widetilde{G})| > |V(G)| > 2km > 4k$, length of $\Gamma > 4k$ and therefore there exists $\alpha \in E(\Gamma)$ which is not adjacent to any of the edges $\widetilde{a}_{j_1}, \widetilde{\alpha}_2, \ldots, \widetilde{\alpha}_k$. These edges together with α form a set of independent representative edges of $E_1^*, \ldots, E_k^*$ and Γ.

Theorem 4. *For each triplet (k, g, r) such that $k \geqslant 1$, $g \geqslant 3$ and $r \geqslant 2k$ there exists a regular graph of degree r and girth g containing a spanning RH_k-subgraph.*

Proof. It follows by induction on k directly from Lemma 1 and Theorem 3.

Theorem 5. *For each pair (k, g) such that $k \geqslant 2$, $g \geqslant 3$ and each $m \geqslant g$ there exists a H_k-graph G with girth g such that $m / |V(G)|$.*

Proof. Take $k' = k - 1$. By Theorem 4 there exists a regular graph of degree $2k'm$ and girth g containing a spanning $H_{k'}$-subgraph. The $H_{k'+1}$-graph $\widetilde{G}$ constructed as in the proof of Theorem 3 has girth g and $m / |V(\widetilde{G})|$.

Theorem 6. *Let X be a bipartite graph containing a spanning RH_k-*

subgraph X_0 *and* B *a regular multigraph of degree* $|V(X)|$ *containing a spanning* H_k*-submultigraph* B_0. *Then there exists a bipartite graph* $G \in Q[B, X]$ *which contains a spanning* RH_k*-subgraph.*

Proof. G is constructed exactly as in the proof of Theorem 2 but some additional conditions are imposed on the bijections θ_u. Let $\{V^0, V'\}$ a partition of $V(X)$ into two independent sets. Since X contains Hamiltonian cycles, one has $|V^0| = |V^1|$ and therefore $|V(X)|$ is even. Give an arbitrary orientation to each cycle γ_i and consider an Eulerian trail P in B such that for every $a \in E(\gamma_i)$ the orientations of a induced by γ_i and P coincide. Give to each edge of B the orientation induced by P. If $a \in E(B)$ is oriented from u to v call (u, a) (resp: (v, a)) a 0-point (resp: 1-point). Clearly each edge of B^* joins a 0-point to a 1-point. Denote by $S^0(u)$ (resp: $S^1(u)$) the set of all 0-points (resp: 1-points) in $S(u)$. Obviously $|S^0(u)| = |S^1(u)|$ for every $u \in V(B)$. Let $V^0(X_u)$ and $V^1(X_u)$ be the sets which correspond in X_u to V^0 and V^1 respectively. Impose on θ_u the conditions $\theta_u(S^0(u)) = V^0(X_u)$, $\theta_u(S^1(u)) = V^1(X_u)$ for every $u \in V(B)$ and continue with the proof as in Theorem 2. It is easy to see that all conditions we have imposed on the θ_u's are compatible and that $\left\{\bigcup_{u \in V(B)} S^0(u), \bigcup_{u \in V(B)} S^1(u)\right\}$ is a partition of $V(G)$ into independent sets.

Lemma 3. *Let* $k_0 \geq 1$ *be an integer. Suppose that for every even number* $g \geq 4$ *there exists a bipartite* RH_{k_0}*-graph with girth* g. *Then for every triplet* (k_0, g, r) *such that* g *is an even number,* $g \geq 4$ *and* $r \geq 2k_0$, *there exists a bipartite regular graph of degree* r *and girth* g *which contains a spanning* RH_{k_0}*-subgraph.*

Proof. Take g_1, r_1 and X like in the proof of Lemma 1. By Theorem 4 there exists a regular graph B of degree $|V(X)|$ and girth $\frac{1}{2} g_1 + 1$ containing a spanning RH_{k_0}-subgraph. By Theorem 6 there exists a bipartite graph $G \in Q[B, X]$ *containing a spanning* RH_{k_0}-subgraph. By Theorem 1 G is regular of degree r_1 and girth g_1. This yields a contradiction.

Theorem 7. *Let* G *be a bipartite regular graph of degree* $2km$

and girth $g \geqslant 4$ *containing a spanning* RH_k*-subgraph. Suppose* $m = 2m_1 + 1 > g$. *Then there exists a bipartite regular* RH_{k+1}*-graph* $\tilde{G}$ *of girth* g *and a covering homomorphism* $\pi: \tilde{G} \to G$ *such that every point of* G *is covered by* m *points of* $\tilde{G}$.

Proof.

(a) This step is exactly the same as in Theorem 3.

(b) Define the cycle Γ (Fig. 1) by $V(\Gamma) = Z_p \times [1, m]$;

$$E(\Gamma) = \{((j, t), (j + 1, t + m_1)) | j \in Z_p,\ 1 \leqslant t \leqslant m_1 + 1\} \cup$$
$$\cup \{((j, t), (j + 1, t + m_1 + 1)) | j \in Z_p,\ 1 \leqslant t \leqslant m_1\}$$

and put $H_j = \{j\} \times [1, m]$, $j \in Z_p$. Let $\theta_{ij}: H_j \to \pi_i^{-1}(x_j)$ $j \in Z_p$, $i = 1, \ldots, k$ be bijections such that $\theta_{1j}^{-1}(w_j) = (j, m_1 + 1)$

$$\theta_{1j}^{-1}(y_j) = \begin{cases} \theta_{1j}^{-1}(w_j) & \text{if} \quad y_j = w_j \\ \left(j, \frac{1}{2} g\right) & \text{if} \quad y_j \neq w_j \end{cases}$$

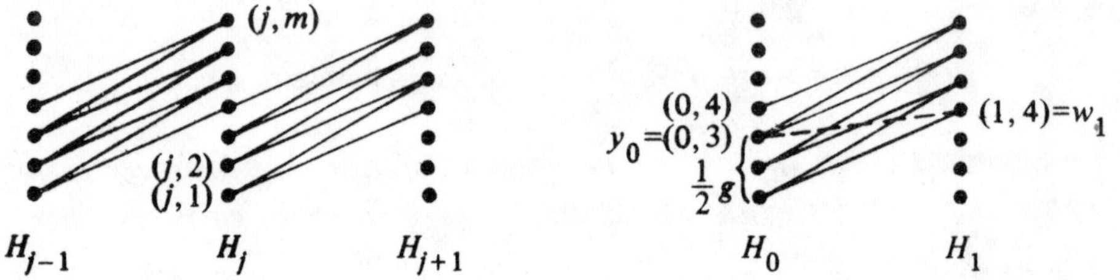

m_1=3, g=6

Figure 1

for $j \in Z_p$. For each j and each $u \in H_j$ identify in $\Gamma \cup \bigcup_{i=1}^{k} \tilde{E}_i$ the elements $u, \theta_{1j}(u), \ldots, \theta_{kj}(u)$. In this way a graph $\tilde{G}$ is obtained. After the identification, Γ and the $\tilde{E}_i$'s become edge-disjoint Hamiltonian cycles of $\tilde{G}$ and the homomorphisms π_i induce a covering homomorphism $\pi: \tilde{G} \to G$. Each point of G is covered by m points of $\tilde{G}$, each edge of G not in C_1 is covered by exactly one edge of $\widetilde{G}$ and each edge a_j by $\widetilde{a}_j$ and the edges of an arc Γ_j of Γ of length m (Fig. 1).

(c) $\widetilde{a}_j$ is a chord of the arc Γ_j. Let λ_j be the cycle formed by $\widetilde{a}_j$ and the subarc of Γ_j determined by $\widetilde{a}_j$. It is easy to prove

(1) $l(\lambda_j) \geqslant g$ for every j;

(2) $l(\lambda_0) = g$ (to prove it notice that $y_0 \neq w_0$ and therefore $\theta_{10}^{-1}(y_0) = \left(0, \frac{1}{2} g\right)$);

(3) $\lambda_0, \ldots, \lambda_{p-1}$ are pairwise edge-disjoint.

The proof of $g(\widetilde{G}) = g$ follows now as in Theorem 3.

(d) As in Theorem 3.

Theorem 8. *For each triplet* (k, g, r) *such that* $k \geqslant 1$, g *is even and* $\geqslant 4$ *and* $r \geqslant 2k$, *there exists a bipartite regular graph of degree* r *and girth* g *containing a spanning* RH_k*-subgraph.*

Proof. It follows by induction on k directly from Lemma 3 and Theorem 7.

Theorem 9. *For each pair* (k, g) *such that* $k \geqslant 2$, g *is even and* $\geqslant 4$ *and each* $m = 2m_1 + 1 > g$ *there exists a bipartite* H_k*-graph with girth* g *such that* $m / | V(G) |$.

Proof. Take $k' = k - 1$. By Theorem 8 there exists a bipartite regular graph of degree $2k'm$ and girth g containing a spanning $H_{k'}$-subgraph. The bipartite $H_{k'+1}$-graph $\widetilde{G}$ constructed as in the proof of Theorem 7 has girth g and $m / | V(\widetilde{G}) |$.

REFERENCES

[1] H. Sachs, Regular graphs with given girth and restricted circuits, *J. London Mat. Soc.*, 38 (1963), 423-429.

[2] H. Sachs, On regular graphs with given girth. *Theory of Graphs and its Applications,* (Proceeding Smolenice 1963), Prag 1964, 91-97.

[3] F. Harary, *Graph Theory,* Addison-Wesley, 1969.

[4] H. Sachs, *Einführung in die Theorie der endlichen Graphen,* Carl Hanser Verlag, München, 1971, 98-108.

COLLOQUIA MATHEMATICA SOCIETATIS JÁNOS BOLYAI
10. INFINITE AND FINITE SETS, KESZTHELY (HUNGARY), 1973.

A THEOREM ON THE THEORY OF THE FOUR-COLOUR PROBLEM

H.-J. PRESIA

By dualisation of the four-colour problem in its original form we obtain – as is well-known – the problem of four-colourability of the vertices of planar graphs without loops; it is this latter problem I am going to deal with in what follows. This paper refers to remarks made by H. Sachs in his paper, particularly.

In our investigation we especially use results which are related to those of Dirac, Aarts – de Groot, Winn – Ore – Stemple, Sachs, and Donec. In particular, a well-known theorem of Dirac, which on certain conditions permits the extension of a partial four-colouring of a planar graph G to the whole of G, is of great relevance to the proof.

This theorem is: *Let* $G = G' \cup G''$ *be partially four-coloured,* G' *denoting the coloured and* G'' *denoting the uncoloured subgraph. Besides, let* G'' *have the following three properties:*

(a) G'' *is a connected graph.*

(b) G'' *has vertices* P *only with* $v(P) \leq 5$ ($v(P)$ *means the valency of* P *with respect to* G).

(c) *In* G'' *there is at least one vertex* P_0 *with* $\nu(P_0) \leqslant 4$.

In that case the partial four-colouring may be extended to $G = G' \cup G''$ *(by an eventual interchange of colours of some vertices already coloured).*

As an immediate application of this theorem, the four-colourability of all connected planar graphs G without loops, which have at least one vertex P_0 with $\nu(P_0) \leqslant 4$ and whose other vertices P satisfy $\nu(P) \leqslant 5$, can be established.

This result could be improved by Aarts – de Groot [1] in such a way that the hypothesis of the existence of a vertex P_0 with $\nu(P_0) \leqslant 4$ is not necessary. Sachs succeeded in proving a further generalization of this theorem; in [4] he showed that also every planar graph G without loops, which has no more than three vertices with valencies greater than 5, is four-colourable. It is now possible to improve the result of Sachs.

The theorem is this: *If the planar graph* G *without loops has no more than four*

(*) *vertices the valencies of which are greater than* 5, *then* G *is colourable in four colours.*

Now some remarks on the proof which will be submitted for publication to the *"Mathematische Nachrichten"* [6].

The theorem will be proved through reductio ad absurdum by means of induction with respect to the number n of vertices, the proof having virtually the same structure as Sachs' proof. Trivially, for small n the proposition is true. Suppose now that the proposition is false: Then there exists a planar graph without loops whose vertices with at most four exceptions all have valencies not greater than 5, which cannot be coloured in four colours. Let G_0 be such a graph having the minimum number n of vertices; then we have to prove that G_0 does not exist.

The proof consists of five steps being partly lengthy:

(1) Similar to the proof of the theorems of Aarts – de Groot and Sachs by using an interchange of colours of Kempe chains we get:

The graph G_0 *is 3-connected (with respect to the vertices) having no loops and no separating triangles; the set of vertices may be denoted* $\{Q_1, Q_2, Q_3, Q_4; P_1, \ldots, P_{n-4}\}$, *the vertices having the following valencies*

$$\nu(Q_\nu) \geqslant 5, \ \nu = 1, 2, 3, 4;$$

$$\nu(P_1) = \nu(P_2) = \ldots = \nu(P_{n-4}) = 5.$$

(2) In connection with these investigations I succeeded in proving the following

Theorem. *If a triangulated planar graph* T *has no more than eight vertices the valencies of which are greater than* 5, *then* T *is four-colourable.*

The proof of this theorem (as well as the proof of the "57-triangle-theorem" quoted in the remark at the end of this paper) will be submitted for publication together with the proof of the theorem I am dealing with in this work [6].The proof makes use of a result of Donec [2] after which the vertices of a planar loopless graph are colourable in four colours unless their number is greater than 44. (Which is an improvement of the results of Winn [5] and Ore – Stemple [3]). From this theorem follows: G_0 is no triangulated graph.

(3) The structure of graphs G_0 with properties as formulated in (1) and (2) has been investigated by Sachs [4]. For that he introduced the term of "associated vertices": Two vertices P' and P'' are called associated via P when there exists in G_0 a vertex P adjacent to both P' and P'' such that both edges (P', P) and (P, P'') are boundary edges of a region Γ of G_0, which is not a triangle (see Fig. 1).

As G_0 is no triangulated graph according to (2), there are at least two pairs of associated vertices. It can be shown that associated vertices are never adjacent.

(4) We can now show that at least one pair of associated vertices P_0', P_0'' can be found such that no subset of $\{P_0', P_0''; Q_1, Q_2, Q_3, Q_4\}$ forms a separating set of vertices; such a pair will be called a "good" pair.

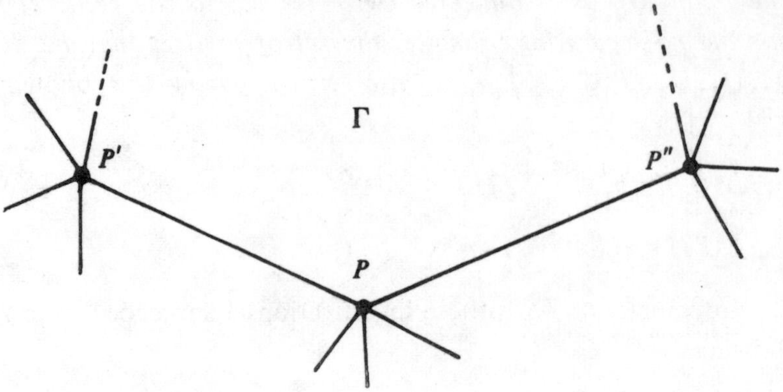

Figure 1

The proof of this proposition constitutes the most important and most involved step of the whole proof, and it requires some cases to be considered separately. This method is based on suitable reductions and interchanges of colours within Kempe chains.

(5) The application of the theorem of Dirac. Let P'_0, P''_0 be any good pair of vertices of G_0 associated via P. If P'_0 and P''_0 are identified thus generating a new vertex $\bar{Q}$ and if subsequently multiple edges are replaced by single ones, from G_0 a new planar graph $\bar{G}$ is derived (see Fig. 2).

In $\bar{G}$ only the valencies of $\bar{Q}, Q_1, Q_2, Q_3, Q_4$ may be greater than 5, and $\nu(\bar{P}) = 4$. Obviously, the five vertices $\bar{Q}, Q_1, Q_2, Q_3, Q_4$ can be coloured by four colours. The uncoloured subgraph $\bar{U} = \bar{G} - \bar{Q} - Q_1 - Q_2 - Q_3 - Q_4$, which evidently coincides with $U_0 = G_0 - P'_0 - P''_0 - Q_1 - Q_2 - Q_3 - Q_4$, is connected according to (4); in addition, it contains the vertex $\bar{P}$ with $\nu(\bar{P}) = 4$ and all the other vertices have valencies being not greater than 5.

Thus, the conditions of the theorem of Dirac being satisfied $\bar{G}$ can be coloured in four colours. From that a four-colouring of the original graph G_0 follows immediately.

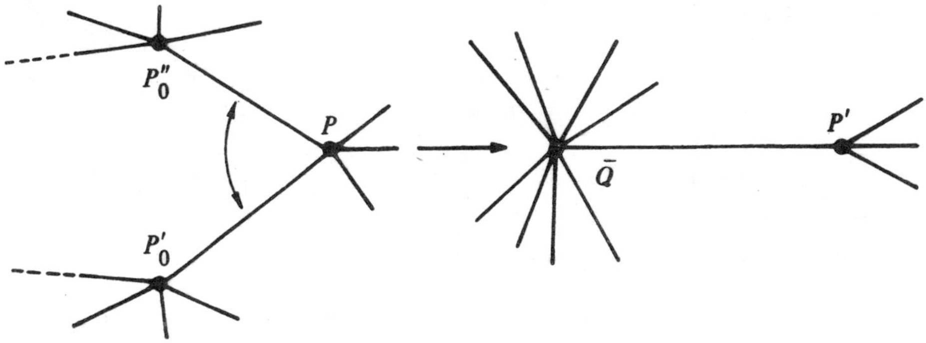

Figure 2

This leads to a contradiction that G_0 cannot be coloured by four colours and this proves the assertion.

Remark. In addition to the statement formulated in (2) I succeeded in improving the "51-triangle-theorem" of Sachs [4] by means of (*) and the result of Donec:

If the planar graph G without loops does not contain more than 57 circuits of length three, then G is four-colourable.

Supplementary Remark.

The result of Donec has in the meantime been improved by Stromquist in the following (unpublished) note:

W. Stromquist, The four-color theorem for small maps, Department of Treasury, Washington, D.C.

He proves that the vertices of a planar loopless graph are colourable in four colours unless their number is greater than 51. By using this new result the theorem on triangulations above can be improved to the theorem:

REFERENCES

[1] J.A. Aarts – J. de Groot, A case of colouration in the four colour problem. *Nieuw Arch. Wisk.*, 11 (1963), 10-18.

[2] G.A. Donec, Issledovanie voprosov raskraski ploskich grafov (*Dissertacija*). Kiev 1970. (Investigation of problems concerning the colouring of planar graphs (*Dissertation*). Kiev 1970).

[3] O. Ore – J. Stemple, Numerical calculations on the four-color problem. *J. Combinatorial Theory*, 8 (1970), 65-78.

[4] H. Sachs, *Einführung in die Theorie der endlichen Graphen*, Teil II. BSB B.G. Teubner Verlagsgesellschaft Leipzig, 1972.

[5] C.E. Winn, On the minimum number of polygons in an irreducible map. *Am. J. Math.*, 62 (1940), 406-416.

[6] H.-J. Presia, Über Färbungen planarer Graphen I, II. *(to appear).*

COLLOQUIA MATHEMATICA SOCIETATIS JÁNOS BOLYAI
10. INFINITE AND FINITE SETS, KESZTHELY (HUNGARY), 1973.

PLANARITY OF TWO-POINT UNIVERSAL GRAPHS

G.B. PURDY

Definition. A graph G is called super-universal if it has the following property: given two distinct points A and B of G, there exists a point F_1 joined to both, a point F_2 joined to neither, a point F_3 joined to A but not B, and a point F_4 joined to B but not A. The object of this note is to prove that such a graph cannot be planar, at any rate not if the number of vertices n is at least 136. The first occurrence of super-universal graphs is [1]. S.H. Hechler and others posed the problem showing the nonexistence of planar super-universal graphs. They are also called two-point universal graphs.

Lemma 1. *Let G be a super-universal graph and let A, B, C be distinct vertices of G. Then there is a path of length ≤ 3 from A to B which does not contain C.*

Proof. There is a point D joined to A and B. If $D \neq C$ we are done. Suppose therefore that $D = C$. Thus C is joined to A and B. There is a point E joined to A and not joined to C. Clearly $E \neq B$ (and of course $E \neq A, C$). There is a point F joined to both E and B.

We have $F \neq E$, since E is not joined to C. Hence $AEFB$ is the de-desired path. (If $F = A$ then there exists a path of length 1).

Theorem 1. *Let G be a planar super-universal graph. Let A be joined to 20 other vertices B, E, and X_i, $(1 \leqslant i \leqslant 18)$. Then the 18 vertices X_i are either all inside or all outside the triangle BAE.*

Proof. Suppose not. Then without loss of generality $X_1, \ldots, X_9$ are outside triangle ABE and $X_{10} = C$ is inside triangle ABE. There exists a point C' joined to C but not A. Hence (see Figure 1) $C' \neq B$, $C' \neq E$, and C' is on the inside of triangle ABE. For each X_i, $(1 \leqslant i \leqslant 9)$, there exists Y_i such that X_i is joined to Y_i and Y_i is joined to C'. Now $Y_i \neq A$, since C' is not joined to A, and C' is on the inside of triangle ABE, the X_i being on the outside.

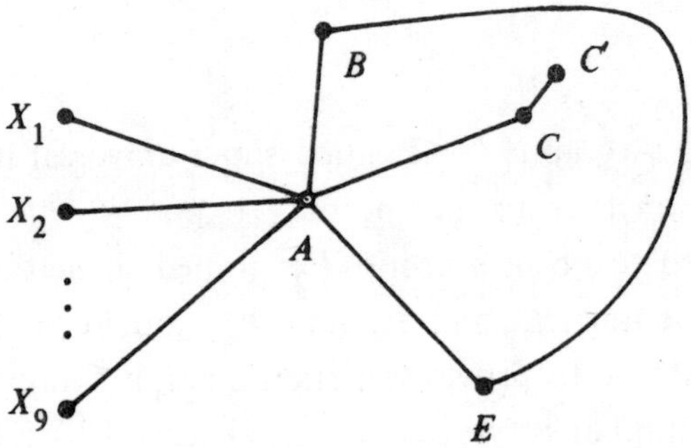

Fig. 1

Hence $Y_i \in \{B, E\}$. By the symmetry between B and E we may assume without loss of generality that $Y_1 = \ldots = Y_5 = B$; hence B is joined to X_1, X_2, X_3, X_4, and X_5. (See Figure 2). There is a point X_3' joined to X_3 but not A; we may suppose the numbering of $X_1, \ldots, X_5$ to be such that X_3' and X_3 are inside quadrilateral AX_4BX_2 which is inside quadrilateral AX_5BX_1. We recall that $X_1, X_2, \ldots, X_5$ are outside triangle AEB and C inside. It will be easier to visualize the situation if we regard $X_1, \ldots, X_5$ as being inside triangle AEB and C as being outside. There is of course no difference in fact between the two.

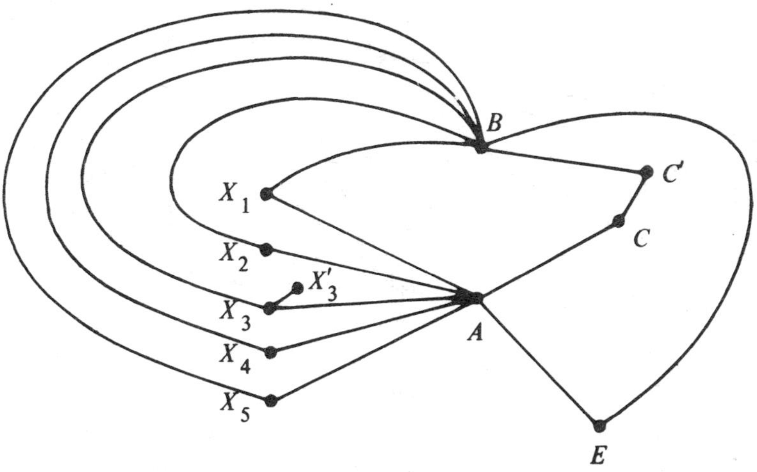

Fig. 2

There is a path of length $\leqslant 3$ from C' to X'_3 avoiding B – this is Lemma 1. Since C' is not joined to A, the path must go through E. Hence there is a path of length $\leqslant 2$, avoiding B, which goes from E to X'_3 (since A is not joined to X'_3). The path $C'EAX'_3$ does not exist because A is not joined to X'_3. The only possible paths remaining are $C'EX_1X'_3$ and $C'EX_5X'_3$. But X_1 and X'_3 are on opposite sides of quadrilateral AX_4BX_2, and are therefore not joined.

Similarly X_5 and X'_3 are on opposite sides of the same quadrilateral AX_4BX_2 and cannot be joined. Hence there is a contradiction and the theorem is proved. (See Frigure 3).

Theorem 2. *Let G be a planar super-universal graph. Let A be joined to B, C, E and X_i, $(1 \leqslant i \leqslant 14)$, let B and E both be joined to F, where F is not joined to A, and suppose that the X_i are on the outside of quadrilateral $BAEF$ and C is on the inside. Then C is joined to F.*

Proof. Suppose that C is not joined to F. Then there exists C' joined to C but not joined to A, and C' must be inside quadrilateral

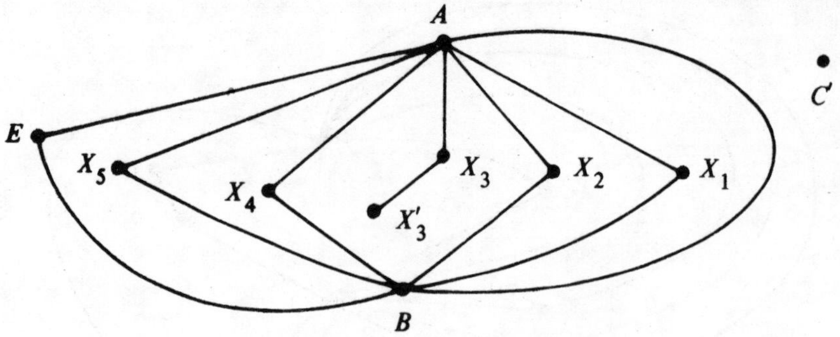

Figure 3

BAEF. For each X_i, $(1 \leqslant i \leqslant 14)$, there is a Y_i joined to both X_i and C'. Since C' is not joined to A, Y_i cannot be A. (See Figure 4.) Hence each $Y_i \in \{B, E, F\}$. Due to the symmetry between B and E, there are only two essentially different possibilities. Either $Y_1 = \ldots = Y_5 = B$ or $Y_1 = \ldots = Y_5 = F$.

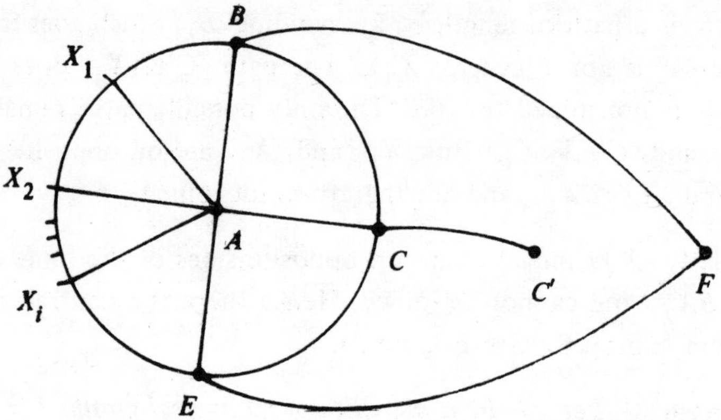

Figure 4

Case 1. Suppose $Y_1 = Y_2 = \ldots = Y_5 = B$. (See Figure 5.) Then B is joined to $X_1, \ldots, X_5$. We may suppose that the X_i are so numbered that X_3 and X'_3 are inside quadrilateral AX_4BX_2 which is inside quadrilateral AX_5BX_1 which is inside quadrilateral $AEFB$ with C' out-

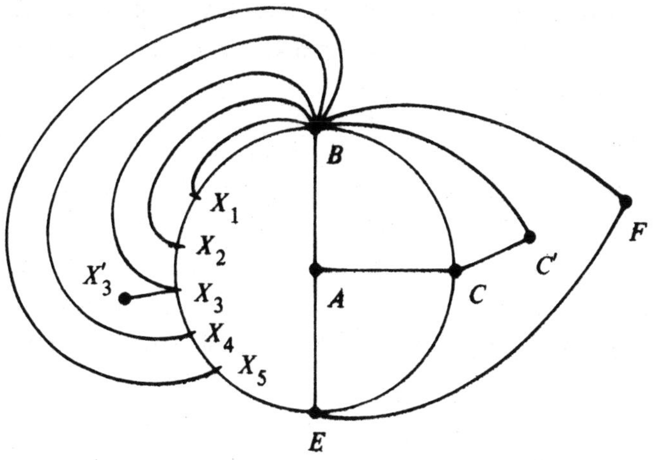

Figure 5

side quadrilateral $AEFB$ (we turned quadrilateral $AEFB$ inside out). Here, of course X_3' is a point joined to X_3 but not joined to A. (See Figure 6.) By Lemma 1, there is a path of length $\leqslant 3$ from C' to X_3' avoiding B. Since C' is not joined to A, the path must go through E or F. Hence, since A is not joined to X_3', there is a path from either E or F to X_3' of length $\leqslant 2$. The only possible paths are $C'EX_1X_3'$, $C'EX_5X_3'$, $C'FX_1X_3'$, and $C'FX_5X_3'$. But X_5 and X_3' are not joined,

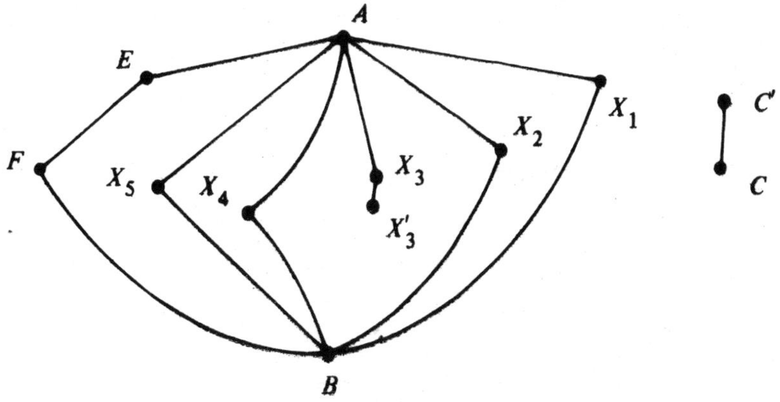

Figure 6

being on opposite sides of quadrilateral AX_4BX_2. Similarly X_1 and X'_3 are not joined, being on opposite sides of the same quadrilateral AX_4BX_2. Thus case 1 ends in a contradiction.

Case 2. Suppose $Y_1 = Y_2 = \ldots = Y_5 = F$. (See Figure 7.) Then F is joined to $X_1, \ldots, X_5$. We may suppose that the X_i are so numbered that X_3 and X'_3 are inside quadrilateral AX_4FX_2 which is inside quadrilateral AX_5FX_1 which is inside quadrilateral $AEFB$ with C' outside quadrilateral $AEFB$ (again we have turned quadrilateral $AEFB$ inside out). Here, of course X'_3 is a point joined to X_3 but not joined

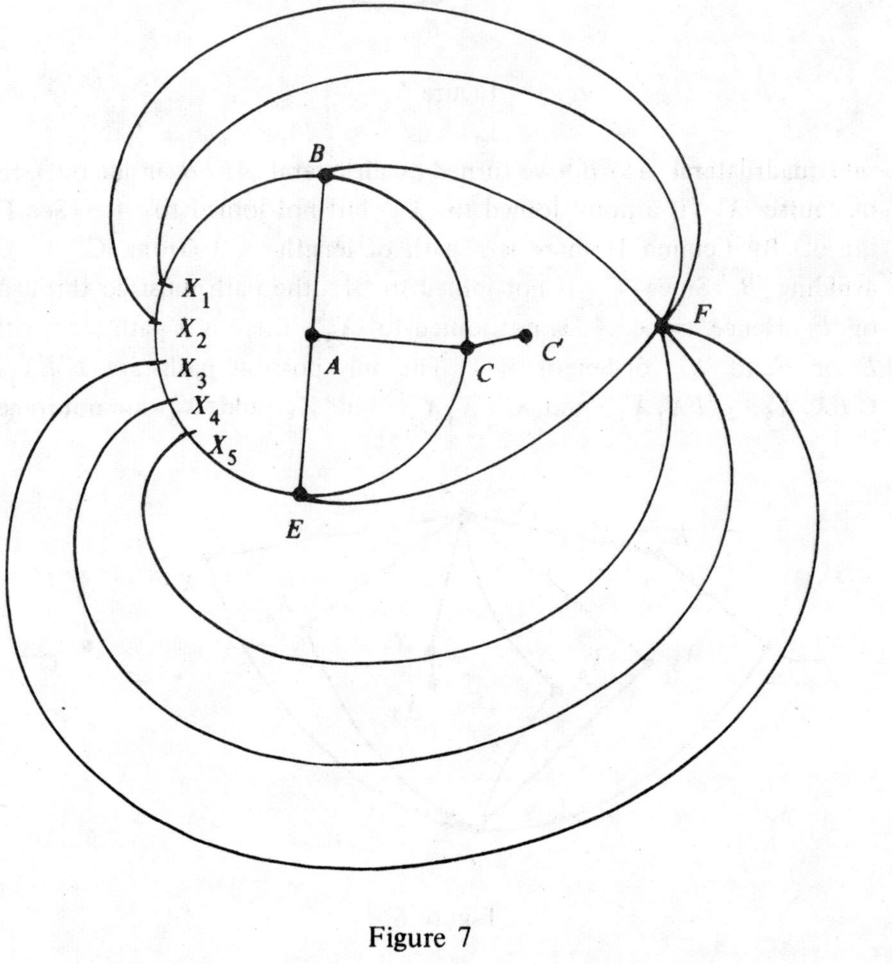

Figure 7

to A. (See Figure 8.) By Lemma 1, there is a path of length $\leqslant 3$ from C' to X'_3 avoiding F. Since C' is not joined to A, the path must go through E or B. Without loss of generality the path goes through E and the argument proceeds as in case 1 giving the desired contradiction.

Lemma 2. *Let G be a planar super-universal graph. Then there exists a vertex with valence at least* $\frac{n+4}{5}$.

Proof. Since G is planar, there is a point A of valence $K \leqslant 5$. Let $X_1, \ldots, X_k$ be the points joined to A. Let $Y_1, Y_2, \ldots, Y_{n-1}$ be the points of $G - A$. For each Y_i there is a point joined to both Y_i and A. Hence each Y_i is joined to some X_j. Hence some X_j must be joined to at least $\frac{n-1}{5}$ of the Y_i and have valence $\geqslant 1 +$ $+ \frac{n-1}{5} = \frac{n+4}{5}$.

Lemma 3. *Let G be a super-universal graph, and let A, B, C, D be four distinct vertices. Then there is a path of length $\leqslant 4$ from A to B which does not pass through C or D.*

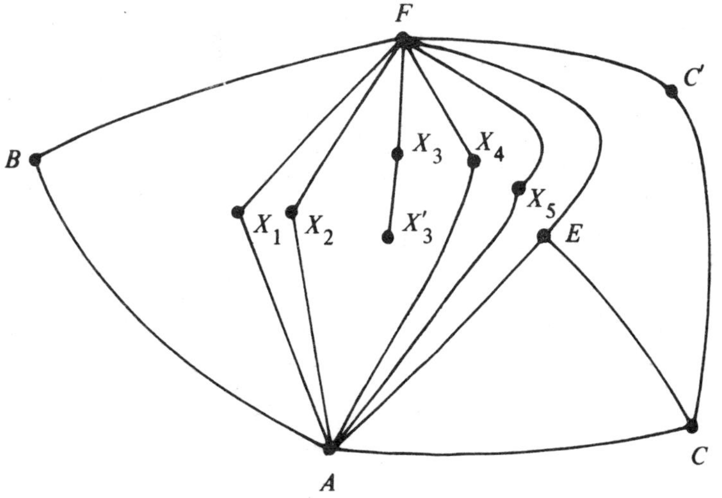

Figure 8

Proof. We may suppose that A is not joined to B. Let E be a point of G joined to neither C nor D. If $E = A$, then let F be a point joined to E and B and we are done. If $E = B$, let F be joined to E and A and we are done. Suppose that $E \neq A, B$. Let F be joined to E and A, and let H be joined to E and B. We are done.

Lemma 4. *Let G be a super-universal planar graph, let A, F, X_i, $(1 \leqslant i \leqslant m)$ be vertices, $m \geqslant 24$ and let A be joined to X_i, $(1 \leqslant i \leqslant m)$ (assume counter-clockwise ordering) but let A not be joined to F. Then F cannot be joined to more than 6 of the points X_i.*

Proof. Suppose that F is joined to 7 points X_i. By Theorem 2 these points must be consecutive modulo m. Without loss of generality they are $X_1, \ldots, X_7$. Now X_4 is inside quadrilateral AX_3FX_5 which is inside quadrilateral AX_2FX_6 which is inside quadrilateral AX_1FX_7 and there exists an X_i outside quadrilateral AX_1FX_7 not joined to either X_1 or X_7, since otherwise X_1 or X_7 must be joined to nine other points contrary to Theorem 1. There is a path of length $\leqslant 4$ from X_i to X_4 avoiding A and F – this is Lemma 3. But this cannot be, and we have the desired contradiction.

Theorem 3. *A super-universal graph G with $n \geqslant 136$ vertices cannot be planar.*

Proof. Suppose that G is planar. By Lemma 2, there is a vertex A with valence $m \geqslant \frac{140}{5} = 28$. Let $X_1, \ldots, X_m$ be joined to A. Let the embedding of G into the plane be fixed, and let the X_i be numbered so that the arcs AX_i emanating from A are in counter-clockwise order. It follows from Theorem 1 that X_i and X_j can be joined only if i and j are consecutive modulo m. Let G' be the graph G restricted to $\{X_1, \ldots, X_m\}$. Any path from X_i to X_j confined to G' has length $\geqslant \min_r |i - j + rm|$, although there may not be a path at all. Since $m \geqslant 28$, there certainly exist four points $X_{i_1}, X_{i_2}, X_{i_3}, X_{i_4}$ in counter-clockwise order such that the distance between any two consecutive ones in G' is at least 7. There is a path of length $\leqslant 3$ between X_{i_1} and X_{i_3} which avoids A. The path must be $X_{i_1}F_1F_3X_{i_3}$, where F_1 and F_3 are not joined to A. Similarly there is a path $X_{i_2}F_2F_4X_{i_4}$ where F_2, F_4 are not joined to A. Theorem 2 and Lemma 4 imply that the F_i are all distinct. Now F_2 and F_4 are on opposite sides of the pentagon $AX_{i_3}F_3F_1X_{i_1}$ and yet they are joined by an edge. This cannot happen in a planar graph.

REFERENCE

[1] S.H. Hechler, Large Super-universal Metric Spaces, *Israel Journal of Mathematics,* 14, 2 (1973), 115-148.

COLLOQUIA MATHEMATICA SOCIETATIS JÁNOS BOLYAI
10. INFINITE AND FINITE SETS, KESZTHELY (HUNGARY), 1973.

ANTI-RAMSEY THEOREMS

R. RADO*

I. INTRODUCTION.

Let us call an equivalence relation $x \equiv y$ on a set S the *maximal equivalence relation* on S if $x \equiv y$ holds for all $x, y \in S$. At the other extreme we have the *minimal equivalence relation* on S, which is such that $x \equiv y$ holds only if $x = y$. Ramsey's theorem [1] can be described roughly as asserting that under suitable conditions every equivalence relation on a sufficiently large set S of sequences induces the maximal equivalence relation on some arbitrarily large subset of S. The present note deals with theorems which establish a conclusion in the opposite direction, by asserting that under certain conditions every equivalence relation on a sufficiently large set S of sequences induces the minimal equivalence relation on some arbitrarily large subset of S. It is interesting to observe that one of our "anti-Ramsey"-theorems, Theorem 4, is proved with the help of Ramsey's theorem itself. Whereas Ramsey imposes an upper bound on

*This research was supported by a Canadian Commonwealth Research Fellowship which the author held at the University of Waterloo, Canada.

the number of equivalence classes, the theorems of this note, naturally, require some hypothesis which, on the contrary, guarantees the existence of many equivalence classes. We achieve this by stipulating the validity of certain cancellation laws, viz. (2), (3) and (5). These laws hold, for instance, whenever the equivalence relation arises in a natural way in connection with a group.

2. NOTATION AND TERMINOLOGY

If nothing is said to the contrary, we use the following conventions. Capital letters denote sets, and $|A|$ denotes the cardinality of A. The relation $A \subset B$ denotes inclusion in the wide sense. Small Roman letters, other than $u, v, w, x, y, z,$ denote cardinal numbers. In particular, r, s, t always denote finite cardinals. An *ordered set* is a pair $(A, \prec)$, where $\prec$ is a total order relation on A. The order type of $(A, \prec)$ is denoted by $\mathrm{tp}\,(A, \prec)$. We put

$$[A]^c = \{X \subset A:\ |X| = c\}\,.$$

We use the *obliterator* $\hat{\ }$, whose effect consists in deleting from a sequence the element or elements above which it is placed. Thus, for $r \leqslant s$ the symbol $(x_r, \ldots, \hat{x}_s)$ denotes the sequence $(x_r, \ldots, x_{s-1})$ if $r < s$, and the empty sequence $(-)$ if $r = s$.

If R is a binary relation on the set* $X = \{x_0, \ldots, \hat{x}_\rho\}$ then the symbol

$$\{x_0, \ldots, \hat{x}_\rho\}_R$$

denotes the set X and, at the same time, expresses the validity of $x_\mu R x_\nu$ for $\mu < \nu < \rho$. Similarly for *sequences* $(x_0, \ldots, \hat{x}_\rho)_R$. *Except in the* proof of Theorem 3, all sequences occurring in this note have a finite number of terms.

An *e-relation* on A is an equivalence relation on the set

$$\{(a_0, \ldots, \hat{a}_t):\ t < \aleph_0;\ a_0, \ldots, \hat{a}_t \in A\}\,.$$

*ρ a finite or infinite ordinal.

The simplest kind of e-relation on A which satisfies the conditions (2), (3) and (5) below, is obtained by embedding A in a group $\Gamma = (G, \cdot)$, so that $A \subset G$, and putting, for $a_0, \ldots, \hat{a}_r, b_0, \ldots, \hat{b}_s \in A$, $(a_0, \ldots, \hat{a}_r) \equiv$ $\equiv (b_0, \ldots, \hat{b}_s)$ whenever $a_0 \ldots a_{r-1} = b_0 \ldots b_{s-1}$. Here $c_0 \ldots c_{t-1}$ is the unit of Γ if $t = 0$. Call this particular relation the e-relation on A *generated* by Γ. An e-relation $\equiv$ is called *commutative* if $(a(0), \ldots, \hat{a}(t)) \equiv$ $\equiv (a(\pi(0)), \ldots, \hat{a}(\pi(t)))$ whenever $\{0, \ldots, \hat{t}\} = \{\pi(0), \ldots, \hat{\pi}(t)\}$.

If φ and ψ are order types then the *partition relation*

$$\varphi \to (\psi)^c_n$$

expresses the condition what whenever $\operatorname{tp}(A, \prec) = \varphi$; $|N| = n$; $[A]^c =$ $= \bigcup_{\nu \in N} I_\nu$, then there is $B \subset A$ and $\nu \in N$ such that $\operatorname{tp}(B, \prec) = \psi$ and $[B]^c \subset I_\nu$.

3. RESULTS

Theorem 1. *Let* $0 < n < \aleph_0$, *and let* $(A, \prec)$ *be an ordered set,*

(1) $$|A| > \frac{1}{3}(2^{2n+1} + 1)\,.$$

Let $\equiv$ *be an* e*-relation on* A *satisfying the following conditions:*

(2) $$\begin{cases} \text{if} & \{x, \ldots, \hat{x}_r\}_\prec,\ \{y_0, \ldots, \hat{y}_s\}_\prec,\ \{z\} \subset A \quad \text{and if} \\ \text{either} & \text{(i)}\ (x_0, \ldots, \hat{x}_r, w) \equiv (y_0, \ldots, \hat{y}_s, w) \quad \text{or} \\ & \text{(ii)}\ (w, x_0, \ldots, \hat{x}_r) \equiv (w, y_0, \ldots, \hat{y}_s), \\ \text{then} & (x_0, \ldots, \hat{x}_r) \equiv (y_0, \ldots, \hat{y}_s); \end{cases}$$

(3) *if* $x, y \in A$ *and* $(x) \equiv (y)$, *then* $x = y$.

Then

(A) $$\begin{cases} \text{there is } B \in [A]^{n+1} \text{ such that} \\ \text{whenever } \{x_0, \ldots, \hat{x}_r\}_\prec,\ \{y_0, \ldots, \hat{y}_s\}_\prec \subset B \text{ and} \\ (x_0, \ldots, \hat{x}_r) \neq (y_0, \ldots, \hat{y}_s), \text{ then} \\ (x_0, \ldots, \hat{x}_r) \not\equiv (y_0, \ldots, \hat{y}_s)\,. \end{cases}$$

Next*, we give a best possible result for commutative e-relations. It shows that the condition (1) is not too far from being best possible in Theorem 1.

Theorem 2. *Let* $0 < n < \aleph_0$.

(i) *If* $|A| > 3^n$ *and* $\equiv$ *is a commutative e-relation on* A *satisfying* (2) *and* (3), *then* (A) *holds.*

(ii) *There is an abelian group* Γ *of order* 3^n, *such that the e-relation generated by* Γ *does not satisfy* (A).

A best possible result for infinite sets:

Theorem 3. *Let* $(A, \prec)$ *be an infinite well-ordered set, and* $\equiv$ *an e-relation on* A *satisfying* (2) *and* (3). *Then there is* $B \in [A]^{|A|}$ *such that whenever* $\{x_0, \ldots, \hat{x}_r\}_\prec, \{y_0, \ldots, \hat{y}_s\}_\prec \subset B$ *and* $(x_0, \ldots, \hat{x}_r) \neq$ $\neq (y_0, \ldots, \hat{y}_s)$ *then*

$$(x_0, \ldots, \hat{x}_r) \not\equiv (y_0, \ldots, \hat{y}_s) .$$

Our last theorem is a generalisation of Theorem 1, except that the modest lower bound in (1) is replaced by a value of the Ramsey function, which is notorious for taking large values. In the conclusion of Theorem 1 the elements of the sequences were arranged in accordance with a fixed ordering of A, and no multiple occurrence of an element was allowed. We now dispense with both these restrictions, but we have to introduce a new condition (6). For a justification of (6) see remark (v) below.

Theorem 4. *Let* $r < \aleph_0$, *and denote by* c_r *the number of equivalence relations on a set of cardinal* $\sum_{s \leq r} (2r)^s$. *Let* $(A, \prec)$ *be an ordered set and* ψ *be an order type such that* $\psi \geq 2r + 1$ *and*

(4) $$\operatorname{tp}(A, \prec) \to (\psi)^{2r}_{c_r} .$$

Let $\equiv$ *be an e-relation on* A. *Suppose that*

*I owe to a conversation with E.C. Milner an improvement to the original, weaker, version of Theorem 2.

$$(5)\quad \begin{cases} \text{whenever} \quad 0 \leqslant s,\ t - s \leqslant r; \\ (x_0, \ldots, \hat{x}_s, y, x_s, \ldots, \hat{x}_t) \equiv (x_0, \ldots, \hat{x}_s, z, x_s, \ldots, \hat{x}_t), \\ \text{then} \quad y = z. \end{cases}$$

Then there is $B \subset A$ *such that* $\mathrm{tp}\,(B, \prec) = \psi$, *and*

$$(x_0, \ldots, \hat{x}_s) \not\equiv (x_s, \ldots, \hat{x}_t)$$

for $0 \leqslant s,\ t - s \leqslant r$ *and* $x_0, \ldots, \hat{x}_t \in B$, *provided that for at least one* $\tau_0 < t$ *we have*

$$(6)\qquad x_{\tau_0} \neq x_0, \ldots, \hat{x}_{\tau_0}, \ldots, \hat{x}_t\,.$$

Remarks.

(i) If ψ is finite then (4) holds provided $|A|$ is finite and sufficiently large.

(ii) If $\psi = \omega$ then (4) holds for $\mathrm{tp}\,(A, \prec) = \omega$.

(iii) If ψ is an arbitrary ordinal number then (4) holds provided $\mathrm{tp}\,(A, \prec)$ is a sufficiently large ordinal number.

(iv) A crude estimate for c_r is $c_r \leqslant 2^{(2r)^{2r+2}}$.

(v) The condition (6) cannot be omitted in Theorem 4.

(vi) The conditions (2), (3) imply (5).

4. PROOFS

Proof of Theorem 1. We may assume $|A| < \aleph_0$. Call a sequence $(x(0), \ldots, \hat{x}(m))$ *good* if $x(0), \ldots, \hat{x}(m) \in A$ and

$$(x(\alpha_0), \ldots, \hat{x}(\alpha_r)) \not\equiv (x(\beta_0), \ldots, \hat{x}(\beta_s))$$

whenever $\{\alpha_0, \ldots, \hat{\alpha}_r\}_<, \{\beta_0, \ldots, \hat{\beta}_s\}_< \subset \{0, \ldots, \hat{m}\}$ and $(\alpha_0, \ldots, \hat{\alpha}_r) \neq (\beta_0, \ldots, \hat{\beta}_s)$. A sequence which is not good is called *bad*. Thus the empty sequence $(-)$ is good. Let $(x_0, \ldots, \hat{x}_m)$ be good and $\alpha < \beta < m$. Then $x_\alpha \neq x_\beta$. For if $x_\alpha = x_\beta$ then $(x_\alpha) \equiv (x_\beta)$ which contradicts the definition of goodness. To prove Theorem 1 it suffices to find a good sequence $(x_0, \ldots, x_n)$. We define inductively, as long ast this is possible, elements $x_0, x_1, \ldots$ of A, using the following rule: if $\mu < \aleph_0$ and $x_0, \ldots, \hat{x}_\mu \in A$, then x_μ is minimal in $(A, \prec)$ such that $(x_0, \ldots, x_\mu)_\prec$ is good. This construction yields a good sequence $(x_0, \ldots, \hat{x}_t)_\prec$ such that x_t does not exist. It suffices to prove $t > n$.

Let $z \in A$. Suppose, for the moment, that the sequence $(x_0, \ldots, \hat{x}_t, z)$ is good. Then $(x_0, \ldots, \hat{x}_t, z)_{\neq}$. If $(x_0, \ldots, \hat{x}_t, z)_\prec$ then x_t exists, which is false. Hence there is $\alpha < t$ such that $(x_0, \ldots$ $\ldots, \hat{x}_\alpha; z, x_\alpha, \ldots, \hat{x}_t)_\prec$. Then $(x_0, \ldots, \hat{x}_\alpha, z)_\prec$ is good, and since $z \prec x_\alpha$ we have a contradiction against the definition of x_α. It follows that $(x_0, \ldots, \hat{x}_t, z)$ is bad.

We have at least one of the following two cases.

Case 1. There is $z \in A$ such that

$$(x_{\alpha_0}, \ldots, \hat{x}_{\alpha_r}, z) \equiv (x_{\beta_0}, \ldots, \hat{x}_{\beta_s}, z)$$

for some sets $\{\alpha_0, \ldots, \hat{\alpha}_r\}_<, \{\beta_0, \ldots, \hat{\beta}_s\}_< \subset \{0, \ldots, \hat{t}\}$ with

$$(\alpha_0, \ldots, \hat{\alpha}_r) \neq (\beta_0, \ldots, \hat{\beta}_s) .$$

Then, by (2) (i), $(x_{\alpha_0}, \ldots, \hat{x}_{\alpha_r}) \equiv (x_{\beta_0}, \ldots, \hat{x}_{\beta_s})$ which contradicts the fact that $(x_0, \ldots, \hat{x}_t)$ a good.

Case 2. For every $z \in A$ we have $(x_{\alpha_0}, \ldots, \hat{x}_{\alpha_r}, z) \equiv (x_{\beta_0}, \ldots, \hat{x}_{\beta_s})$ for some sets $\{\alpha_0, \ldots, \hat{\alpha}_r\}_<, \{\beta_0, \ldots, \hat{\beta}_s\}_< \subset \{0, \ldots, \hat{t}\}$. Here $r, s,$ $\alpha_0, \ldots, \hat{\alpha}_r, \beta_0, \ldots, \hat{\beta}_s$ depend on z. For each z make a choice of these numbers for which r is minimal, and put

$$f(z) = ((\alpha_0, \ldots, \hat{\alpha}_r), (\beta_0, \ldots, \hat{\beta}_s)) .$$

Then it follows from (2) (ii) that $\alpha_0 \neq \beta_0$ whenever $r, s \geqslant 1$. If $z, z' \in A$ and $f(z) = f(z')$, then

$$(x_{\alpha_0}, \ldots, \hat{x}_{\alpha_r}, z) \equiv (x_{\beta_0}, \ldots, \hat{x}_{\beta_s}) \equiv (x_{\alpha_0}, \ldots, \hat{x}_{\alpha_r}, z')$$

and hence, by (2) (ii) and (3), $z = z'$. An easy counting argument gives

$$\frac{1}{3}(2^{2n+1}+1) < |A| = |\{f(z)\colon z \in A\}| \leqslant$$

$$\leqslant |\{((\alpha_0, \ldots, \hat{\alpha}_r)_<, (\beta_0, \ldots, \beta_s)_<)\colon \text{ if } r, s > 0 \text{ then}$$

$$\alpha_0 \neq \beta_0\}| = 2^t \cdot 2^t - \sum_{\lambda < t} 2^{t-\lambda-1} \cdot 2^{t-\lambda-1} = \frac{1}{3}(2^{2t+1}+1)$$

and completes the proof.

Proof of Theorem 2 (i). The proof is similar to that of Theorem 1. Define good and bad sequences as in the proof of Theorem 1. The terms of a good sequence are pairwise distinct. Hence there is a good sequence $(x(0), \ldots, \hat{x}(m))$ for which m is maximal. Let $z \in A$. Then there are sets

$$\{\alpha_0, \ldots, \hat{\alpha}_r\}_<, \{\beta_0, \ldots, \hat{\beta}_s\}_< \subset \{0, \ldots, \hat{m}\}$$

with $(x(\alpha_0), \ldots, \hat{x}(\alpha_r), z) \equiv (x(\beta_0), \ldots, \hat{x}(\beta_s))$. Choose $r, s, \alpha_0, \ldots, \hat{\alpha}_r$, $\beta_0, \ldots, \hat{\beta}_s$ so that r is minimal. Then, by (2) and (3) and commutativity, $(\alpha_0, \ldots, \hat{\alpha}_r, \beta_0, \ldots, \hat{\beta}_s)_{\neq}$.

Put $f(z) = (t_0, \ldots, \hat{t}_m)$, where, for $\lambda < m$,

$$t_\lambda = \begin{cases} 0 & \text{if } \lambda \in \{\alpha_0, \ldots, \hat{\alpha}_r\} \\ 1 & \text{if } \lambda \in \{\beta_0, \ldots, \hat{\beta}_s\} \\ 2 & \text{if } \lambda \in \{0, \ldots, \hat{m}\} - \{\alpha_0, \ldots, \hat{\alpha}_r, \beta_0, \ldots, \hat{\beta}_s\}. \end{cases}$$

By (2) and (3), $f(z) = f(z')$ implies $z = z'$. Hence

$$3^n < |A| = |\{f(z)\colon z \in A\}| \leqslant 3^m ,$$

and the conclusion follows.

Proof of Theorem 2 (ii). Let $(A, +, \cdot)$be the Galois field $GF(3^n)$. Consider the e-relation on A which is generated by the group $(A, +)$. Let $x_0, \ldots, x_n \in A$. Then $\sum_\nu \nu_\nu x_\nu = 0$ for some $\nu_0, \ldots, \nu_n \in \{0, 1, 2\}$, not all zero. This follows since $GF(3^n)$ is a vector space of dimension n. Thus, in the terminology used above, the sequence $(x_0, \ldots, x_n)$ is bad for the relation $\equiv$. Since $x_0, \ldots, x_n$ are arbitrary, (ii) follows.

Proof of Theorem 3. In this proof sequences are not restricted to be of finite length. Again, the argument is ismilar to that establishing Theorem 1. A sequence $(x(0), \ldots, \hat{x}(\mu))$ is *good* if and only if $x(0), \ldots, \hat{x}(\mu) \in A$, and $(x(\alpha_0), \ldots, \hat{x}(\alpha_r)) \not\equiv (x(\beta_0), \ldots, \hat{x}(\beta_s))$ whenever $\{\alpha_0, \ldots, \hat{\alpha}_r\}_<, \{\beta_0, \ldots, \hat{\beta}_s\}_< \subset \{0, \ldots, \hat{\mu}\}$ and $(\alpha_0, \ldots, \hat{\alpha}_r) \neq \neq (\beta_0, \ldots, \hat{\beta}_s)$. The empty sequence is good, and the terms of a good sequence are pairwise distinct. Define inductively, as long as possible, elements $x_0, x_1, \ldots$ of A as follows. If μ is a finite or infinite ordinal, and if $x_0, \ldots, \hat{x}_\mu$ have already been defined then x_μ is the least element of $(A, \prec)$ such that $(x_0, \ldots, x_\mu)_\prec$ and $(x_0, \ldots, x_\mu)$ is good. By transfinite construction we obtain a good sequence $(x_0, \ldots, \hat{x}_\lambda)_\prec$ which is such that x_λ does not exist. Then $|\lambda| \leqslant |A|$. Let $z \in A$. Suppose that $(x_0, \ldots, \hat{x}_\lambda, z)$ is good. Then $(x_0, \ldots, \hat{x}_\lambda, z)_{\neq}$. If $(x_0, \ldots, \hat{x}_\lambda, z)_\prec$ then we have a contradiction against the definition of λ. Hence there is a least ordinal $\alpha < \lambda$ such that $z \prec x_\alpha$. Then the sequence $(x_0, \ldots, \hat{x}_\alpha, z)_\prec$ is good, which contradicts the definition of x_α. Thus we have shown that $(x_0, \ldots, \hat{x}_\lambda, z)$ is bad. We have two cases:

Case 1. There is $z \in A$ such that $(x_{\alpha_0}, \ldots, \hat{x}_{\alpha_r}, z) \equiv \equiv (x_{\beta_0}, \ldots, \hat{x}_{\beta_s}, z)$ for some $r, s, \alpha_0, \ldots, \hat{\alpha}_r, \beta_0, \ldots, \hat{\beta}_s$ such that $\{\alpha_0, \ldots, \hat{\alpha}_r\}_<, \{\beta_0, \ldots, \hat{\beta}_s\}_< \subset \{0, \ldots, \hat{\lambda}\}$ and $(\alpha_0, \ldots, \hat{\alpha}_r) \neq \neq (\beta_0, \ldots, \hat{\beta}_s)$. Then, by (2) (i), $(x_{\alpha_0}, \ldots, \hat{x}_{\alpha_r}) \equiv (x_{\beta_0}, \ldots, x_{\beta_s})$ which contradicts the fact that $(x_0, \ldots, \hat{x}_\lambda)$ is good.

Case 2. If $z \in A$ then $(x_{\alpha_0}, \ldots, \hat{x}_{\alpha_r}, z) \equiv (x_{\beta_0}, \ldots, \hat{x}_{\beta_s})$ for some $\{\alpha_0, \ldots, \hat{\alpha}_r\}_<, \{\beta_0, \ldots, \hat{\beta}_s\}_< \subset \{0, \ldots, \hat{\lambda}\}$. We make a choice of $r, s, \alpha_0, \ldots, \hat{\alpha}_r, \beta_0, \ldots, \hat{\beta}_s$ and put $f(z) = ((\alpha_0, \ldots, \hat{\alpha}_r), (\beta_0, \ldots, \hat{\beta}_s))$. If $f(z) = f(z')$ then, by (2) (ii) and (3), $z = z'$. Therefore

$$|A| = |\{f(z) \colon z \in A\}| \leqslant$$

$$\leqslant |\{((\alpha_0, \ldots, \hat{\alpha}_r)_<, (\beta_0, \ldots, \hat{\beta}_s)_<) \colon$$

$$r, s < \aleph_0;\ \alpha_0, \ldots, \hat{\beta}_s < \lambda\}|\,.$$

If $|\lambda| < \aleph_0$ then $|A| < \aleph_0$, which is false. Hence $|\lambda| \geqslant \aleph_0$ and so

$|\lambda| \leqslant |A| \leqslant \sum_{r,s<\aleph_0} |\lambda|^{r+s} = |\lambda|$. Thus the set $B = \{x_0, \ldots, \hat{x}_\lambda\}_\prec$ has the required property.

Proof of Theorem 4. We begin by justifying the remarks (i)-(vi) made after the statement of the theorem. (i) and (ii) follow from Ramsey's theorem [1]. For (iii) see [2], Theorem 4. It should be remarked here that the proof of Theorem 4 in [2] assumes the General Continuum Hypothesis. This assumption may be omitted at the cost of increasing $\mathrm{tp}\,(A, \prec)$. We obtain (iv) by observing that a binary relation on A is a subset of $A \times A$. To justify (v) consider the e-relation on an infinite set A which is generated by a group $\Gamma = (A, \cdot)$. Suppose that the orders of the elements of Γ do not exceed r. Then there is no non-empty set $B \subset A$ which has the property required in Theorem 4. For if $x \in B$ then there is $s \in \{1, \ldots, r\}$ such that $(x_0, \ldots, x_{s-1}) \equiv (-)$ if $x_0 = \ldots = \hat{x}_s = x$. Alternatively, a counter example is provided by every infinite abelian group. The assertion (vi) is clearly true.

We now assume the hypotheses of Theorem 4. For $X = \{x(0), \ldots, \hat{x}(2r)\}_\prec \subset A$ put $f(X) = \{((\alpha_0 \ldots, \hat{\alpha}_s), (\alpha_s, \ldots, \hat{\alpha}_t)):$ $0 \leqslant s, t - s \leqslant r;\ \alpha_0, \ldots, \hat{\alpha}_t < 2r;\ (x(\alpha_0), \ldots, \hat{x}(\alpha_s)) \equiv (x(\alpha_s), \ldots, \hat{x}(\alpha_t))\}$. Then

(7) $\qquad |\{f(X)\colon X \in [A]^{2r}\}| \leqslant c_r$.

In order to see this we note that the set

$$P = \{(\alpha_0, \ldots, \hat{\alpha}_s)\colon s \leqslant r;\ \alpha_0, \ldots, \hat{\alpha}_s < 2r\}$$

has cardinality $\sum_{s \leqslant r} (2r)^s$, and that the pairs $((\alpha_0, \ldots, \hat{\alpha}_s), (\alpha_s, \ldots, \alpha_t))$ which occur in the definition of $f(X)$ define an equivalence relation on P. It now follows from (7) and (4) that there is a set $B \subset A$ with $\mathrm{tp}\,(B, \prec) = \psi$, such that $f(X)$ is constant on $[B]^{2r}$. Let us now assume that $0 \leqslant s,\ t - s \leqslant r;\ x_0, \ldots, \hat{x}_t \in B;$

$$\tau_0 < t;\ x_{\tau_0} \neq x_0, \ldots, \hat{x}_{\tau_0}, \ldots, \hat{x}_t$$

and that, moreover, $(x_0, \ldots, \hat{x}_s) \equiv (x_s, \ldots, \hat{x}_t)$. We have to deduce a contradiction. We may assume $\tau_0 < s$. We can find a set B' such that

$x_0, \dots, \hat{x}_t \in B' = \{(0), \dots, \hat{y}(2r)\}_{\prec} \subset B$. Then $x_\tau = y(\gamma(\tau))$ for $\tau < t$, where $\gamma(0), \dots, \hat{\gamma}(t) < 2r$. By (6), $\gamma(\tau_0) \neq \gamma(\tau)$ for $\tau_0 \neq \tau$. Since $|B| \geqslant 2r + 1$ we can choose a set $B^* \in [B]^{2r+1}$. We can write

$$B^* = \{z(0), \dots, \hat{z}(\gamma(\tau_0)), z', z'', \hat{z}(\gamma(\tau_0)), \dots, \hat{z}(2r)\}_{\prec} .$$

By definition of B, $f(B^* - \{z''\}) = f(B^* - \{z'\})$. We conclude, in view of our assumptions, that, with $z(\gamma(\tau)) = w_\tau$ for $\tau < t$,

$$(w_0, \dots, \hat{w}_{\tau_0}, z', \hat{w}_{\tau_0}, \dots, \hat{w}_s) \equiv (w_s, \dots, \hat{w}_t) .$$

Similarly,

$$(w_0, \dots, \hat{w}_{\tau_0}, z'', \hat{w}_{\tau_0}, \dots, \hat{w}_s) \equiv (w_s, \dots, \hat{w}_t) .$$

But now (5) gives $z' = z''$, which is the required contradiction.

REFERENCES

[1] F.P. Ramsey, On a problem in formal logic, *Proc. London Math. Soc.*, 2, 30 (1930), 264-286.

[2] P. Erdős – A. Hajnal – R. Rado, Partition relations for cardinal members, *Acta Math. Acad. Sci. Hungar.*, 16 (1965), 93-196.

COLLOQUIA MATHEMATICA SOCIETATIS JÁNOS BOLYAI
10. INFINITE AND FINITE SETS, KESZTHELY (HUNGARY), 1973.

ON PARTITIONAL MATROIDS WITH APPLICATIONS

A. RECSKI

Let S denote a finite set. $\mathcal{S} = (S_1, S_2, \ldots, S_p)$ is called a *partition* of S if $\bigcup_{i=1}^{p} S_i = S$ and if $S_i \cap S_j \neq \phi$ iff $i = j$. $|\mathcal{S}|$ will denote the number of subsets of $\mathcal{S}$. A partial ordering of the set of the partitions of S is defined: $\mathcal{S}_1 \leqslant \mathcal{S}_2$ iff $\mathcal{S}_1$ is a *refinement* of $\mathcal{S}_2$ that is if each element of $\mathcal{S}_1$ is contained in a single element of $\mathcal{S}_2$. $\mathcal{S}_0 = (\mathcal{S}'; \mathcal{S}'')$ will denote the *"roughest common refinement"* of $\mathcal{S}'$ and $\mathcal{S}''$ that is $\mathcal{S}_0 \leqslant \mathcal{S}'$ and $\mathcal{S}_0 \leqslant \mathcal{S}''$ and if $\mathcal{S} \leqslant \mathcal{S}'$ and $\mathcal{S} \leqslant \mathcal{S}''$ then $\mathcal{S} \leqslant \mathcal{S}_0$.

A partition $\mathcal{S}$ is considered and let $\mathcal{A} = (a_1, a_2, \ldots, a_p)$ be an ordered set of integers such that $0 \leqslant a_i \leqslant |S_i|$, $(i = 1, 2, \ldots, p)$. The system $[\mathcal{S}, \mathcal{A}]$ corresponds to a matroid on S, namely a subset $X \subseteq S$ is independent in this matroid iff $|X \cap S_i| \leqslant a_i$ for $i = 1, 2, \ldots, p$ (see Berge [1]). The matroids of this type are called *partitional matroids.* The matroid $[\mathcal{S}, \mathcal{A}]$ is said to be defined *over* the partition $\mathcal{S}$ in the following sense: The matroid is uniquely determined by the system $[\mathcal{S}, \mathcal{A}]$ but this system is not unique for a given partitional matroid unless for $i = 1, 2, \dots, p$, $0 \neq a_i \neq |S_i|$. In order to establish a one-one correspondence

(*) *all of the subsets* S_i *with* $a_i = 0$ *or with* $a_i = |S_i|$ *are supposed to consist of a single element only.*

For a partitional matroid $[\mathcal{S}, \mathcal{A}]$ one can easily see that $|\mathcal{S}| = |S|$ iff the matroid has a single basis only. The rank of a matroid $\mathcal{M}$ will be denoted by $\rho(\mathcal{M})$. Obviously, $\rho([\mathcal{S}, \mathcal{A}]) = \sum_{i=1}^{|\mathcal{S}|} a_i$.

The system of the matroids on a set S has also a partial ordering: $\mathcal{M}_1 \subseteq \mathcal{M}_2$ iff any independent subset in the matroid $\mathcal{M}_1$ is independent in $\mathcal{M}_2$ too. If $\mathbf{1}$ and $\mathbf{0}$ denote the matroids with the property that all of the subsets of S are independent and that none of the nonempty subsets is independent respectively then for an arbitrary matroid $\mathcal{M}$ on S $\mathbf{0} \subseteq \mathcal{M} \subseteq \mathbf{1}$. This partially ordered set is not lattice-ordered if $|S| > 2$ (see Recski [4]), and it is an open question whether there exists a function of dimension on it (I conjecture yes).*

This partial ordering $\subseteq$ of the matroids will be considered on the set of the partitional matroids over S only. The relation between the partial ordering of the partitional matroids and that of the corresponding partitions will be presented and the structure of the partially ordered set of the partitional matroids will thus be treated.

The importance of the partitional matroids can be emphasized by two facts. On the one hand, a matroid is transversal iff it is the sum of partitional matroids (see Mirsky [3]). On the other hand, the partitional matroids can be applied in the electric network theory, see Appendix II.

I.

Lemma 1. *Let* $\mathcal{M} = [\mathcal{S}, \mathcal{A}]$ *and* $\mathcal{M}' = [\mathcal{S}', \mathcal{A}']$ *and* $\mathcal{M} \subseteq \mathcal{M}'$. *One can always find a partitional matroid* $\mathcal{N}$ *over a partition* $\mathcal{T}$ *such that* $\mathcal{M} \subseteq \mathcal{N} \subseteq \mathcal{M}'$ *and* $\mathcal{T} = (\mathcal{S}, \mathcal{S}')$.

*A counterexample was given by A.W. Ingleton (private communication, 31st July 1974). Using his construction a maximal chain of length $n(n + 1)/2$ can be given.

Proof. Let $\mathcal{S} = (S_1, S_2, \ldots, S_p)$; $\mathcal{S}' = (S'_1, S'_2, \ldots, S'_q)$ and $\mathcal{T} = (T_1, T_2, \ldots, T_r)$. For $i = 1, 2, \ldots, r$, $T_i = S_{\sigma(i)} \cap S'_{\sigma'(i)}$. Let $b_i = \min\{|T_i|, a_{\sigma(i)}\}$ and $\mathcal{B} = (b_1, b_2, \ldots, b_r)$, and, finally, $\mathcal{N} = [\mathcal{T}, \mathcal{B}]$. Obviously, $\mathcal{M} \subseteq \mathcal{N}$. One has to prove that $\mathcal{N} \subseteq \mathcal{M}'$. Let the subset $X \subseteq S$ be independent in $\mathcal{N}$ but not independent in $\mathcal{M}'$. X may be supposed to be contained totally in one of the subsets of $\mathcal{S}'$, say in S'_1. Let $I \subseteq \{1, 2, \ldots, r\}$ such that $i \in I$ iff $|X \cap T_i| > 0$. If i and j are different elements of I then $\sigma(i) \neq \sigma(j)$, since $\mathcal{T}$ is the "*roughest* common refinement" of $\mathcal{S}$ and $\mathcal{S}'$. $X \cap T_i$ is independent in $\mathcal{N}$ for any $i \in I$, since X is independent in $\mathcal{N}$. But these subsets are contained in different subsets of $\mathcal{S}$ and X is therefore independent in $\mathcal{M}$. In this case $\mathcal{M} \subseteq \mathcal{M}'$ implies that X is independent in $\mathcal{M}'$ too, a contradiction.

Lemma 2. *Let* $\mathcal{M} = [\mathcal{S}, \mathcal{A}]$ *and* $\mathcal{M}' = [\mathcal{S}', \mathcal{A}']$ *and* $\mathcal{M} \subseteq \mathcal{M}'$. *If* $\mathcal{S} > \mathcal{S}'$ *then* $\rho(\mathcal{M}) < \rho(\mathcal{M}')$.

Proof. Let $S'_1 \in \mathcal{S}'$ be a proper subset of $S_i \in \mathcal{S}$. According to (*) $0 < a_i < |S_i|$. If a'_1 denotes the integer corresponding to the subset S'_1 in $\mathcal{M}'$ then $a'_1 \geqslant \min\{a_i, |S'_1|\}$, since $\mathcal{M} \subseteq \mathcal{M}'$. The same reasoning holds for the other subset (or subsets) of $\mathcal{S}'$ contained in S_i, and since $\rho(\mathcal{M}') = \sum_{i=1}^{|\mathcal{S}'|} a'_i$, the statement follows.

Lemma 3. *Let* $\mathcal{M} = [\mathcal{S}, \mathcal{A}]$ *and* $\mathcal{M}' = [\mathcal{S}', \mathcal{A}']$ *and* $\mathcal{M} \subseteq \mathcal{M}'$. *If* $\rho(\mathcal{M}) = \rho(\mathcal{M}')$ *then* $\mathcal{S} \leqslant \mathcal{S}'$.

Proof. Lemma 2 implies that $\mathcal{S}' < \mathcal{S}$ is not possible. Let us suppose that $\mathcal{S}$ and $\mathcal{S}'$ cannot be compared in this partial ordering. Let $\mathcal{N}$ be a partitional matroid on S over $\mathcal{T}$ such that $\mathcal{M} \subseteq \mathcal{N} \subseteq \mathcal{M}'$ and $\mathcal{T} = (\mathcal{S}, \mathcal{S}')$ (see Lemma 1). Since $\mathcal{S}$ and $\mathcal{S}'$ cannot be compared, $\mathcal{T}$ is a proper refinement of $\mathcal{S}$. Thus, Lemma 2 implies that $\rho(\mathcal{M}) < \rho(\mathcal{N})$, a contradiction, since $\rho(\mathcal{M}) \leqslant \rho(\mathcal{N}) \leqslant \rho(\mathcal{M}') = \rho(\mathcal{M})$ implies that $\rho(\mathcal{M}) = \rho(\mathcal{N})$.

If x and y are different elements of a partially ordered set then x is said to be *covered* by y ($x \prec y$) if $x \subset y$ yet $x \subset t \subset y$ for no t. The function $d(x)$ is called a *function of dimension* of the partial ordering if, for any pair x, y, $x \prec y$ iff $x \subseteq y$ and $d(x) + 1 = d(y)$. [$d(x)$ is sometimes called the "rank-function" of the partially ordered set. Throughout, the rank has the matroid-theoretical meaning only and $d(x)$ will be called the function of dimension in order to avoid misunderstanding.]

Theorem 1. *Let* $|S| = n$ *and* $1 \leqslant k \leqslant n-1$. *The partitional matroids on* S *with rank* k *form a partially ordered subset with the following function of dimension* d_k: *If* $\mathcal{M}$ *is a partitional matroid over* $\mathcal{S}$ *and with rank* k *then* $d_k(\mathcal{M}) = n - |\mathcal{S}|$.

Proof. Obviously, $d_k(\mathcal{M}) = 0$ iff $\mathcal{M}$ consists of a single basis only. Similarly, $d_k(\mathcal{N}) = n - 1$ iff each k-element subset of S is a basis in $\mathcal{N}$. If $\mathcal{M}_1 \subseteq \mathcal{M}_2$ and $d_k(\mathcal{M}_1) + 1 = d_k(\mathcal{M}_2)$ then $\mathcal{M}_1 \prec \mathcal{M}_2$ is obviously implied by Lemma 3. On the other hand let $\mathcal{M}_2$ cover $\mathcal{M}_1$. The relation $\mathcal{M}_1 \subseteq \mathcal{M}_2$ is evident and $d_k(\mathcal{M}_1) < d_k(\mathcal{M}_2)$ is also implied by Lemma 3. Let the partitions of $\mathcal{M}_1$ and $\mathcal{M}_2$ be denoted by $\mathcal{S}_1$ and $\mathcal{S}_2$ respectively. If $d_k(\mathcal{M}_2) - d_k(\mathcal{M}_1)$ were greater than one then *either* at least one of the members, say S_0, of $\mathcal{S}_2$ would contain at least 3 members of $\mathcal{S}_1$ *or* at least two of the members, say S_1 and S_2, of $\mathcal{S}_2$ would contain each at least two members of $\mathcal{S}_1$. In the first case let $S_0 = T_1 \cup T_2 \cup T_3$, $(T_1 \in \mathcal{S}_1, T_2 \in \mathcal{S}_1)$ and let $\mathcal{S}_0$ be a partition of S such that $T_1 \in \mathcal{S}_0$, $T_2 \cup T_3 \in \mathcal{S}_0$ and $\mathcal{S}_1 \leqslant \mathcal{S}_0 \leqslant \mathcal{S}_2$. In the second case let $S_1 = V_1 \cup V_2$, $(V_1 \in \mathcal{S}_1, V_2 \in \mathcal{S}_1)$ and let $\mathcal{S}_0$ be a partition of S such that $V_1 \in \mathcal{S}_0$, $V_2 \in \mathcal{S}_0$ and $\mathcal{S}_1 \leqslant \mathcal{S}_0 \leqslant \mathcal{S}_2$. One can easily find a partitional matroid $\mathcal{M}_0$ over $\mathcal{S}_0$ is both cases such that $\mathcal{M}_1 \subset \mathcal{M}_0 \subset \mathcal{M}_2$, a contradiction, since $\mathcal{M}_1 \prec \mathcal{M}_2$.

Q.E.D.

II.

The partially ordered set of the partitional matroids on the set S consists of $n + 1$ subsets $(n = |S|)$; each subset contains the partitional matroids of a fixed rank k, $(k = 0, 1, \dots, n)$. There is a single matroid if $k = 0$ or if $k = n$ (0 and 1 respectively); all the other subsets possess a function of dimension d_k. Let $\mathcal{M}_{n,k}$ denote the matroid with $d_k(\mathcal{M}_{n,k}) = n - 1$ (that is, all the k-element subsets of S are bases in $\mathcal{M}_{n,k}$). Similarly, let $\mathcal{M}^{(k)}$ denote the set of all partitional matroids $\mathcal{N}$, $\left[\binom{n}{k} \text{ in number}\right]$ with rank k and with the property that $d_k(\mathcal{N}) = 0$ (that is, they possess a single basis each).

The structure of the partially ordered set of the partitional matroids on S is to be characterized by the relation among these $n + 1$ subsets.

Lemma 4. *Let two subsets be considered, containing all the partitional matroids of rank x and y respectively. These subsets are isomorphic (with respect to the ordering relation) iff $x + y = n$.*

Proof. If $x + y \neq n$ then for example $|\mathcal{M}^{(x)}| \neq |\mathcal{M}^{(y)}|$. If $x + y = n$ then the required one-one correspondence is the duality.

Note that $\mathcal{M}_1 \subseteq \mathcal{M}_2$ implies that $\mathcal{M}_1^* \subseteq \mathcal{M}_2^*$ iff $\rho(\mathcal{M}_1) = \rho(\mathcal{M}_2)$.

Lemma 5. *Let $\mathcal{N}_1 \in \mathcal{M}^{(k)}$ and $\mathcal{N}_2 \in \mathcal{M}^{(k+1)}$ and $\mathcal{N}_1 \subset \mathcal{N}_2$. Then any chain between $\mathcal{N}_1$ and $\mathcal{N}_2$ has length $k + 1$.*

Proof. Let $\mathcal{N}_1 \subseteq \mathcal{N} \subset \mathcal{N}_2$. If X_1 and X_2 denote the single basis of $\mathcal{N}_1$ and $\mathcal{N}_2$ respectively then $X_1 \subset X_2$. The rank of $S - X_2$ is zero and that of X_2 is k in the matroid $\mathcal{N}$. X_2 must therefore contain a single circuit C such that $C \cap X_1 \neq \phi$. Let $s = |C|$. Obviously, $s = 1$ iff $\mathcal{N} = \mathcal{N}_1$. All the possible values $s = 1, 2, \ldots, k + 1$ result some matroids and the statement follows. Fig. 1a visualizes symbolically all the matroids greater than or equal to one of the matroids $\mathcal{M}^{(k)}$ and less than $\mathcal{N}_2$: adding the broken lines would result the lattice of the $k + 1$-dimensional cube.

Lemma 6. *Any chain of partitional matroids between $\mathcal{M}_{n,k}$ and $\mathcal{M}_{n,k+1}$ has length 2.*

Proof. Let $\mathcal{M}_{n,k} \subset [\mathcal{S}, \mathcal{A}] \subset \mathcal{M}_{n,k+1}$ and $\mathcal{S} = (S_1, S_2, \ldots, S_p)$. Let $k > 1$ at first. If $|S_i| \geqslant k$ then $a_i = k$. Since $\sum_{i=1}^{|\mathcal{S}|} a_i = k + 1$, at most one of the elements of $\mathcal{S}$ may have cardinality greater than or

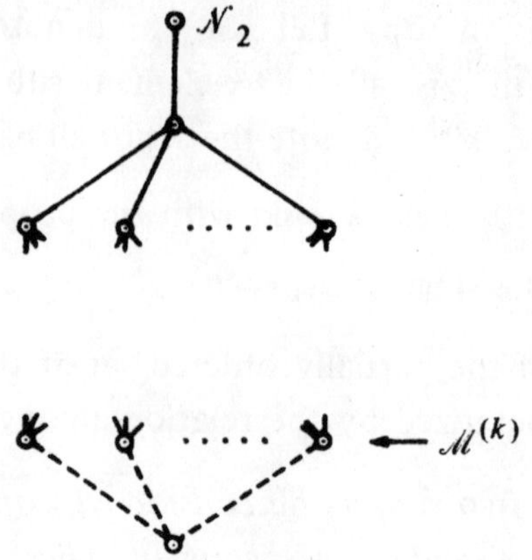

Fig. 1a

equal to k. If none of them had, $[\mathcal{S}, \mathcal{A}]$ would equal to 1. Therefore, all but one of the elements of $\mathcal{S}$ must have cardinality one (see (*)) and each corresponding number a_i equals to 1. Therefore, $\mathcal{S} = (S_1, S_2)$; $\mathcal{A} = (k, 1)$ and $|S_1| = n - 1$ since $\rho([\mathcal{S}, \mathcal{A}]) = k + 1$. Fig. 1b visualizes all the partitional matroids (n in number) between $\mathcal{M}_{n,k}$ and $\mathcal{M}_{n,k+1}$. If $k = 1$ then $|\mathcal{S}| = 2$ follows in the same way, and the matroids $[(S_1, S_2), (1, 1)]$ are pairwise incomparable, by Lemma 1.

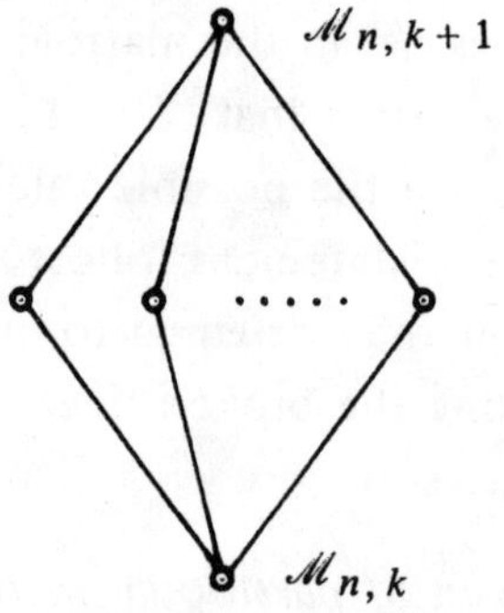

Fig. 1b

Theorem 2. *The partially ordered set of the partitional matroids on the set* S *does not possess any function of dimension if* $|S| \geqslant 4$.

Proof. The statement is trivially implied by Theorem 1 and Lemmata 5 and 6. Let us add that two maximal chains of partitional matroids between **0** and **1** can easily be given with lengths $3n - 3$ and $\frac{n(n+1)}{2}$ respectively.

APPENDICES

I.

"he . . . flung himself upon his horse
and rode madly off in all directions"
(S. Leacock: Gertrude the Governess)

Figure 2 presents the diagram of the partially ordered set of the partitional matroids on a set of cardinality 4, in order to visualize the above statements. Dotted lines connect pairs of matroids of the *same* rank.

II.

The application of the partitional matroids in the theory of electric networks with controlled sources is briefly presented.

If a network consists of independent (voltage and/or current) generators and passive one-port elements only then it can be visualized by a graph and described by the aid of the circuits and minimal cut-sets of this graph (see Kirchhoff [2]). If the (2-connected) graph consists of e edges and v vertices then $e - v + 1$ independent circuits and $v - 1$ independent cut-sets can be found in the graph (by the aid of a fixed but arbitrary spanning tree of the graph), yielding together e Kirchhoff-type equations. (The Kirchhoff Voltage (or Current) Laws state that the sum of the voltages (or currents) of the elements along any circuit (or cut-set respectively) is zero.) Further e equations state some connections between the voltage and the current of the same element (Ohm's Laws).

More complex networks (e.g. those containing semiconductor elements like transistors) can be modelled by one-port elements as well, but the concept of the controlled sources is required, that is, the voltage (or the current) of the generator will be the function of the voltage or the current

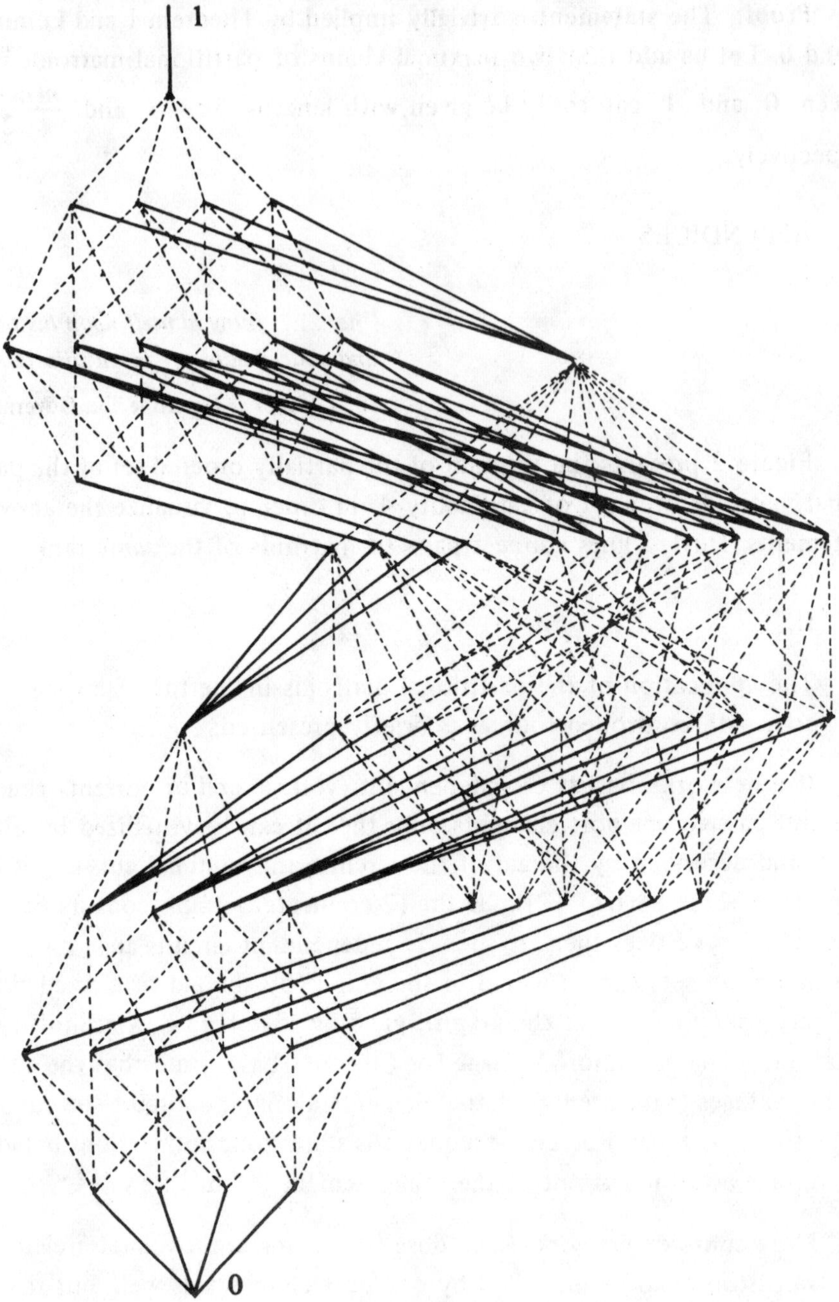

Fig. 2

of (one or more) other elements. The circuits and the cut-sets of the graph of the network cannot alone describe all the dependence-properties of this network.

For brevity's sake controlled current generators are considered only. (If the network consists of controlled voltage generators only, the model is obtained by duality. We refer to [4] for the case if controlled sources of both type are required.) The dependence relations of the controlled sources can be described by a partitional matroid. E.g. if the current of a resistance controls the current of a generator then these two elements are joined in a two-element subset of $\mathcal{S}$ and the corresponding element of $\mathcal{A}$ is one. Generally, if certain elements control certain generators then all these elements together form a subset of $\mathcal{S}$ and the corresponding element of $\mathcal{A}$ is the "degree of freedom". (This latter may be greater than one too. E.g. if the sum of the currents of p elements control a current-source then the subset is of cardinality $p + 1$ and the corresponding integer is p.)

This matroid $[\mathcal{S}, \mathcal{A}]$ together with the circuit-matroid $\mathcal{C}$ of the graph of the network will be the model of the network. The independent circuits and cut-sets of $[\mathcal{S}, \mathcal{A}] \vee \mathcal{C}$ present a system of equations. (If two matroids $\mathcal{M}_1$ and $\mathcal{M}_2$ are given on the same set S then their *sum* $\mathcal{M}_1 \vee \mathcal{M}_2$ is a matroid on S such that $X \subseteq S$ is independent iff $X = X_1 \cup X_2$ where $X_1 \in \mathcal{M}_1$ and $X_2 \in \mathcal{M}_2$.)

As an example, the d.c. feedback amplifier of Fig. 3a is considered. If both I_1 and I_2 were independent current generators then the graph of Fig. 3b would show the dependence relations. If, for example, $\{4, 5\}$ is the fixed tree then the following equations can be found:

$$i_1 + i_2 + i_3 + i_4 = 0$$

$$i_1 + i_3 = i_5$$

$$R_3 i_3 + R_5 i_5 = R_4 i_4$$

and one can easily find that the output voltage is

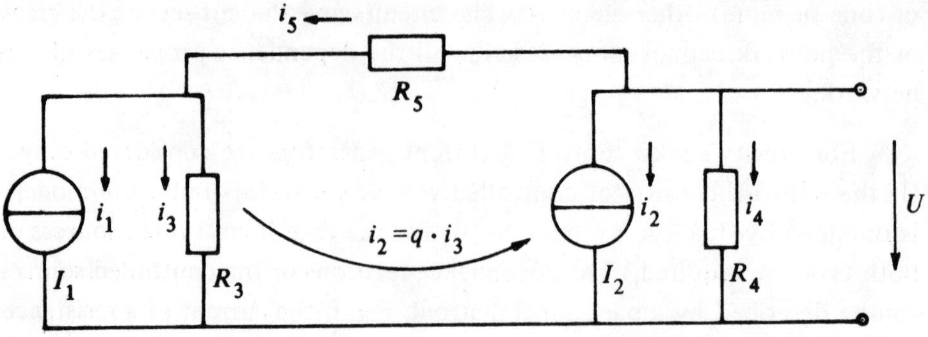

Fig. 3a

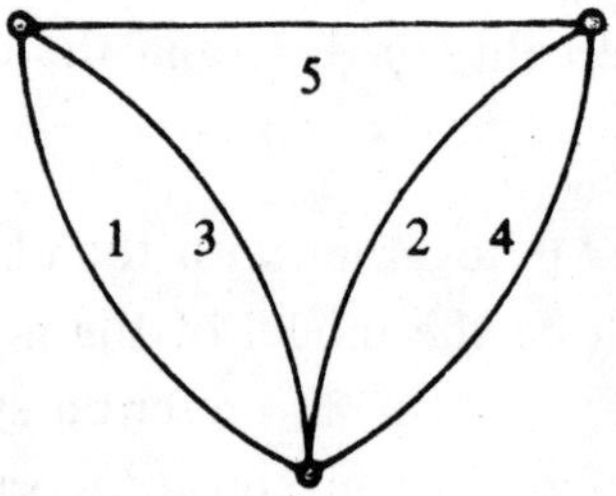

Fig. 3b

$$U = -\frac{R_3R_4i_1 + (R_3 + R_5)R_4i_2}{R_3 + R_4 + R_5} .$$

If I_2 is controlled by the current of R_3, $(i_2 = q \cdot i_3)$ then the appropriate combinatorial structure must contain a cut-set $\{2, 3\}$ and all the cut-sets of the graph of Fig. 3b still have to be contained in a cut-set. The system is described by $\mathscr{C} \vee [\mathscr{S}, \mathscr{A}]$ where $\mathscr{C}$ is the circuit-matroid of the graph, $\mathscr{A} = (0, 1, 0, 0)$ and $\mathscr{S} = (\{1\}, \{2, 3\}, \{4\}, \{5\})$. Let, e.g. $\{2, 4, 5\}$ be the fixed basis of this matroid, then the following equations are yielded:

$$i_1 + (1+q)i_3 + i_4 = 0$$

$$i_1 + i_3 = i_5$$

$$i_2 = q \cdot i_3$$

$$R_3 i_3 + R_5 i_5 = R_4 i_4$$

in accordance with the fact that the independent cut-sets corresponding to the basis $\{2, 4, 5\}$ are $\{2, 3\}$, $\{1, 3, 5\}$ and $\{1, 3, 4\}$. The output voltage can easily be computed:

$$U = - \frac{qR_4 R_5 - R_3 R_4}{R_3 + (1+q)R_4 + R_5} i_1 .$$

The description of the presented network can trivially be given in the traditional way too, but these traditional (and sometimes rather intuitive) methods cannot be applied for solving the independence-problems of a typical electronic circuit (consisting of about 100 controlled sources, 300 resistances and many other elements). Therefore a combinatorial approach is required in order to develop an algorithm for solving such network problems (by a digital computer).

REFERENCES

[1] C. Berge, *Graphes et hypergraphes,* Dunod, Paris, 1970.

[2] G. Kirchhoff, Über die Auflösung der Gleichungen, auf welche man bei der Untersuchungen der linearen Verteilung Galvanischer Ströme geführt wird, *Ann. Physik,* 72 (1847), 497-508.

[3] L. Mirsky, *Transversal Theory,* Academic Press, 1971.

[4] A. Recski, On the sum of matroids with applications in electric network theory, *doctoral dissertation,* Budapest, 1972.

COLLOQUIA MATHEMATICA SOCIETATIS JÁNOS BOLYAI
10. INFINITE AND FINITE SETS, KESZTHELY (HUNGARY), 1973.

COVERING OF VERTICES OF A UNIFORM HYPERGRAPH BY EDGES AND STABLE SETS

M. RIVIERE

§1. DEFINITIONS

Let H be a uniform, simple, and finite hypergraph of rank $s \geqslant 2$

$$H = (X, \mathscr{E}),$$

$$X = \{x_1, x_2, \ldots, x_n\}, \quad \mathscr{E} = \{E_1, E_2, \ldots, E_m\},$$

$$\forall i \quad \text{if} \quad 1 \leqslant i \leqslant m \quad \text{then} \quad E_i \subseteq X \quad \text{and} \quad |E_i| = s .$$

The results obtained also apply to graphs $(r = 2)$. The definitions are those of Berge [1].

A stable set (in the weak sense) of H is a subset of X not containing any edge. Thus a subset of cardinality $< s$ of X or a proper subset of an edge is always a stable set.

A clique of H is a subset of X such that all its subsets of s elements are edges of H.

The (weak) *chromatic number* $\chi(H)$ of H is the minimal cardinality of a partition of X into stable sets.

The aim of this paper is to study coverings of vertices of H by edges and stable sets. The minimal cardinality of such a covering will be denoted by $\chi'(H)$. The most interesting results are obtained in the case when a minimal number of edges is used in the covering, so that it is as close to a coloring as possible.

This work originated with the study "Indices de *P*-recouvrement d'un graphs" by Michel Chein and Michel Riviere [3], in which χ' appeared as a natural upper bound of a class of indices. This index differs from the classical ones in that two kinds of objects occur in the coverings.

Definition. With each subset X' of X we associate the pair $H|X' = (X', \mathscr{E}')$, where $\mathscr{E}'$ is the set of edges included in X'.

This concept will be used in §4. $H|X'$ is a uniform hypergraph of order s, which may have isolated vertices.

§2. λ-COVERINGS

Let $\lambda \geqslant 1$ be an integer. A λ-covering of H is defined as an arbitrary set

$$R_\lambda = \{A_1, A_2, \ldots, A_q;\ S_1, S_2, \ldots, S_r\}$$

such that

$$q \geqslant 0\,, \quad r \geqslant 0\,, \quad q + r = \lambda$$

$$\forall i \quad \text{if} \quad 1 \leqslant i \leqslant q \quad \text{then } A_i \text{ is an edge of } H,$$

$$\forall j \quad \text{if} \quad 1 \leqslant j \leqslant r \quad \text{then } S_j \text{ is a nonempty stable set of } H.$$

The union of the A_i's and S_j's is equal to X.

For a value of λ, q and r may possibly take several values; denote by q_λ the minimal value of q, and by r_λ the maximal value of r. Clearly

$$q_\lambda + r_\lambda = \lambda .$$

The index $\chi'(H)$ (abbreviated as χ') is defined as the minimal value of λ for which there exists a λ-covering.

Examples of λ-coverings. The sets consisting of a single vertex $(\lambda = n)$, the stable sets occuring in a coloring $(\lambda = \chi)$, the edges $(\lambda = m)$, the edges of a covering of minimal cardinality in the sense of Berge $(\lambda = \rho)$ are examples of λ-coverings. Hence χ' exists and we have

$$\chi' \leqslant \chi \leqslant n \quad \text{and} \quad \chi' \leqslant \rho \leqslant m .$$

Proposition 1. *If $R_\lambda = \{A_1, \ldots, A_{q_\lambda}; S_1, \ldots, S_{r_\lambda}\}$ is a λ-covering with a minimal number of edges, then we have $A_i \cap A_{i'} = \phi$ for every i and i' with $i \neq i'$ and $A_i \cap S_j = \phi$ for every i and j.*

Proof. Assume, on the contrary, that there is a vertex x such that $x \in A_i \cap A_{i'}$ or $x \in A_i \cap S_j$ holds. x is covered by $A_{i'}$ or S_j, and so one obtains a new λ-covering by replacing the edge A_i in R_λ by the nonempty stable set $A_i - \{x\}$, which contradicts the minimality of q_λ. (This argument will systematically be used without being detailed.) The proof is complete.

Proposition 2. *For every λ with $\chi' \leqslant \lambda \leqslant n$, if there is a λ-covering at all, then there is one which partitions X and uses a minimal number of edges; such a covering will be called fine.*

Proof. Take a λ-covering $R_\lambda = \{A_1, \ldots, A_{q_\lambda}; S_1, \ldots, S_{r_\lambda}\}$ which minimizes the number of edges. In view of Proposition 1, only the stable sets can be nondisjoint. Assume that $S_i \cap S_j \neq \phi$ holds for some $i < j$.

(a) If $S_i - S_j \neq S_j \neq \phi$, then replace S_i with $S_i - S_j$ in R_λ.

(b) If $S_i - S_j = \phi$ and $S_j - S_i \neq \phi$, then replace S_j with $S_j - S_i$.

(c) If $S_i = S_j$ and it contains at least two vertices one of which is, say, x, then replace S_i and S_j with $\{x\}$ and $S_i - \{x\}$.

Every time when it can be applied, this process replaces nondisjoint stable sets with smaller disjoint ones. When it is not applicable any more, the only nondisjoint stable sets that may possibly remain are of form $S_i = S_j = \{x\}$. Assume that there exist such stable sets.

In this case we must have $q_\lambda = 0$; otherwise, choosing a vertex t of A_1, we could obtain a new λ-covering containing fewer edges by replacing A_1, S_i, and S_j with $\{t\}$, $A_1 - \{t\}$, and S_i, which is a contradiction.

Moreover, there must exist a stable set S_k with $|S_k| > 1$ in this case; otherwise the vertices would be covered by λ one-element stable sets ($q_\lambda = 0$ and $r_\lambda = \lambda$), and so, x being covered twice, we would have $\lambda > n$, in contradiction with the hypothesis $\lambda \leqslant n$.

So choose an $x' \in S_k$ and replace S_j and S_k with $S'_j = \{x'\}$ and $S'_k = S_k - \{x'\}$; noting that $x \notin S_k$ holds, one can see that the sets S_i, S'_j, and S'_k are pairwise disjoint.

By iterating this second procedure as long as it is applicable, finally we obtain a λ-covering with a minimal number of edges that partitions X, i.e. a fine λ-covering. It consists of a partial matching and a partial coloring of the graph H. The proof is complete.

With r_λ colors one can color at least $n - sq_\lambda$ vertices in H. This result will be improved in §4.

An example of covering. The case of cliques.

Proposition 3. *If H is an s-clique with n vertices, where $n \geqslant 0$, then*

$$\chi(H) = \left[\frac{n}{s-1}\right]^* \quad \text{and} \quad \chi'(H) = \left[\frac{n}{s}\right]^*$$

Moreover $\chi(H) = \chi'(H)$ if and only if there is an integer $a \geqslant 0$ with $sa - s < n < sa - a$. ($[x]^*$ denotes the least integer exceeding or equal to x.)

Proof. The stable sets are the sets consisting of at most $s - 1$ vertices, and so

$$\chi = \left[\frac{n}{s-1}\right]^* .$$

The edges are the sets consisting of s vertices, and if we take $[n/s]$ pairwise disjoint ones from among them, there remains a (possibly empty) stable set, and so we have

$$X' = [n/s]^* .$$

As for the last assertion, we clearly have

$$\chi = \chi' \Leftrightarrow \left[\frac{n}{s-1}\right]^* = \left[\frac{n}{s}\right]^*$$

$$\Leftrightarrow \exists a \geqslant 0\left[a - 1 < \frac{n}{s-1} \leqslant a \quad \text{and} \quad a - 1 < \frac{n}{s} \leqslant a\right]$$

$$\Leftrightarrow \exists a \geqslant 0[sa - a - s + 1 < n \leqslant sa - a \quad \text{and}$$

$$sa - a < n \leqslant sa]$$

$$\Leftrightarrow \exists a \geqslant 0[sa - s < n \leqslant sa - a] ,$$

which completes the proof.

§3. FINE λ-COVERINGS

Theorem 1. *(Existence theorem) There exists a fine* λ*-covering for every* λ *with* $\chi' \leqslant \lambda \leqslant n$.

Proof. The assertion is valid for $\lambda = \chi'$. We are going to show that, for every λ with $\chi' \leqslant \lambda < n$, if there exists a fine λ-covering, then there also exists a fine $\lambda + 1$-covering.

(a) If $q_\lambda > 0$ then let

$$R_\lambda = \{A_1, \ldots, A_{q_\lambda}; S_1, \ldots, S_{r_\lambda}\}$$

be a fine λ-covering. Choose a vertex $x \in A_1$; the sets $\{x\}$ and $A_1 - \{x\}$ are stable sets of H, and so we obtain a $\lambda + 1$-covering of form

$$R_{\lambda+1} = \{A_2, \ldots, A_{q_\lambda}; S_1, \ldots, S_{r_\lambda}, \{x\}, A_1 - \{x\}\} .$$

It contains $q_\lambda - 1$ edges and $r_\lambda + 2$ stable sets. $R_{\lambda+1}$ is not necessarily a fine $\lambda+1$-covering, but in view of Proposition 2 there exists one that is fine, and we have

$$q_{\lambda+1} \leqslant q_\lambda - 1, \quad \text{i.e.} \quad q_{\lambda+1} < q_\lambda,$$

and

$$r_{\lambda+1} \geqslant r_\lambda + 2, \quad \text{i.e.} \quad r_{\lambda+1} > r_\lambda + 1.$$

(b) If $q_\lambda = 0$ and, consequently $r_\lambda = \lambda$, then let

$$R_\lambda = \{S_1, \ldots, S_\lambda\}$$

be a fine λ-covering. One of the stable sets S_j contains at least two elements, since otherwise one would have $\lambda = n$. Therefore it can be partitioned into two nonempty stable sets, e.g. $\{x\}$ and $S_j - \{x\}$, and

$$R_{\lambda+1} = \{S_1, \ldots, S_{j-1}, \{x\}, S_j - \{x\}, S_{j+1}, \ldots, S_\lambda\}$$

is a fine $\lambda+1$-covering. We have

$$q_{\lambda+1} = 0, \quad \text{i.e.} \quad q_{\lambda+1} = q_\lambda,$$

and

$$r_{\lambda+1} = \lambda + 1, \quad \text{i.e.} \quad r_{\lambda+1} = r_\lambda + 1.$$

The theorem is proved, and so is the following

Proposition 4. *We have the following inequalities for any* λ *with* $\chi' \leqslant \lambda < \chi$

$$q_\lambda \geqslant 1 \quad \textit{and} \quad q_{\lambda+1} < q_\lambda,$$

$$r_\lambda \leqslant \chi - 2 \quad \textit{and} \quad r_{\lambda+1} > r_\lambda + 1.$$

Moreover, we have $q_\lambda = 0$ *and* $r_\lambda = \lambda$ *for any* λ *with* $\chi \leqslant \lambda \leqslant n$.

From here one immediately obtains the following

Corollary 1. $\chi' = \chi$ *holds if and only if* $q_{\chi'} = 0$.

The result given in Proposition 4 saying that $\lambda < \chi$ implies $q_{\lambda+1} < q_\lambda$

cannot be improved, as we can have $q_{\lambda+1} < q_\lambda - 2$. This is shown by the graph H presented in Fig. 1. For this graph we have

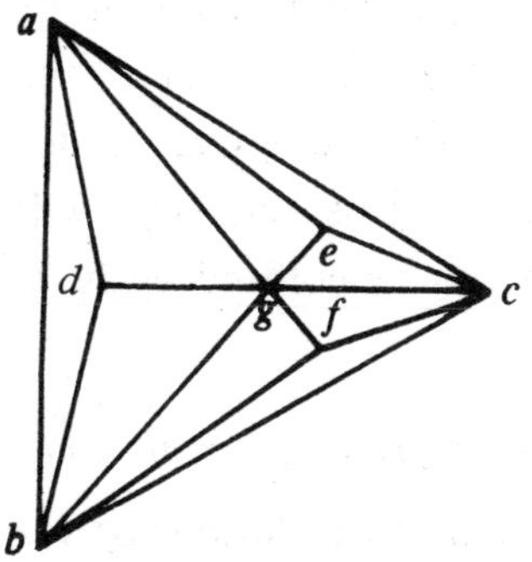

$$\chi = 4 \quad q_4 = 0 \quad r_4 = 4$$

$$\chi' = 3 \quad q_3 = 2 \quad r_3 = 1$$

$$R_{\chi'} = \{\{a, b\}, \{g, c\}; \{d, e, f\}\}$$

$$q_4 = q_3 - 2$$

Fig. 1

Proposition 5. *The union of the edge of a fine* λ*-covering*

$$R_\lambda = \{A_1, \ldots, A_{q_\lambda}; S_1, \ldots, S_{r_\lambda}\}$$

is a clique of H. *We shall use the notation*

$$C(R_\lambda) = \bigcup_{1 \leqslant i \leqslant q_\lambda} A_i \,.$$

Proof. $C(R_\lambda)$ is a subset of sq_λ elements of X. Let E_1 be a subset of s elements of $C(R_\lambda)$, and consider a partition of $C(R_\lambda)$ into q_λ classes each containing s elements, the first of which is E_1: $E_1, E_2, \ldots, E_{q_\lambda}$. If E_1 is not an edge of H then it is a stable set; in this case one obtains a new λ-covering formed by $E_2, \ldots, E_{q_\lambda}; S_1, \ldots, S_{r_\lambda}, E_1$ containing less than q_λ edges, which is a contradiction. So every subset of s elements of $C(R_\lambda)$ is an edge of H, showing that $C(R_\lambda)$ is a clique. The proof is complete.

The maximal matchings of $C(R_\lambda)$ provide the edges of the fine λ-coverings that use the same stable sets as R_λ.

§4. PARTIAL COLORINGS OF H AND AN ALGORITHM FOR DETERMINING χ'

We saw in §2 that the values of q_λ for $\lambda < \chi$ have irregular jumps which are significant indicators of the structure of the graph. That is, the sequence of pairs (q_λ, r_λ) is of some interest; we are now going to describe this sequence in terms of partial colorings. To this end, φ_k will denote the maximal number of vertices of H that can be colored with k colors.

Proposition 6. *For every λ with $\chi' \leqslant \lambda < n$ we have $\varphi_{r_\lambda} = n - sq_\lambda$. Furthermore, every coloring with r_λ colors of a maximal number of vertices provides the stable sets of a fine λ-covering.*

Proof. First assume that $\chi \leqslant \lambda \leqslant n$. In this case we have $q_\lambda = 0$ and $r_\lambda = \lambda$, and all the n vertices of H can be colored with λ colors; i.e. $\varphi_{r_\lambda} = n$.

Assume now that $\chi' \leqslant \lambda < \chi$. In this case we have $q_\lambda > 0$, and the r_λ stable sets of a fine λ-covering constitute a coloring of $n - sq_\lambda$ vertices with r_λ colors; i.e. $\varphi_{r_\lambda} \geqslant n - sq_\lambda$ holds.

Now let us be given a coloring of φ_{r_λ} vertices with r_λ colors. Denote by X' the set of the $n - \varphi_{r_\lambda}$ vertices that are not colored. Put $H' = H | X'$ (see §1). Choose a maximal matching of H' consisting of a edges; the set S of vertices not belonging to this matching is obviously a stable set. The a edges, the r_λ stable sets of the coloring, and the set S (if nonempty) form a λ'-covering of H, where $\lambda' \leqslant a + r_\lambda + 1$; equality holds here if and only if $S \neq \phi$.

We are going to show that $\varphi_{r_\lambda} = n - sq_\lambda$. In fact, assuming the contrary, we should have $\varphi_{r_\lambda} > n - sq_\lambda$. In this case,

$$sa \leqslant n - \varphi_{r_\lambda} < sq_\lambda$$

holds, and so

$$a < q_\lambda \quad \text{and} \quad \lambda' < q_\lambda + r_\lambda + 1 = \lambda + 1 \, .$$

The λ'-covering just found contains a edges, and so $q_{\lambda'} \leqslant a < q_\lambda$. I.e. we have $\lambda' \leqslant \lambda$ and $q_{\lambda'} < q_\lambda$, which contradicts Proposition 3 about the decreasing of q_λ. The equality $\varphi_{r_\lambda} = n - sq_\lambda$ is established.

Next we show that S is empty. In fact, assuming the contrary, we have

$$sa < n - \varphi_{r_\lambda} = sq_\lambda \, , \quad \text{i.e.} \quad a < q_\lambda \, ;$$

hence $\lambda' = a + r_\lambda + 1 < \lambda + 1$, and so $q_{\lambda'} \leqslant a < q_{\lambda'}$ which is a contradiction. Thus we can also see that the stable sets of any fine λ-covering are the stable sets occuring in a coloring of a maximal number of vertices with r_λ colors. The proof is complete.

We are now going to deduce a condition for the existence of a λ-covering, which will provide us with a procedure to obtain the sequence of pairs (q_λ, r_λ) when λ varies from χ' to n.

Theorem 2. *For an integer* λ *with* $0 \leqslant \lambda \leqslant n$ *consider the set*

$$E_\lambda = \{u \colon \, 0 \leqslant u \leqslant n, \; \varphi_n \geqslant n - s(\lambda - u)\} \, ,$$

where u *runs over integers.*

The necessary and sufficient condition in order that H *possess a fine* λ*-covering is that* E_λ *is nonempty. If* E_λ *is nonempty then we have* $r_\lambda = \sup E_\lambda$.

Proof. Assuming that there exists a fine λ-covering for a λ with $0 \leqslant \lambda \leqslant n$, we have

$$0 \leqslant r_\lambda \leqslant n \quad \text{and} \quad \varphi_{r_\lambda} = n - sq_\lambda = n - s(\lambda - r_\lambda) \, ,$$

which shows that $r_\lambda \in E_\lambda$, and so E_λ is not empty.

Assume now that E_λ is not empty. First consider the case $\chi \leqslant \lambda \leqslant n$. Then H possesses a fine λ-covering without edges (cf. Theorem 1) and so $r_\lambda = \lambda$ holds. In this case we have $\lambda \in E_\lambda$, since

$$0 \leqslant \lambda \leqslant n \quad \text{and} \quad \varphi_\lambda = n = n - s(\lambda - \lambda) ;$$

moreover, if $\lambda < \lambda' \leqslant n$, then $\lambda' \notin E_\lambda$, holds, as $\varphi_{\lambda'} = n < n - s(\lambda - \lambda')$. Consequently, we have $r_\lambda = \lambda = \sup E_\lambda$. The theorem is established for $\chi \leqslant \lambda \leqslant n$.

Consider now the case $\lambda < \chi$ (and $E_\lambda \neq \phi$). Then we have $\varphi_\lambda < n = n - s(\lambda - \lambda)$, and so $\lambda \notin E_\lambda$. Moreover, for every $u > \lambda$ we have $\varphi_u \leqslant n < n - s(\lambda - u)$, and so $u \notin E_\lambda$ holds as well. Put $u_\lambda = \sup E_\lambda$; we have $u_\lambda < \lambda < \chi$. Therefore, we obtain

$$\varphi_{u_\lambda} < n \quad \text{as} \quad u_\lambda < \chi ;$$

(1) $$\varphi_{u_\lambda + 1} < n \quad \text{as} \quad u_\lambda + 1 < \chi ;$$

(2) $$\varphi_{u_\lambda} \geqslant n - s(\lambda - u_\lambda) , \quad \text{as} \quad u_\lambda \in E_\lambda ;$$

(3) $$\varphi_{u_\lambda + 1} < n - s(\lambda - u_\lambda - 1) \quad \text{as} \quad u_\lambda + 1 \notin E_\lambda .$$

Suppose we are given a coloring of φ_{u_λ} vertices with u_λ colors. Denote by X' the set of $n' = n - \varphi_{u_\lambda}$ vertices that are not colored. Put $H' = H | X'$ (see § 1).

(α) First we show that $\varphi_{u_\lambda} = n - s(\lambda - u_\lambda)$. Assuming the contrary, we have

$$\varphi_{u_\lambda} > n - s(\lambda - u_\lambda)$$

in view of (2). We distinguish two cases:

(i) If we have $n' < s - 1$, then H' is a stable set in H; hence its vertices can be colored with a single color, and so we have $\varphi_{u_\lambda + 1} = n$, which contradicts (1).

(ii) Assume now $n' \geqslant s - 1$. In this case one can choose $s - 1$ vertices of H' that form a stable subset of H; these vertices can be colored with a single color, and so we have

$$\varphi_{u_\lambda+1} \geqslant \varphi_{u_\lambda} + s - 1 > n - s(\lambda - u_\lambda) + s - 1 ,$$

i.e.

$$\varphi_{u_\lambda+1} \geqslant n - s(\lambda - u_\lambda) + s \geqslant n - s(\lambda - u_\lambda - 1) ,$$

which contradicts (3). So we have established that $n' = n - \varphi_{u_\lambda} = s(\lambda - u_\lambda)$.

Choose now a maximal matching of H', using a edges. The vertices not belonging to this matching form a set S that is stable in H', and so in H.

(β) We show that S is empty, and so $a = \lambda - u_\lambda$. Assuming the contrary, the a edges of the matching, the u_λ stable sets of the coloring, and the set S form a λ'-covering $R_{\lambda'}$ of H with $\lambda' = a + u_\lambda + 1$. Furthermore, we have $sa < n' = s(\lambda - u_\lambda)$, and so $a < \lambda - u_\lambda$, and $\lambda' < \lambda + 1$, i.e. $\lambda' \leqslant \lambda$.

The λ'-covering $R_{\lambda'}$ may possibly not be fine, but in any case we have

$$q_{\lambda'} \leqslant a, \; r_{\lambda'} \geqslant u_\lambda + 1 , \quad \text{i.e.} \quad r_{\lambda'} > u_\lambda ,$$

and, moreover, $0 \leqslant r_{\lambda'} \leqslant n$ and

$$\varphi_{r_{\lambda'}} = n - sq_{\lambda'} = n - s(\lambda' - r_{\lambda'}) \geqslant n - s(\lambda - r_{\lambda'}) .$$

Thus $r_{\lambda'} \in E_\lambda$ and $r_{\lambda'} > u_\lambda$; this contradicts the choice of $u_\lambda = \sup E_\lambda$. Hence S must be empty, and we have $sa = n' = s(\lambda - u_\lambda)$, and so $a = \lambda - u_\lambda$.

(γ) It is easy to see that H has a λ-covering. In fact, the $a = \lambda - u_\lambda$ edges of the matching and the u_λ stable sets of the coloring form a λ-covering R_λ of H. We claim that this is a fine λ-covering. To this end, assume the contrary. In this case we have $q_\lambda < a$ and $r_\lambda > u_\lambda$. Since $0 \leqslant r_\lambda \leqslant n$ and $\varphi_{n_\lambda} = n - sq_\lambda = n - s(\lambda - r_\lambda)$ hold, we have $r_\lambda \in E_\lambda$; so $r_\lambda > u_\lambda$ contradicts the definition of u_λ. We have thus established that R_λ is a fine λ-covering such that $r_\lambda = u_\lambda = \sup E_\lambda$ and $q_\lambda = \lambda - u_\lambda$. The proof is complete.

Corollary 2. *Put*

$$E'_\lambda = \{\nu\colon\ 0 \leqslant \nu \leqslant n,\ \varphi_\nu = n - s(\lambda - \nu)\}\,,$$

where ν *runs over integers.* H *has a fine* λ*-covering if and only if* $E'_\lambda \neq \phi$, *and in this case we have* $r_\lambda = \sup E'_\lambda$.

Proof. We have $E'_\lambda \subset E_\lambda$. Moreover, in view of Theorem 2, if $E_\lambda \neq \phi$, then $\sup E_\lambda \in E'_\lambda$ holds. This establishes the corollary.

Next we state a generalization of a theorem of Berge [1] (cf. Theorem 13, p. 365). This will enable us to reduce the evaluation of φ_u to the evaluation of a weak stability number.

Let $H = (X, \mathscr{E})$ *and* $L = (Y, \mathscr{F})$, *be two hypergraphs. Denote by* $H + L$ *the hypergraph* $(Z, \mathscr{G})$ *defined as follows:*

$$Z = X \times Y$$

$$\mathscr{G} = \{\{x_i\} \times Y_j\colon\ x_i \in X,\ Y_j \in \mathscr{T}\} \cup \{X_k \times \{y_l\}\colon\ X_k \in \mathscr{E},\ y_l \in Y\}\,.$$

Then we have

$$\varphi_u(H) = \alpha(H + K_u)\,,$$

where K_u *denotes the* 2*-clique* (i.e. *complete graph*) *having* u *vertices, and* $\alpha(G)$ *denotes the weak stability number of the hypergraph* G.

Method of calculation of χ'. We clearly have

$$\chi' \leqslant \chi\,,$$

$$E_\chi \supset E_{\chi-1} \supset \ldots \supset E_{\chi-i} \supset \ldots.$$

and $\chi' = \chi - i$ iff $E_{\chi-i} \neq \phi$ and $E_{\chi-i-1} = \phi$. Moreover, $r_\lambda = \sup E_\lambda$ and $r_{\lambda-1} \leqslant r_\lambda - 2$, which enables one to spare some trials. So the calculation can proceed as follows

(1) Compute χ (note that $r_\chi = \chi$).

(2) Compute $\varphi_{\chi-2}, \varphi_{\chi-3}, \ldots, \varphi_u$, $(u \leqslant \chi - 2)$

until you find a u such that $\varphi_u = n - s(\chi - 1 - u)$.

If there is no such u then $\chi' = \chi$.

If there is one, then $r_{\chi-1} = u$.

(3) More generally, if $r_{\chi-1}, r_{\chi-2}, \ldots, r_\lambda$ have already been found, then compute $\varphi_{r_\lambda - 2}, \varphi_{r_\lambda - 3}, \ldots, \varphi_u$, $(u \leqslant r_\lambda - 2)$ until you find a u such that $\varphi_u = n - s(\lambda - 1 - u)$.

If there is no such u then $\chi' = \lambda$.

If there is one, then $r_{\lambda-1} = u$.

§5. λ-COVERINGS OF GRAPHS

If H is a graph, then one can obtain more complete results. As in §3, let $C(R_\lambda)$ denote the set of endpoints of the edges belonging to the fine λ-covering R_λ. Following Berge, we shall use the notation $\gamma = \gamma(H)$ for the chromatic number of graphs instead of $\chi(H)$.

Proposition 7. *Let* $R_\lambda = \{A_1, \ldots, A_{q_\lambda}; S_1, \ldots, S_{r_\lambda}\}$ *be fine a* λ-*covering, where* $\chi' \leqslant \lambda \leqslant \gamma$. *Then* R_λ *contains at least one edge, and for every pair of vertices* x, y *belonging to* $C(R_\lambda)$ *and for every stable set* S_j, *there exists a vertex* $t \in S_j$ *that is adjacent to* x *and* y.

Proof. In view of Proposition 5, we can assume that x and y are the endpoints of an edge of R_λ, say A_1. Given $S_j \in R_\lambda$, define the set

$$S = \{t\colon\ t \in S_j \text{ and } t \text{ is adjacent to } x \text{ in } H\}\ .$$

We are going to show that S contains a vertex adjacent to y. Assuming the contrary, the sets

$$S' = \{x\} \cup (S_j - S) \quad \text{and} \quad S'' = \{y\} \cup S$$

are nonempty stable sets of H covering $A_1 \cup S_j$, and

$$R'_\lambda = \{A_2, \ldots, A_{q_\lambda}; S_1, \ldots, S_{j-1}, S_{j+1}, \ldots, S_{r_\lambda}, S', S''\}$$

is a new λ-covering containing less edges, which is a contradiction. The proof is complete.

Theorem 3. *Assume the same as in Proposition 7* $(\chi' \leqslant \lambda < \gamma)$. *Then for every pair of vertices* x, y *of* $C(R_\lambda)$ *and for every collection of* l *stable sets belonging to* R_λ *(for the sake of simplicity, we assume that this collection consists of the first* l *stable sets)* $S_1, S_2, \ldots, S_l$, *there exist vertices* $u_1 \in S_1$, $u_2 \in S_2, \ldots, u_l \in S_l$ *belonging to the same connected component of the subgraph* H' *of* H *spanned by the set* $X' = \bigcup_{1 \leqslant j \leqslant l} S_j$ *such that each of them is adjacent to* x *and* y.

Proof. Denote by $L_1, L_2, \ldots, L_p$ the connected components of the graph H' and put

$$L_{k,j} = L_k \cap S_j \quad \text{for} \quad 1 \leqslant k \leqslant p\,,\ 1 \leqslant j \leqslant l\,.$$

The sets $L_{k,j}$ are pairwise disjoint. Consider the following hypothesis:

(1) For every k with $1 \leqslant k \leqslant p$ there exists a j_k such that L_{k,j_k} does not contain any vertex adjacent to x and y.

We are going to show that this hypothesis is absurd. Put

$K_1 = \{k$: $L_{k,1}$ does not contain any vertex adjacent to x and $y\}$,

$K_2 = \{k$: $L_{k,1}$ contains at least one vertex adjacent to x and $y\}$.

(1) Assuming $k \in K_2$, we have $j_k > 1$ in hypothesis (1). In this case we are going to verify two assertions.

(a) The sets $S_1 - L_{k,1}$ and L_{k,j_k} cannot both be empty. In fact, assuming the contrary, no vertex of $S_1 = L_{k,1}$ is adjacent to any vertex of S_{j_k}, and so the set $T = S_1 \cup S_{j_k}$ is a nonempty stable subset of H.

Hence we can obtain a new λ-covering by replacing $A_1 = \{x, y\}$, S_1, and S_{j_k} with $\{x\}, \{y\}$, and T, which is a contradiction.

(b) We can obtain a new fine λ-covering of H by "exchanging" $L_{k,1}$ and L_{k,j_k} that is, by replacing the sets S_1 and S_{j_k} with

$$T_1 = (S_1 - L_{k,1}) \cup L_{k,j_k}$$

and

$$T_2 = (S_{j_k} - L_{k,j_k}) \cup L_{k,1} .$$

In fact, T is not empty in view of (a). Moreover, the sets $S_1 - L_{k,1}$ and L_{k,j_k} are stable in H. In fact, if $u \in S_1 - L_{k,1}$ and $v \in L_{k,j_k}$, then u and v cannot be adjacent in H, since otherwise one would have $u \in L_k$, i.e. $u \in L_{k,1}$, which is absurd. This shows that T_1 is a nonempty stable subset of H.

T_2 is not empty, since $k \in K_2$ implies that $L_{k,1}$ is not empty. Moreover, T_2 is stable in H; this can be shown similarly as it was done for T_1.

(2) Note that the exchange described under (1) (b) between $L_{k,1}$ and L_{k,j_k} takes place between two subsets of L_k, and it does not modify $L_{k',j}$ for $k' \neq k$. Therefore, if $k \in K_2$, $k' \in K_2$, and $k \neq k'$, then the corresponding exchanges can be carried out in L_k and in $L_{k'}$, since they concern pairwise disjoint subsets and do not counteract each other.

Carry out these exchanges for all $k \in K_2$. In the fine λ-covering obtained, the stable set replacing S_1 will be the following

$$S'_1 = \Big(\bigcup_{k \in K_1} L_{k,1}\Big) \cup \Big(\bigcup_{k \in K_2} L_{k,j_k}\Big) ,$$

and, in view of hypothesis (1), this stable set does not contain any vertex adjacent to x and y. This contradicts Proposition 7. Therefore, hypothesis (1) is false, and so there exists an integer k with $1 \leqslant k \leqslant p$ such that there exist vertices $u_1 \in L_{k,1}, u_2 \in L_{k,2}, \ldots, u_l \in L_{k,l}$ belonging to the same connected component L_k of H' each of which is adjacent to x and y. The theorem is proved.

Proposition 8. *Where K_n denotes the complete graph with n vertices, we have $\chi(K_n) = [n/2]^*$. Moreover, for every graph H we have $\gamma(H)/2 \leqslant \chi'(H) \leqslant \gamma(H)$, and $\chi'(H) = \gamma(H)/2$ holds if and only if H is a complete graph.*

Proof. The first claim follows from Proposition 3. As for the second, consider a fine covering $R_{\chi'}$, of cardinality χ'. Then $C(R_{\chi'})$ is a clique with $2q_{\chi'}$ vertices. Consider a coloring of $C(R_{\chi'})$ with $2q_{\chi'}$ colors and extend this coloring to a coloring of H with the aid of the $r_{\chi'}$, stable sets belonging to $R_{\chi'}$; we obtain the relation

$$\gamma(H) \leqslant 2q_{\chi'} + r_{\chi'} \leqslant 2q_{\chi'} + 2r_{\chi'} = 2\chi'(H)\,,$$

which establishes the second claim.

As for the third one, assume that $\gamma(H)/2 = \chi'(H)$ holds. Then we must have equality everywhere in the centered line above, i.e. we must have $r_{\chi'} = 0$, and so H must be equal to the complete graph $C(R_{\chi'})$. This establishes the "only if" part of the last claim. The "if" part is included in the first claim.

Proposition 9. *The completely triangulated planar graphs with chromatic number $\gamma \leqslant 4$ satisfying $\chi' < \gamma$ are the following:*

(1) $\chi' = 1 + 0 = 1\,,$ $\gamma = 2$

(2)  $\chi' = 1 + 1 = 2\,,$ $\gamma = 3$

(3) 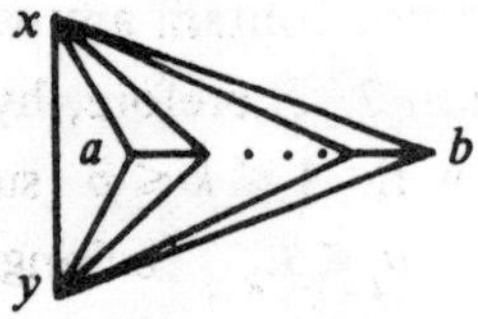

$\chi' = 1 + 2 = 3\,,$ $\gamma = 4$

(4) 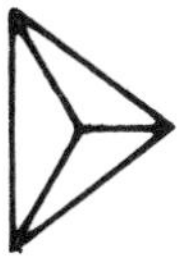 $\chi' = 2 + 0 = 2$, $\gamma = 4$

(5) 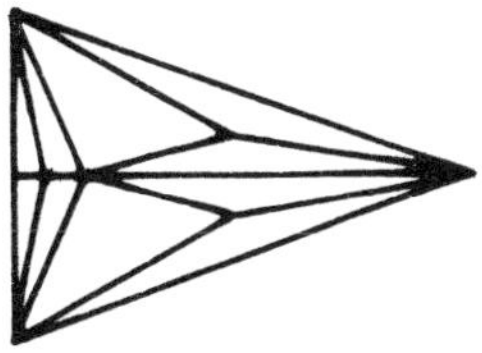 $\chi' = 2 + 1 = 3$, $\gamma = 4$

(6) 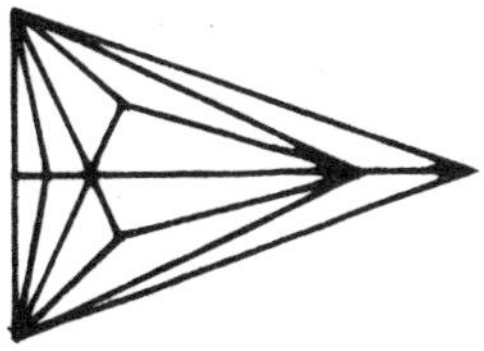 $\chi' = 2 + 1 = 3$, $\gamma = 4$

Here scheme (3) represents on infinite number of graphs such that a and b are connected with a path of length $\geqslant 2$ each vertex of which is adjacent to x and y.

Proof. Study each case by using Propositions 5, 7 and 8.

REFERENCES

[1] C. Berge, *Graphes et hypergraphes,* Dunod, Paris, 1970. (English translation: *Graphs and hypergraphs,* North-Holland – American Elsevier, Amsterdam – London – New York, 1973.)

[2] M. Chein, Indice de *P*-recouvrement d'un graphe, *C.R. Acad. Sc. Paris,* 272, 772-775.

[3] M. Chein – M. Riviere, Indice de *P*-recouvrement d'un graphe, (in this proceedings).

COLLOQUIA MATHEMATICA SOCIETATIS JÁNOS BOLYAI
10. INFINITE AND FINITE SETS, KESZTHELY (HUNGARY), 1973.

SUBDIRECT REPRESENTATIONS OF GRAPHS

G. SABIDUSSI

1. INTRODUCTION

It has been, and largely still is, customary to treat graphs as strictly combinatorial structures. In many ways this has been a fruitful point of view; quite often, however, it has given to graph-theoretical problems and methods a suspicious flavor of being somewhat *ad hoc*, and it has certainly impeded the throwing of bridges from graph theory to other parts of mathematics. Our purpose here is to establish one such bridge by exploiting the fact that graphs are relational systems, and to show that integration of graph theory into this wider and more widely known framework provides a new setting for certain time-honored graph-theoretical concepts, and at the same time leads to purely combinatorial problems that even a very classical graph theorist might find interesting.

A graph X may be considered as a relational system consisting of exactly one binary relation E (= adjacency) on a base set V (= vertices) subject to two conditions:

$$(x, y) \in E \rightarrow (y, x) \in E \quad \text{(the graph is non-oriented)}$$

and

$$(x, y) \in E \rightarrow \neg\, x = y \quad \text{(there are no loops)} .$$

Being definable by first-order sentences the class of graphs is thus analogous to an equational class of algebras, and one may therefore transfer to graphs many standard concepts of universal algebra.

One of the most fundamental results in universal algebra is Birkhoff's theorem which says that any algebra belonging to a given equational class is a subdirect product of subdirectly irreducible algebras belonging to that class ([1], p. 140). Fawcett ([2] chapter 1) has shown that Birkhoff's theorem also holds for graphs. Regrettably, Fawcett's identification ([2] (1.33)) of the subdirectly irreducible graphs as the complete graphs with one edge missing ("almost complete" graphs) is not entirely correct. Complete graphs are likewise subdirectly irreducible and it is precisely the existence of two species of subdirectly irreducible graphs which gives rise to the combinatorial problems to which this paper is primarily devoted.

Section 2 gives an account of Fawcett's proof (unpublished) of Birkhoff's theorem. Section 3 deals with "pure" subdirect representations (in which either all factors are complete or all factors are almost complete). The principal result here is that any graph has a pure subdirect representation of at least one of the two kinds. Finally, in Section 4 we consider those graphs all of whose subdirectly irreducible representations are pure complete. It is in dealing with this question that representation theory touches on, and gives a more algebraic setting to, certain aspects of classical coloring theory.

For a graph X we shall denote by $V(X)$ its set of vertices and by $E(X)$ its set of edges. Since the relation of adjacency is symmetric edges will usually (but not always) be written as unordered pairs $[x, y]$. For $x \in V(X)$, $V(x; X)$ will denote the neighborhood of x, i.e.,

$$V(x; X) = \{y \in V(X) \colon [x, y] \in E(X)\} .$$

The complement of X will be denoted by $-X$.

If R is an equivalence on $V(X)$ we define the *quotient graph* X/R by

$$V(X/R) = V(X)/R\ ,$$

$$E(X/R) = \{[R[x], R[y]]:\ R[x] \neq R[y]$$

$$\text{and} \quad [x', y'] \in E(X) \quad \text{for some} \quad x'Rx, y'Ry\}.$$

A *homomorphism* from X to Y is a map Φ: $V(X) \to V(Y)$ such that $[x, y] \in E(X)$ implies $[\Phi x, \Phi y] \in E[Y]$. We shall write Φ: $X \to Y$ for short. Graphs and homomorphisms form a category in which the monomorphisms [epimorphisms] are simply the injective [surjective] homomorphisms. However for our purposes it will be convenient to adopt the convention that an *epimorphism* Φ: $X \to Y$ is a surjective homomorphism such that for any $[y, y'] \in E(Y)$ there is an edge $[x, x'] \in E(X)$ with $[y, y'] = [\Phi x, \Phi x']$, i.e., we require that an epimorphism be *full*. An *embedding* $X \to Y$ is an isomorphism of X onto a restriction (= full subgraph) of Y.

A *congruence* R on a graph X is the kernel of a homomorphism of X into some graph Y. Since a homomorphism can never identify two adjacent vertices we obtain the following trivial (but extremely useful) intrinsic characterization of a congruence: R is a congruence if and only if it is an equivalence on $V(X)$ contained in the complement of the adjacency relation $E(X)$. An important consequence of this simple observation is that the set of all congruences on a graph is inductive, i.e., any graph possesses maximal congruences. The following lemma (whose proof we leave to the reader) characterizes maximal congruences.

Lemma 1.1. *A congruence* R *on* X *is maximal if and only if* X/R *is complete.*

2. SUBDIRECT REPRESENTATIONS

In the category of all graphs and homomorphisms the product P of a non-empty family of graphs $(X_\alpha)_{\alpha \in A}$ always exists, and is obtained as follows:

$$V(P) = \prod_{\alpha \in A} V(X_\alpha),$$

and

$$[x, y] \in E(P) \quad \text{if and only if} \quad [\mathrm{pr}_\alpha x, \mathrm{pr}_\alpha y] \in E(X_\alpha)$$

for each $\alpha \in A$.

Definition 2.1. Let X and X_α, $\alpha \in A$, be graphs. X is called a *subdirect product* of the family $(X_\alpha)_{\alpha \in A}$ if and only if

(i) there exists an embedding $\Phi: X \to P$, where $P = \prod_{\alpha \in A} X_\alpha$, or, in other words, X is isomorphic to a restriction of P; and

(ii) for each $\alpha \in A$, $\mathrm{pr}_\alpha \circ \Phi: X \to X_\alpha$ is an epimorphism.

Example 2.2. The complete bipartite graphs $K_{2,2}$, $K_{1,3}$ and $K_{1,4}$ are subdirect products of two copies of $K_{1,2}$. In Figure 1 the heavy lines form a copy of $K_{2,2}$, the light lines form $K_{1,4}$, Each of the four 3-stars contained in $K_{1,2} \times K_{1,2}$ provides a subdirect representation of $K_{1,3}$.

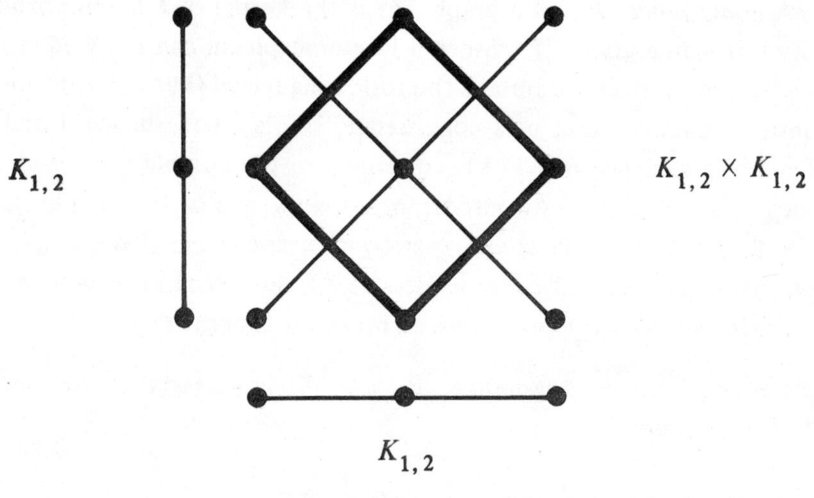

Figure 1

Suppose X is a subdirect product of the family $(X_\alpha)_{\alpha \in A}$. For each $\alpha \in A$ we can define an equivalence C_α on $V(X)$ by

$$xC_\alpha y \quad \text{if and only if} \quad (\mathrm{pr}_\alpha \circ \Phi)x = (\mathrm{pr}_\alpha \circ \Phi)y\,.$$

Since the projections pr_α are homomorphisms, C_α is a congruence on X.

Note that $X_\alpha = (\mathrm{pr}_\alpha \circ \Phi)X \cong X/C_\alpha$ by the natural isomorphism which maps $C_\alpha[x] \mapsto (\mathrm{pr}_\alpha \circ \Phi)x$. One may therefore say that X is a subdirect product of the graphs X/C_α, $\alpha \in A$.

In order to obtain the characteristic properties of the family of congruences $(C_\alpha)_{\alpha \in A}$ we need the following definition.

Definition 2.3. Let X be a graph with vertex-set V, R an equivalence on V. In this context it will be convenient to consider $E = E(X)$ as a binary relation rather than as a set of unordered pairs. We define

$$ER = R \circ E \circ R\,,$$

where "$\circ$" is the usual relational product, i.e., $[x, y] \in ER$ if and only if there exist $x', y' \in V$ with $(x, x') \in R$, $(x', y') \in E$, and $(y', y) \in R$.

Note that always $E \subset ER$ (the only property of R which is used here is reflexivity), and that since E is symmetric, so also is ER.

By $\bar{X}$ we denote the graph with $V(\bar{X}) = V$, $E(\bar{X}) = ER$. In other words, $\bar{X}$ is obtained from X by adding to X all possible edges joining two vertices belonging to two different congruence classes, provided that X already contains one such edge.

Lemma 2.4. *If* $(C_\alpha)_{\alpha \in A}$ *is the family of congruences associated with a subdirect representation of a graph* X, *then*

(i) $\bigcap_{\alpha \in A} C_\alpha = I$ *(the identity on* $V(X)$*); and*

(ii) $\bigcap_{\alpha \in A} EC_\alpha = E(X)$.

Proof. The proof of (i) is obvious since Φ is one-one.

(iii) Since $E(X) \subset \bigcap_{\alpha \in A} EC_\alpha$ it suffices to prove only the reverse inclusion. Let $(x, y) \in EC_\alpha$ for all $\alpha \in A$. This implies that for each

$\alpha \in A$ there exist $x_\alpha, y_\alpha \in V(X)$ with $x_\alpha C_\alpha x$, $y_\alpha C_\alpha y$, and $[x_\alpha, y_\alpha] \in E(X)$. Hence $\mathrm{pr}_\alpha(\Phi x_\alpha) = \mathrm{pr}_\alpha(\Phi x)$, $\mathrm{pr}_\alpha(\Phi y_\alpha) = \mathrm{pr}_\alpha(\Phi y)$, and $e_\alpha = [\Phi x_\alpha, \Phi y_\alpha] \in E(\Phi X) \subset E(P)$ for each $\alpha \in A$. Projection into X_α yields

$$[\mathrm{pr}_\alpha(\Phi x), \mathrm{pr}_\alpha(\Phi y)] = [\mathrm{pr}_\alpha(\Phi x_\alpha), \mathrm{pr}_\alpha(\Phi y_\alpha)] \in E(X_\alpha)$$

for each $\alpha \in A$, and hence $[\Phi x, \Phi y] \in E(P)$. Since ΦX is a restriction of P this means $[\Phi x, \Phi y] \in E(\Phi X)$, and finally, since Φ is an isomorphism between X and ΦX, $[x, y] \in E(X)$.

A family of congruences $(S_\alpha)_{\alpha \in A}$ with the property that $\bigcap_{\alpha \in A} S_\alpha = I$ is called *point-separating.* It is clear that a family is point-separating if and only if for any pair of distinct non-adjacent vertices x, y, i.e., for any edge $[x, y] \in E(-X)$, there is an $\alpha \in A$ such that $(x, y) \notin S_\alpha$.

The converse of the preceding lemma is also true.

Lemma 2.5. *If* $(S_\alpha)_{\alpha \in A}$ *is a family of congruences on a graph* X *such that*

(i) $\bigcap_{\alpha \in A} S_\alpha = I$ *and*

(ii) $\bigcap_{\alpha \in A} ES_\alpha = E(X)$,

then X *is a subdirect product of the family* $(X/S_\alpha)_{\alpha \in A}$.

Proof. Put $P = \prod_{\alpha \in A} X_\alpha$, where $X_\alpha = X/S_\alpha$, $\alpha \in A$. For each $x \in V(X)$ denote the family of congruence classes $(S_\alpha[x])_{\alpha \in A}$ by $\bar{x}$, and define a mapping $\Phi \colon V(X) \to V(P)$ by $x \mapsto \bar{x}$. That Φ is a homomorphism is obvious, since $[x, y] \in E(X)$ implies $[S_\alpha[x], S_\alpha[y]] \in E(X_\alpha)$ for each $\alpha \in A$, hence $[\bar{x}, \bar{y}] \in E(P)$. To get that Φ is one-one, suppose that $\bar{x} = \bar{y}$. Then $S_\alpha[x] = S_\alpha[y]$ or, equivalently, $(x, y) \in S_\alpha$ for each $\alpha \in A$. By condition (i), $x = y$.

Finally, to show that Φ is full, let $[\bar{x}, \bar{y}] \in E(P)$. This implies $[S_\alpha[x], S_\alpha[y]] \in E(X/S_\alpha)$ for each $\alpha \in A$. By the definition of a quotient graph there exist $x_\alpha, y_\alpha \in V(X)$ such that $x_\alpha S_\alpha x, y_\alpha S_\alpha y$, and $[x_\alpha, y_\alpha] \in E(X)$. But this means $(x, y) \in ES_\alpha$ for each $\alpha \in A$, whence by (ii), $[x, y] \in E(X)$. Therefore $[\bar{x}, \bar{y}]$ is the image of an edge in X.

For any $\alpha \in A$, $\mathrm{pr}_\alpha \circ \Phi$ is the natural projection $X \to X_\alpha$, and hence trivially an epimorphism. This completes the proof.

The two preceding lemmas say that the representations of X as a subdirect product are in one-one correspondence with the families $(S_\alpha)_{\alpha \in A}$ of congruences on X which satisfy conditions (i) and (ii). These two conditions are trivially satisfied if one of the congruences S_α is the identity I. The graphs for which this always happens, i.e., for which it is the case that any family of congruences satisfying (i) and (ii) contains the identity, are of particular importance. They have the property that in any representation as a subdirect product at least one factor is the graph itself.

Definition 2.6. A graph X is called *subdirectly irreducible* if and only if for any family $(S_\alpha)_{\alpha \in A}$ of congruences on X the relations

$$\bigcap_{\alpha \in A} S_\alpha = I \quad \text{and} \quad \bigcap_{\alpha \in A} ES_\alpha = E(X)$$

imply that $S_{\alpha_0} = I$ for some $\alpha_0 \in A$.

Subdirectly irreducible graphs are easy to characterize.

Proposition 2.7. *A graph is subdirectly irreducible if and only if it is complete or a complete graph with exactly one edge missing. Graphs of the second type will be called almost complete.*

It is clear that two almost complete graphs are isomorphic if and only if their orders are the same.

For the proof of 2.7 and also for later use we need:

Lemma 2.8. *Let $(S_\alpha)_{\alpha \in A}$ be a family of congruences on X such that for each pair of non-adjacent vertices x, y of X there is a congruence S_α which identifies x and y (such a family will be called point-identifying). Then*

$$\bigcap_{\alpha \in A} ES_\alpha = E(X).$$

Proof. Let $(x, y) \in ES_\alpha$ for each $\alpha \in A$. This means there are vertices x_α, y_α with $[x_\alpha, y_\alpha] \in E(X)$ and $(x, x_\alpha) \in S_\alpha$, $(y, y_\alpha) \in S_\alpha$. If

$[x, y] \notin E(X)$, then by hypothesis $(x, y) \in S_{\alpha_0}$ for some $\alpha_0 \in A$. Together with $(x, x_{\alpha_0}) \in S_{\alpha_0}$ and $(y, y_{\alpha_0}) \in S_{\alpha_0}$ this gives $(x_{\alpha_0}, y_{\alpha_0}) \in \in S_{\alpha_0}$, contrary to $[x_{\alpha_0}, y_{\alpha_0}] \in E(X)$. Hence $[x, y] \in E(X)$.

Proof of Proposition 2.7. Sufficiency: In a complete graph X any two distinct vertices are adjacent. Hence the identity is the only congruence on X. If X is almost complete there is exactly one pair x, y of distinct non-adjacent vertices. Thus the smallest equivalence which identifies x and y is the only non-identity congruence on X. Hence any family of congruences whose intersection is the identity must contain the identity as a member, i.e., X is subdirectly irreducible.

Necessity: Suppose X contains two pairs of distinct non-adjacent vertices, say a, b, and a', b'. Put $A = E(-X)$, and for each $\alpha = [x, y] \in \in A$ let S_α be the smallest equivalence on X which identifies x and y. Each S_α is a congruence, and by lemma 2.8, $\bigcap_{\alpha \in A} ES_\alpha = E(X)$. Also note that no S_α is the identity. To complete the proof that X is not subdirectly irreducible we have to show $\bigcap_{\alpha \in A} S_\alpha = I$. But this is obvious since already $S_{\alpha_0} \cap S_{\alpha_1} = I$, where $\alpha_0 = [a, b]$ and $\alpha_1 = [a', b']$.

The key role in the proof of Birkhoff's theorem is played by those congruences whose quotients turn out to be subdirectly irreducible. They were introduced by Fawcett [2] (1.31).

Definition 2.9. A congruence R on X is called *relatively maximal* if and only if there is a pair of distinct non-adjacent vertices x, y of X which are separated by R (i.e., $(x, y) \notin R$), and R is maximal with this property. If R is relatively maximal without being maximal it will be called *almost maximal.*

The existence of such congruences follows trivially from Zorn's lemma.

Proposition 2.10. *If* R *is relatively maximal, then* X/R *is subdirectly irreducible.*

Proof. Suppose x, y are two distinct non-adjacent vertices of X separated by R. Let $(S_\alpha)_{\alpha \in A}$ be a family of congruences on X/R

whose intersection is the identity I on X/R, and assume that $S_\alpha \neq I$ for each $\alpha \in A$. This implies that for each $\alpha \in A$, $(R[x], R[y]) \in S_\alpha$. But then $R[x] = R[y]$, contrary to R separating x and y.

Lemma 2.11. *If X/R is almost complete, then R is almost maximal. Hence X/R is subdirectly irreducible if and only if R is either relatively maximal or maximal.*

Proof. By hypothesis there is a unique pair of distinct classes $R[a]$, $R[b]$ such that $[R[a], R[b]] \notin E(X/R)$. In particular, R does not identify a and b. To show that it is maximal with this property take a congruence $R' \supset R$ and suppose that there is an $(x, y) \in R' - R$. Since X/R is almost complete, $(x, y) \notin R$ implies that either

$$[R[x], R[y]] \in E(X/R) \quad \text{or} \quad [R[x], R[y]] = [R[a], R[b]] .$$

The first alternative leads to a contradiction against R having a maximality property, whereas the second yields

$$R'[x] = R'[y] = R'[x] \cup R'[y] \supset R[x] \cup R[y] = R[a] \cup R[b] ,$$

whence $aR'b$.

Lemma 2.12. *If $[x, y]$ and $[a, b]$ are two distinct edges of $-X$, there exists a relatively maximal congruence on X which identifies x and y and separates a and b.*

Proof. Consider the graph $Y = X \cup (e)$, where $e = [a, b]$. Trivially, $V(Y) = V(X)$, and any congruence on X which does not identify a and b is a congruence on Y, and conversely. Hence the maximal congruences on Y are identical with the relatively maximal congruences on X which separate a and b. It follows that any maximal congruence on Y which identifies x and y has the required property. Such a congruence exists since $[x, y] \notin E(Y)$.

This completes the preliminaries and we are now in a position to prove the analogue of Birkhoff's theorem.

Theorem 2.13. *Every graph X is a subdirect product of subdirectly irreducible graphs.*

Proof. All we have to do is to find a point-separating and point-identifying family of relatively maximal congruences on X.

If X is complete or almost complete, then it is subdirectly irreducible, and hence there is nothing to prove. We may therefore assume that $|E(-X)| \geqslant 2$. Let

$$A = \{([x, y], [a, b]): [x, y], [a, b] \in E(-X), [x, y] \neq [a, b]\} .$$

By 2.12, for each $\alpha = ([x, y], [a, b])$ there is a relatively maximal congruence S_α on X which identifies x and y and separates a and b. Clearly, this family has the required properties.

3. PURE SUBDIRECTLY IRREDUCIBLE REPRESENTATIONS

In view of the fact that there are two types of subdirectly irreducible graphs, viz. the complete ones and the almost complete ones, the question now arises which graphs have "pure" representations in which either all factors are complete or all factors are almost complete. Several variations of the question are possible:

1. Which graphs have pure complete representations?
2. Which graphs have pure almost complete representations?
3. Which graphs have *only* pure complete represention?

The straight analogue of question 3 for almost complete representations is meaningless, for if X is any non-empty graph one may take a maximal congruence R on X, and then proceed to add enough relatively maximal congruences to obtain a point-separating and point-identifying family $(S_\alpha)_{\alpha \in A}$. The resulting subdirect representation will have at least one complete factor, viz. X/R. In other words, there is no non-empty graph having only pure almost complete representations. The "correct" almost complete version of question 3 therefore is:

4. Which graphs have the property that any subdirectly irreducible representation can be reduced to a pure almost complete representation?

The main result concerning questions 1 and 2 is Theorem 3.6 which says that every graph has a pure complete or a pure almost complete representation. The two situations are, however, not mutually exlusive (for example any product of at least two copies of $K_{1,2}$ has both types of pure representation). Graphs with a pure complete representation are easily characterised (3.2); for those with a pure almost complete representation a characterization is still lacking. Question 3 is also largely open. The known partial results will be dealt with in Section 4. Nothing is known concerning question 4.

We begin with those graphs which are subdirect products of *complete* graphs.

Lemma 3.1. *A n.a.s.c. that a graph X be a subdirect product of complete graphs is that the intersection of all maximal congruences on X be the identity.*

Proof. Let M be the intersection of all maximal congruences on X.

If X is a subdirect product of a family $(X_\alpha)_{\alpha \in A}$ of complete graphs, consider the associated congruences C_α on X. Since $X/C_\alpha \cong X_\alpha$ it follows from 1.1 that each C_α is maximal. Hence, using 2.4 (i),

$$M \subset \bigcap_{\alpha \in A} C_\alpha = I\,.$$

Conversely, suppose that $M = I$. Let $(M_\alpha)_{\alpha \in A}$ be the family of all maximal congruences on X. For each $[x, y] \in E(-X)$ there is a congruence on X which identifies x and y. Hence the maximal congruences are point-identifying, whence by Lemma 2.5, X is a subdirect product of the graphs X/M_α, $\alpha \in A$, and by 1.1 these are complete.

Our problem is, then, to characterize those graphs for which M is the identity. First of all we notice that if X is a discrete graph, then X has exactly one maximal congruence, viz. $V(X) \times V(X)$, and unless $|X| \leqslant 1$, this congruence does not separate vertices.

Lemma 3.2. *Let X be a non-discrete graph (i.e., $E(X) \neq \phi$). If x and y are two distinct vertices with xMy, then*

(i) $V(x; X) = V(y; X)$; *and*

(ii) *every edge of* X *is incident with some vertex belonging to* $V(x; X)$.

Conversely, if two distinct vertices x, y *satisfy* (i) *and* (ii), *then* xMy.

Note that since $E(X) \neq \phi$ condition (ii) implies that $V(x; X) \neq \phi$. Hence neither x nor y is an isolated vertex. In other words, in a non-discrete graph any two distinct isolated vertices are separated by some maximal congruence.

Proof.

(i) Suppose there is an $a \in V(x; X) - V(y; X)$. Since $[a, y] \notin E(X)$, there is a maximal congruence R which identifies a and y, i.e., $a \in \in R[y]$. On the other hand, $[a, x] \in E(X)$ implies $a \notin R[x]$. Hence $R[x] \neq R[y]$, contrary to xMy. This means $V(x; X) \subset V(y; X)$, and by symmetry the two sets are equal.

(ii) Suppose there is an edge $[a, b] \in E(X)$ with $a, b \notin V(x; X) = = V(y; X)$. This means $[a, x], [b, y] \notin E(X)$, and hence there is a maximal congruence R on X with aRx and bRy. $[a, b] \in E(X)$ implies $(a, b) \notin R$, and hence $(x, y) \notin R$, contrary to xMy.

To prove the converse, put $A = V(X) - V(x; X)$, and let R be any maximal congruence on X. Since R is a congruence, $R[z] \subset V(X) - - V(z; X)$ for every $z \in V(X)$. Hence by (i), $R[x]$ and $R[y]$ are contained in A. By (ii) at least one end of every edge of X is not in A, hence A is independent. Since R is maximal this says $R[x] = R[y]$.

It follows that any graph which violates at least one of the two conditions of Lemma 3.2 is a subdirect product of complete graphs. Hence we have (from condition (i)):

Corollary 3.3. *Any graph for which the relation of equal neighborhoods is the identity is a subdirect product of complete graphs.*

Condition (ii) of 3.2 is of a very different nature. It says that the vertices of a graph with a non-trivial M-class $M[x]$ are "very close together", in the sense that if a vertex z of X is not isolated, then it is adjacent to some vertex in $\{x\} \cup V(x; X)$. This means that if X has no isolated vertices, and if M is non-trivial, then X is connected and its diameter is $\leqslant 4$. Hence we obtain

Corollary 3.4. *If a graph without isolated vertices is either disconnected, or is connected and has diameter* $\geqslant 5$, *then it is a subdirect product of complete graphs.*

We leave the details of the proof to the reader (see Figure 2).

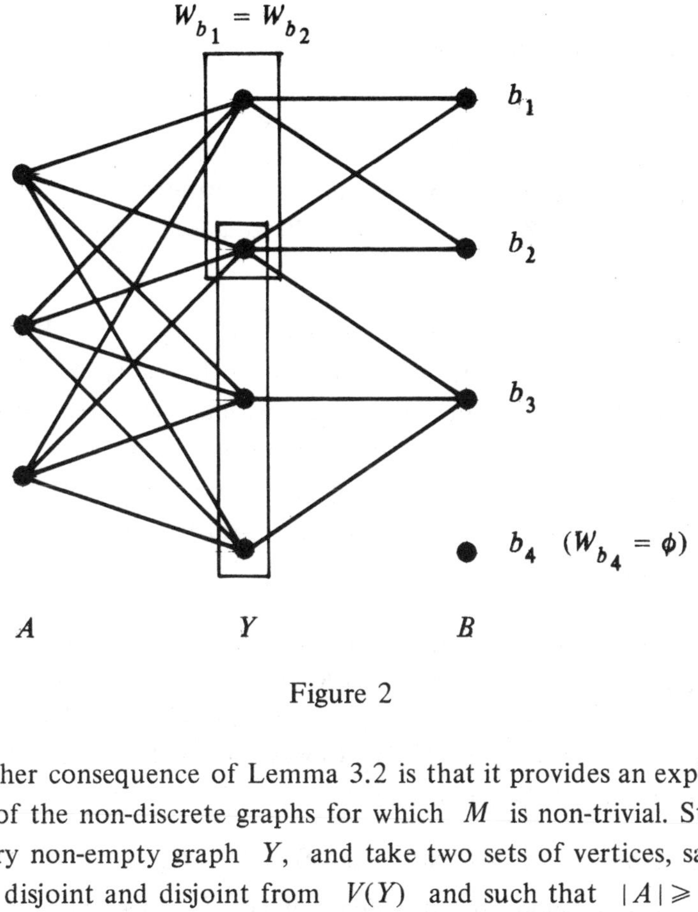

Figure 2

Another consequence of Lemma 3.2 is that it provides an explicit construction of the non-discrete graphs for which M is non-trivial. Start with an arbitrary non-empty graph Y, and take two sets of vertices, say A, B, which are disjoint and disjoint from $V(Y)$ and such that $|A| \geqslant 2$. For each $b \in B$ choose a proper subset W_b of $V(Y)$. Join every $a \in A$ with every $y \in V(Y)$, and every $b \in B$ with every $w \in W_b$ (see Figure 2).

In the resulting graph X the set A is precisely one of the M-classes. Also it is easy to verify that $M[b] = \{b\}$ for each $b \in B$; hence any other non-trivial M-class of X is contained in $V(Y)$. From this last remark we obtain immediately:

Corollary 3.5. *If two vertices a, x of a graph X belong to different M-classes, where $|M[a]| \geq 2$, $|M[x]| \geq 2$, then $[a, x] \in E(X)$. In other words, the restriction of X to the union of its non-trivial M-classes is complete multipartite.*

The most important consequence of 3.2 is the following.

Theorem 3.6. *Every graph has a pure complete or a pure almost complete representation.*

Proof. By 3.1 we may assume that M is non-trivial. We use the notation of Figure 2, i.e., we let A be a non-trivial M-class of X, Y the restriction of X to $V(a; X)$, where $a \in A$, and $B = V(X) - (A \cup V(a; X))$.

Let I be the set of all ordered pairs $\alpha = (e, e')$ of distinct edges of $-X$ such that the ends of e do not both belong to A. For each $\alpha = (e, e') \in I$ we shall construct an almost maximal congruence R_α such that $(x, y) \in R_\alpha$ and $(u, v) \notin R_\alpha$, where $e = [x, y]$, $e' = [u, v]$. Since in the pairs belonging to I, e' may run through all edges of $-X$ it is clear that the family $(R_\alpha)_{\alpha \in I}$ is point-separating. It also has the property that $\bigcap_{\alpha \in I} ER_\alpha = E(X)$.

This is based on the observation that if $x, y \in A$ and if R is any congruence on X, then $(x, y) \notin ER$. Otherwise there exist x', y' such that xRx', yRy' and $[x', y'] \in E(X)$. From xRx' we have $[x, x'] \notin E(X)$; similarly $[y, y'] \notin E(X)$. In view of 3.2 this means $x', y' \in A \cup B$, contrary to x', y' being adjacent.

It follows from this that if $(x, y) \in \bigcap_{\alpha \in I} ER_\alpha$ then at least one of x, y does not belong to A. Hence if $e = [x, y] \notin E(X)$, then $\alpha = (e, e') \in I$ for any edge $e' \in E(-X)$ different from e, so that $xR_\alpha y$ and the proof can be completed as for 2.8.

The construction of the family $(R_\alpha)_{\alpha \in I}$ is a little tedious because many cases must be distinguished. We list them all, but carry out the proof for only three of them, leaving the others to the reader.

Given $e = [x, y]$, $e' = [u, v]$, $e \neq e'$ one has the following possibilities

(i) $x, y \in B$ or $x \in A$, $y \in B$ (i) $u, v \in A \cup B$

(ii) $x \in B$, $y \in V(Y)$ (ii) $u \in B$, $v \in V(Y)$

(iii) $x, y \in V(Y)$ (iii) $u, v \in V(Y)$

Case (i.i). Let S_α be any maximal congruence on Y. Partition $A \cup B$ into two non-empty subsets C, D such that $A \cap C \neq \phi$, $A \cap D \neq \neq \phi$, and $x, y, u \in C$, $v \in D$. Since $|A| \geqslant 2$ and not both of x, y are in A this is possible. Now extend S_α to a congruence R_α on X by adding C and D as new congruence classes. Clearly R_α is almost maximal, C and D being the two non-adjacent vertices of X/R_α.

Case (i.ii). Same as case (i.i) except that here one does not (and cannot) require that $v \in D$.

Case (i.iii). Here let S_α be a relatively maximal congruence on Y which separates u and v. If S_α is almost maximal, extend it to R_α on X by adding $A \cup B$ as a new congruence class. If S_α is maximal, partition $A \cup B$ into C and D with $x, y \in C$, and extend S_α by adding C and D as extra congruence classes.

We conclude this section with some brief remarks concerning the third question raised above, viz. the characterization of those graphs all of whose subdirectly irreducible representations are pure complete. The complete description of these graphs is still a wide-open problem. The partial results obtained so far are closely tied in with some fundamental notions of coloring theory and constitute the subject of the next section. The link between the two sections is provided by Proposition 3.8 which, incidentally, also disposes of the converse of 2.8. For its proof we need the following lemma.

Lemma 3.7. *Let* $(R_\alpha)_{\alpha\in A}$ *be a family of maximal congruences on* X *for which* $\bigcap_{\alpha\in A} ER_\alpha = E(X)$. *Then* $(R_\alpha)_{\alpha\in A}$ *is point-identifying.*

Proof. If $(R_\alpha)_{\alpha\in A}$ is not point-identifying, then there is an edge $[x, y] \in E(-X)$ such that $(x, y) \notin R_\alpha$ for each $\alpha \in A$. X/R_α being complete this implies $(x, y) \in ER_\alpha$ for each $\alpha \in A$, i.e., $[x, y] \in E(X)$, a contradiction.

Proposition 3.8. *For any graph* X *the following statements are equivalent:*

(i) *Every subdirectly irreducible representation of* X *consists of complete graphs.*

(ii) *Every relatively maximal congruence on* X *is maximal.*

(iii) *A family* $(R_\alpha)_{\alpha\in A}$ *of relatively maximal congruences on* X *is point-identifying if and only if*

$$\bigcap_{\alpha\in A} ER_\alpha = E(X) .$$

(iv) *There is no epimorphism of* X *onto an almost complete graph.*

Proof. The pattern of the proof is as follows: (i) ⇒ (ii) ⇒ (iv) ⇒ (i), (ii) ⇔ (iii).

(i) ⇒ (ii). Take any relatively maximal congruence R on X, and any point-separating, point-identifying family $(S_\alpha)_{\alpha\in A}$ of relatively maximal congruences on X of which R is a member. By (i) each X/S_α is complete, hence X/R is complete, i.e., R is maximal.

(ii) ⇒ (iv). Suppose there is an epimorphism $\Phi: X \to Y$ onto an almost complete graph. Consider the kernel R of Φ. By 2.11, R is relatively maximal, hence maximal, whence $Y \cong X/R$ is complete, a contradiction.

(iv) ⇒ (i). Let $(X_\alpha)_{\alpha\in A}$ be a subdirectly irreducible representation of X. Then $\mathrm{pr}_\alpha \circ \Phi: X \to X_\alpha$ is an epimorphism, hence by (iv), X_α cannot be almost complete. Since X_α is subdirectly irreducible it must therefore be complete.

(ii) ⇒ (iii) is immediate from Lemma 3.7.

(iii) ⇒ (ii). Suppose X has an almost maximal congruence R. Then X/R is almost complete, i.e., there exist two distinct R-classes $R[a]$, $R[b]$ such that $[a', b'] \notin E(X)$ for every $a' \in R[a]$ and $b' \in R[b]$. In particular this implies that $(a, b) \notin ER$.

Let $(R_\alpha)_{\alpha \in A}$ be the family of all relatively maximal congruences which separate a and b. Clearly this family is not point-identifying and R is one of its members. To complete the proof we have to show that $\bigcap_{\alpha \in A} ER = E(X)$. This will provide a contradiction to (iii).

Let $(x, y) \in \bigcap_{\alpha \in A} ER_\alpha$. Since $(a, b) \notin ER$, we have $(x, y) \neq (a, b)$. Moreover, for each $\alpha \in A$ there exist x_α, y_α such that $(x, x_\alpha) \in R_\alpha$, $[x_\alpha, y_\alpha] \in E(X)$, $(y_\alpha, y) \in R_\alpha$. If $[x, y] \notin E(X)$, then by 2.12 there is a $\beta \in A$ with $xR_\beta y$, and hence $x_\beta R_\beta y_\beta$, a contradiction to $[x_\beta, y_\beta] \in$ $\in E(X)$. Thus $[x, y] \in E(X)$.

The existence of graphs (other than complete graphs) satisfying the conditions of 3.8 is shown by the graph Q_r of Figure 3.

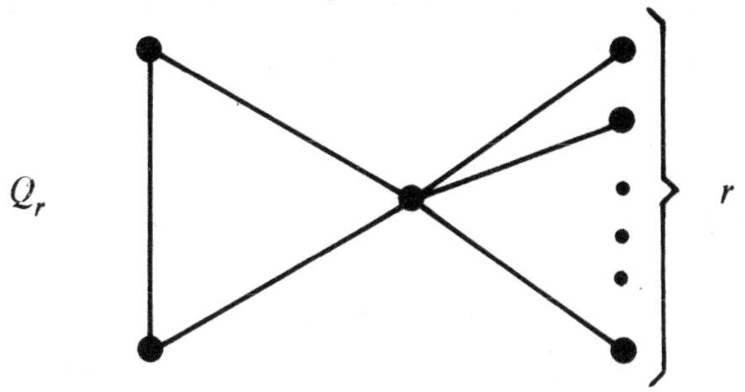

Figure 3

It is clear that every homomorphic image of Q_r is isomorphic to Q_s, $0 \leqslant s \leqslant r$, and consequently Q_r has no epimorphism onto an almost complete graph. The triangle in Q_r can be replaced by any complete graph.

4. GRAPHS WITHOUT MIXED REPRESENTATIONS

The preceding sections make it clear that the key position in the theory of subdirectly irreducible representations is enjoyed by maximal congruences and almost maximal congruences. Corresponding to these two types of congruences are epimorphisms onto complete graphs and almost complete graphs. The former have been known and studied – under the name of colorings – long before the general notion of graph homomorphisms appeared in the literature. Epimorphisms onto almost complete graphs are new, for the simple reason that so far there did not exist any motivation to consider them. Motivation is now provided primarily by the problem of determining those graphs all of whose subdirectly irreducible representations are pure complete. This problem appears to be of considerable difficulty, and it has been solved so far only in the case where the graphs in question are at most 3-chromatic (Theorem 4.2). The method we employ does not seem to be extendable even to chromatic number 4.

For the sake of simplicity we shall assume that all graphs in this section are *finitely colorable.* To fix our terminology, an n-coloring of X is a homomorphism of X into K_n, the complete graph on the vertices $0, \ldots, n-1$. By K_n^- we shall denote the almost complete graph on the same set of vertices as K_n, more precisely,

$$K_n^- = \begin{cases} K_n \setminus e_0, & \text{where} \quad e_0 = [0,1] \quad \text{if} \quad n \geqslant 2 \\ \phi, & \text{if} \quad n = 1 . \end{cases}$$

For $n \geqslant 2$ there is a natural epimorphism $\kappa_n \colon K_n^- \to K_{n-1}$ defined by $0 \mapsto 0$, and $\alpha \mapsto \alpha - 1$ for $0 < \alpha < n$. In order to maintain the existence of this epimorphism even in the case $n = 1$ we have chosen to define K_1^- as the empty graph.

Definition 4.1. Let X be a graph, n a positive integer. An *almost n-coloring* is an epimorphism $X \to K_n^-$. X is *almost n-colorable* [*almost colorable*] if and only if it has an almost n-coloring [for some n].

In connection with this definition it should be recalled that the key to the characterization problem of graphs without mixed subdirectly irreducible representations is Proposition 3.8 which says that the graphs in question are *exactly those which are not almost colorable.* With this in mind note that, in contradistinction to colorings, *almost colorings must be defined as epimorphisms.* Were one to define an almost n-coloring simply as a homomorphism *into* K_n^-, then since the restriction of K_{n+1}^- to $\{1, \ldots, n\}$ is isomorphic to K_n, every n-coloring would automatically be an almost $(n+1)$-coloring, and hence every graph would be almost colorable. In view of Proposition 3.8 it is precisely this circumstance which we want to avoid.

We now state the main result of this section and then give the necessary definitions and auxiliary results which lead up to its proof.

Theorem 4.2. *Let* X *be an* n*-chromatic graph without isolated vertices;* $n \leqslant 3$. *Then the following are equivalent*

(i) X *has no almost coloring;*

(ii) X *has no almost* $(n+1)$*-coloring;*

(iii) *if* $n = 2$, *then* $X \cong K_2$;

if $n = 3$, *then* X *is isomorphic to one of the following:* K_3, $K_3 \oplus K_2$, Q_r *for any* $r > 0$ *(see Figure 3), and the 5-circuit* C_5.

The case $n = 4$ is presently being studied by Mme Reine Fournier of the Université de Montréal and her preliminary results indicate a similar situation as for $n = 3$: there is a finite number of easily described infinite families of graphs which may or may not have K_4 as a restriction, as well as a finite number of additional "peculiar" graphs.

The structure of the proof of 4.2 is as follows: 4.2 (i) $\Rightarrow$ 4.2 (ii) $\Rightarrow$ $\Rightarrow$ 4.4 $\Rightarrow$ 4.5 $\Rightarrow$ 4.2 (iii) $\Rightarrow$ 4.2 (i).

Definition 4.3. Let X be n-chromatic, Φ a coloring of X, $\alpha \in \{0, \ldots, n-1\}$. A vertex $x \in V(X)$ is called an α*-full* vertex of Φ if and only if $\Phi x = \alpha$ and for each $\beta \in \{0, \ldots, n-1\} - \{\alpha\}$ there is a $y \in V(x; X)$ with $\Phi y = \beta$. For convenience we shall say that x is a *full* vertex of Φ if it is 0-full. The choice of 0 as the color of x is, of course, arbitrary; what is essential is that all full vertices of Φ have the same color.

Note that a coloring Φ of X has at least one α-full vertex for every $\alpha \in \{0, \ldots, n-1\}$. Otherwise one could change the color of every vertex of color α to some color not occurring in its neighborhood, and this would result in an $(n-1)$-coloring of X.

Lemma 4.4. *Let X be an n-chromatic graph. Then X has no almost $(n+1)$-coloring if and only if every coloring of X has exactly one full vertex.*

Proof. Necessity: Suppose X has an almost $(n+1)$-coloring Φ_0. By an argument which is similar to that which establishes the existence of a full vertex for every coloring of a graph, one can prove the existence of an $x_0 \in \Phi_0^{-1}[0]$ such that for each $\alpha \in \{2, \ldots, n\}$ there is a $y \in V(x_0; X)$ with $\Phi_0 y = \alpha$. Similarly, by considering the almost $(n+1)$-coloring $\Phi_1 = \tau \circ \Phi_0$, where τ is the transposition $0 \leftrightarrow 1$, one obtains an $x_1 \in \Phi_1^{-1}[0] = \Phi_0^{-1}[1]$ with the same property, i.e., that for each $\alpha \in \{2, \ldots, n\}$ there is a $z \in V(x_1; X)$ with $\Phi_1 z = \alpha$. It is then clear that x_0 and x_1 are two distinct full vertices of the n-coloring $\Phi = \kappa_{n+1} \circ \Phi_0 = \kappa_{n+1} \circ \Phi_1$.

Sufficiency: Suppose a coloring Φ of X has two distinct full vertices x_0, x_1. Define $\psi \colon X \to K_{n+1}^-$ by $x_0 \mapsto 0$ and $x \mapsto \Phi x + 1$ for all other vertices.

Let $[\alpha, \beta] \in E(K_{n+1}^-)$, $\alpha < \beta$. If $\alpha = 0$ or 1, then $\beta \geqslant 2$, hence by fullness of x_α there is a vertex $y_\alpha \in V(x_\alpha; X)$ with $\Phi y_\alpha = \beta - 1$. Thus

$$[\psi x_\alpha, \psi y_\alpha] = \begin{cases} [0, \Phi y_0 + 1] = [0, \beta] & \text{if} \quad \alpha = 0 \\ [\Phi x_1 + 1, \Phi y_1 + 1] = [1, \beta] & \text{if} \quad \alpha = 1 . \end{cases}$$

Finally, if $\alpha \geqslant 2$, then since Φ is an epimorphism there is an edge $e = [x, y] \in E(X)$ with $\Phi x = \alpha - 1$, $\Phi y = \beta - 1$. Since both $\alpha - 1$ and $\beta - 1$ are $\geqslant 1$, and $\Phi x_\alpha = 0$, it follows that neither end-vertex of e coincides with x_0 or x_1. Hence $[\psi x, \psi y] = [\Phi x + 1, \Phi y + 1] = [\alpha, \beta]$. Thus ψ is an epimorphism, i.e., an almost $(n+1)$-coloring.

Lemma 4.5. *If an n-chromatic graph* X *has no almost* $(n+1)$*-coloring, then the (unique) full vertex corresponding to each coloring of* X *is critical.*

Proof. Let Φ be a coloring of X, x_0 the full vertex corresponding to Φ, Φ_0 the restriction of Φ to $X_0 = X - x_0$. If x_0 is not critical, then X_0 is n-chromatic and Φ_0 is a coloring of X_0. Hence X_0 has a full vertex x_1 relative to Φ_0. Evidently, $x_1 \neq x_0$. Now note that x_1 is a full vertex of X with respect to Φ, since Φ and Φ_0 are the same mapping on X_0 (and, incidentally, since $\Phi x_0 = \Phi x_1 = 0$, $x_0 \notin V(x_1; X)$, whence $V(x_0; X) = V(x_0; X_0)$). This means that X has two full vertices, a contradiction to 4.4.

Lemma 4.5 can be considerably strengthened. However, since we will not use this result we state it without proof.

Lemma 4.6. *Let* X *be an n-chromatic graph which is not* $(n+1)$*-almost colorable. Then every critical vertex of* X *is adjacent to* $n-1$ *other critical vertices.*

Proposition 4.7. *Any non-degenerate shortest path* W *in a bipartite graph* X *is a retract of* X.

This is a special case of a much more general theorem (due to P. Hell) concerning retractions of bipartite graphs (see [3], 6.3.1.A).

Proof. We use induction on n, the length of W.

Since X is 2-chromatic there exists an epimorphism $\sigma: X \to K_2$ such that $\sigma x_0 = 0$. Hence for $n = 1$ one can define $\Phi: X \to W = (x_0, x_1)$ by $x \mapsto x_\alpha$ if $x \in \sigma^{-1}[\alpha]$, $\alpha = 0, 1$.

Now suppose the statement true for $n-1$. Let $W = (x_0, \ldots, x_n)$ be a shortest path in X. $V(x_n; X)$ is contained in one of the two color-classes of X, hence the equivalence R whose classes are $V(x_n; X)$ and $\{x\}$, $x \in V(X) - V(x_n; X)$, is a congruence and $\chi(X/R) = 2$. For $x \in V(X)$ we shall denote the class $R[x]$ by x'.

Note that $W' = (x'_0, \ldots, x'_{n-1})$ is a shortest path in X/R. Hence by induction hypothesis there is a retraction $\Phi'\colon X/R \to W'$. Define $\Phi\colon X \to W$ by

$$x_n \longmapsto x_n ,$$

$$x \longmapsto x_i, \quad \text{where} \quad x \neq x_n \quad \text{and} \quad \Phi' x' = x_i' .$$

This is the desired retraction.

Corollary 4.8. *Let X be a bipartite graph, C a shortest circuit in X. Then C is a retract of X.*

Proof. Let $e \in E(C)$, $Y = X \setminus e$, $W = C \setminus e$. Y is bipartite and since $|C|$ is minimal, W is a shortest path in Y (joining the two ends of e). Hence by 4.7 there is a retraction $\Phi\colon Y \to W$. Note that $V(Y) = V(X)$ and $V(W) = V(C)$. Hence considered as a mapping $V(X) \to V(C)$, Φ is a retraction $X \to C$.

This completes the preliminaries of the proof of 4.2.

Proof of Theorem 4.2. The implication (i) ⇒ (ii) is trivial. We remark here that for chromatic number $\geqslant 4$ the converse is false; a counterexample will be given after this proof.

(ii) ⇒ (iii): We suppose first that X is 2-chromatic. By 4.5, X has a critical vertex x_0. $X_0 = X - x_0$ is 1-chromatic, and hence discrete. By hypothesis, X has no isolated vertex, hence $V(x_0; X) = V(X_0)$, (i.e., X is a star with center x_0). We wish to show that $|X_0| = 1$. If not, choose any $x_1 \in V(X_0)$ and map $X \to K_3^-$ as follows:

$$x_0 \mapsto 0,\ x_1 \mapsto 1,\ x \mapsto 2 \quad \text{for any} \quad x \in V(X_0),\ x \neq x_1 .$$

Since we assume that X_0 has more than one vertex, this mapping is an epimorphism, which contradicts hypothesis (ii). But $|X_0| = 1$ means that X consists of the edge $[x_0, x_1]$, i.e., $X \cong K_2$.

The case $n = 3$ is considerably more involved. Again we choose a critical vertex x_0; this time $X_0 = X - x_0$ is 2-chromatic. We fix an arbitrary coloring of X. A vertex $x \in V(X_0)$ will be called an *α-vertex* if its color with respect to the chosen coloring is α (where $\alpha = 0$ or 1).

We distinguish two cases.

Case (i). $d(x_0; X) \geqslant 3$.

Without loss of generality, x_0 is adjacent to two distinct 0-vertices of X_0, say y_0, y_2, and to at least one 1-vertex, y_1 (for if all members of $V(x_0; X)$ belong to the same color-class of X_0, then X would be 2-chromatic). y_1 is not an end-vertex of X, for if so, then it is an isolated vertex of X_0, and these can all be colored with color 0. Moreover, y_0 and y_2 can be so chosen that at least one of them, say y_0, is incident with an edge of X_0. For if all 0-vertices of X_0 belonging to $V(x_0; X)$ are isolated vertices of X_0, then by re-coloring them with color 1, and assigning color 0 to x_0, we would obtain a 2-coloring of X. Thus, there exist $z_i \in V(X_0)$ such that $[y_i, z_i] \in E(X_0)$, $i = 0, 1$. This implies that z_i is a $(1 - i)$-vertex. Let R_0 be the equivalence on $V(X_0)$ with classes $a_1 = \{y_1, z_0\}$, $a_2 = \{y_2, z_1\}$, and $\{x\}$ otherwise.

Since y_i and z_{i-1} have the same color, R_0 is a congruence, and $Y_0 = X_0/R_0$ is 2-chromatic. We extend R_0 to a congruence R on X be adding the equivalence class $\{x_0\}$. Let $Y = X/R$.

Note that $W = (\{y_0\}, a_1, a_2)$ is a path in Y_0, and since Y_0 contains no triangle (being 2-chromatic), W is a shortest path. Let Z be the restriction of Y to $\{\{x_0\}, a_1, a_2, \{y_0\}\}$. Since W is shortest, Z is isomorphic to K_4^- (see Figure 4), and (by 4.7) there is a retraction Φ_0: $Y_0 \to W$ which can be extended to a retraction Φ: $Y \to Z$ by mapping $\{x_0\} \mapsto \{x_0\}$. Thus we have the following epimorphisms:

$$X \xrightarrow[\text{proj.}]{} X/R = Y \xrightarrow[\Phi]{} Z \xrightarrow[\cong]{} K_4^-$$

which contradicts the assumption that X is not almost 4-colorable.

Case (ii). $d(x_0; X) = 2$. Then $V(x_0; X) = \{y_0, y_1\}$, and the colors of y_0, y_1 in X_0 are different, say y_α is an α-vertex, $\alpha = 0, 1$ (otherwise X would be 2-chromatic).

If there is an epimorphism σ_0 of X_0 onto a 4-circuit $C_4 = (y_0, y_1, y_2, y_3)$, then this can evidently be extended to an epimorphism σ: $X \to H$, where

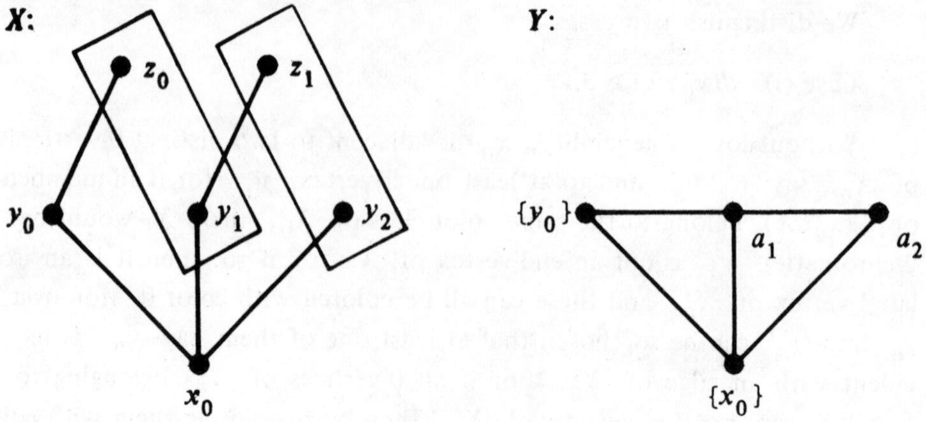

Figure 4

$$V(H) = \{x_0\} \cup V(C_4), \quad E(H) = \{[x_0, y_\alpha]: \alpha = 0, 1\} \cup E(C_4)\,.$$

Moreover, there is an epimorphism $H \to K_4^-$ defined by $x_0 \mapsto 3$, $y_i \mapsto i$ $i = 0, \ldots, 3$. Hence we conclude that no epimorphism of X_0 onto a 4-circuit can exist.

This implies that the length of any path in X_0 cannot exceed 3. Otherwise, take any path of length 4 in X_0, say $W = (w_0, \ldots, w_4)$, and identify w_0 and w_4. Since w_0 and w_4 have the same color in X_0, the quotient graph Z_0 is 2-chromatic and contains a 4-circuit C (resulting from W). C is a shortest circuit in Z_0, hence by 4.8 it is a retract of Z_0. Consequently, X_0 has an epimorphism onto a 4-circuit, but we already know that this is impossible.

Since y_0 and y_1 must have different colors in X_0 they belong to the same component of X_0, say X_1, and $E(X_1) \neq \phi$. Next note that X_0 has at most one other component. To see this let $\{X_i: i \in I\}$ be the set of all components of X different from X_1. Then $E(X_i) \neq \phi$, otherwise $X_i = (x_i)$, and hence each x_i would be an isolated vertex of X, contrary to our hypothesis. Now suppose that $|I| \geqslant 2$. Choose an $i_0 \in I$ and define a mapping $\tau: X \to K_4^-$ by

$$x \mapsto \begin{cases} 0, & \text{if } x = x_0, \\ \alpha + 2, & \text{if } x \text{ is an } \alpha\text{-vertex of } X_1, \quad \alpha = 0, 1, \\ \alpha + 1, & \text{if } x \text{ is an } \alpha\text{-vertex of } X_{i_0}, \quad \alpha = 0, 1, \\ 1, & \text{if } x \text{ is a } 0\text{-vertex of } X_i, \quad i \in I - \{i_0\}, \\ 3, & \text{if } x \text{ is a } 1\text{-vertex of } X_i, \quad i \in I - \{i_0\}. \end{cases}$$

It is straightforward to verify that τ is an epimorphism, a contradiction. Henceforth we shall say that X_0 has two components X_1, X_2, premitting the case that $X_2 = \phi$.

X_2 has at most one edge. For if $|E(X_2)| \geqslant 2$, then since X_2 is connected it has two adjacent edges e_1, e_2, and by 4.7, X_2 can be retracted onto the path $U = (e_1) \cup (e_2)$.

In other words, X_2 has an epimorphism onto any path of length 2. But this means that we can obtain an epimorphism $X \to K_4^-$ by mapping X'_1, the component of X which contains x_0, onto the triangle $0, 2, 3$ (X'_1 is 3-chromatic), and X_2 onto the path $(2, 1, 3)$, again a contradiction.

We now distinguish further cases.

Case (ii.a). X_1 has a single edge, i.e., $X_1 \cong K_2$. Here either $X = X'_1 \cong K_3$, if $X_2 = \phi$; or $X \cong K_3 \oplus K_2$, if $X_2 \cong K_2$.

Case (ii.b). X_1 contains a path of length 2 but none of length 3. This means that X_1 is a star, i.e., $X_1 \cong K_{1,r}$ with $r \geqslant 2$, and y_0, y_1 are adjacent in X_1. Thus X'_1 has the form indicated in Figure 5. It follows that $X_2 = \phi$. For if not, map X onto K_4^- as follows: $x_0 \mapsto 0$, $y_0 \mapsto 2$, $y_1 \mapsto 3$, $x \mapsto 1$ for any $x \in V(y_0; X) - \{x_0, y_1\}$, and X_2 onto the edge $[1, 3]$ (see Figure 5). But $X_2 = \phi$ implies that $X \cong Q_s$, where $s = r - 1 \geqslant 1$.

Case (ii.c). X_1 contains a path of length 3. Since 4-circuits are forbidden, X_1 must have the form depicted in Figure 6.

Moreover, since y_0 and y_1 have different colors in X_1 we must have that (in the notation of Figure 6), either $y_0 = x$ and $y_1 = y$, or $y_0 \in A$

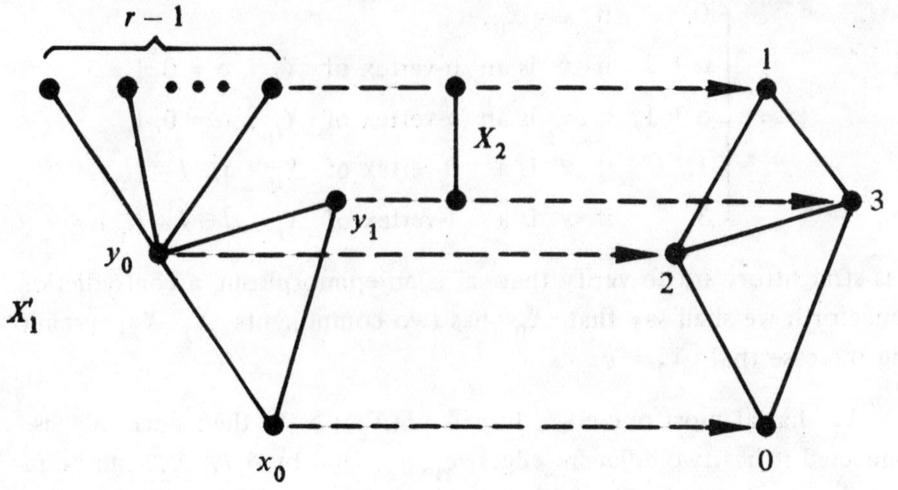

Figure 5

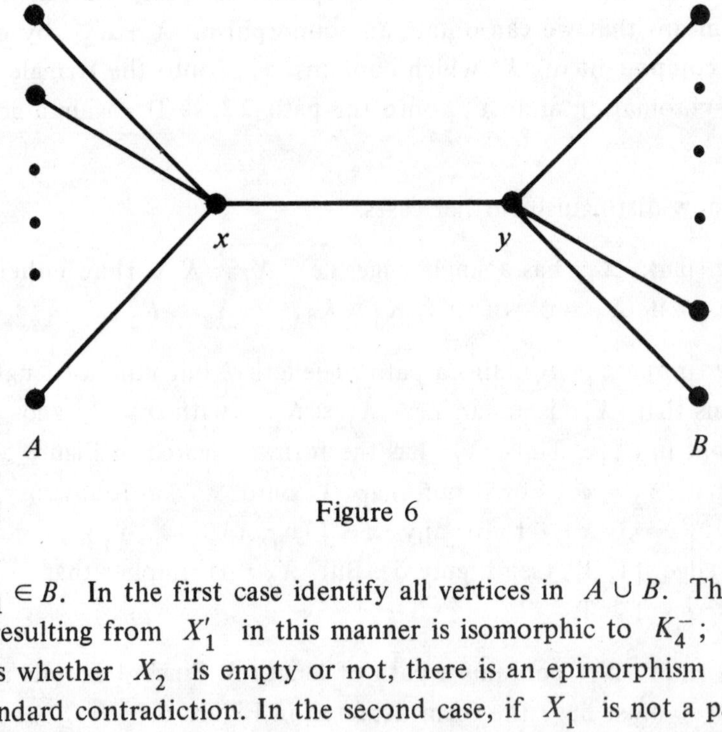

Figure 6

and $y_1 \in B$. In the first case identify all vertices in $A \cup B$. The quotient graph resulting from X'_1 in this manner is isomorphic to K_4^-; hence, regardless whether X_2 is empty or not, there is an epimorphism $X \to K_4^-$, the standard contradiction. In the second case, if X_1 is not a path, identify all vertices in $(A \cup B) - \{y_1\}$. The resulting quotient Z of X is isomorphic to the graph H introduced at the beginning of the proof of case

(ii) (again, the presence or absence of X_2 is immaterial), and hence there is an epimorphism $Z \to K_4^-$. The last remaining alternative is that X_1 is a path (of length 3), i.e., X_1' is a 5-cirucit, and $X_2 = \phi$ (for if X_2 is non-empty, X can be mapped onto the graph Z of the preceding case). It follows that $X \cong C_5$.

(iii) ⇒ (i). For K_3 and $K_3 \oplus K_2$ this is completely obvious. For Q_r see the remark after 3.8. Concerning the 5-circuit, if there were an epimorphism $C_5 \to K_n^-$ for some n, then since $|E(C_5)| = 5$, we must have $n \leqslant 4$. Since C_5 is 3-chromatic, $n = 2$ or 3 is impossible, and it is easily verified that the only homomorphic images of C_5 are C_5 itself, Q_1 and K_3. This completes the proof.

It will be observed that all that was used in order to prove Theorem 4.2 was the fact that the given n-chromatic graph has no almost $(n + 1)$-coloring. For $n \geqslant 4$ this is no longer equivalent to the total absence of almost colorings as is shown by the 4-chromatic graph of Figure 7 (due to P. Hell, oral communication).

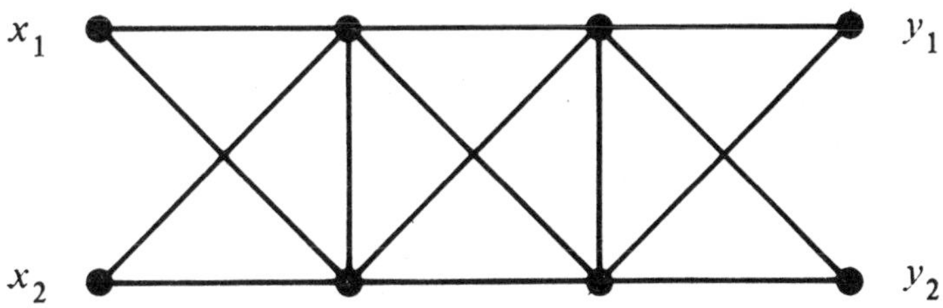

Figure 7

Identification of x_i and y_i, $i = 1, 2$, provides an epimorphism onto K_6^-. However, no epimorphism of the graph onto K_5^- exists.

Added in proof. Mme Reine Fournier has kindly pointed out that the condition of Lemma 4.4 is not necessary. The lemma remains, however, applicable to the proof of Theorem 4.2.

REFERENCES

[1] G. Birkhoff, *Lattice theory;* AMS Colloquium Publications, 25, Providence R.I., 1967.

[2] B. Fawcett, Graphs and ultrapowers; *Ph. D. thesis,* McMaster University, 1969.

[3] P. Hell; Rétractions de graphes; *Thèse de doctorat,* Université de Montréal, 1972.

COLLOQUIA MATHEMATICA SOCIETATIS JÁNOS BOLYAI
10. INFINITE AND FINITE SETS, KESZTHELY (HUNGARY), 1973.

ON SOME INVESTIGATIONS IN THE THEORY OF THE FOUR-COLOUR PROBLEM

H. SACHS

Only recently Y. Shimamoto (cf. Haken [6]) tried to establish the validity of the four-colour conjecture by means of computers. As is well-known, his attempt failed but nevertheless caused wide and encouraging activities in the field of the four-colour problem: Heesch's [7] reduction theory with its many results became widely known, possible means for proving the conjecture were critically scrutinized – especially by the papers of Tutte and Whitney [15], [17] – and by all this a deeper understanding of the underlying relations and of the particular difficulties of the four-colour problem was arrived at.

I would like to express a few naive thoughts connected with this field in order to – perhaps – make a modest contribution towards clearing up the question of what might reasonably be expected and what sort of phenomena, paradoxical as these may appear, may occur. Certainly, whenever a difficult central problem is being attacked, on the one hand, a great deal of optimism is absolutely necessary but, on the other hand, a simultaneous critical estimate of the scope of one's means will always be useful and will help to avoid disappointment: My contribution is rather meant to lie in this latter (critical) direction. But I do not intend to formulate a new conjecture or a particular assertion – I would just like to say that, in may opinion, the arguments in favour of the four-colour conjecture and the arguments in favour of its negation are rather counterbalancing each other. The possibility that the four-colour problem may not be solvable at all (which principially must be taken into account) will be excluded from my considerations, though.

First of all a remark of a more general nature: Optimists convinced of the truth of the four-colour theorem will of course object pointing out that there are numerous partial results in favour of the four-colour conjecture, that a great number of reductions is already known, and particularly that every planar graph with fewer than 45 vertices is four-colourable – still, I do not hold this arguments to be convincing. In this connection, let us consider an example taken from a different field. For small n, the cyclomatic polynomials have only coefficients 0, 1, and -1, and for a long time, this was supposed to be true for all n. Emma Lehmer, however, provided a method for constructing counter-examples, using numbers n containing at least three distinct odd primes, the smallest of these numbers being equal to $3 \cdot 5 \cdot 7 = 105$. It is true, however, that for this, a new idea, a deeper insight into the structure of the object in question was necessary, and the same will certainly be true in the case of the four-colour problem: If there exists a counter-example at all, it will then hardly be found by "trial and error" – there exist indeed too many reductions – but by systematic construction methods. Such a method would theoretically, too, be of great value so that even a – possibly – negative outcome would mean an essential theoretical progress. Moreover, it would not be surprising at all if in a sense, "almost all" planar graphs should turn out to be five-chromatic though possibly the irreducible ones among them are very rare: Every non-four-colourable planar graph G defines a class of infinitely many reducible planar graphs, all of which can be reduced to G and among which there are possibly many non-four-colourable graphs. It is not very difficult to imagine that for a large number of vertices, the sets of non-four-colourable graphs contained in these classes may cover almost the entire region. If planar graphs with sufficiently many vertices are considered, it may turn out that the probability for a randomly selected graph to be non-four-colourable will perhaps be not too small; in this case, however, we are faced with the difficulty of being forced to test for non-colourability rather than for non-reducibility, and that, in its turn, cannot be realized in practice, even when using computers.

In this respect G. Grohmann [4] (a student of mine) has made some interesting investigations: Combining various procedures with an algorithm of A.A. Zykov et al. [19] based on the recoloration (interchange of colours) of defective Kempe chains, he developed a computer program for finding counter-examples; it turned out that the procedures to be performed (which, in cases of 15 or 20 vertices, will not take more than a few minutes even when performed by hand) will require computing times to be measured in hours, days, or even years if graphs of some interesting size (i.e., with more than 60 vertices, say) are taken into consideration. Without the application of new theoretical results, the search for counter-examples seems to be absolutely hopeless (even if there should be any).

Let us now consider a bit more in detail those methods being used in order to obtain a confirmation of the four-colour conjecture. These can be subdivided into two classes as follows:

(1) *Recoloration methods,* i.e., methods based on interchanging colours within Kempe chains (cf. [8]).

The interchange of colours per se constitutes, as it were, the "combinatorial component" of such a method whereas the homotopy property of the sphere (the fact that every simply closed curve on the sphere can be continually contracted to a point) guaranteeing the (eventual) existence of a suitable recoloration constitutes the "topological component". Therefore, interchanging colours within Kempe chains is indeed a very plausible near-at-hand procedure.

(2) *All other methods.* These will not be itemized here; I would just like to point out that numerous Hungarian mathematicians, amongst them Paul Erdős playing a prominent rôle, have made substantial contributions, explicitly or implicitly. It may suffice here to mention only a few of

the more important theorems which are of particular significance for the theory of the four-colour problem:

The combinatorial characterization of planar graphs due to Kuratowski [9];

The theorem of Tutte [14] stating that every four-connected planar graph is Hamiltonian;

The theorem of Grötzsch [5] stating that every schlicht planar graph containing no circuit of length three is three-colourable.

All three of these theorems are proved by means of induction; Grötzsch's procedure provides an excellent example for a proof making use of (substantially) only a finite number of reduction figures.

Before returning to the discussion of the recoloration methods, I may be permitted a brief remark concerning the theorem of Tutte. One might try to extend Tutte's theorem by showing that every cyclically 5-connected (with respect to the edges) normal map is Hamiltonian because this proposition would already imply the validity of the four-colour conjecture. The latter statement, however, does not hold true, as has been proved by H. Walther [16] and E.J. Grinberg [3] in 1965: Therefore, the Tutte theorem cannot be improved in this direction. The Grinberg – Walther result is not in favour of the four-colour conjecture: Here again we meet with the typical phenomenon observed so many times, that investigations showing great promise will – up to a certain point – produce results in favour of the four-colour conjecture, but that they will then fail, and fail in principle, thus rather providing quite a strong argument against the four-colour conjecture.

Let us now consider the recoloration methods. As is well-known, by means of recolouring Kempe chains various reduction methods were developed already in the classical period (by Birkhoff, Errera, Franklin, Winn, et al.). For a long time the result of Winn [18] stating that every planar graph with fewer than 36 vertices is four-colourable was the best statement available until some years ago when the number of vertices could be raised to 40 in a voluminous paper by O. Ore and J. Stemple (cf. [11]). Recently, G.A. Donec (Kiev) [2] succeeded in pushing this bound as far as 45, his paper of only 25 typewritten pages being surprisingly short. Donec makes use of methods very similar to those used by his predecessors, and all of them are based on the central idea of recolouring Kempe chains.

The doubtlessly most significant contribution towards a systematic theory of "Kempe reductions" (i.e., reductions by means of recolouring suitably chosen Kempe chains) has been made by H. Heesch [7] who introduced the notions of A, B, C and D-reducibility and found a great number of new reduction figures: It is to be hoped that he will succeed in complementing his catalogue of reduction figures so that every planar graph may be recognized as either four-colourable or reducible, thus confirming the four-colour conjecture. Whether this will be possible or not remains, of course, an open question; it may be taken for granted, however, that by means of Heesch's results the bounds established by Winn, Ore – Stemple, and Donec (mentioned above) can in principle be improved considerably: In searching for counter-examples, one should not start with less than 100 vertices, say.

In concluding I should like to mention some investigations of my own. Aarts and de Groot [1] proved that every planar graph G containing no vertex of valency greater than 5 is four-colourable. The proof, again, is based on recolouring Kempe chains and can be brought into a particularly elegant form by making use of a well-known theorem of Dirac (concerning the eventuality of extending a "partial coloration" of a planar graph; cf. [10], p. 110, or [13], p. 179) which, in a way, contains the "recoloration argument" in a highly concentrated form. Here the question arises whether the hypothesis that G should contain no vertex of valency greater than 5 might be replaced by a less stringent one. This is indeed possible: If G is allowed to have at most two vertices of valency greater than 5, the four-colourability may be proved fairly easily, and if G is allowed to have at most three such vertices, the proof is a bit more involved but in principle still runs along similar lines (cf. [13], pp. 182-203). H.-J. Presia

[12] succeeded in settling the case of four exceptional vertices, the expenditure for the proof in this case being noticeably greater; this result together with some immediate consequences is discussed in Dr. Presia's paper so that it will not be necessary for me to go into details.

The assessment of the result may differ widely: On the one hand, only little seems to be gained by allowing G to have (no more than) four exceptional vertices, and on the other hand, the fact that there are no bounds prescribed for the valencies of the exceptional vertices still comprises a new point of view – besides presenting a new infinite class of four-colourable planar graphs.

It should be noted that in this case, again, the main means of proof lastly consists in recolouring Kempe chains.

One might be inclined to believe that this result is nothing else but the commencement of an eventual induction with respect to the number of vertices having valencies greater than 5 – but I am afraid that would be entirely false! Even the analogous statement for (no more than) five exceptional vertices would – if provable at all – necessitate new arguments and insights far beyond our present knowledge. Here again we arrive at the well-known barrier: Up to the number 4 there are no major difficulties, but the transition from 4 to 5 is barred by insurmountable obstacles. I for my part am convinced that an investigation of the circumstances – if somebody would take the trouble of carrying out such an analysis – would clearly show that the deeper reason for the phenomenon that the way of reasoning successful in the case of (no more than) 4 exceptional vertices will fail if applied to the case of 5 exceptional vertices, is, in fact, exactly the same reason which renders the reduction of the pentagon by means of recoloruing Kempe chains impossible*. In other words: If it were possible to prove the four-colour conjecture in the way pointed out above, then there would also exist a simpler proof, viz., it would be possible to overcome (or to evade) the difficulties of the non-reducibility of the pentagon – thus complementing Kempe's original idea.

*The author here refers to the fact that – in terms of Heesch's reduction theory of triangulations the "5-wheel" is not D-reducible.

Here we meet again with the phenomenon already mentioned above: Up to a certain point we obtain results all in favour of the four-colour conjecture but, finally, the "proof" collapses thus supplying a fairly strong argument against the four-colour conjecture; or, in order to qualify the last assertion: ... against the possibility of settling the four-colour problem by means of recoloration methods without, in addition, using some substantially new idea.

REFERENCES

[1] J.M. Aarts – J. de Groot, A case of colouration in the four colour problem. *Nieuw Arch. Wisk.,* 11 (1963), 10-18.

[2] G.A. Donec, Issledovanie voprosov raskraski ploskich grafov, *(Dissertacija).* Kiev, 1970. (Investigation of problems concerning the colouring of planar graphs. *(Dissertation).* Kiev, 1970.)

[3] E.Ja. Grinberg, O ploskich odnorodnych grafach stepeni tri bez gamil'tonovych ciklov, *Latv. mat. ezhegodnik,* 4, Riga (1968), 51-57. (On planar homogeneous graphs of degree three without Hamilton circuites. *Latvian math. yearbook,* 4, Riga (1968), 51-57.)

[4] G. Grohmann, Ein Verfahren für die Suche nach Gegenbeispielen für die Vierfarbenvermutung mit Hilfe von Rechenautomaten. *Dissertation,* 1972, TH Ilmenau.

[5] H. Grötzsch, Zur Theorie der diskreten Gebilde, 7. Mitteilung: Ein Dreifarbensatz für dreikreisfreie Netze auf der Kugel. Wiss. Z. Martin-Luther-Univ. Halle – Wittenberg, *Math.-Nat. Reihe,* 8 (1958/59), 109-120.

[6] W.R.G. Haken, Haken on Shimamoto's construction (4-color-seminar, October 1971; *hectographed*).

[7] H. Heesch, *Untersuchungen zum Vierfarbenproblem.* Mannheim – Wien – Zürich, 1969.

[8] A. B. Kempe, On the geographical problem of the four colours. *Amer. J. Math.*, 2 (1879), 193-200.

[9] C. Kuratowski, Sur le problème des courbes gauches en Topologie. *Fund. Math.*, 15 (1930), 271-283.

[10] O. Ore, *The Four-Color Problem.* New York – London, 1967.

[11] O. Ore – J. Stemple, Numerical calculations on the four-color problem. *J. Comb. Theory,* 8 (1970), 65-78.

[12] H.-J. Presia, Über Färbungen planarer Graphen. I und II. *Math. Nachr.*, (to appear).

[13] H. Sachs, *Einführung in die Theorie der endlichen Graphen,* Teil II. Leipzig, 1972.

[14] W. T. Tutte, A theorem on planar graphs. *Trans. Amer. Math. Soc.*, 82 (1956), 99-116.

[15] W. T. Tutte, Shimamoto's attack on the four- colour problem. (University of Waterloo; *hectographed*).

[16] H. Walther, Über das Problem der Existenz von Hamiltonkreisen in planaren, regulären Graphen. *Math. Nachr.*, 39 (1969), 277-296.

[17] H. Whitney – W. T. Tutte, Kempe chains and the four colour problem. *Utilitas Mathematica,* (Winnipeg), 2 (1972), 241-281.

[18] C. E. Winn, On the minimum number of polygons in an irreducible map. *Amer. J. Math.*, 62 (1940), 406-416.

[19] A. A. Zykov, (A. A. Sykow – D. J. Kesselman – J. I. Neimark – W. N. Podkorytow), Über ein Verfahren zur Färbung ebener Triangulationen. *Math. Nachr.*, 40 (1969), 51-60.

COLLOQUIA MATHEMATICA SOCIETATIS JÁNOS BOLYAI
10. INFINITE AND FINITE SETS, KESZTHELY (HUNGARY), 1973.

AN EXTREMAL PROBLEM IN FINITE GRAPH THEORY

J. SHEEHAN

1. INTRODUCTION

It is well-known and easily proved that a regular graph of degree n and girth 4 has at least $2n$ vertices and that the only such graph with exactly $2n$ vertices is the bipartite graph $K_{n,n}$. Let $F(k, d)$ be the set of non-bipartite regular graphs of degree k with diameter d and girth $2d$. Let $f(k, d) = \min \{|V(G)| : G \in F(k, d)\}$. If $F(k, d) = \phi$ we write $f(k, d) = \infty$. In this talk we discuss the function $f(k, d)$. Our results are mainly for the case $d = 2$.

2. THE FUNCTION $f(k, 2)$

Let $P(n)$ be the set of integers modulo n. Let k be a positive integer $(k > 1)$ and let p be the smallest prime dividing k. Let $k = pq$. The graph $G_k(p, q)$ (see Fig. 1) is defined as follows: $V(G_k(p, q)) = P((3p - 1)q)$ and vertices i and j are adjacent if and only if $i - j \equiv 1 + 3n \pmod{3p - 1}$ for some $n \in P(p)$. It is easy to prove that $G_k(p, q) \in F(k, 2)$. Hence

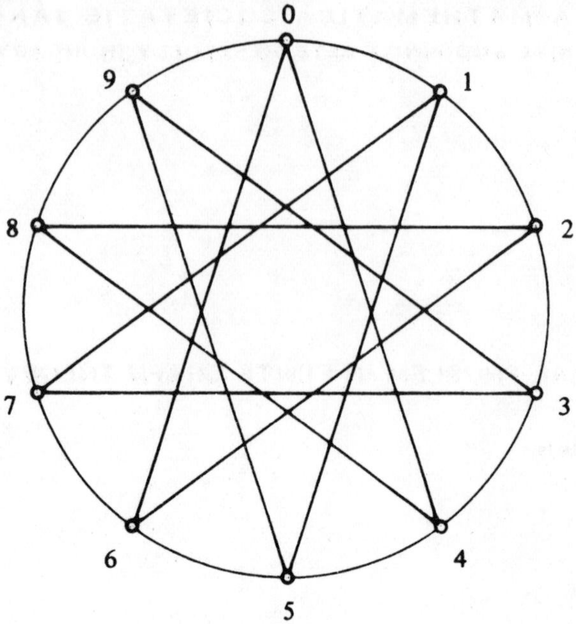

$G_4(2, 2)$

Figure 1

(1) $$f(k, 2) \leqslant 3k - \frac{k}{p}.$$

Theorem. *Let* p *be the smallest prime dividing* k *then, if* $p \in \{2, 3\}$, $f(k, 2) = 3k - \frac{k}{p}$ *and* $G_k(p, q)$ *is the only member of* $F(k, 2)$ *with exactly* $3k - \frac{k}{p}$ *notices.*

Proof. Let $p = 2$. Let G be a minimal member of $F(k, 2)$ and let $n = |V(G)|$. It is easy to prove that G must contain a pentagon. Let X be the set of vertices of a pentagon contained in G. Since there are no edges across X the number of edges leaving X is $5(k-2)$. However each vertex of $G - X$ can be joined to at most 2 vertices in X and hence the number of edges leaving $G - X$ is at most $2(n-5)$. Hence $5(k-2) \leqslant 2(n-5)$ i.e., $n \geqslant 3k - \frac{k}{2}$. Therefore $f(k, 2) \geqslant 3k - \frac{k}{2}$ and so, from (1), $f(k, 2) = 3k - \frac{k}{2}$.

The case when $p = 3$ (see [1]) seems more difficult to prove as does the uniqueness of $G_k(p, q)$, $p \in \{2, 3\}$, and we do not include the proofs here.

Conjecture. Let p be the smallest prime dividing k then

$$f(k, 2) = 3k - \frac{k}{p}$$

and $G_k(p, q)$ is the only member of $F(k, 2)$ with exactly $3k - \frac{k}{p}$ vertices. (Incidentally $f(k, 2) > 3k - \frac{k}{3}$ if $p \notin \{2, 3\}$.)

3. REMARKS

When $d > 2$ our information on $f(k, d)$ is negligible. We do not know even if $f(k, d) \neq \infty$ for any k. We do know that $f(3, 3) \in \{18, \infty\}$.

We have considered the same problem without any diameter restrictions. Thus let $H(k, d)$ be the set of non-bipartite regular graphs of degree k and girth $2d$. Let $h(k, d) = \min\{|V(G)|: G \in H(k, d)\}$. It is easy to prove that

$$(2) \qquad h(k, 2) \geqslant \frac{5k}{2}.$$

We define the graph $G(k)$ (see Fig. 2) as follows: Let $\left\{\frac{2k}{3}\right\}$ be the smallest even integer greater than or equal to $\frac{2k}{3}$. Write $2s = \left\{\frac{2k}{3}\right\}$. Let $V(G(k)) = \bigcup_{i=1}^{8} X_i$ be a disjoint union of independent sets with cardinalities $|X_i| = k - 2s$ $(i = 1, 2, 3, 4)$, $|X_i| = 4s - k$ $(i = 5, 6)$, $|X_i| = s$ $(i = 7, 8)$. Then, assuming the X_i's to be independent sets of vertices, $E(G(k))$ is defined as follows. The induced subgraphs of $G(k)$ with vertex sets $X_i \cup X_j$, indicated by unbroken lines in Figure 2, are complete bipartite graphs and the induced subgraph with vertex set $X_5 \cup X_6$, indicated by a broken line in Figure 2, is regular, bipartite of degree s.

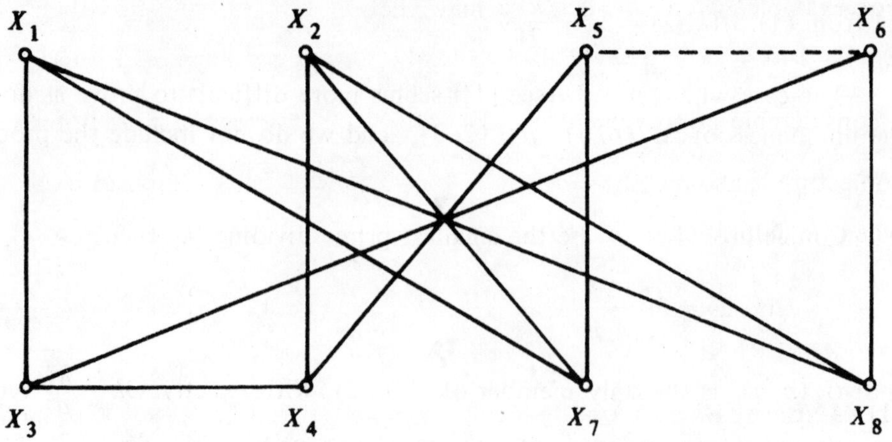

$G(k)$

Figure 2

It is easy to show that $G(k) \in H(k, 2)$. Hence

(3) $$h(k, 2) \leqslant 2k + \left\{\frac{2k}{3}\right\}.$$

For certain values of k this upper bound for $h(k, 2)$ can be improved. However the improvement is quite messy as we shall see. We define $\Theta(k)$, assuming its existence, as follows. $\Theta(k) = 2k + \dfrac{2\alpha\beta}{k}$ where α, β are integers such that

(1) $\dfrac{k}{3} \leqslant \beta \leqslant \alpha \leqslant \dfrac{2k}{3}$;

(2) $\dfrac{k^2}{4} \leqslant \alpha\beta \leqslant \dfrac{k^2}{3}$;

(3) $k \mid \alpha\beta$ and

(4) $\alpha\beta$ is minimal subject to (1), (2) and (3).

It is easy to prove that, providing $\Theta(k)$ exists,

(4) $$h(k, 2) \leqslant \Theta(k) \leqslant 2k + \left\{\frac{2k}{3}\right\}.$$

For example when $k = 55$ we may choose $\alpha = 33$, $\beta = 30$ then $\Theta(k) \leqslant 146 < 2k + \left\{\frac{2k}{3}\right\} = 148$. Clearly for some values of k, $\Theta(k)$ is a considerable improvement on $2k + \left\{\frac{2k}{3}\right\}$. Finally when k is prime we conjecture that $h(k, 2) = 2k + \left\{\frac{2k}{3}\right\}$ (obviously $\Theta(k)$ does not exist when k is prime).

REFERENCE

[1] J. Sheehan, Non-bipartite graphs of girth 4, *Journal of Discrete Mathematics,* 8 (1974), 383-402.

COLLOQUIA MATHEMATICA SOCIETATIS JÁNOS BOLYAI
10. INFINITE AND FINITE SETS, KESZTHELY (HUNGARY), 1973.

GRAPHS WITH PRESCRIBED ASYMMETRY AND MINIMAL NUMBER OF EDGES

S. SHELAH*

§0. INTRODUCTION

We shall deal with non-directed graphs, without loops and double edges, and having a finite number of vertices.

A graph is symmetric if it has a non-trivial automorphism = a permutation of its vertices, such that a pair of vertices is connected iff their images are connected. The asymmetry of a graph is the minimal number of changes (i.e. adding and deleting of edges) which is necessary to make the graph symmetric. Erdős and Rényi [1] defined and investigated this notion, and defined, $F(n, k)$ $[C(n, k)]$ for $k \geqslant 1$, $n > 1$ as the minimal number of edges in a [connected] graph, with n vertices, whose asymmetric is k; if there is no such graph the value of the function will be ∞ (If n is too small, this happens). (see [1], §5 p. 311): They proved that $C(6, 1) = 6$, $C(1, 1) = 0$, $C(n, 1) = n - 1$ for $n \geqslant 7$; also $C(n, 2) > n + 1$ for

*Research done in summer 1968, paper written in summer 1970. The preparation of this paper was supported in part by NSF Grant #GP-22937, while the author was supported in part by NSF Grant #GP-27994.

$n \geqslant 7$ and $F(n, 3) \geqslant 4n/3 - 3/2$. It is obvious that $F(n, k) \leqslant C(n, k)$. They also show that $C(n, 1) = \infty$ for $1 < n \leqslant 5$, and $C(n, k) = \infty$ for $n < 2k + 1$.

We shall compute $C(n, k)$ *and* $F(n, k)$ *for* $k > 1$ *and* n *sufficiently larger than* k. For $k > 2$, n sufficiently larger than k, we affirm the conjecture in [1] that $C(n, k) = F(n, k)$. (See [1] p. 314. before the remarks.) It will be interesting to know for any k, from what n our formulas are correct. From the proof a bound can be found, but seemingly it will be far from the exact value.

First we shall formulate the results. Then, in § 1, we prove that $F(n, k)$ is not smaller than the values mentioned in the theorems, by generalizing a proof from [1]. In § 2 we describe examples of connected graphs with n vertices and asymmetry k, whose number of edges is the number appearing in the theorems. Finally, in § 3 we shall prove for the case $k \geqslant 41$, that the graphs described in § 2, have the required asymmetry. (For $3 \leqslant k \leqslant 40$, the proof is messy and with the same central idea).

The results are the following:

Theorem 0.1. *For* n *sufficiently large*

$$F(n, 2) = n + 1, \quad C(n, 2) = n + 2 .$$

Remark. This was independently found by Nesetril in his M. Sc. thesis.

Theorem 0.2. *For odd* $k > 2$, *and* n *sufficiently larger than* k

$$F(n, k) = C(n, k) = [(k + 3)n/4 - 0.5[2n/(k + 3)] + 1/2] .$$

Theorem 0.3. *For even* $k > 2$ *and* n *sufficiently larger than* k

$$F(n, k) = C(n, k) = [(k + 2)n/4 + 1/2] .$$

Notations. Let G denote a graph, P, Q, R, S vertices of the graph, $N = N(G)$ the number of vertices of G, $E = E(G)$ the number of edges of G. Let v_P be the valence of P (= the number of edges incident to P), and v^i the valence of P_i. Also V_k will denote the number of vertices (in G) whose valence is k, $V_{\geqslant k}$ the number of vertices (in G) whose valence is $\geqslant k$, etc. Thus, $2E = \sum_P' v_P = \sum_k kV_k$. Let m, n, k, l denote natural numbers, and i, j, r integers. $A(G)$ will stand for the asymmetry of G.

We say that $P_1, P_2, \ldots, P_m$ is a path, if $P_1P_2, P_2P_3, \ldots, P_{m-1}P_m$ are edges, and $P_1P_2 \ldots P_m$ is a circle, if $P_1P_2, P_2P_3, \ldots, P_{m-1}P_m, P_mP_1,$ are edges. $[x]$ is the integral part of x.

§1. PROOF OF THE LOWER BOUNDS

First we shall observe some facts, which, in fact, appear in [1].

If P, Q are vertices of G, which are not connected to any other vertices (but PQ may be an edge) then the permutation interchanging them is an automorphism of G. Hence

Observation 1. $A(G) \leqslant v_P + v_Q$ if P, Q are distinct vertices of G.

Observation 2. $A(G) \leqslant v_P + v_Q - 2$ if P, Q are vertices of G, and PQ is an edge.

If P, Q, R are vertices of G, such that RP, RQ are edges, and there are no other edges containing P or Q, except possibly PQ, then the permutation interchanging P and Q is an automorphism of G, hence G is symmetric.

Observation 3. $A(G) \leqslant v_P + v_Q - 2$, if P, Q, R are distinct vertices of G, and RP, RQ are edges of G.

Lemma 1.1. *$F(n, 2) \geqslant n + 1$ for $n > 7$, and $C(n, 2) \geqslant n + 2$ for $n \geqslant 7$.*

Proof. By [1], $C(n, 2) \geqslant n + 2$, for $n > 6$.

Suppose $N(G) = n$, $A(G) = 2$; we should prove $E(G) \geqslant n + 1$. Let $G_1, \ldots, G_k$ be the components of G. By [1] there are, up to isomorphism, only two asymmetric connected graphs G with $E(G) \leqslant N(G)$. One, G^1 is the graph with one point, and the other G^2 have six vertices and six edges. Now $E(G) = \sum_{l=1}^{k} E(G_l)$ and $N(G) = \sum_{l=1}^{k} N(G_l)$. Now clearly no two of the G_k's can be isomorphic, hence the worst case is when, say, $G_1 = G^1$, $G_2 = G^2$. As $N(G) > 7$, $k > 2$. Hence

$$E(G) = \sum_{l=1}^{k} E(G_l) = E(G_1) + E(G_2) + \sum_{l=3}^{l} E(G_l) \geqslant$$

$$\geqslant 0 + 6 + \sum_{l=3}^{k} (N(G_l) + 2) =$$

$$= 0 + 6 + \sum_{l=3}^{k} N(G_l) + 2(k-2) =$$

$$= 6 + N(G) - 7 + 2(k-2) =$$

$$= N(G) - 1 + 2(k-2) \geqslant N(G) + 1 .$$

Lemma 1.2. *If* $k > 2$ *is even, then*

$$F(n,k) \geqslant [(k+2)n/4 + 1/2] .$$

Proof. Let G be a graph with n vertices, $A(G) \geqslant k$. We should prove that $E = E(G) \geqslant [(k+2)n/4 + 1/2]$, or, as $E(G)$ is an integer, $E(G) \geqslant (k+2)n/4$, or $2E(G) \geqslant (k+2)n/2$.

If the valency of every vertex is $\geqslant (k+2)/2$, then

$$2E = \sum_l lV_l \geqslant ((k+2)/2) \sum_l V_l = (k+2)n/2 ,$$

so let R_0 be a vertex with valency $< (k+2)/2$, that is $\leqslant k/2$. Then, for any other vertex P we have $v_P \geqslant k/2$, because by observation 1

$$k \leqslant A(G) \leqslant v_P + v_Q \leqslant v_P + k/2 .$$

Now if $v_Q \leqslant k/2$, and PQ is an edge, then v_P is $\geqslant k/2 + 2$ as by observation 2

$$k \leqslant A(G) \leqslant v_P + v_Q - 2 = v_P + k/2 - 2 .$$

Similarly if Q, P, S, are vertices of G, QS, PS are edges, then $v_Q \leqslant k/2$ implies $v_P \geqslant k/2 + 2$ (by observation 3).

Assume first $\nu_{R_0} = k/2$. As we have shown that for every other P, $\nu_P \geqslant k/2$, clearly $V_{<k/2} = 0$. As every vertex of valence $\leqslant k/2$ is connected only with vertices of valence $\geqslant k/2 + 2$, and no vertex is connected with two vertices of valency $k/2$, *clearly* $V_{\geqslant (k/2+2)} \geqslant V_{k/2}$.

Hence

$$2E = \sum l V_l \geqslant (k/2)V_{k/2} + (k/2+1)V_{k/2+1} +$$
$$+ (k/2+2)V_{\geqslant (k/2+2)} =$$
$$= (k/2)V_{k/2} + (k/2+1)(n - V_{k/2} - V_{\geqslant (k/2+2)}) +$$
$$+ (k/2+2)V_{\geqslant (k/2+2)} =$$
$$= - V_{k/2} + (k/2+1)n + V_{\geqslant (k/2+2)} \geqslant$$
$$\geqslant (k/2+1)n = (k+2)n/2 .$$

Now assume $\nu_{R_0} < k/2$. Then by observation 1, the valency of any other vertex P is $\geqslant (k+2)/2$, as $k \leqslant A(G) \leqslant V_P + k/2 - 1$. If $\nu_{R_0} \neq 0$, and P is connected with R_0, then $\nu_P \geqslant k - \nu_{R_0} + 2$. Hence

$$2E = \sum_Q \nu_Q \geqslant (k/2+1)(n-2) + \nu_{R_0} + \nu_P \geqslant$$
$$\geqslant (k+2)n/2 - (k+2) + \nu_{R_0} + k - \nu_{R_0} + 2 = (k+2)n/2 .$$

So we are left with the case $\nu_{R_0} = 0$. Then for every $P \neq R_0$, $\nu_P \geqslant k$ (by observation 1). Hence

$$2E \geqslant \sum_Q \nu_Q \geqslant k(n-1) = kn - k =$$
$$= (k+2)n/2 + (k-2)n/2 - k \geqslant$$
$$\geqslant (k+2)n/2 + (4-2)n/2 - k =$$
$$= (k+2)n/2 + n - k \geqslant (k+2)n/2$$

(we use the assumption that k is even and > 2, hence $\geqslant 4$; and that $n \geqslant 2k+1 > k$ (for $n < 2k+1 > k$ implies $F(n,k) = \infty$).

Lemma 1.3. *If k is odd and > 2, then*

$$F(n, k) \geqslant [(k+3)n/4 - 0.5[2n/(k+3)] + 1/2] .$$

Remark. For $k = 3$, this slightly improves Th. 8 p. 314 [1].

Proof. Let $l = (k+1)/2$. Let $N(G) = n$, $A(G) \geqslant k$.

If R is a vertex of G with valency $< l$, then for every other vertex P we have $\nu_P \geqslant l$, because by observation 1

$$k \leqslant A(G) \leqslant \nu_P + \nu_R \leqslant \nu_P + l - 1$$

$$\nu_P \geqslant k - (l-1) = k - (k+1)/2 + 1 =$$

$$= k/2 - 1/2 + 1 = k/2 + 1/2 = l .$$

Hence there is at most one vertex with valency $< l$. Now if $\nu_P \leqslant l$, and P, Q are connected, then $\nu_Q \geqslant l + 1$ (by observation 2), and similarly if PR, QR are edges then $\nu_Q \geqslant l + 1$ (by observation 3). Hence if $\nu_P \leqslant l$ and PQ are connected, then $\nu_Q \geqslant l + 1$, and P is the only vertex connected with Q with a valency $\leqslant l$. Hence $V_{\geqslant (l+1)} \geqslant \sum_{m \leqslant l} mV_m \geqslant lV_l$.

Case I. Let us assume first $V_{<l} = 0$.

Then $n = V_l + V_{>l} \geqslant V_l + lV_l = (l+1)V_l$, or $V_l \leqslant n/(l+1)$.

$$2E(G) = \sum mV_m \geqslant lV_l + (l+1)V_{\geqslant (l+1)} =$$

$$= lV_l + (l+1)(n - V_l) = (l+1)n - V_l \geqslant$$

$$\geqslant (l+1)n - [n/(l+1)] = (k+3)n/2 - [2n/(k+3)] .$$

So, if $V_{<l} = 0$, the lemma holds. Suppose $V_{<l} \neq 0$, hence $V_{<l} = 1$, as noted in the beginning of the proof, and let R_0 be the only vertex with valency $< l$.

Case II. Assume now $\nu_{R_0} = 0$. Then, by observation 1, every $P \neq R_0$ has valency $\geqslant k$. Hence $2E \geqslant k(n-1)$. For $k > 3$, as $k \geqslant 5$

$$2E - (k+3)n/2 \geqslant k(n-1) - (k+3)n/2 =$$

$$= kn - k - kn/2 - 3n/2 = n(k - k/2 - 3/2) - k =$$

$$= n(k/2 - 3/2) - k \geqslant n - k > 0 .$$

This clearly implies the required inequality. For $k = 3$, $n \geqslant 9$, the required inequality also holds.

$$2E \geqslant k(n-1) = 3n - 3 = (k+3)n/2 - 3 =$$

$$= (k+3)n/2 - [2 \cdot 9/6] \geqslant (k+3)n/2 - [2n/(k+3)] .$$

As $\infty = F(n,k)$ for $n < 2k+1 = 7$, the remaining cases are $k = 3$, $n = 7$, $k = 3$, $n = 8$. If we remove R_0, we get a graph G_1, $N(G_1) = n - 1$, $E(G_1) = E(G)$, $A(G_1) \geqslant 3$, and the valency of every vertex is $\geqslant 3$. For $n = 7$, we get a graph with six vertices and asymmetry 3, contradicting Theorem 1.1 in [1], according to which

$$A(G) \leqslant (N(G) - 1)/2 .$$

So we are left with the case $n = 8$. As $\sum v_P$ is even, there is in G_1 at least one vertex with valency $\geqslant 4$. If there are two such vertices, or one with valency > 4, we get $E(G) = E(G_1) \geqslant 12$ which is the required inequality. So let P be the only vertex of valency four, Q_1, Q_2, Q_3, Q_4 the vertices connected with it, and S_1, S_2 the two other vertices. As S_1 has valency three, and it is not connected with P, it is connected with two of the Q's, say Q_1, Q_2. Now clearly in order to make the permutation interchanging Q_1 and Q_2 to an automorphism of G, it is sufficient to remove two edges. This is a contradiction. So have finished the case $v_{R_0} = 0$.

Case III. $l > v_{R_0} > 0$

By observations 2 and 3 it is clear that if P is connected with R_0, or connected with a vertex which is connected with R_0, then $v_P \geqslant m = k - v_{R_0} + 2$, and hence $V_{\geqslant m} \geqslant m$. Clearly $0 < v_{R_0} < l = (k+1)/2$ implies $m > (k+3)/2 = l + 1$. As noted before in Case I

$$V_{\geqslant (l+1)} \geqslant \sum_{m \leqslant l} mV_m \geqslant lV_l, \text{ hence}$$

$$n = 1 + V_l + V_{\geqslant l} \geqslant 1 + V_l + lV_l = 1 + (l+1)V_l ,$$

$$V_l \leqslant (n-1)/(l+1) \leqslant n/(l+1) - 1$$

whence

$$V_l \leqslant [n/(l+1)] - 1 = [2n/(k+3)] - 1 .$$

Now

$$2E = \sum_r rV_r \geqslant v_{R_0} \cdot 1 + lV_l +$$

$$+ (l+1)(V_{\geqslant (l+1)} - V_{\geqslant m}) + mV_{\geqslant m} =$$

$$= v_{R_0} + lV_l + (l+1)V_{\geqslant (l+1)} + (m-l-1)V_{\geqslant m} =$$

$$= v_{R_0} + lV_l + (l+1)(n-1-V_l) + (m-l-1)V_{\geqslant m} =$$

$$= v_{R_0} - V_l + (l+1)n - (l+1) + (m-l-1)V_{\geqslant m} \geqslant$$

$$\geqslant (v_{R_0} - V_l) + (l+1)n - (l+1) +$$

$$+ (m-l-1)m \geqslant$$

$$\geqslant - [2n/(k+3)] + (k+3)n/2 +$$

$$+ (-(l+1) + (m-l-1)m) \geqslant$$

$$\geqslant (k+3)n/2 - [2n/(k+3)]$$

(the last inequality holds, as $m > l + 1$, implies

$$(m-l-1)m \geqslant m > (l+1)) .$$

So the required inequality holds in case III, and Lemma 1.3 is proved.

§2. THE DESCRIPTIONS OF THE EXAMPLES

Example 2.1. We will show that $C(n, 2) \leq n + 2$.

Description. Let $n_1, \ldots, n_6$ be different natural numbers $> 0'$. The vertices of G will be $R_1, \ldots, R_4$ (of valency three), and P^i_k, $i = 1, \ldots, 6$, $k = 1, \ldots, n_i$ (of valency 2). Now

$$R_1 P^1_1 \ldots P^1_{n_1} R_2, R_1 P^2_1 \ldots P^2_{n_2} R_3, R_1 P^3_1 \ldots$$

$$R_1 P^3_1 \ldots P^3_{n_3} R_4, R_2 P^4_1 \ldots P^4_{n_4} R_3,$$

$$R_2 P^5_1 \ldots P^5_{n_5} R_4, R_3 P^6_1 \ldots P^6_{n_6} R_4$$

will be paths, (and every edge of G appears in one of them)

Example 2.2. Proof of $F(n, 2) \leq n + 1$.

It is the same as the previous one, if we add one isolated vertex.

Example 2.3. We show the upper bound for $F(n, k)$, for even $k > 2$.

Remark. Every pair of vertices which will not be said to be connected, will be considered unconnected. We shall concentrate on the case $k > 40$.

Construction. Clearly, by Lemma 1.2, every vertex will have a valency $l = (k + 2)/2$, except one vertex if n is odd. Let us choose numbers $r_1, \ldots, r_l$ such that

1. r_i is odd, and $0 < r_1 < -r_2 < r_3 < -r_4 < \ldots < (-1)^l r_l$.

2. If $r_{i_1} + r_{i_2} = r_{m_1} + r_{m_2}$ then $\{i_1, i_2\} = \{m_1, m_2\}$.

3. If $r_{i_1} + r_{i_2} + r_{i_3} = r_{m_1} + r_{m_2} + r_{m_3}$ then $\{i_1, i_2, i_3\} = \{m_1, m_2, m_3\}$, and $i_1 = i_2 = m_1$ implies $m_1 = m_2$ or $m_1 = m_3$.

4. No sum of ≤ 5 of the numbers $\{\pm r_i \colon 1 \leq i \leq l\}$ is > 0 and $\leq k$.

Clearly $r_i = (-2)^{i+1}(k + 1) + 1$ satisfy the conditions, but we can easily find much smaller r_i's; for example defining r_i by induction as the first satisfying the conditions.

Let us first define a graph $G(n, k)$.

Case I. n is even.

The vertices are $P_1, \ldots, P_n$; and let $P(i) = P_i$, and $P_i = P_j$ if $i = j$ (mod n). Now for even i, P_i is connected with $P(i + r_j)$ for $j = 1, \ldots, l$.

Case II. n is odd.

The points will be $P_1, \ldots, P_{n-1}$ and Q. As before $P_i = P(i)$ $P_i = P_j$ if $i = j$ (mod $n - 1$). For even i, P_i is connected with $P(i + r_j)$ $j = 1, \ldots, l$; except if $j = 1$ and i belongs to $\{2r: 1 \leqslant r \leqslant \leqslant (l + 1)/2\}$. Q is connected to P_{2r}, P_{2r+r_1} for $1 \leqslant r \leqslant (l + 1)/2$.

Let $L_1 = 6|r_l|$, $L = 12|r_l|$, $(|r_l| = (-1)^l r_l)$. (For $k > 40$ this is more then sufficient, but for $3 \leqslant k \leqslant 40$, greater values can be more convenient).

Now we shall define the required graph $G^*(n, k)$, by slightly modifying $G(n, k)$. For $m = 1, \ldots, k + 15$ we omit the edges

$$P(2[n/4] + 2mL)P(2[n/4] + 2mL + r_1)$$

and $P(2[n/4] + 2mL + 2L_1 + 2m)P(2[n/4] + 2mL + 2L_1 + 2m + r_1)$ and add the edges $P(2[n/4] + 2mL)P(2[n/4] + 2mL + 2L_1 + 2m + r_1)$ and $P(2[n/4] + 2mL + r_1)P(2[n/4] + 2mL + 2L_1 + 2m)$.

Notice that the r_i's and L, L_1 depend only on k and not on n.

Example 2.4. Proof of the upper bound of $F(n, k)$ for odd $k > 3$.

Clearly here the valencies of the vertices will be $l = (k + 1)/2$ or $l + 1$. Let n_1 be such that $((l + 1)n_1 + l(n - n_1))/2$ is the number appearing in Theorem 0.3; clearly there is such a number. We define the r_i's as in example 2.3, and also L, L_1. Clearly $ln_2 \leqslant n_1$, where $n_2 = n - n_1$. Now we define $G(n, k)$.

Case I. n_1 is even.

The vertices will be $P_1, \ldots, P_{n_1}, R_1, \ldots, R_{n_2}$, where $n_2 = n - n_1$. As before $P_i = P(i)$, and $P_i = P_j$ if $i = j \pmod{n_1}$. For even i, we connect P_i with $P(i + r_j)$ for $j = 1, \ldots, l$. We also connect R_1 with $P_1, \ldots, P_l$; R_2 with $P_{l+1}, \ldots, P_{2l}; \ldots ; R_m$ with $P_{lm-l+1}, \ldots$ $\ldots, P_{lm}; \ldots$. Finally, if $n_1 > ln_2$ we connect P_{ln_2+1} with $P_{ln_2+2}, \ldots, P_{ln_2+2m}$ with $P_{ln_2+2m-1}, \ldots$. (Note that this is possible as $n_1 - ln_2$ is even, because n_1 is even, and as $(l+1)n_1 + ln_2$ is even, ln_2 is even). Notice that the valency of the P_i's is $l+1$, and that of the R_i's is l.

Case II. n_1 is odd.

Note that as $(l+1)n_1 + ln_2$ is even, l cannot be even. So l is odd and n_2 is even. Now the vertices will be $P_1, \ldots, P_{n_1-1}, R_1, \ldots$ $\ldots, R_{n_2}$ and Q. As usual $P_i = P(i)$ and $P_i = P_j$ if $i = j \pmod{n_1 - 1}$. For even i we connect P_i with $P(i + r_j)$, $j = 1, \ldots, l$. We alco connect for $m = 1, \ldots, n_2$, P_m with $P(ml - l + j)$ for $j = 1, \ldots, l$. Now for $m = 1, \ldots, (l+1)/2$, we "disconnect" $P(2Lm)P(2Lm + r_1)$ and connect each of them with Q. Note that the P_i's and Q have valency $l+1$, whereas the R_i's have valency l. Now the definition of $G^*(n, k)$ from $G(n, k)$ is the same as in example 3.3.

Example 2.5. Proof of the upper bound for $F(n, 3)$.

Let n_1 be a number such that $(3n_1 + 2(n - n_1))/2$ is the number in Theorem 0.3, and $n_2 = n - n_1$. Clearly n_1 is even, and $2n_2 \leqslant n_1$; and so $n_1 - 2n_2$ is even, hence it is zero or two. Let us define $G(n, k)$.

Case I. $n_1 = 2n_2$.

The vertices of the graph will be $P_1, \ldots, P_{n_1}, R_1, \ldots, R_{n_2}$. $P_1, \ldots, P_{n_1}$ will be a circle. R_m will be connected with P_{2m} and $P_{(2m+17)}$ (where as usual $P_i = P_j$ if $i = j \pmod{n_1}$).

Case II. $n_1 = 2n_2 + 2$.

The vertices will be $P_1, \ldots, P_{n_1-2}$, $R_1, \ldots, R_{n_2}, S_1, S_2$.

$S_1P_1 \ldots P_{L-1}S_2P_L \ldots P_{n_1-2}$ is a circle, S_1S_2 is an edge, and R_m is connected with P_{2m}, P_{2m+17}.

The definition of $G^*(n, k)$ is as in 2.3, taking $r_1 = 1$, $L_1 = 100$, $L = 200$.

§3. PROOF THAT THE EXAMPLES HAVE THE REQUIRED ASYMMETRY

We shall prove it only for $k > 40$. Let us mention some properties of the graphs $G^*(n, k)$ we shall need. We assume implicitly that n is always sufficiently large.

Property A. There is no square in the graph.

Proof. By property (4) of the r_i's, a square cannot contain as a vertex one of the R_i's in 2.4, nor Q in 2.3 II. By the definition of L it cannot contain Q from 2.4 II. By the definition of $G^*(n, k)$ and L, L_1, it suffices to prove that in $G(n, k)$, there is no circle $P(i_1)P(i_2)P(i_3)P(i_4)$. Now if there is a such a circle, i_1 is odd iff i_2 is even. So assume i_1 is even, hence i_2, i_4 are odd, i_3 is even. Moreover; $i_2 - i_1, i_4 - i_1, i_2 - i_3, i_4 - i_3 \in \in \{r_m : m = 1, \ldots, l\}$. As $(i_4 - i_1) + (i_2 - i_3) = (i_2 - i_1) + (i_4 - i_3)$ by property (2) of the r_i's

$$\{i_4 - i_1, i_2 - i_3\} = \{i_2 - i_1, i_4 - i_3\}.$$

Hence $i_4 - i_1 = i_2 - i_1$ or $i_4 - i_1 = i_4 - i_3$ so $i_4 = i_2$ or $i_1 = i_3$, and so this is not a square.

Property B. If $P(i_1) \ldots P(i_6)$ is a circle in $G^*(n, k)$ then $i_2 - i_1 = = i_4 - i_5$; and similarly $i_3 - i_2 = i_5 - i_6$, $i_4 - i_3 = i_6 - i_1$. Moreover $i_2 \equiv i_4 \equiv i_6$, $i_1 \equiv i_3 \equiv i_5 \pmod 2$.

The proof is similar to that of (A).

Property C. For every P_i, the number of vertices among $\{P(i-j): r_1 < j \leqslant r_l\}$ adjacent to P_i is $\geqslant [(l-1)/2]$.

Proof. Suppose i is even. Then clearly P_i is connected with

$P(i - r_{2m+1})$ for $m = 1, \ldots, [(l-1)/2]$. Now if i is odd, it is connected with $P(i - r_{2m})$ for $m = 1, \ldots, [(l-1)/2]$.

Now we shall prove the theorem itself.

Theorem 3.1. *$A[G^*(n, k)] = k$, for $k > 40$, and n sufficiently large relative to k. (Clearly it is $\leqslant k$).*

Proof. We shall prove a stronger result: if θ is a permutation of the vertices of $G^*(n, k)$, then the number of edges PQ, such that $\theta(P), \theta(Q)$ are not connected, is $\geqslant k$.

Suppose G was obtained from $G^*(n, k)$ by $< k$ changes, and θ is an automorphism of G. We should prove θ is the identity.

We shall first prove

(*) there are m_1, m_2, r such that $(r_l)^2 \leqslant m_2$ and for every i, $m_1 \leqslant i \leqslant m_1 + m_2$, $\theta(P_i) = P(i + r)$, or for every i, $m_1 \leqslant i \leqslant m_1 + m_2$, $\theta(P_i) = P(-i + r)$.

Proof of (*). Let A be the set of all vertices of $G^*(n, k)$ satisfying at least one of the following conditions:

(1) An edge which contains it, was removed or added in the change of $G(n, k)$ to $G^*(n, k)$ or from $G^*(n, k)$ to G.

(2) It is connected to Q or is Q (when there is a vertex named Q in the graph).

(3) Its image by θ satisfies (1) or (2).

Now the number of vertices satisfying (1) is $\leqslant 4(k + 15) + 2(k - 1) \leqslant \leqslant 8k$, the number of vertices satisfying (2) is $\leqslant 1 + (l + 1) \leqslant k/2 + 4 \leqslant \leqslant 2k$; and the number of vertices satisfying (3) cannot be more than $8k + 2k$. So $|A| \leqslant 20k$. Hence there are m_1, m_2; $(r_l)^2 < m_2$ such that: for every i, $m_1 \leqslant i \leqslant m_1 + m_2$, P_i and $\theta(P_i)$ do not belong to A. (If there are $> 20k(r_l)^2$ P_i's, this clearly holds, and we have assumed n is sufficiently large). Now clearly P_i and $\theta(P_i)$ have the same valency in $G^*(n, k)$ (as they both $\notin A$) and also they are not Q. So $\theta(P_j)$ is not

Q, and not an R_i, hence it is a P_j let $\theta(P_j) = P_{\theta(j)}$. Clearly for $m_1 \leqslant i$, $j \leqslant m_1 + m_2$, P_i, P_j are connected iff $P_{\theta(i)} P_{\theta(j)}$ are connected. If i is even, $P_i P_j$ are connected if $j - i \in \{r_m : 1 \leqslant m \leqslant l\}$, and similarly if $\theta(i)$ is even, $P_{\theta(i)} P_{\theta(j)}$ are connected iff $\theta(j) - \theta(i) \in \{r_m : 1 \leqslant m \leqslant l\}$. Between the ordered pairs from $\{i\colon m_1 \leqslant i \leqslant m_1 + m_2\}$ we define a relation E_1: $(j_1, j_2) E_1 (j_3, j_4)$ holds iff there are k_1, k_2 in this interval such that $P_{j_1}, P_{j_2} P_{k_1} P_{j_4} P_{j_3} P_{k_2}$ is a circle. Let E be the minimal equivalence relation which extends E_1. By property (B), $(j_1, j_2) E_1 (j_3, j_4)$ implies $j_2 - i_1 = j_4 - j_3$, and $j_1 = j_3 \pmod 2$, $j_2 = j_4 \pmod 2$. Clearly also $(j_1, j_2) E (j_3, j_4)$ implies the same. If we restrict ourselves to pairs (i, j) such that P_i, P_j is an edge, we have exactly $2l$ equivalence classes ($(\langle i, j\rangle$ and $\langle j, i\rangle$ belongs to different equivalence classes), $\langle j_1 j_2\rangle E \langle i_1, i_2\rangle$ holds iff $j_1 = j_2 \pmod 2$ and $j_1 - j_2 = i_1 - i_2 \in \{\pm r_m : 1 \leqslant m \leqslant l\}$. We can define similarly E'_1 and E' among $\{\theta(i)\colon m_1 \leqslant i \leqslant m_1 + m_2\}$. On the one hand we can easily see that E'_1 and E' are the images of E_1 and E under θ. That is $\langle j_1, j_2\rangle E \langle j_3, j_4\rangle$ holds iff $\langle \theta(j_1), \theta(j_2)\rangle E' \langle \theta(j_3), \theta(j_4)\rangle$ holds, where $j_1, j_2, j_3, j_4 \in \{i\colon m_1 \leqslant i \leqslant m_1 + m_2\}$. On the other hand repeating the proof for E, and using the fact that E', restricted to pairs $\langle i, j\rangle$ for which $P_i P_j$ is an edge, has exactly $2l$ equivalence classes (as an image of E) *we can conclude* that: $\langle j_1, j_2\rangle E' \langle j_3, j_4\rangle$ holds iff $j_1 = j_3 \pmod 2$ and $j_1 - j_2 = j_3 - j_4 \in \{\pm r_m : 1 \leqslant l\}$ (where of course $j_1, j_2, j_3, j_4 \in \{\theta(i)\colon m_1 \leqslant i \leqslant m_1 + m_2\}$.

For $2 \leqslant m \leqslant l$, let $\theta(r_m) = \theta(j + r_m) - \theta(j)$ for any even j_1, $m_1 \leqslant j \leqslant m_1 + m_2$. (Clearly $\theta(r_m)$ is independent of the choice of j). It is easy to prove that either for every m, $\theta(r_m) \in \{r_i\colon 1 \leqslant i \leqslant l\}$, or for every m, $\theta(r_m) \in \{-r_i\colon 1 \leqslant i \leqslant l\}$. Now we shall prove that in the first case $\theta(r_m) = r_m$ and in the second $\theta(r_m) = -r_m$. This is done by considering circles whose edges are from four equivalence classes. Clearly, this implies $(*)$.

Now we shall prove that

$(**)$ either for every i, $\theta(P_i) = P(i + r)$ or for every i. $\theta(P_i) = = P(-i + r)$.

Suppose this is not true. We prove first that there are less than nine

i's for which this fails. Otherwise let $i_1 < i_2 < \ldots < i_9$ be the first nine i's $> m_1$. Each i_j, by property C, is connected with $\geqslant [(l-1)/2]$ of the vertices $\{P(i_j - m): r, < m < r_l\}$ in $G^*(n,k)$ and $G(n,k)$. It is easy to observe that there are no $k_1, k_2 < i_j$, $k_1, k_2 \neq i_1, \ldots, i_{j-1}, i_j - r_1$ such that $P_{k_1}P_{i_j}, P_{k_2}P_{i_j}$ are connected in $G^*(n,k)$ and $G(n,k)$, and also their images are connected in $G^*(n,k)$ and $G(n,k)$. (Otherwise $P(\theta(k_1))P(\theta(k_2))P(\pm i_j + r)$ is a square in $G^*(n,k)$, contradiction.) Hence the number of edges (in $G^*(n,k)$) $P_iP_{i_j}$, $i \leqslant i_j$ such that $P_{\theta(i)}P_{\theta(ij)}$ is not an edge in $G^*(n,k)$, is $\geqslant [(l-1)/2] - j$. Hence the number of changes is

$$\geqslant \sum_{j=1}^{9} ([(l-1)/2] - j) =$$

$$= 9[l-1)/2] - \sum_{j=1}^{9} j \geqslant 9(l-2)/2 - 10 \cdot 9/2 =$$

$$= 9l/2 - 9 - 5 \cdot 9 = 9l/2 - 54 \geqslant 9(k+1)/4 - 54 =$$

$$= 9k/4 - 9/4 - 54 = k + 5k/4 - 51.75 =$$

$$= (k-1) + (5k-203)/4 > k-1$$

(if $k > 203/5 = 40.6$) contradiction.

So we have proved that (**) holds except perhaps for $\leqslant 8P_i$'s. Noticing the edges we add to $G(n,k)$ to create $G^*(n,k)$, clearly, for every i, except possibly the nine mentioned above, $\theta(P_i) = P_i$.

Suppose there are $m > 2$, i's for which $\theta(P_i) \neq P_i$, and let $i_1, \ldots, i_m$ be such indices. Then as before, for each i_j, if $\theta(P_i) = P_i$, $P_iP_{i_j}$ are connected then $\theta(P_i)\theta(P_{i_j})$ are not connected, except possibly one i. Hence the number of edges $P_iP_{i_j}$ such that $\theta(P_i)\theta(P_{i_j})$ are not connected, is $\geqslant l - m$. Hence the number of changes made in $G^*(n,k)$ to create G is $\geqslant m(l-m)$; a contradiction, for $2 < m < 9$, $k > 40$.

So except possibly for two i's $\theta(P_i) = P_i$. Now it is not hard to see that also the number of vertices not transferred to themselves is $\leqslant 2$.

So if θ is not the identity, it interchanges only two vertices. As in $G^*(k, n)$ there is no square or triangle, clearly this also leads to contradiction. So we proved theorem 3.1.

Hint to the proof for $k < 41$. The same way, as we choose $i_1 < \ldots < i_9$, we can choose $i^1 > \ldots > i^9$, which are $< m_1$ and are the greatest nine i's for which $(**)$ fails. Also we can have several intervals satisfying $(*)$, and we can prove their number is not too large, and that the number of i's not in any of them is also not too large. Then we should examine many cases separately, until at least if follows that $(**)$ holds, and the rest is similar to the proof that appears here. For $k = 3$ $(*)$ should include the R_j's as well as the P_i's.

REFERENCES

[1] P. Erdős – A. Rényi, Asymmetric graphs, Magyar Tudományos Akadémia Budapest, *Acta Math.*, 14 (1963), 295-317.

[2] P. Hell – J. Nesetril, Rigid and inverse rigid graphs, *Combinatorial structures and their applications.* Gordon and Breach, N.Y., London, Paris, 1970, 169-171.

[3] R. Frucht – A. Gewirtz, *Asymetric graphs; Recent trends in graph theory,* ed. Dold, Springer-Verlag 186.

COLLOQUIA MATHEMATICA SOCIETATIS JÁNOS BOLYAI
10. INFINITE AND FINITE SETS, KESZTHELY (HUNGARY), 1973.

NOTES ON PARTITION CALCULUS

S. SHELAH

§0. INTRODUCTION

We deal here with some separate problems which appeared in the problem list of Erdős and Hajnal [1].

In §1 we consider problem 3 of [1] asked by Erdős, Hajnal and Rado, which was the only open case (for infinite cardinals) of $\lambda \to (\mu)^2_2$, and we solve it affirmatively. Thus if $\aleph_\omega < 2^{\aleph_{n(0)}} < \ldots < 2^{\aleph_{n(k)}} < \ldots$, then $\sum'_{n<\omega} 2^{\aleph_n} \to (\aleph_\omega, \aleph_\omega)^2$. We prove a canonization lemma for it.

In §2 we deal with problem 32 of [1] asked by Erdős and Hajnal, which asks whether there is a graph G with $\aleph_1$ vertices and with no $[[\aleph_0, \aleph_1]]$ subgraph for which $\aleph_1 \not\to (G, G)^2$. We provide a wide class of such graphs, assuming CH. If $V = L$ is assumed we show that $\aleph_1 \to (G, G)^2$ iff G has coloring number $\leqslant \aleph_0$.

In §3 we deal with problems 48, 50 of [1] (asked by Erdős and Hajnal): partition relations concerning coloring numbers.

In §4 we deal with problem 42 of [1] asked by Erdős and Hajnal (part A, B) and Gustin (part C). We get compactness and incompactness results on the existence of transversals and on property B. We also find sufficient and necessary conditions for the existence of transversals.

§1

Canonization Lemma 1.1. *Suppose* κ, λ_i, $(i<\kappa)$ *are regular,* $i<j \rightarrow \lambda_i < \lambda_j$ *and* $\lambda^j \overset{\text{def}}{=} \prod_{i<j} \lambda_i^{\mu(i)} < \lambda_j$. *Suppose* $|A_i| = \lambda_i$ *and* F_i $(i<\chi)$ *is an* n_i*-place function from* $A = \bigcup_{i<\kappa} A_i$ *into* χ, $2^{\chi+\kappa} < \lambda_0$, *so* $2^{\chi+\kappa} < \lambda_j$.

Suppose that for every $B_i \subseteq A_i$ $(i<\alpha<\kappa)$, $|B_i| \leqslant \mu(i)$ *and* $a_i \in A_i$ $(\alpha<i<\kappa)$ *and* $C \subseteq A_\alpha$, $|C| = \lambda_\alpha$ *there is* $B_\alpha \subseteq C$, $|B_\alpha| \leqslant \mu_\alpha$ *such that* $P_\alpha(\langle B_i\colon i \leqslant \alpha\rangle, \langle a_i\colon \alpha<i<\kappa\rangle)$ *holds, for specified properties* P_α.

Then there are $a_i^* \in A_i$, $B_i \subseteq A_i$ *such that*

(1) *for any* $a_1, \ldots \in \bigcup_{i<\alpha} B_i$, $b, b' \in B_\alpha$, $c, c' \in B_\beta$ $(\alpha<\beta)$ *and* $i<\chi$,

(1A) $F_i(b, a_1, \ldots) = F_i(b', a_1, \ldots)$;

(1B) $F_i(b, c, a_1, \ldots) = F_i(b', c', a_1, \ldots) = F_i(b', a_\beta^*, a_1, \ldots)$;

(2) *for any* $\alpha<\kappa$, $P_\alpha(\langle B_i\colon i \leqslant \alpha\rangle, \langle a_i^*\colon \alpha<i<\kappa\rangle)$.

(3) *If* F_i *is three-place,* $2^{\chi+\kappa} < \mathrm{cf}\,[\mu(i)]$ *for every* i, P_α *hereditary for the* B_i*'s for subsets of the same cardinality then:*

$$a, a' \in B_\alpha,\ b, b' \in B_\beta,\ c, c' \in B_\gamma,\ \alpha<\beta<\gamma<\kappa \quad \textit{implies}$$

$$F_i(a, b, c) = F_i(a', b', c').$$

Remark. We could refine the lemma along the lines of [7] §5, but there is no application of it. We can assume that the range of F_i is 2^χ.

Proof. We may assume without loss of generality that the set of functions F_i is closed under permutations and identifications of variables.

Define for any $B \subseteq A$, $\bar{a} \in A$, $\mathrm{tf}(\bar{a}, B)$ as the following set of equations

$$\{F_i(\bar{x}, \bar{b}) = c \colon i < \chi,\ F_i(\bar{a}, \bar{b}) = c,\ \bar{b} \in B,\ c \in \chi\}\,.$$

Clearly $|\{\mathrm{tf}(\bar{a}, B) \colon \bar{a} \in A\}| \leqslant 2^{|B|+\chi}$. Hence for every $\alpha < \kappa$, any $B_i \subseteq A_i$ $(i < \alpha)$, $|B_i| \leqslant \mu_i$,

$$\left|\left\{\mathrm{tf}\left(a, \bigcup_{i<\alpha} B_i\right) \colon a \in A_\alpha\right\}\right| \leqslant$$

$$\leqslant 2^{\chi + \sum_{i<\alpha} \mu(i)} \leqslant 2^\chi \cdot \prod_{i<\alpha} 2^{\mu(i)} \leqslant \lambda^\alpha\,.$$

Now the number of such $\langle B_i \colon i < \alpha\rangle$ is $\prod_{i<\alpha} \lambda_i^{\mu(i)} \leqslant \lambda^\alpha$.

So the set C_α of $a \in A_\alpha$ such that for some $B_i \subseteq A_i$ $(i < \alpha)$, $|B_i| \leqslant \mu(i)$, $\left|\left\{a' \in A_\alpha \colon \mathrm{tf}\left(a', \bigcup_{i<\alpha} B_i\right) = \mathrm{tf}\left(a, \bigcup_{i<\alpha} B_i\right)\right\}\right| < \lambda_\alpha$ is the union of $\leqslant \lambda^\alpha \cdot \lambda^\alpha < \lambda_\alpha$ sets each of cardinality $< \lambda_\alpha$; hence is of cardinality $< \lambda_\alpha$. Choose $a_i^* \in A_i - C_i$ for each $i < \kappa$. Now define inductively $B_i \subseteq A_i$, $|B_i| \leqslant \mu(i)$. Suppose we have defined B_i for $i < \alpha$. As $a_\alpha^* \notin C_\alpha$, the set

$$B_\alpha^1 = \left\{a \in A_\alpha \colon \mathrm{tf}\left(a, \bigcup_{i<\alpha} B_i\right) = \mathrm{tf}\left(a_\alpha^*, \bigcup_{i<\alpha} B_i\right)\right\}$$

has cardinality λ_α. As $\left\{\mathrm{tf}\left(a, \bigcup_{i<\alpha} B_i \cup \{a_i^* \colon i < \kappa\}\right) \colon a \in B_\alpha^1\right\}$ has cardinality $\leqslant 2^{\chi + \sum_{i<\alpha} \mu(i) + \kappa} < \lambda_\alpha = |B_\alpha^1|$; there are $B_\alpha^2 \subseteq B_\alpha^1$, $|B_\alpha^2| = \lambda_\alpha$ and t_α such that for every $a \in B_\alpha^2$

$$t_\alpha = \mathrm{tf}\left(a, \bigcup_{i<\alpha} B_i \cup \{a_i^* \colon i < \kappa\}\right)\,.$$

Now by assumption there is $B_\alpha \subseteq B_\alpha^2$ such that $|B_\alpha| \leqslant \mu(\alpha)$ and $P_\alpha(\langle B_i \colon i \leqslant \alpha\rangle, \langle a_i^* \colon \alpha < i < a\rangle)$ holds.

Now clearly (2) holds by the choice of B_α. (1A) holds as $B_\alpha \subseteq B_\alpha^1$; as for (1B), $B_\alpha \subseteq B_\alpha^2$ implies that $F_i(b, a_\beta^*, a_1, \ldots) = F_i(b', a_\beta^*, a_1, \ldots)$, and $B_\beta \subseteq B_\beta^1$ implies

$$F_i(b, c, a_1, \ldots) = F_i(b, a^*_\beta, a_1, \ldots) = F_i(b, c', a_1, \ldots),$$

$$F_i(b', c, a_1, \ldots) = F_i(b', a^*_\beta, a_1, \ldots) = F_i(b', c', a_1, \ldots),$$

and combining the equalities we get (1B). In order to get (3) we should replace the B_α by a subset of the same cardinality.

Theorem 1.2. *If* $\kappa \to (\kappa)^2_2$, $\kappa = \mathrm{cf}\,\lambda$, $\langle 2^\mu : \mu < \lambda \rangle$ *is not eventually constant, but is eventually* $\geqslant \kappa$, *then*

$$\chi = \sum_{\mu<\lambda} 2^\mu \to (\lambda)^2_2$$

(and in fact even $\chi \to (\lambda, \lambda, \omega)^2$*).*

Proof. Let f be a function from $[\chi]^2$ into $2 = \{0, 1\}$. Choose $\mu(i) < \lambda$ $(i < \kappa)$ such that $\sum_{i<\kappa} \mu(i) = \lambda$ and $2^{\mu(i)}$ is strictly increasing and $2^{\mu(i)} \geqslant \lambda$, and let $\lambda_i = (2^{\mu(i)})^+$ and $A_i = \{\alpha : \bigcup_{j<i} \lambda_j \leqslant \alpha < \lambda_i\}$. If for some $i < \kappa$ there is a $B \subseteq A_i$, $|B| \geqslant \lambda$ such that f is constant on $[B]^2$, then we are ready; so assume there is no such B. As (by [4]) $\lambda_i \to (\lambda_i, \mu(i))^2$ and $\lambda_i \to (\mu(i), \lambda_i)^2$ hold for every $A'_i \subseteq A_i$, $|A'_i| = \lambda_i$, there are sets $B_{i,0}, B_{i,1} \subseteq A_i$ of cardinality $\mu(i)$ such that f has the constant value 0 (1) on $B_{i,0}$ $(B_{i,1})$. Define $P_\alpha(\langle B_i : i \leqslant \alpha \rangle, \langle a^*_i : \alpha < i < \kappa \rangle) \overset{\text{def}}{=\!=} [B_\alpha = B_{\alpha,0} \cup B_{\alpha,1}$, f has the constant value 0 (1) on $B_{\alpha,0}$ $(B_{\alpha,1})$, and $|B_{\alpha,0}| = |B_{\alpha,1}| = \mu(\alpha)]$.

So by Lemma 1.1 there are $B_\alpha \subseteq A$, $|B_\alpha| = \mu(i)$, and (by (1B) in the lemma) there is a two-place function g from κ to $\{0, 1\}$ such that $f(a, b) = g(i, j)$ for $a \in B_i$, $b \in B_j$, $i < j$. As $\kappa \to (\kappa)^2_2$ there is $I \subseteq \kappa$, $|I| = \kappa$ and $\delta \in 2$ such that for any $i < j \in I$, $g(i, j) = \delta$. Let $B = \bigcup_{\alpha \in I} B_{\alpha,\delta}$. Then clearly $|B| = \sum_{i \in I} \mu(i) = \lambda$, and f has on $[B]^2$ the constant value δ.

Corollary 1.3. *If* $\aleph_\omega < 2^{\aleph_{n(0)}} < 2^{\aleph_{n(1)}} < \ldots$ *then* $\sum_{n<\omega} 2^{\aleph_n} \to (\aleph_\omega, \aleph_\omega)^2$.

Remark. This answers problem 3 of [1], and Theorem 2 completes the answer to the question "when $\lambda \to (\mu)^2_2$" for infinite λ, μ.

Conjecture 1A (Hajnal). *If* $\aleph_\omega < 2^{\aleph_{n(0)}} < 2^{\aleph_{n(1)}} < \ldots$ *then* $\sum_{n<\omega} 2^{\aleph_n} \to (\aleph_\omega, 4)^3$.

Remark. For all previously known proved cases of $\lambda \to (\mu, \mu)^2$ also $\lambda \to (\mu, 4)^3$ holds; on the other hand $\sum_{n<\omega} 2^{\aleph_n} \not\to (\aleph_\omega, 5)^3$.

§ 2. NOTATION

Notation. For a graph G, let $V(G)$ be its set of vertices, and $E(G)$ its set of edges. We write $a \in G$ instead of $a \in V(G)$, etc.

Definition 2.1. $\lambda \to (G_i)^2_{i<\alpha}$ $(\lambda \not\to [G_i]^2_{i<\alpha})$ (G_i graphs) if for every α-colouring f of λ (i.e. a function $f\colon [\lambda]^2 \to \{i\colon i < \alpha\}$) there is an $i < \alpha$ and a one-to-one function F from $V(G_i)$ into λ such that $a \neq b \in G_i \wedge \{ab\} \in E(G_i) \Rightarrow f(F(a), F(b)) = i$, $(f(F(a), F(b)) \neq i)$.

Definition 2.2. $V_\lambda(A, G) = |\{a \in G\colon |\{b \in A\colon a, b$ are connected$\}| \geqslant \lambda\}|$.

Theorem 2.1.

(A) (CH) $\Rightarrow \aleph_1 \not\to (G)^2_2$ *if* $V_{\aleph_0}(A, G) = \aleph_1$ *for some countable* $A \subseteq G$.

(B) $(2^\lambda = \lambda^+)$ $\lambda^+ \not\to [G_i]^2_{i<\lambda^+}$ *if for every* i *there is* A_i *such that* $V_\lambda(A_i, G) = \lambda^+$ *where* $A_i \subseteq G_i$, $|A_i| = \lambda$.

Proof. Clearly it suffices to prove (B). Let $\{\langle\alpha_i, F_i\rangle\colon \lambda \leqslant i,\ i < \lambda^+\}$ be a list of the pairs $\langle\alpha, F\rangle$, where $\alpha < \lambda^+$ and F is a one-to-one function from A_α into λ^+; we may assume without loss of generality that the range of F_i is $\subseteq \{j\colon j < i\}$.

Let $V(G_{\alpha_i}) = \{a^i_j\colon j < \lambda^+\}$ and $B^i_j = \{F_i(a^i_\beta)\colon a^i_\beta \in A_{\alpha_i},\ \beta < \lambda^+,\ (a^i_\beta, a^i_j) \in E(G_{\alpha_i})\}$, We may choose the enumeration of $V(G_{\alpha_i})$ so that $|B^i_j| = \lambda$ holds for limit j.

Now we define by induction on i the coloring f on $[i]^2$. For $i, j < \lambda$ we define $f(i, j)$ arbitrarily. Suppose f is defined on $[i]^2$.

Now we define $f(i,j)$ for $j<i$. It is well-known that if $\{C_j: j<\lambda\}$ is a family of λ sets, $|C_i|=\lambda$, then we can find pairwise disjoint $C'_i \subseteq C_i$, $|C'_i|=\lambda$, hence we can define $f(i,\alpha)$ such that if $j,\beta \leqslant i$, $|B^j_\beta| = \lambda$, $B^j_\beta \subseteq i$ then

$$\{f(i,\gamma)\colon \gamma \in B^j_\beta\} = i = \{\gamma\colon \gamma < i\}\,.$$

Now the coloring f is defined. Suppose F is an embedding of G_α into λ^+ contradicting our claim, then for some $i<\lambda^+$, $\alpha=\alpha_i$, $F_i = F \restriction A_\alpha$ and for some limit δ we have $i, \lambda < \delta < \lambda^+$; $a^i_\beta \in A_\alpha \Rightarrow \beta < \delta$; and $\beta < \delta \Leftrightarrow F(a^i_\beta) < \delta$. But then $f(F(a^i_\delta), b) \neq \alpha < \delta$ for every $b \in B^i_\delta$, contradicting the definition of f.

Corollary 2.2. *There exists a graph G with $\aleph_1$ vertices which does not have a subgraph of type $[[\aleph_0, \aleph_1]]$ (bipartite graph) but $\aleph_1 \nrightarrow (G,G)$.*

Remark. This answers problem 32 from [1] affirmatively.

Proof. Let $\{A_\alpha\colon \alpha<\omega_1\}$ be a set of subsets of ω such that $\alpha \neq \beta$ implies $|A_\alpha \cap A_\beta| < \aleph_0$. G will have ω_1 as the set of vertices, and α, β are connected iff $\alpha < \omega \leqslant \beta$, $\alpha \in A_\beta$. Then by 2.1 $\aleph_1 \nrightarrow (G)^2_2$, but G satisfies the other requirements by the construction.

Definition 2.2. The coloring number of a graph G, $\mathrm{cl}(G)$ is the minimal cardinal λ such that we can list its set of vertices $\{a_i\colon i<l_0\}$ such that each a_i is connected to $<\lambda$ a_j's for $j<i$;

Theorem 2.3. *If G has coloring number $\aleph_0$ and $\leqslant \lambda$ vertices then $\lambda \to (G)^2_n$ for every $n<\aleph_0$.*

Proof. By [3] we can assume the set of vertices of G is $\{a_i\colon i<\mu\}$, where $\mu \leqslant \lambda$, such that each a_i is connected to $<\aleph_0$ a_j's with $j<i$.

Let $f\colon [\lambda]^2 \to \{0,\ldots,n-1\}$ be an n-coloring of λ.

Let D be a uniform ultrafilter over λ and define $g(\alpha) =$ the $i \in n$ such that $A_i = \{\beta<\lambda\colon f(\alpha,\beta)=i\} \in D$. ($g$ is well defined because if $\lambda = A_0 \cup \ldots \cup A_{n-1}$, the A_i's are disjoint, and so $A_i \in D$ for exactly one $i<n$.) Let $i_0<n$ be such that $\{\alpha<\lambda\colon g(\alpha)=i_0\} \in D$. Now define $b_j<\lambda$ by induction on j, such that $g(b_j)=i_0$, $f(b_k,b_j)=i_0$, if a_k, a_j is connected, and $k \neq j \Rightarrow b_k \neq b_j$. If for $k < j$, $b_k < \lambda$ is defined, let $\{k_1, \ldots, k_m\}$ be the set of $k < j$ such that a_k, a_j are connected in G. Then $A_{k_1}, \ldots, A_{k_n} \in D$ hence $A_{k_1} \cap \ldots \cap A_{k_m} \in D$, hence it has cardinality λ so there is $b_j \in A_{k_1} \cap \ldots \cap A_{k_m} - \{b_k : k < j\}$. Clearly $a_i \to b_i$ is the embedding we seek.

Theorem 2.4 (V = L).

(A) *If G is a graph with $\aleph_1$ vertices which has coloring number $> \aleph_0$ (that is $\aleph_1$) then $\aleph_1 \not\to (G)^2_2$.*

(B) *If each G_i, $(i < \omega_1)$ has $\aleph_1$ vertices and coloring number $> \aleph_0$ then $\aleph_1 \not\to [G_i]^2_{i<\omega_1}$.*

Remarks.

(1) I first claimed the theorem incorrectly without V = L, and A. Hajnal and A. Máté, who tried to reconstruct the proof, also proved this theorem.

(2) We prove only (A). The proof of (B) is similar.

Proof. (A) We may assume without loss of generality that ω_1 is the set of vertices of G. We first show that

(*) the set $I \subseteq \omega_1$ of $\alpha < \omega_1$ such that, for some $j = j_\alpha \geqslant \alpha$, j is connected to infinitely many $\beta < \alpha$ is stationary.

In fact, assuming the contrary, let $C \subseteq \omega_1$ be closed and unbounded and disjoint to I, and write $C = \{c_i : i < \omega_1\}$ $(c_0 = 0)$. Now, for each $i < \omega_1$, rearrange $[c_i, c_{i+1})$ in a sequence of type ω; this rearrangement shows that G has coloring number $\aleph_0$, which is a contradiction. So (*) holds.

Clearly, we may assume that if $\alpha \in I$ then α is limit and $j_\alpha = \alpha$ holds. By Jensen [5] there are functions $F_\alpha : \alpha \to \alpha$, $(\alpha \in I)$ such that the set $\{\alpha \in I : F \upharpoonright \alpha = F_\alpha\}$ is stationary for every function $F: \omega_1 \to \omega_1$. We are about to define the coloring function $f: [\omega_1]^2 \to \{0, 1\}$. We define $f(i, j)$ by induction on $\max\{i, j\}$. f is defined arbitrarily on $[\omega]^2$.

If f is defined on $[\beta]^2$, $\beta < \omega_1$, then define $f(\beta, i)$ for every $i < \beta$ such that $\{f(\beta, F_\alpha(j))\colon \alpha \leqslant \beta,\ \alpha \in I,\ \text{and}\ j \in A_\alpha\} = \{0, 1\}$, where $A_\alpha = \{\gamma < \alpha\colon \{\gamma, \alpha\} \in G\}$ (note that A_α is infinite).

Assume that $F\colon \omega_1 \to \omega_1$ is an embedding of G into ω_1 such that if i and j are connected then $f(F(i), F(j)) = \delta$ for a fixed δ ($\delta = 0$ or 1). Clearly, $C = \{\alpha < \omega_1\colon F(j) < \alpha \text{ iff } j < \alpha\}$ is closed and unbounded. By the definition of the F_α's, the set $\{\alpha \in I\colon F \restriction \alpha = F_\alpha\}$ is stationary; hence there is an $\alpha \in C \cap I$ such that $F \restriction \alpha = F_\alpha$. Then the definition of f with $\beta = F(\alpha)$ shows that there are $\gamma, \gamma' < \alpha$ such that $\{\gamma, \alpha\}, \{\gamma', \alpha\} \in G$ and $f(F(\alpha), F(\gamma)) = 0$ and $f(F(\alpha), F(\gamma')) = 1$, which contradicts our assumption on F, completing the proof.

Conclusion 2.5. (V = L). *For graphs G with $\aleph_1$ vertices, $\aleph_1 \nrightarrow (G)^2_2$ iff G has coloring number $\leqslant \aleph_0$.*

§3.

Definition 3.1. $(\lambda, \mu) \to (\kappa, \chi)$ holds if every graph G with λ vertices all whose subgraphs spanned by a set of $< \mu$ vertices have colouring number $\leqslant \kappa$ has coloring number $\leqslant \chi$.

See [7] for more material on this. Just as Erdős and Hajnal [2] notice that V = L implies a positive anwer to problem 42c of [1], we can notice:

Lemma 3.1 (V = L).

(1) *Assume λ is regular. Then $(\lambda, \lambda) \to (\aleph_0, \aleph_0)$ iff λ is weakly compact.*

(2) *If λ is not weakly compact, then there is a graph of cardinality λ showing $(\lambda, \lambda) \nrightarrow (\aleph_0, \aleph_0)$ that has chromatic number $\aleph_1$ and every subgraph of smaller cardinality has chromatic number $\aleph_0$.*

Remark. In fact we can replace $\aleph_0$ by any $\mu < \lambda$; this partially answers 48A [1] (when μ is regular).

Proof.

(1) If λ is weakly compact, it is immediate that $(\lambda, \lambda) \to (\aleph_0, \aleph_0)$ (formulate a suitable set of $L_{\lambda,\lambda}$-sentences such that every subset of power $< \lambda$ has a model by the assumption of $(\lambda, \lambda) \to (\aleph_0, \aleph_0)$, and a model of it gives the conclusion).

Suppose now λ is not weakly compact. By Jensen [5] there is a stationary $C \subseteq \lambda$, $\alpha \in C \Rightarrow \mathrm{cf}\,\alpha = \omega$, and we may assume that $\beta < \alpha$, $\alpha \in C \Rightarrow \beta + \omega < \alpha$, but for every limit ordinal $\delta < \lambda$, $C \cap \delta$ is not stationary. Choose $A_\alpha \subseteq \alpha$ for $\alpha \in C$ such that the order type of A_α is ω and $\sup A_\alpha = \alpha$. Define a graph G with set of vertices $\{\alpha\colon \alpha < \lambda\}$ and set of edges $\{(i, \alpha)\colon i \in A_\alpha,\ \alpha \in C\}$. Now we prove by induction on α that the restriction of G to $\{i\colon i < \alpha\}$ has coloring number $\aleph_0$. For $\alpha = 0$, or α successor it is immediate; and if α is limit, choose a continuous increasing unbounded sequence $\alpha_i < \alpha$, $i < \mathrm{cf}\,\alpha$, $\alpha_i \notin C$. By the induction hypothesis the restriction of G to $[\alpha_i, \alpha_{i+1})$ has coloring number $\aleph_0$, so let $<_i$ be a suitable order. Define an order $<^*$ on $[0, \alpha)$: $a <^* b$ iff $a < \alpha_i \leqslant b$ for some i or $\alpha_i \leqslant a, b < \alpha_{i+1}$, $a <_i b$ for some i.

For any $a < \alpha$, let $\alpha_i \leqslant a < \alpha_{i+1}$; then

$$\{b <^* a\colon a, b \text{ are connected}\} =$$

$$= \{b < \alpha_i\colon a, b \text{ are connected}\} \cup \{b\colon a, b \text{ are connected},$$

$$b <_i a,\ \alpha_i < b < \alpha_{i+1}\} = \{b < \alpha_i\colon b \in A_a,\ a \in C\} \cup$$

$$\cup \{b\colon b <_i a,\ \alpha_i \leqslant b < \alpha_{i+1},\ a, b \text{ are connected}\}.$$

Both sets are finite (the first one since A_a has order type ω and $\alpha_i \notin C$, the second by the definition of $<_i$). Hence the coloring number of G restricted to $\{i\colon i < \alpha\}$ is $\aleph_0$.

Suppose the coloring number of G is $\leqslant \aleph_0$. By [3] there is an order $<^*$ of $\{\alpha\colon \alpha < \lambda\}$ of order-type λ such that $|\{\beta <^* \alpha\colon \beta, \alpha$ are connected in $G\}| < \aleph_0$ for every α. It is well known that $S = \{\alpha < \lambda\colon \beta < \alpha \Rightarrow \beta <^* \alpha$ for any $\beta < \lambda\}$ is a closed unbounded subset of λ, hence there is $\alpha \in S \cap C$, and so $A_\alpha \subseteq \{\beta\colon \beta <^* \alpha\}$; a contradiction.

(2) By $\diamondsuit_\lambda$ of Jensen [5], there are partitions $\langle B_n^\alpha : n < \omega \rangle$ of α such that the set $\{\alpha < \lambda : \operatorname{cf} \alpha = \omega, \forall n < \omega [B_n \cap \alpha = B_n^\alpha]\}$ is stationary for any partition $\langle B_n : n < \omega \rangle$ of λ. Choose A_α in the construction above so that if B_n^α is unbounded in α, then $A_\alpha \cap B_n^\alpha \neq \phi$.

Definition 3.2. $\operatorname{Col}(\lambda, \mu, \kappa, \chi)$ holds if every graph G with $|V(G)| = |V(G)| = \lambda$, $\operatorname{cl}(G) > \mu$, contains a $[[\kappa, \chi]]$ subgraph.

Theorem 3.2 (G.C.H.). *The following are equivalent:*

(A) *not* $\operatorname{Col}(\aleph_{\omega+1}, \aleph_1, \aleph_1, \aleph_0)$

(B) *not* $\operatorname{Col}(\aleph_{\omega+1}, \aleph_1, \aleph_2, \aleph_0)$

(C) $(\aleph_{\omega+1}, \aleph_3) \nrightarrow (\aleph_1, \aleph_1)$

(D) *there are a stationary set* $C \subseteq \{\alpha < \aleph_{\omega+1} : \operatorname{cf} \alpha = \aleph_1\}$ *and sets* $S_\alpha \subseteq \alpha$, $\operatorname{tp}(S_\alpha) = \omega_1$, $\sup S_\alpha = \alpha$, *such that* $\alpha, \beta \in C$, $\alpha \neq \beta \Rightarrow$ $\Rightarrow |S_\alpha \cap S_\beta| < \aleph_0$.

Remark. This gives a partial answer to problem 5.7 of [3], which is between (A) and (B).

Proof.

(D) $\Rightarrow$ (A). Define G by $V(G) = \aleph_{\omega+1}$, $E(G) = \{(\alpha, \beta) : \alpha \in S_\beta, \beta \in C\}$. Suppose $A \times B \subseteq E(G)$, $|A| = \aleph_0$, $|B| = \aleph_1$. As $|A| \neq |B|$, we may assume $A < B$ (i.e. $a \in A, b \in B \Rightarrow a < b$) or $B < A$. If $A < B$ then choose $b_1 \neq b_2 \in B$, so $S_{b_1} \cap S_{b_2} \supseteq A$ is infinite, a contradiction. If $B < A$, the contradiction is similar. So G does not have an $[[\aleph_0, \aleph_1]]$ subgraph.

On the other hand, trivially, $\operatorname{cl} G \leqslant \aleph_2$, as the natural ordering of ordinals shows. Suppose $\operatorname{cl}(G) \leqslant \aleph_1$, and $<^*$ is an order of $\aleph_{\omega+1}$ confirming this; we may assume by [3] that $<^*$ has order-type $\omega_{\omega+1}$. It is well known that $\{\delta < \omega_{\omega+1} : (\forall \alpha)(\alpha < \delta \Rightarrow \alpha <^* \delta)\}$ is a closed and unbounded subset of $\omega_{\omega+1}$, so some $\delta \in C$ belongs to it, hence it is connected to $\aleph_1$ of its predecessors, a contradiction. Hence $\operatorname{cl}(G) = \aleph_2$; but we have proved that G has no $[[\aleph_0, \aleph_1]]$ subgraph, so (A) holds.

(A) $\Rightarrow$ (B). Trivial.

(B) $\Rightarrow$ (C). Suppose G shows (B), that is $|V(G)| = \aleph_{\omega+1}$, $\mathrm{cl}(G) > \aleph_1$ but G has no $[[\aleph_2, \aleph_0]]$ subgraph; so by [3] 5.5, every subgraph G' of G with $\leqslant \aleph_\omega$ vertices has coloring number $\leqslant \aleph_1$. Hence this G shows that (C) holds.

(C) $\Rightarrow$ (D). Let G show (C), and assume $V(G) = \aleph_{\omega+1}$: For any $\alpha < \aleph_{\omega+1}$, choose A_n^α, $\bigcup_{n<\omega} A_n^\alpha = \{i\colon i < \alpha\}$, $|A_n^\alpha| \leqslant \aleph_n$, and let $F(\alpha) < \aleph_{\omega+1}$ be the first ordinal such that if $n < \omega$, $B \subseteq A_n$ and

$$c(B, A_n) = \{\gamma < \aleph_{\omega+1}\colon (\forall a \in A_n)[(a, \gamma) \in E(G) \equiv a \in B]\}$$

has cardinality $\leqslant \aleph_\omega$ then $c(B, A_n) < F(\alpha)$, and if $|c(B, A_\alpha)| = \aleph_{\omega+1}$ then $c(B, A_n) \cap \{i\colon i < F(\alpha)\}$ has cardinality $\aleph_\omega$; and also if $a \leqslant \alpha$, $A = \{b < a\colon (b, a) \in E(G)\}$, $|A| < \aleph_\omega$, $B \subseteq A$, $|c(B, A)| < \aleph_{\omega+1}$, then $c(B, A) \leqslant F(\alpha)$. Now let $C_1 = \{\delta < \aleph_{\omega+1}\colon \alpha < \delta \to F(\alpha) < \delta\}$. We may assume that if $\delta \in C_1$ and $|\{a < \delta\colon (a, c) \in E(G)\}| \geqslant \mathrm{cf}\,\delta$ for some $c > \delta$, *then* $|\{a < \delta\colon (a, \delta) \in E(G)\}| \geqslant \mathrm{cf}\,\delta$.

By our choice, for every $G' \subseteq G$ with $|V(G')| < \aleph_3$ we have $\mathrm{cl}(G') \leqslant \aleph_1$, hence it is easy to see that G has no $[[\aleph_1, \aleph_2]]$ subgraph. So if $|\{a < \alpha\colon (a, c) \in E(G)\}| \geqslant \aleph_1$ *then* $c < F(\alpha)$.

Let $C = \{\delta \in C_1\colon |\{\alpha < \delta\colon (\alpha, \delta) \in E(G)\}| \geqslant \aleph_1\}$.

Let us show that C is stationary. If not, let $C_2 \subseteq C_1$ be closed, unbounded and disjoint to C, and let

$$C_2 = \{\delta_i\colon i < \aleph_{\omega+1}\}.$$

Clearly for any i, the subgraph of G spanned by $[\delta_i, \delta_{i+1})$, G_i, has coloring number $\leqslant \aleph_1$. By our construction and definition of C_2, if $a \in [\delta_i, \delta_{i+1})$ then $|\{c\colon c < \delta_i, (c, a) \in E(G)\}| \leqslant \aleph_0$. So it is easy to see that $\mathrm{cl}(G) \leqslant \aleph_1$, a contradiction. So C is stationary. For $\delta \in C$ let $S_\delta = \{\alpha < \delta\colon (\alpha, \delta) \in E(G)\}$.

By definition of C, $|S_\delta| \geqslant \aleph_1$. If for some $\alpha < \delta$, $S_1 = S_\delta \cap \{i\colon i < \alpha\}$ has cardinality $\aleph_1$, then for some n, $|A_n^\alpha \cap S| \geqslant \aleph_1$, hence

$\delta < F(\alpha)$, contradicting $C \subseteq C_1$. So $|S_\delta| = \aleph_1$, and $\text{tp}(S_\delta) = \omega_1$, and $\sup S_\delta = \delta$, so $\text{cf}\,\delta = \omega_1$. Suppose $\delta_1 < \delta_2 \in C$, $S_{\delta_1} \cap S_{\delta_2}$ is infinite. (Note that $\text{cf}\,\delta_1 = \text{cf}\,\delta_2 = \omega_1$.)

Let $a = \delta_1$, $A = S_{\delta_1}$, $B = S_{\delta_1} \cap S_{\delta_2}$; then by the definition of F, $F(\delta_1) > \delta_2$, a contradiction. So C and the S_α's show that (D) holds.

Lemma 3.3. *If the coloring number of G is $< \mu$, then there are no sets A, B of vertices such that:*

$$|B| > |A| \geqslant \mu, \text{ and for every } b \in B$$

$$|\{a \in A\colon (a, b) \in E(G)\}| \geqslant \mu .$$

Proof. Easy.

This enables us to eliminate G.C.H. from some results in this section and from similar results after suitable changes.

§4.

Definition 4.1. Let $\lambda^+ \geqslant \kappa$.

(A) $\text{PT}(\lambda, \kappa)$ holds if there is an indexed family S of λ sets, each of cardinality $< \kappa$, such that S has no transversal, but every $S' \subseteq S$, $|S'| < \lambda$ has a transversal.

$\text{PT}^*(\lambda, \kappa)$ is defined in the same way except that $A \in S \Rightarrow |A| = \kappa$; $\lambda \geqslant \kappa$. A transversal of S is a one-to-one function f such that for any $A \in S$, $f(A) \in A$.

(B) $\text{PB}(\lambda, \kappa)$, $\text{PB}^*(\lambda, \kappa)$ are defined similarly replacing "has a transversal" by having property B. A family S has property B if there is a set C such that $A \in S \Rightarrow A \cap C \neq \phi$, $A - C \neq \phi$.

(C) $\text{PD}(\lambda, \mu)$ hold if there is a graph G with λ vertices such that G does not have the property $D(\mu)$, but every subgraph with $< \lambda$ vertices has. G has property $D(\mu)$ if we can direct its edges so that the number of directed edges emanating from any vertex is $< \mu$.

These properties arise from problem 42 [1] ($\lambda = \aleph_2$ there); PT from a question of Gustin; PB, PD from questions of Erdős and Hajnal. We give partial answers.

Lemma 4.1.

(A) $\mathrm{PT}^*(\lambda, \kappa)$ *implies* $\mathrm{PT}(\lambda, \kappa^+)$; $\mathrm{PB}^*(\lambda, \kappa)$ *implies* $\mathrm{PB}(\lambda, \kappa^+)$.

(B) *If* $\lambda > \kappa$ *then* $\mathrm{PT}^*(\lambda, \kappa)$ *iff* $\mathrm{PT}(\lambda, \kappa^+)$.

(C) $\mathrm{PT}(\mu, \mu^+)$, $\mathrm{PT}^*(\kappa^+, \kappa)$, $\mathrm{PB}^*(2^\kappa, \kappa)$, *and* $\mathrm{PD}(\kappa^+, \kappa)$ *hold, but* $\mathrm{PB}(\lambda, \aleph_0)$, $\mathrm{PT}(\lambda, \aleph_0)$, $\mathrm{PT}^*(\kappa, \kappa)$, $\mathrm{PB}^*(\kappa, \kappa)$, $\mathrm{PD}(\kappa, \kappa)$ *do not hold.*

(D) *If* $\kappa_2 \geqslant \kappa_1$ *then* $\mathrm{PT}(\lambda, \kappa_1)$ *implies* $\mathrm{PT}(\lambda, \kappa_2)$; $\mathrm{PB}(\lambda, \kappa_1)$ *implies* $\mathrm{PB}(\lambda, \kappa_2)$.

(E) $\mathrm{PT}(\lambda, \kappa) \Rightarrow \mathrm{PT}(\lambda^+, \kappa)$ *for regular* λ, *hence* $\mathrm{PT}(\aleph_{\alpha+n}, \aleph_\alpha)$ *holds.*

(F) *If* κ *is singular, then* $\mathrm{PT}(\kappa, \kappa)$.

(G) *If* $\mathrm{cf}\,\lambda < \mathrm{cf}\,\mu$ *then* $\mathrm{PD}(\lambda, \mu)$ *fails.*

Proof.

(A) Immediate by the definitions: use the same family.

(B) Let the family S exemplify $\mathrm{PT}(\lambda, \kappa^+)$; we may assume that the elements of each $A \in S$ are ordinals. Let

$$S' = \{\kappa \cup (A \times \{\alpha\}) \colon A \in S,\ \alpha < \kappa^+\}.$$

It is easy to check that S' exemplifies $\mathrm{PT}^*(\lambda, \kappa)$

(C) For $\mathrm{PT}(\mu, \mu^+)$ take

$$S = \{\mu\} \cup \{\{\alpha\} \colon \alpha < \mu\},$$

for $\mathrm{PT}^*(\kappa^+, \kappa)$ take $S = \{\alpha \colon \kappa \leqslant \alpha < \kappa^+\}$ (these examples are well known). For $\mathrm{PB}^*(2^\kappa, \kappa)$ take a maximal family S of subsets of κ of power 2^κ such that $A \neq B \in S \Rightarrow A \not\subseteq B$, and S is closed under complements (there is such family of power 2^κ, and we can extend it to a maximal one). For PD (κ^+, κ) take the complete graph with κ^+ vertices.

It is easy to show that PT* (κ, κ), PB* (κ, κ), and PD (κ, κ) fail. PT $(\lambda, \aleph_0)$ fails by Hall's theorem, and PB $(\lambda, \aleph_0)$ fails by compactness arguments.

(D) Immediate by the definitions (use the same family).

(E) See [9].

(F) Let $\kappa = \sum_{i<\mu} \kappa_i$, $\mu < \kappa_0 < \kappa_1 < \ldots$, $\kappa_\delta = \bigcup_{i<\delta} \kappa_i$ for limit δ. Let

$$S = \{\{\alpha\}: \alpha < \kappa,\ \alpha \neq \kappa_i \text{ for any } i < \mu\} \cup$$
$$\cup \{\{\alpha: \kappa_i \leqslant \alpha < \kappa_{i+1}\}: i < \mu\} \cup \{\{\kappa_i: i < \mu\}\}.$$

(G) Immediate.

Lemma 4.2. *Assume* V = L *and let* λ, μ *be regular,* $\lambda > \mu$. *Then the following are equivalent:*

(A) λ *is not weakly compact.*

(B) PT* (λ, μ)

(C) PB* (λ, μ)

(D) PD (λ, μ)

(E) *If* λ *is inaccessible there is a family* S *exemplifying* PB (λ, λ) *(and also* PT (λ, λ)*) such that no two members of* S *have the same cardinality.*

Remark. Erdős and Hajnal [2] already noticed (A) ⇒ (B) and the proof of 3.1 is similar to it. Therefore, we only give it here concisely. Parts (B)-(D) give answers to problem 42 [1]. (E) is a privately communicated problem of Erdős.

Proof. Clearly if λ is weakly compact then (B)-(E) fail. So suppose λ is not weakly compact. Then by Jensen [5] if $\mu = \aleph_0$, and by a slight improvement of A. Beler otherwise, there is a stationary $C \subseteq \subseteq \{\alpha < \lambda\colon \operatorname{cf} \alpha = \operatorname{cf} \mu,\ \beta + \mu \leqslant \alpha \text{ for } \beta < \alpha\}$ such that for every $\delta < \lambda$, $C \cap \delta$ is not stationary. For each $\alpha \in C$, choose $A_\alpha \subseteq \alpha$, $\sup A_\alpha = \alpha$, $\operatorname{tp} A_\alpha = \mu$; then $S = \{A_\alpha\colon \alpha \in C\}$ proves (B). Let G be a graph whose set of vertices is λ, and its set of edges $\{(\alpha, \beta)\colon \alpha \in A_\beta\}$; this proves (D). Now by Jensen [5], there are sets $T^\alpha \subseteq \alpha$ such that for any $A \subseteq \lambda$ of cardinality λ, $\{\alpha \in C\colon A \cap \alpha = T^\alpha\}$ is stationary, and $|T^\alpha| = |\alpha|$. Choose $A^\alpha \subseteq T^\alpha$, $|A^\alpha| = \mu$ and then $\{A^\alpha\colon \alpha \in C\}$ proves (C). Now (E) is proved by $S = \{T^\alpha\colon \alpha \in C,\ \alpha \text{ is a limit cardinal}\}$ (any $S' \subseteq S$, $|S'| < |S|$ has property B because $S' \cup S$ (which is $\neq S'$, as it is an indexed family) has a transversal by [9]).

Notation. For a family S, and sets A, B

$$S(A) = \{C\colon C \in S,\ C \subseteq A\}$$

$$S(A, B) = S_B^A = \{C - B\colon C \in S,\ C \subseteq A,\ C \nsubseteq B\}.$$

Definition 4.2. Define $m(S, \kappa)$ recursively, where S is a family of sets of cardinality $< \kappa$, as follows:

Case I. $|S| < \kappa$, then $m(S, \kappa)$ is 0 if S has a transversal, and -1 otherwise.

Case II. $\lambda = |S| \geqslant \kappa$ is regular. Then $m(S, \kappa)$ is $-\lambda$ if there is a continuous increasing sequence of sets A_α, $\alpha < \lambda$, $|A_\alpha| < \lambda$, and $A \in S \Rightarrow A \subseteq \bigcup_{\alpha<\lambda} A_\alpha$, and C_1 or C_2 is stationary where

$$C_1 = \Big\{\alpha\colon S(A_\alpha) - \bigcup_{\beta<\alpha} S(A_\beta) \neq \phi\Big\}$$

$$C_2 = \{\alpha\colon m[S(A_{\alpha+1}, A_\alpha), \kappa] < 0\}$$

or $|\bigcup_{\alpha<\lambda} A_\alpha| < \lambda$; and $m(S, \kappa) = 0$ otherwise.

Case III. $\lambda = |S| \geqslant \kappa$ has cofinality $\aleph_0$, $\lambda = \aleph_{\alpha+\beta}$, where $\kappa = \aleph_\alpha$ and $0 < \beta < \omega_1$. Then $m(S, \kappa) = 0$.

Remark. The definition is interesting only if κ is regular $\kappa = \aleph_\beta$, $|S| < \aleph_{\beta+\omega_1}$.

Lemma 4.3.

(A) *If for some* A, $m(S(A), \kappa) < 0$, *then* S *has no transversal.*

(B) *If* F *is a transversal of* S, *and* $m(S^A_B, \kappa) = -\mu$ *then* $\{b \in B$: *for some* $C \in S^A_B$, $F(C) = b)\}$ *has cardinality* μ.

Proof. *Immediate.*

Theorem 4.4. *Suppose* $\kappa = \aleph_\alpha$ *is regular,* $|S| < \aleph_{\alpha+\omega_1}$. *Then* S *has a transversal iff* $m[S(A), \kappa] \geqslant 0$ *for every* A.

Proof. The only if part is 4.3 (A). So we now prove by induction on $|S|$ that

(*) if for every A, $m[S(A), \kappa] \geqslant 0$, then S has a transversal. Note that if $|S(A)| > |A|$, $S(A)$ has no transversal.

Case I. $|S| < \kappa$, there is nothing to prove, by definition.

Case II. $\lambda = |S|$ is regular.

We may assume that S is a family of subsets of λ and $|S(\beta)| < \lambda$ for $\beta < \lambda$. Let

$$C_1 = \{\alpha < \lambda\text{: there is } A \in S,\ A \subseteq \alpha,\ \sup A = \alpha\}$$

$$C_2 = \{\alpha < \lambda\text{: there is } A \subseteq \lambda,\ \alpha \subseteq A,\ |A| < \lambda \text{ such that } m(S^A_\alpha, \kappa) < 0\}.$$

If C_1 is stationary, put $A_\alpha = \alpha$; then $m[S(\lambda), \kappa] = -\lambda$, a contradiction.

If C_2 is stationary, define A_α by induction. If $A_\alpha = \alpha \in C_2$, $A_{\alpha+1}$ is the A mentioned in C_2, otherwise $A_{\alpha+1} = \beta$ where β is the first ordinal bigger than any $\gamma \in A_\alpha$ and than α. $A_0 = \phi$, $A_\delta = \bigcup_{\alpha<\delta} A_\alpha$ for limit δ. This shows $m[S(\lambda), \kappa] = -\lambda$, a contradiction. Hence there is a closed unbounded $C \subseteq \lambda$ disjoint to $C_1 \cup C_2$. Let $C = \{\alpha(i)\colon i < \lambda\}$, $S_i = S^{\alpha(i+1)}_{\alpha(i)}$. As $C \cap C_1 = \phi$, $S = \bigcup_{i<\lambda} \{A\colon A - \alpha_i \in S_i\}$ and as $C \cap C_2 = \phi$ for every A, $m[S_i(A), \kappa] \geqslant 0$. So by the induction hypothesis each S_i has a transversal F_i, and F defined by $F(A) = F_i(A - \alpha_i)$ if $A \subseteq \alpha_{i+1}$, $A \not\subseteq \alpha_i$ is a transversal of S.

Case III. $|S| = \aleph_{\alpha+\delta}$, $\aleph_\alpha = \kappa$, $\mathrm{cf}\, \delta = \omega$, $\delta < \omega_1$.

We may assume that S is a family of subsets of $\aleph_{\alpha+\delta}$, and let $\lambda_n < \aleph_{\alpha+\delta}$, $\sum_{n<\omega} \lambda_n = \aleph_{\alpha+\delta}$. It suffices to prove

(**) for any $A_0 \subseteq \aleph_{\alpha+\delta}$ with $|A_0| < \aleph_{\alpha+\delta}$, there is an A_1 with $A_0 \subseteq A_1 \subseteq \aleph_{\alpha+\delta}$, $|A_1| < \aleph_{\alpha+\delta}$ such that for any $A_2 \subseteq \aleph_{\alpha+\beta}$ we have $m[S^{A_2}_{A_1}, \kappa] \geqslant 0$.

Indeed, assuming this, define $A(n)$ $(n < \omega)$ recursively such that $|A(n)| < \aleph_{\alpha+\delta}$, $\lambda_n \subseteq A(n)$, and for every A, $m(S^A_{A(n)}, \kappa) \geqslant 0$. Then by the induction hypothesis $S[A(0)]$, $S[A(n+1), A(n)]$ have transversals, and combining them we get a transversal of S.

Suppose that A_0 contradicts (**), and let $\mu = |A_0| + \kappa$. Define A_α, $\alpha \leqslant \mu^+$ by recursion so that $|A_\alpha| \leqslant \mu$. A_0 is already defined, and $A_\delta = \bigcup_{i<\delta} A_i$ for limit δ. If A_α is defined, then, by the definition of A_0, there is an A with $|A| < \aleph_{\alpha+\delta}$, $m[S^A_{A_\alpha}, \kappa] < 0$, and, as is easily seen, with $A \supseteq A_\alpha$, and choose such A with minimal cardinality. If $|A| > \mu$, then we get by definition 4.2 that $m(S^A_{A_\alpha}, \kappa) = -|S^A_{A_\alpha}| \leqslant \leqslant -|A| < -\mu$, and so, by Lemma 4.3 (B), $S(A)$ has no transversal, but $|S(A)| < \aleph_{\alpha+\delta}$, so by the induction hypothesis we get a contradiction. Thus $|A| \leqslant \mu$; let $A_{\alpha+1} = A$. Now clearly $m[S(A_{\alpha+1}, A_\alpha), \kappa) < 0$, hence $m[S(A_{\mu^+}), \kappa] = -\mu^+$, a contradiction. So (**) holds, completing the proof.

Corollary 4.5. *If* β *is limit,* $0 < \beta < \omega_1$, $\kappa < \aleph_{\alpha+\beta}$ *then* PT $(\aleph_{\alpha+\beta}, \kappa)$ *does not hold.*

A similar theorem is

Theorem 4.6. *If* λ *is a strong limit cardinal of cofinality* $\aleph_0$, $\kappa < \lambda$ *then* PT (λ, κ) *does not hold.*

Proof. We may assume that S is a family of subsets of λ of cardinality $< \kappa$ such that each $S' \subseteq S$, $|S'| < \lambda$ has a transversal. We must have $|S(A)| < \lambda$ for any $A \subseteq \lambda$, $|A| < \lambda$ (as λ is strong limit), hence $|S(A)| \leqslant |A|$. Similarly to the proof of 4.5, it suffices to prove

(***) for any $A \subseteq \lambda$ with $|A| < \lambda$ there is a $B, A \subseteq B \subseteq \lambda$, and a transversal F of $S(B)$ such that if C is the range of F then the family $S' = \{D - C\colon D \in S,\ D \notin S(B)\}$ satisfies: for every $S'' \subseteq S'$, $|S''| < \lambda$, S'' has a transversal.

Suppose $A \subseteq \lambda$, $|A| < \lambda$ is given. If taking $B = A$, there is a transversal F of $S(A)$ satisfying (***), then we are ready. Otherwise, for each transversal F of $S(A)$ there is a subfamily S_F of $S - S(A)$, such that $|S_F| < \lambda$, and $S_F^1 = \{C - \text{Range}\, F\colon C \in S_F\}$ has no transversal. Let $\lambda = \sum_{n<\omega} \lambda_n$, $\lambda_n < \lambda$, and $S^n = \bigcup \{S_F\colon F$ is a transversal of $S(A)$, $|S_F| \leqslant \lambda_n\}$. Clearly $|S^n| \leqslant 2^{|A|} + \lambda_n < \lambda$, so let F^n be a transversal of $S^n \cup S(A)$. Let B be the smallest set with $A \subseteq B$ such that we have $F^n(C) \in B \Rightarrow C \subseteq B$ for any $C \in S$ and $n < \omega$. Clearly B exists and $|B| \leqslant |A| + \kappa + \aleph_0 < \lambda$, hence $S(B)$ hence a transversal F_0, and let F_1 be its restriction to $S(A)$, and let $\lambda_n \geqslant |S_{F_1}|$. Now we shall show that $S_{F_1}^1$ has a transversal F, and so we get a contradiction: if $C \in S_{F_1}$ and $C \not\subseteq B$, then let $F(C - \text{Range}\, F_1) = F^n(C)$ (which $\notin B$) and if $C \in S_{F_1}$ and $C \subseteq B$, then let $F(C - \text{Range}\, F_1) = F_0(C)$ (note that $C \not\subseteq \text{Range}\, F_1$). It is easy to check that the F is a transversal, a contradiction. So (***) holds, and the proof is complete.

Open Problems.

(A) In 4.4, what can we say about singular cardinals λ of cofinality $> \omega$?

(B) In 4.6, can the strong limitness of λ be dropped?

(C) Can we generalize the definition of $m(S, \kappa)$ to all possible $|S|$ so that 4.4 holds?

We can deal with PB as with PT and prove, e.g.

Theorem 4.7. *If* $\mathrm{cf}\,\lambda = \aleph_0 < \lambda$, λ *is strong limit or* $\lambda = \aleph_{\alpha+\beta}$, $0 < \beta < \omega_1$ *then* $\mathrm{PB}(\lambda, \mu)$ *does not hold for any* $\mu < \lambda$.

Added in proof. The answer to all questions is yes; the proofs will appear in [10]; there we defined $m(S, \kappa) = 0$ whenever $|S|$ was singular.

We noticed long ago that there was a trivial solution of problem B: in case $\lambda = |S|$ is singular, $m(S, \kappa) = -\lambda$ if no $S' \subseteq S$ with $|S - S'| < \lambda$ has a transversal, and $m(S, \kappa) = 0$ otherwise.

REFERENCES

[1] P. Erdős – A. Hajnal, Unsolved problems in set theory, *Proceedings of Symposia in Pure Math.* XIII, Part I, A.M.S. Providence, R.I., (1971), 17-48.

[2] P. Erdős – A. Hajnal, Unsolved and solved problems in set theory, *Proceedings of the Symp. in honor of Tarksi's seventieth birthday in Berkeley,* 1971, to appear.

[3] P. Erdős – A. Hajnal, On chromatic number of graphs and set systems, *Acta Math. Acad. Sci. Hungar.*, 17 (1966), 61-99.

[4] P. Erdős – A. Hajnal – R. Rado, Partition relations for cardinals, *Acta. Math. Acad. Sci. Hungar.*, 16 (1965), 93-196.

[5] R.L. Jensen, The fine structure of the constructible universe, *Ann. of Math. Logic,* 4 (1972), 229-308.

[6] R.L. Jensen – K. Kunen, Some combinatorial properties of L and V, notes.

[7] S. Shelah, Notes in combinatorial set theory, *Israel J. of Math.*, 14 (1973), 262-277.

[8] E.C. Milner – S. Shelah, Two theorems on transversals, appearing in this volume.

[9] E. C. Milner – S. Shelah, Sufficiency conditions for the existence of transversals, *Canadian J. of Math.*, 26 (1974), 948-961.

[10] S. Shelah, to appear.

COLLOQUIA MATHEMATICA SOCIETATIS JÁNOS BOLYAI
10. INFINITE AND FINITE SETS, KESZTHELY (HUNGARY), 1973.

COMMAND GRAPHS*

G.J. SIMMONS

1. INTRODUCTION

A command graph $G(k, l, n)$ is defined to be an undirected graph without loops on $n+1$ vertices, one of which is designated as the source (command post) and the other n as outputs (outposts), in which cutting k edges isolates at most l of the outputs from the source. Parallel edges are permitted. This paper is concerned with characterizing minimal command graphs, i.e. those command graphs having the least possible number, $L(k, l, n)$, of edges.

Surprisingly, there appears to be no literature on the problem in this general form, although on extensive literature [1], [2], [3], [4], [5], [6], [7], [8] exists on the following closely related problem. For minimal command graphs $G(k, 0, n)$ in which "everyone gets the word" in spite of up to k failures of the communication links, it happens to be true that an arbitrary vertex can be designated to be the source. It is this symmetric form,

*This work was supported by the United States Atomic Energy Commission.

i.e. of indistinguishable nodes, that has been treated. This is not the case when $l > 0$ as the following minimal command graph $G(5, 1, 4)$ shows

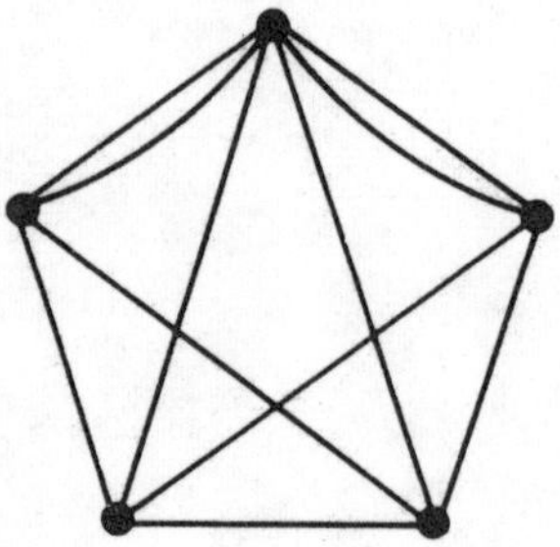

where clearly the source must be vertex s otherwise four cuts would suffice to isolate all output vertices from the source. For this reason we refer to command graphs as asymmetric and to the invulnerable communication networks of Boesch, Felzer et al as symmetric.

2. BOUNDS FOR THE NUMBER OF EDGES

A command graph is characterized by the property that the subgraph G' on every set of $l + \alpha$ output vertices, $\alpha > 0$, is connected to its complement by at least $k + 1$ edges: $\deg(G') \geqslant k + 1$.

Theorem 1. (Superposition). *Given two graphs* $G(k_1, l, n)$ *and* $G(k_2, l, n)$ *having* L_1 *and* L_2 *edges respectively, their union with an arbitrary one-to-one association of the output vertices is a* $G(k_1 + k_2 + 1, l, n)$ *graph with* $L_1 + L_2$ *edges.*

Proof. Since the vertices of the two component graphs are one-to-one associated, the subgraph on every $l + \alpha$ output vertices, $\alpha > 0$, has at least $k_1 + 1$ connections to its complement from $G(k_1, l, n)$ and $k_2 + 1$ from $G(k_2, l, n)$.∎

Corollary. $L(k, l, n) \leqslant \min_{k_1} \{L(k_1, l, n) + L(k - k_1 - 1, l, n)\}$ where $1 \leqslant k_1 \leqslant [\frac{k-1}{2}]$. *The extension to more than two terms on the right is obvious.*

The bound for the number of edges in a minimal command graph $G(k, l, n)$ provided by the Corollary to Theorem 1 generally isn't the smallest achievable number. For example, a minimal $G(9, 1, 6)$ graph has 23 edges while the best bound given by the Corollary is 24 edges obtained either by superimposing a minimal $G(1, 1, 6)$ and a $G(7, 1, 6)$ or a $G(3, 1, 6)$ and $G(5, 1, 6)$. On the other hand the least number of edges for a $G(7, 1, 4)$ is 14 which can be realized by superimposing two minimal $G(3, 1, 4)$ graphs. The following theorem bounds $L(k, l, n)$ from below.

Theorem 2. *The least number of edges,* $L(k, l, n)$ *which a* $G(k, l, n)$ *graph can have satisfies the following;*

(1) $L(k, l, n) = 0 \qquad l \geqslant n$

(2) $L(k, l, n) = n - l + k \qquad n > l \geqslant k$

(3) $L(k, l, n) \geqslant \left\lceil \frac{n(k+1)}{l+1} \right\rceil^{*} \qquad 2l + 1 \geqslant n > l, \; l < k$

(4) $L(k, l, n) \geqslant \left\lceil \frac{(n + 2l + 1)(k + 1)}{2(l + 1)} \right\rceil \qquad n > 2l + 1, \; l < k$

Proof.

(1) is obvious.

(2) follows by first noting that for $n > l \geqslant k$, if α vertices are assumed to be originally isolated, then

$$L(k, l, n) \leqslant \min_{\alpha} L(k, l - \alpha, n - \alpha) \leqslant L(k, k, n - l + k) = \\ = n - l + k\,.$$

But this says that at most $n - l + k$ of the vertices can have edges incident on them, i.e. $l - k + \gamma$ vertices are already isolated before any of the k cuts are made; where $\gamma \geqslant 0$. Therefore, each permissible $G(k, l, n)$ graph must be the union of $l - k + \gamma$ isolated vertices and a $G(k, k - \gamma, n - l + k - \gamma)$ graph since at most $k - \gamma$ of the $n - l + k - \gamma$ connected

* $\lceil x \rceil$ denotes the least integer greater than or equal to x.

vertices can be isolated by k cuts if at most l vertices of the $G(k, l, n)$ graph are to be isolated: i.e.

$$L(k, l, n) = \min_{\gamma} L(k, k - \gamma, n - l + k - \gamma) .$$

But, by considering γ isolated vertices;

$$L(k, k - \gamma, n - l + k - \gamma) \geqslant L(k, k, n - l + k) = n - l + k$$

for $0 \leqslant \gamma \leqslant k$, hence $L(k, l, n) = n - l + k$. To prove (3) and (4), start with any $G(k, l, n)$ graph for which $l < n$ and $l < k$ and note that the valence of the source, ν, must be at least $k + 1$. Denote by $L(k, l, n, \alpha)$ the number of edges in such a $G(k, l, n)$ graph in which $\nu = k + 1 + \alpha$; $\alpha \geqslant 0$.

The exact number of edges linking subgraphs on $l + 1$ output vertices to their complements, when summed over all possible such subgraphs is,

$$\nu \binom{n-1}{l} + (2L(k, l, n, \alpha) - 2\nu) \binom{n-2}{l} .$$

On the other hand, from the definition of a $G(k, l, n)$ graph, every subgraph on $l + 1$ output vertices must be connected to its complement by at least $k + 1$ edges, hence.

$$\nu \binom{n-1}{l} + (2L(k, l, n, \alpha) - 2\nu) \binom{n-2}{l} \geqslant \binom{n}{l+1} (k + 1) . \tag{5}$$

Solving (5) for $L(k, l, n, \alpha)$, we obtain:

$$L(k, l, n, \alpha) \geqslant \frac{n(n-1)}{2(l+1)(n-l-1)} (k + 1) + \frac{n - 2l - 1}{2(n - l - 1)} \nu = B \tag{6}$$

or, upon replacing ν by $k + 1 + \alpha$,

$$L(k, l, n, \alpha) \geqslant \frac{n + 2l + 1}{2(l + 1)} (k + 1) + \frac{n - 2l - 1}{2(n - l - 1)} \alpha = B . \tag{7}$$

If $n \leqslant 2l + 1$, then

$$L(k, l, n) \geqslant \min L(k, l, n, \alpha) \geqslant \min_{\alpha} B \tag{8}$$

i.e. $L(k, l, n)$ is bounded below by the least value the right hand term can assume as α varies.

Case 1. If $B \geqslant \nu$, then

$$\frac{n+2l+1}{2(l+1)}(k+1)+\frac{n-2l-1}{2(n-l-1)}\alpha \geqslant k+1+\alpha$$

or

$$\alpha \leqslant \frac{n-l-1}{l+1}(k+1) \tag{9}$$

and (8) is minimized by assuming equality in (9) and substituting for α in (7):

$$L(k,l,n) \geqslant \frac{n(k+1)}{l+1}\,. \tag{10}$$

Case 2. If $B \leqslant \nu$ then from (6)

$$\frac{n(n-1)}{2(l+1)(n-l-1)}(k+1)+\frac{n-2l-1}{2(n-l-1)}\nu \leqslant \nu$$

and

$$\nu \geqslant \frac{n(k+1)}{l+1}\,.$$

but the total number of edges is at least as large as the valence of the source;

$$L(k,l,n,\alpha) \geqslant \frac{n(k+1)}{l+1}\,. \tag{11}$$

Since α does not appear in the right hand term of (11), equation (10) holds in this case also, and (3) follows by taking the least integer greater than or equal to this bound.

If $n \geqslant 2l+1$, the right hand term in (7) is non negative for all α, so that

$$L(k,l,n) \geqslant \frac{n+2l+1}{2(l+1)}(k+1) \tag{12}$$

and (4) follows by taking the least integer greater than or equal to this bound.∎

The bounds of Theorem 2 are tight in the sense that equality holds for infinitely many n for arbitrary, but fixed, k and l. Theorem 3 constructs one such class of minimal command graphs.

Theorem 3. *If* $n > 2l + 1$ *the graph consisting of the complete graph on* n *vertices, each of which is linked to the source by* $\alpha = l + 1$ *parallel edges, is a minimal* $G(n\alpha - 1, l, n)$ *graph which realizes the lower bound of Theorem* 2.

Proof.* Let G be a complete graph on n vertices each of which is linked to the source vertex by α parallel edges. The total number of edges in G is

$$(13) \qquad L = \frac{n(n-1)}{2} + n\alpha = \frac{n(n + 2\alpha - 1)}{2}$$

L equals the lower bound of (4) if $k = n(l + 1) - 1$.

Since deleting the $\nu = n(l + 1)$ edges incident on the source isolates all output vertices, it is enough to prove that no set of $k < n(l + 1)$ cuts of edges can isolate $l + \beta$ vertices where

$$(14) \qquad 1 \leqslant \beta \leqslant n - l\,.$$

The sum of the valences at $l + \beta$ output vertices is $(l + \beta)(n + \alpha - 1)$ and the number of edges in the subgraph on these $l + \beta$ vertices is $\frac{1}{2}(l + \beta)(l + \beta - 1)$ so that the number of edges, K, which must be cut to isolate this subset is

$$(15) \qquad K = (l + \beta)(n + \alpha - 1) - (l + \beta)(l + \beta - 1) = (l + \beta)(n + 1 - \beta)\,.$$

But $(l + \beta)(n + 1 - \beta) < n(l + 1)$ holds only if

$$(16) \qquad \text{either} \quad \beta < 1 \quad \text{or else} \quad \beta > n - l$$

neither of which can be true by (14).∎

The minimal $G(7, 1, 4)$ graph has at least 14 edges by Theorem 3 and, as was noted earlier, at most 14 by the Corollary to Theorem 1 since

*The author is indebted to the reviewer for this proof.

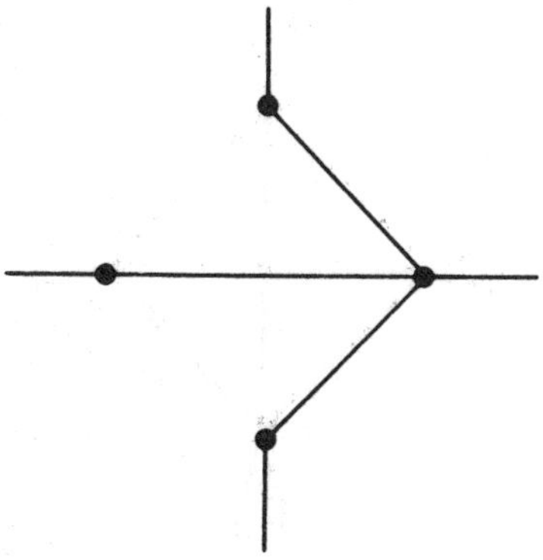

is a minimal $G(3, 1, 4)$ with 7 edges.* Hence any association of output points between two replicas of this minimal $G(3, 1, 4)$ graph would be a minimal $G(7, 1, 4)$ graph.

The two possibilities are

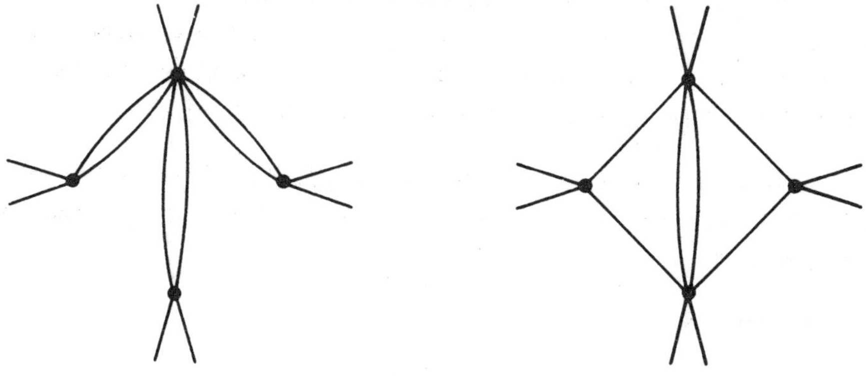

both of which have the same number of edges, 14, as the graph constructed by Theorem 3

*The drawing convention is that open edges terminate at the source vertex.

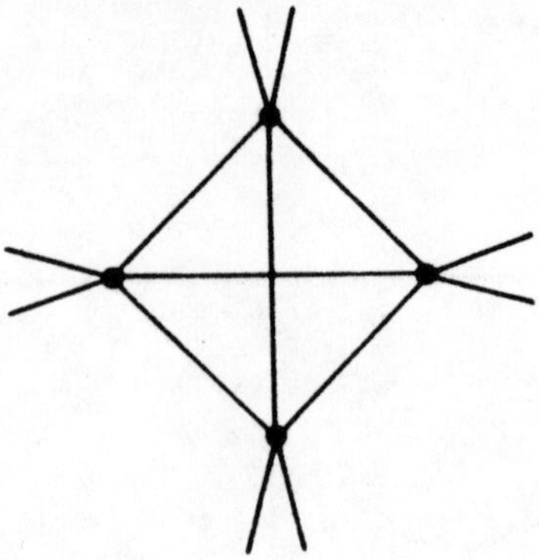

which is not the result of superimposing any lower order graphs.

3. THE CASE $l = 0$

As mentioned in the introductory remarks, the case $l = 0$, i.e. the case in which the performance of the communication network is not degraded by k or fewer link failures, has received extensive attention. We discuss this case only because we have achieved a novel constructive characterization of the minimal command graphs $G(k, 0, n)$.

Obviously, in a minimal $G(k, 0, n)$ graph, the valence at every vertex $\nu_i \geqslant k + 1$. In fact equality holds "almost everywhere" as the following theorem shows.

Theorem 4.

$$L(k, 0, n) = \left\lceil \frac{(k+1)(n+1)}{2} \right\rceil .$$

Proof. The inequality

$$L(k, 0, n) \geqslant \left\lceil \frac{(k+1)(n+1)}{2} \right\rceil$$

follows from Theorem 2. The proof is completed by the following construction [7] of a family of command graphs realizing the lower bound.

If $k = 2j - 1$, let $G(k, 0, n)$ be the superposition of j cycles on $n + 1$ points. If $k = 2j$, let $G(k, 0, n)$ be the superposition of j cycles with $[\frac{n+2}{2}]$ cross links between pairs of vertices such that $[\frac{n-1}{2}]$ vertices of the cycle lie on one side of them and at most one vertex has two cross links incident on it. Clearly, both of these constructions are $G(k, 0, n)$ graphs which achieve the lower bound for the number of edges.∎

For example, the minimal $G(4, 0, 4)$ graph constructed by the method of Theorem 4 would be;

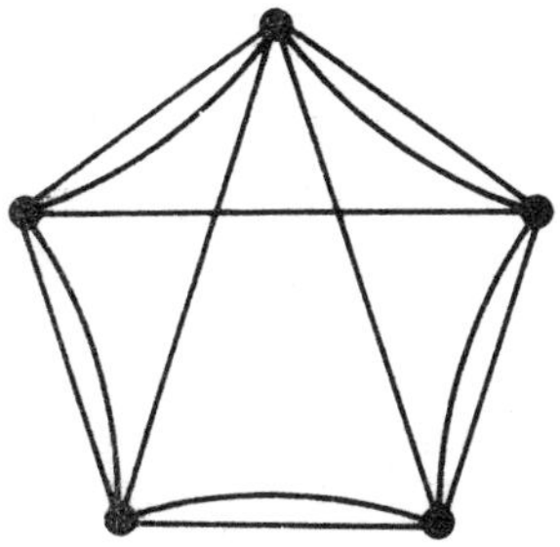

where, as was noted earlier, any vertex could be chosen to be the source.

Although the construction of Theorem 4 shows that for $l = 0$ the lower bound on the number of edges of Theorem 2 is always achieved, there are many minimal command graphs not realizable by this construction – nor even surprisingly, by the superposition theorem on lower order components as we shall show after first characterizing the command graphs $G(1, 0, n)$.

Theorem 5. *A $G(1, 0, n)$ graph, $n \geqslant 2$, is minimal if and only if it is a cycle on the source and the n output vertices.*

Proof. If ν_i is the valence of the i-th vertex in a minimal graph

$$\sum_{i=1}^{n+1} \nu_i = 2L(1, 0, n) = 2(n + 1)$$

by Theorem 4. But $\nu_i \geqslant 2$ for all i in a $G(1, 0, n)$ graph. Therefore, $\nu_i = 2$ and every component is a cycle. The theorem follows since a

$G(k, 0, n)$ graph must be connected.∎

It is interesting to note that there are numerous minimal $G(k, 0, n)$ graphs which are not the result of superimposing lower order minimal command graphs as the following minimal $G(3, 0, 9)$ graph shows

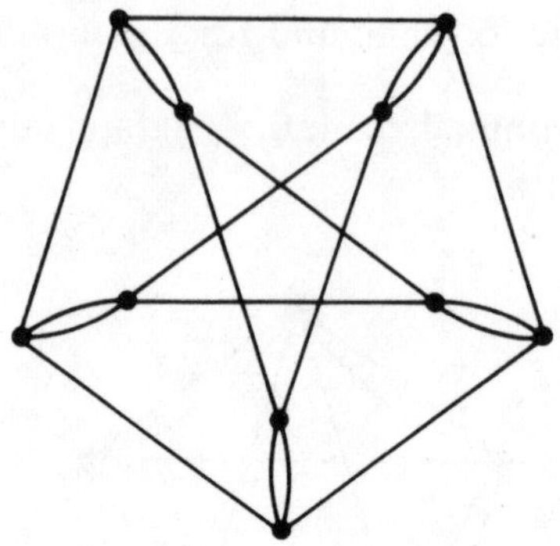

where obviously there is no spanning cycle, i.e. no minimal $G(1, 0, 9)$ component. The parallel edges are a non-essential part of the failure since each such pair of edges could be replaced by the subgraph

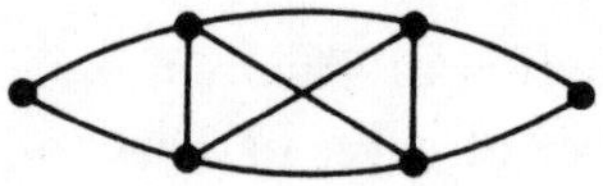

without affecting the result.

Theorem 5 established that the minimal $G(1, 0, n)$ graph was uniquely determined to be a cycle on $n + 1$ vertices. On the other hand, a minimal $G(2, 0, n)$ graph must be a cubic graph on $n + 1$ vertices; however, a cubic graph may not be a command graph as the following two examples show. If parallel edges are allowed the subgraph

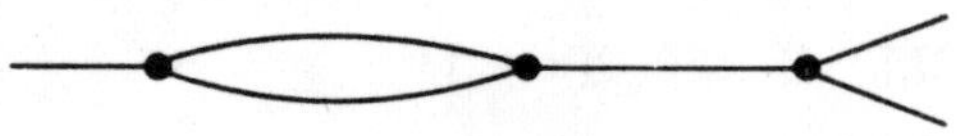

whose smallest containing cubic graph is

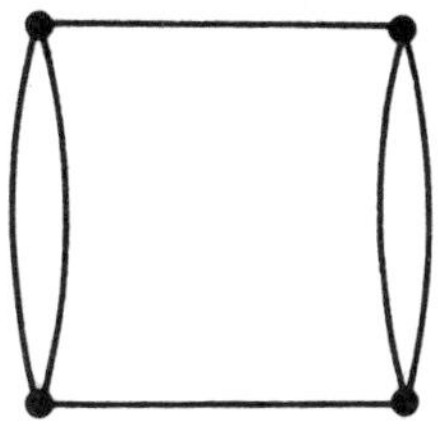

or if parallel edges are disallowed, the subgraph

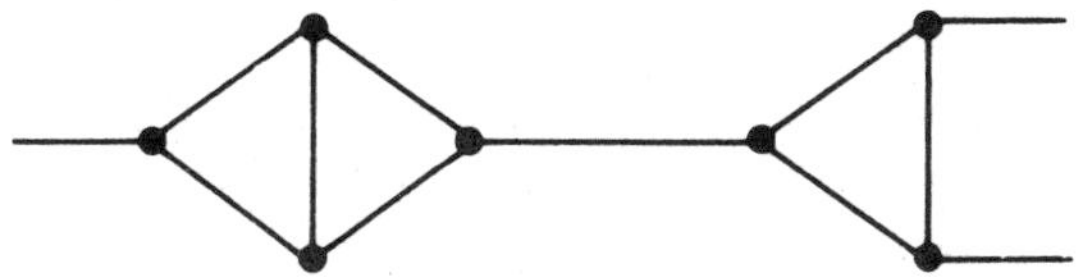

whose smallest containing cubic graph is

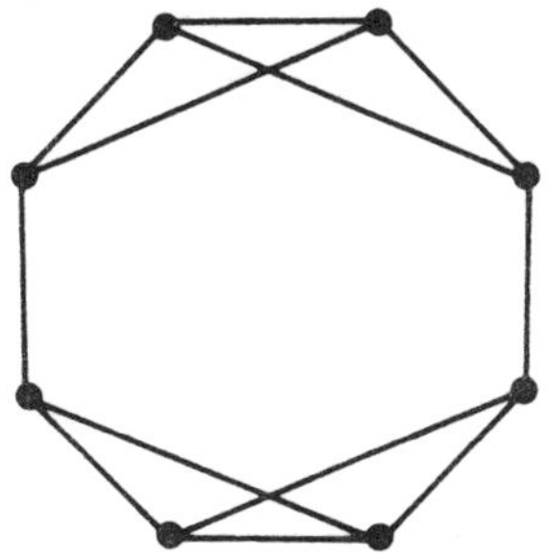

demonstrate the existence of cubic graphs which are not $G(2, 0, n)$ command graphs.

The unique minimal command graphs on 1, 3, and 4 vertices corresponding to $n = 1, 2$ and 3 are:

$G(2, 0, 1)$

$G(2, 0, 2)$

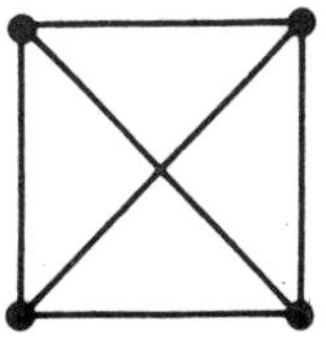

$G(2, 0, 3)$

and the two minimal command graphs on 5 vertices are:

$G_1(2, 0, 4)$

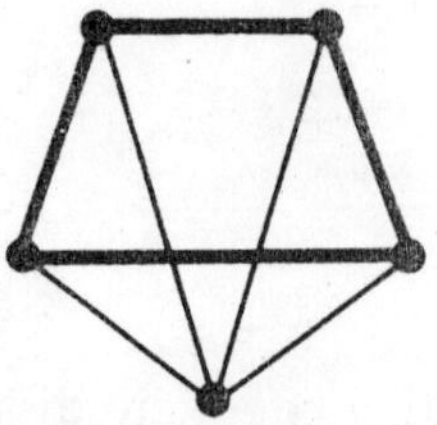

$G_2(2, 0, 4)$

If the midpoints of any pair of edges in the $G(2, 0, 1)$ graph are connected by adding a new edge, the $G(2, 0, 3)$ graph is obtained. Similarly if the midpoints of any pair of edges in the $G(2, 0, 2)$ graph are connected with a new edge one of the two $G(2, 0, 4)$ graphs is obtained.

We shall show that in a minimal $G(2, 0, n)$, $n \geqslant 3$, graph there is always at least one edge which can be deleted by reversing the extension procedure to yield a minimal $G(2, 0, n-2)$ graph. The following smallest such example shows that the choice of an edge for such a reduction, if possible at all cannot be made indiscriminately.

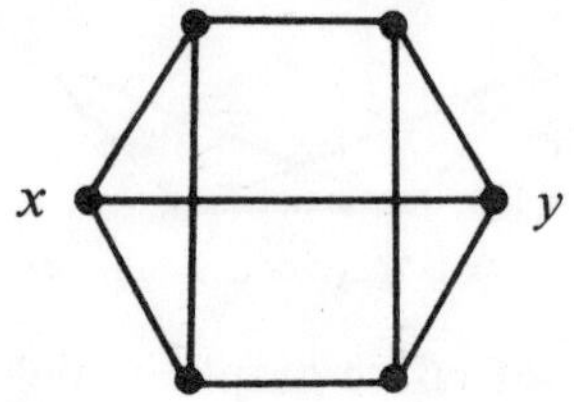

If edge xy is removed, the graph

is obtained, which is clearly not a command graph. Consequently, only the bold edges in the $G_1(2, 0, 4)$ and $G_2(2, 0, 4)$ graphs can be the extending edge of the $G(2, 0, 2)$ graph. The procedure for extending minimal command graphs when $k = 2$ suggested by this observation completely characterizes the minimal $G(2, 0, n)$ graphs as is shown in the following theorem.

Theorem 6. *A $G(2, 0, n)$ graph, $n \geq 3$, is minimal if and only if it is an extension of some minimal $G(2, 0, n-2)$ graph through adding an edge joining the midpoints of a pair of edges.*

Proof. Clearly, such an extension of a minimal $G(2, 0, n-2)$ graph has the proper number of edges to be a minimal $G(2, 0, n)$ graph. To see that it is, consider an arbitrary subgraph of the extended graph. If this subgraph consists of only one or both of the vertices generated by the extension, then it is connected to its complement by 3 or 4 edges respectively. On the other hand if the subgraph includes vertices of the original $G(2, 0, n-2)$ graph – it will be connected to its complement in the extended graph by at least as many edges as the subgraph restricted to those vertices in the $G(2, 0, n-2)$ graph would have been connected to its complement in that graph. Therefore, the extension is a $G(2, 0, n)$ graph and hence a minimal $G(2, 0, n)$ graph.

The proof that every minimal $G(2, 0, n)$ graph is an extension of some $G(2, 0, n-2)$ graph is considerably more delicate. In a minimal $G(2, 0, n)$ graph define a subgraph to be a 3-subgraph if it has more than a single vertex, no parallel edges, is triply connected to its complement, and contains no triply connected subgraph on more than one vertex itself. For reasons of logical consistency, since there are $G(2, 0, n)$ graphs which have no proper 3-subgraphs, for example,

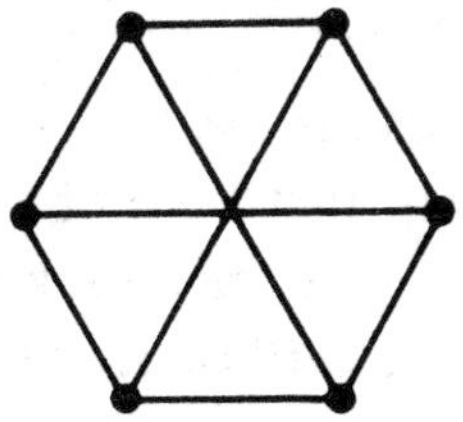

we define $G(2, 0, n)$ to be triply connected to its complement – the null set – so that a minimal $G(2, 0, n)$ can itself be a 3-subgraph.

If $G(2, 0, 2m)$ contains a pair of parallel edges, then it necessarily contains the subgraph H

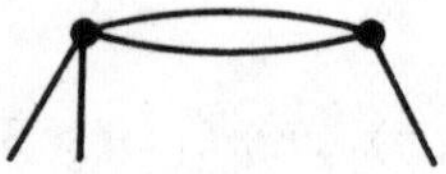

H

which is always triply connected to its complement. Either the complement of H is a 3-subgraph or else it fails to be because it contains a 3-subgraph. In either case, there is always a 3-subgraph (not necessarily proper) of a minimal $G(2, 0, 2m)$ graph. Therefore, from the definition of a 3-subgraph and the result just given, every minimal $G(2, 0, n)$ graph contains a 3-subgraph.

The proof of the theorem depends on the fact that an arbitrary edge, x, in a 3-subgraph, both of whose endpoints have valence 3, can be removed to reduce the minimal $G(2, 0, 2)$ graph to a minimal $G(2, 0, n - 2)$ graph. This line of proof required that one first show that every 3-subgraph has such an edge and then, that its removal results in a minimal $G(2, 0, n - 2)$ graph. Obviously the reduction results in the right number of vertices with the proper valences and the required number of edges for the reduced graph to be a minimal $G(2, 0, n - 2)$ graph.

Let G_1 designate an arbitrary 3-subgraph of a minimal $G(2, 0, n)$ graph. G_1 has at least three vertices since it contains two by definition and a subset of two vertices is always connected to its complement by at least four edges. If G_1 has s vertices, then it contains $\frac{3(s-1)+4-3}{2} = \frac{3s-2}{2}$ edges if it contains a vertex of valence 4 and $\frac{3s-3}{2}$ edges if not. In the first case G_1 contains at least five edges and in the second at least three. It is easy to verify that both of the endpoints of at least two of the contained edges in the first case (and obviously all of them in the second) have valence 3. This is illustrated by the following pair of minimal 3-subgraphs

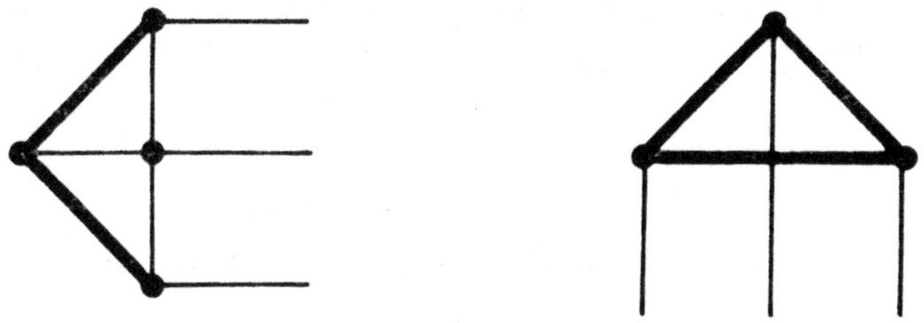

for the two possible cases.

We have shown that every minimal $G(2, 0, n)$ graph contains at least one 3-subgraph and that every 3-subgraph contains an edge (in fact at least two edges), both of whose endpoints have valence 3. If x is such an edge – clearly its removal cannot change the valence of any of the remaining $n - 2$ vertices. Also any subgraph of the $G(2, 0, n)$ graph which contained x (or contained neither of its endpoints) will still be connected to its complement by the same number of edges in the reduced graphs as before. Therefore, we need only consider subgraphs of the $G(2, 0, n)$ graph containing more than one vertex which are linked to their complement by x. If there are four or more linking edges, removing x leaves the subgraph at least triply connected and hence does not contradict the hypothesis that the reduced graph is a minimal $G(2, 0, n - 2)$ graph.

Therefore, we have only to show that x cannot be one of the three edges triply linking some subgraph of the $G(2, 0, n)$ graph with more than one vertex, to its complement. Assume that x is such a link and that G_2 is a subgraph having x as one of three links with its complement so that the graph configuration is of the form:

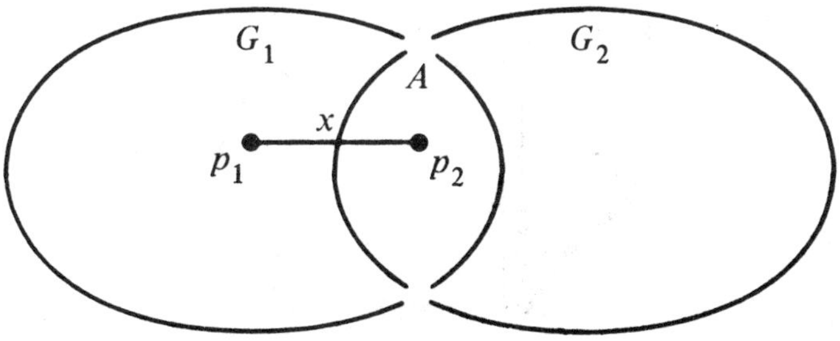

The proof will be completed by showing that p_1 is the only point in the complement of G_2. For then, after the removal of x, G_2 with the one remaining edge which passed through p_1 becomes $G(2, 0, n-2)$.

G_1 is minimal so that G_2 cannot be a subgraph of G_1 (since G_2 was assumed to be triply connected with its complement), therefore $G_2 \setminus A$ contains at least one vertex.

An edge linking A to its complement can have its other endpoint in either $G_1 \cap \bar{G}_2$, $\bar{G}_1 \cap G_2$ or $\bar{G}_1 \cap \bar{G}_2$. Let a, b and c be the number of edges of each type respectively. The subgraphs $G_1 \setminus A$ and $G_2 \setminus A$ are connected to their complements by $3 + a - b - c$ and $3 + b - a - c$ edges. Since these are both nonempty subgraphs in a $G(2, 0, n)$ graph

$$3 + a - b - c \geqslant 3$$

and

$$3 - a + b - c \geqslant 3$$

which implies $a = b$ and $c = 0$. But this says that $G_1 \setminus A$ is triply connected to its complement which is possible for a proper subset of a 3-subgraph only if the subgraph is a single vertex – p_1. Clearly $a = b \neq 1$, otherwise the subgraph A could be isolated by two cuts contrary to the assumption that the entire graph is a $G(2, 0, n)$ graph. But if $a = b = 2$ or 3, since the subgraph $G_1 \setminus A$ contains only the single vertex p_1, $G_1 \cup G_2$ must be the entire $G(2, 0, n)$ graph. It should be noted that there are graphs of this type, for example,

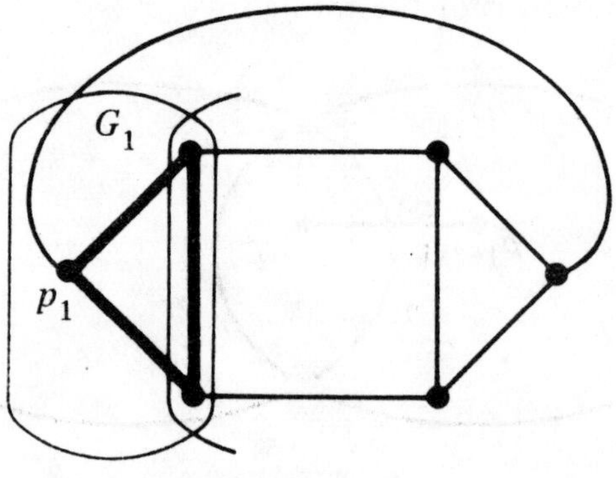

or

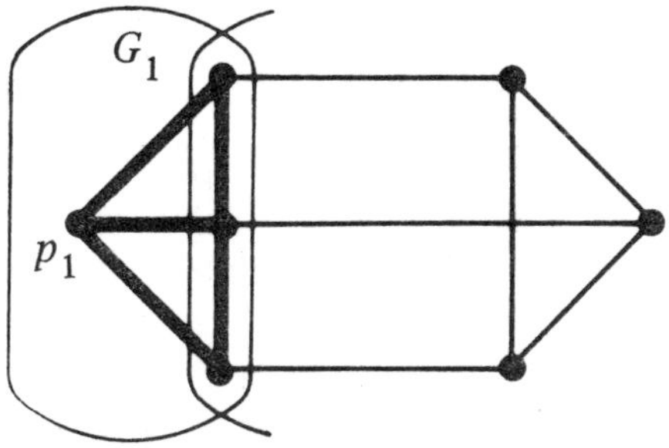

The removal of an edge linking p_1 with a vertex of valence 3 in A must also remove p_1 from the reduced graph. This means that G_2 with the one remaining edge which passed through p_1 becomes $G(2, 0, n-2)$, so that the reduction cannot result in any subgraph which was triply connected to its complement in its original $G(2, 0, n)$ graph being doubly connected in the reduced $G(2, 0, n-2)$ graph. This proves that the reduced graph is a minimal $G(2, 0, n-2)$ graph.∎

The unique minimal command graphs, $G(3, 0, n)$, on 2 and 3 vertices corresponding to $n = 1$ and 2 are

$G(3, 0, 1)$

$G(3, 0, 2)$

and the two minimal command graphs on 4 vertices are

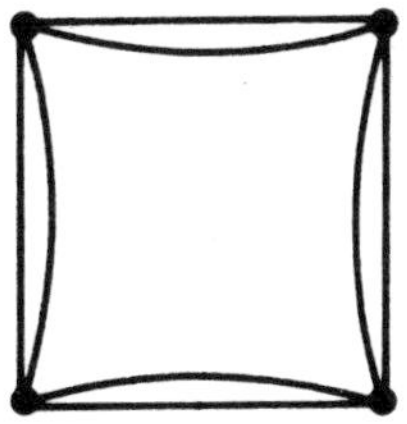

$G_1(3, 0, 3)$

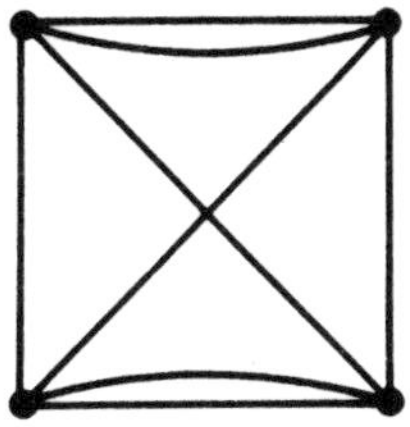

$G_2(3, 0, 3)$

If the midpoints of any pair of edges in the $G(3, 0, 1)$ graph are superimposed to form one new vertex and two new edges, the $G(3, 0, 2)$ graph is obtained. Similarly, if the midpoints of any pair of the parallel edges in the $G(3, 0, 2)$ graph are superimposed the $G_1(3, 0, 3)$ graph is obtained, while superimposing the midpoints of any pair of non-parallel edges yields the $G_2(3, 0, 3)$ graph.

We shall show in fact that in a minimal $G(3, 0, n)$ graph any vertex may be deleted by reversing the extension procedure to yield a minimal $G(3, 0, n - 1)$ graph. The following smallest such example shows that such a reduction, if possible at all, cannot be made without care as to how the reduced edges are selected.

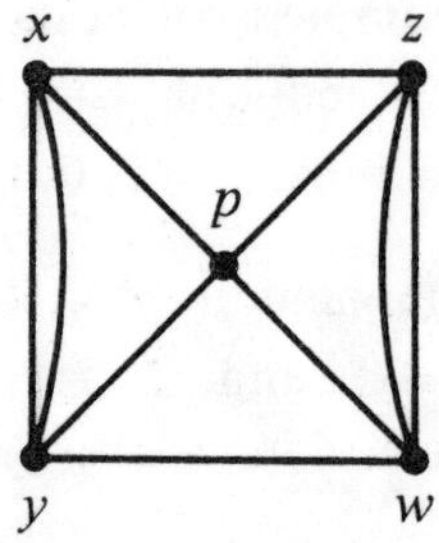

If, when p is removed, the reduced edges are taken to be xz and yw, the graph $G_1(3, 0, 3)$ is obtained, or if xw and yz are taken $G_2(3, 0, 3)$ is obtained; either of which is a minimal command graph on four vertices. Taking the reduced edges to be xy and zw, however, leads to be graph

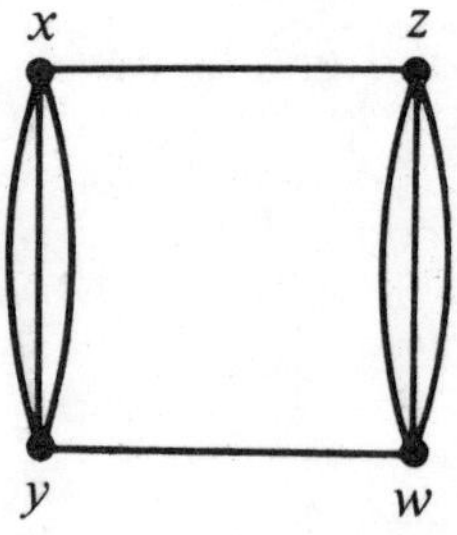

which clearly is not a command graph.

The procedure for extending minimal command graphs, when $k = 3$, suggested by this observation completely characterizes the minimal $G(3, 0, n)$ graphs as is shown in the following theorem.

Theorem 7. *A $G(3, 0, n)$ graph, $n \geqslant 2$, is minimal if and only if it is an extension of some minimal $G(3, 0, n-1)$ graph through superimposing the midpoints of a pair of edges to form a new vertex.*

Proof. Consider the graph resulting from such an extension of a minimal $G(3, 0, n-1)$ graph. A subgraph of this extended graph which does not include the extending vertex, p, is connected to its complement in the extended graph by at least as many edges as in the $G(3, 0, n-1)$ graph. On the other hand, if a subgraph includes p it is connected to its complement in the extended graph by at least as many edges as the subgraph with p deleted was connected to its complement in the $G(3, 0, n-1)$ graph. Hence, the extended graph is a command graph and since it has the minimal number of edges, it is a minimal $G(3, 0, n)$ command graph.

The proof proceeds by showing that for every vertex, p, in G there is always at least one choice for the pair of reduced edges resulting from the deletion of p such that the reduced graph will be a command graph and hence a minimal $G(2, 0, n-1)$ graph.

There are two ways in which a reduction of a minimal command graph can fail to be a command graph. First, if p has a pair of parallel edges incident on it, both of these edges cannot belong to the same edge in the reduced graph since, by definition, command graphs do not have loops. Second, if p is connected by two edges to a subgraph linked to its complement in the $G(3, 0, n)$ graph by either four of five edges, the union of these two edges cannot be an edge in the reduced graph. For example, in the following subgraphs,

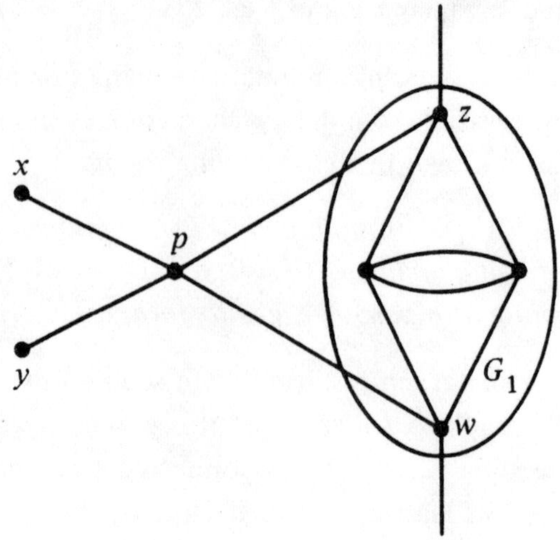

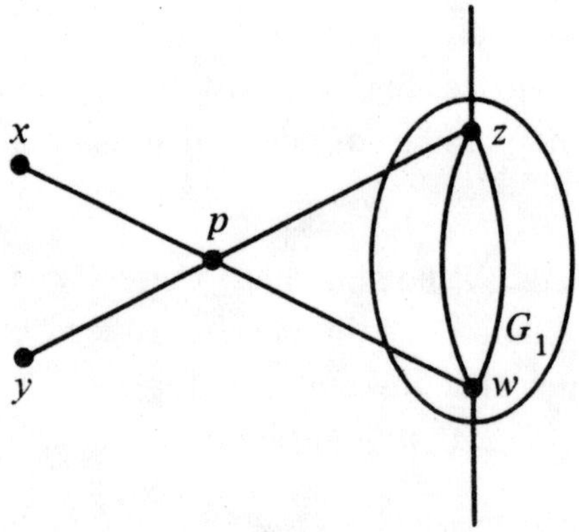

taking the reduced edge to be zw causes the subgraph G_1' on the vertices of G_1 to be linked to its complement by only two edges in the reduced graph.

If the vertex, p, to be deleted has two pairs of parallel edges incident on it

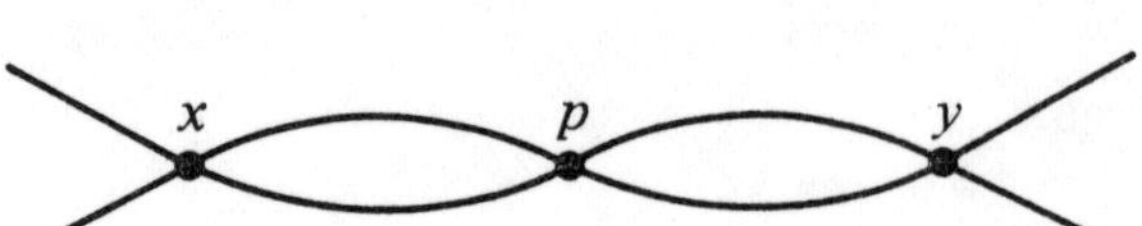

then the reduced graph will be of the form

The only graphs whose connectivity to their complements are affected by the reduction are those including both vertices x and y but not p in $G(3, 0, n)$. However, such a subgraph has the same number of connections to its complement in the reduced graph as the subgraph extended to include vertex p has in the $G(3, 0, n)$ graph. Hence the reduced graph is a $G(3, 0, n-1)$ graph.

If p has only one pair of parallel edges incident on it

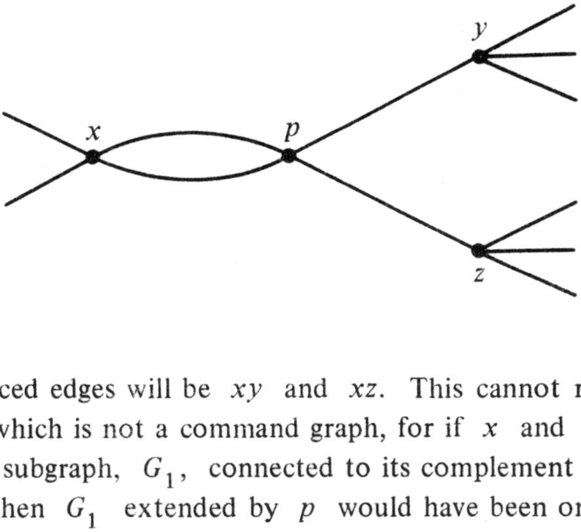

then the reduced edges will be xy and xz. This cannot result in a reduced graph which is not a command graph, for if x and y were vertices of some subgraph, G_1, connected to its complement by five or fewer edges, then G_1 extended by p would have been only triply connected in a $G(3, 0, n)$ graph which is a contradiction.

If p does not have a pair of parallel edges incident on it, then as was pointed out earlier, some choices for the reduced edges may not result in the reduced graph being a command graph. The proof proceeds by showing that there is always at least one (in fact two) acceptable choices for the reduced edges at any p.

First, it cannot be the case that either 3 or 4 of the endpoints of the edges incident on p all belong to a subgraph, G_1, where $\deg(G_1) \leqslant \leqslant 5$, since this would imply that $\deg(G(3, 0, n)/(G_1 \cup p)) \leqslant 3$ which is a contradiction. Second, if two of the edges incident on p, say x and y, belong to a subgraph, G_1, where $\deg(G_1) \leqslant 5$ then $\deg(G(3, 0, n)/(G_1 \cup p)) \leqslant 5$;

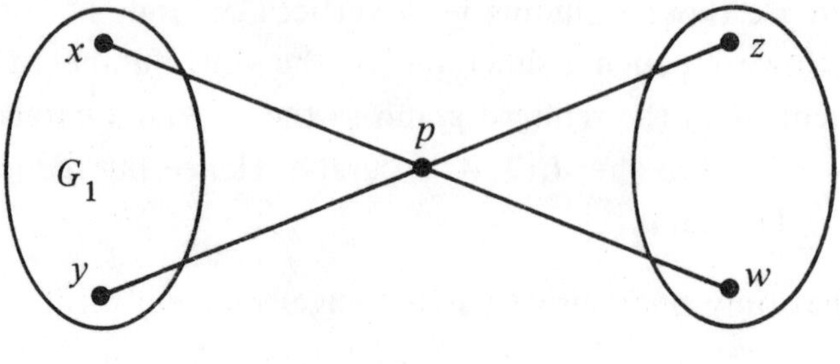

but the choice of the reduced edges as either xz and yw or xw and yz would not change the degree of the reduced subgraphs. The reduced graph would fail to be a command graph only if neither of these choices was possible, i.e., only if vertices x and w belonged to a subgraph, G_2, $\deg(G_2) \leqslant 5$ so that the reduced edge could not be chosen to xw and similarly for vertices x and z. To show that this cannot be the case, assume that x and y are vertices in a subgraph G_1 and x and z vertices in a subgraph G_2 where $\deg(G_1) \leqslant 5$ and $\deg(G_2) \leqslant 5$ so that neither xy nor xz would be a permissible edge in the reduced graph. Let α, β and γ be the number of edges linking $G_1 \cap G_2$ with $G_2/(G_1 \cap G_2)$, $G_1/(G_1 \cap G_2)$ and $G(3, 0, n)/(p \cup G_1 \cup G_2)$, respectively. Also let δ and ϵ be the number of edges linking $G_2/(G_1 \cap G_2)$ and $G_1/(G_1 \cap G_2)$ with $G(3, 0, n)/(p \cup G_1 \cup G_2)$, respectively. Finally, let φ be the number of edges linking $G_1/(G_1 \cap G_2)$ and $G_2/(G_1 \cap G_2)$. We then have by the fact that $G(3, 0, n)$ is a command graph and by the previously made assumptions:

(17) $\deg(G_1 \cap G_2) \geqslant 4 \Rightarrow \alpha + \beta + \gamma \geqslant 3$

(18) $\deg(G_1) \leqslant 5 \Rightarrow \alpha + \gamma + \epsilon + \varphi \leqslant 3$

(19) $\deg(G_2) \leqslant 5 \Rightarrow \beta + \gamma + \delta + \varphi \leqslant 3$

(20) $\deg (G_1/(G_1 \cap G_2)) \geqslant 4 \Rightarrow \beta + \epsilon + \varphi \geqslant 3$

(21) $\deg (G_2/(G_1 \cap G_2)) \geqslant 4 \Rightarrow \alpha + \delta + \varphi \geqslant 3 .$

Forming the sums of (18) and (19) and of (20) and (21) we get

$$\alpha + \beta + 2\gamma + \delta + \epsilon + 2\varphi \leqslant 6$$

and

$$\alpha + \beta + \delta + \epsilon + 2\varphi \geqslant 6$$

which implies $\gamma = 0$, and hence that

$$\alpha + \beta + \delta + \epsilon + 2\varphi = 6 .$$

Therefore

(22) $\beta + \epsilon + \varphi = \alpha + \delta + \varphi = 3$

and

(23) $\alpha + \epsilon + \varphi = \beta + \delta + \varphi = 3$

so that

(24) $\alpha = \beta$ and $\delta = \epsilon .$

From (17) and (24) $\alpha + \beta \geqslant 3$ hence

$$\alpha = \beta \geqslant 2$$

but from (18) $\alpha + \epsilon + \varphi \leqslant 3$, so that $\epsilon + \varphi = 1$ or 0. If $\epsilon = 0$, then $\deg (G_1 \cup G_2 \cup p) = 1$ which is impossible, therefore $\epsilon = 1$ and $\alpha = \beta = 2$, but then $\deg (G_1 \cup G_2 \cup p) = 3$ which is again impossible. Hence, if x and one other vertex belongs to a subgraph G_1 such that $\deg (G_1) \leqslant 5$, then x does not belong to any other such subgraph.

The preceding arguments have shown that it is possible to choose the reduced edges at any vertex, p, of a minimal command graph $G(3, 0, n)$ in such a way that the removal of p results in a minimal command graph $G(3, 0, n-1)$.∎

4. CONCLUSIONS AND OPEN PROBLEMS

The constructive characterization of the minimal $G(1, 0, n)$, $G(2, 0, n)$ and $G(3, 0, n)$ graphs, given in Theorems 5, 6 and 7 respectively, raises the question of whether a comparable characterization might be possible for all minimal $G(k, 0, n)$ graphs. We have been able to partially answer this question affirmatively.

Conjecture 1*. A $G(2m - 1, 0, n)$ graph, $n \geqslant 2$, is minimal if and only if it is an extension of some minimal $G(2m - 1, 0, n)$ graph through superimposing the midpoints of m edges to form one new vertex and m new edges.

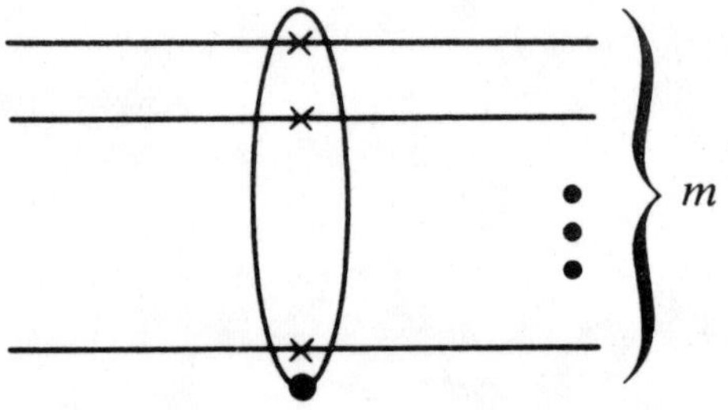

Conjecture 2. A $G(2m, 0, n)$ graph, $n \geqslant 3$, is minimal if and only if it is an extension of some minimal $G(2m, 0, n - 2)$ graph through superimposing the midpoints of two sets of m edges and joining the resultint pair of vertices with a new edge to form two new vertices and $2m + 1$ new edges.

*Conjecture 1 has recently been proven for graphs without parallel edges by L. Lovász.

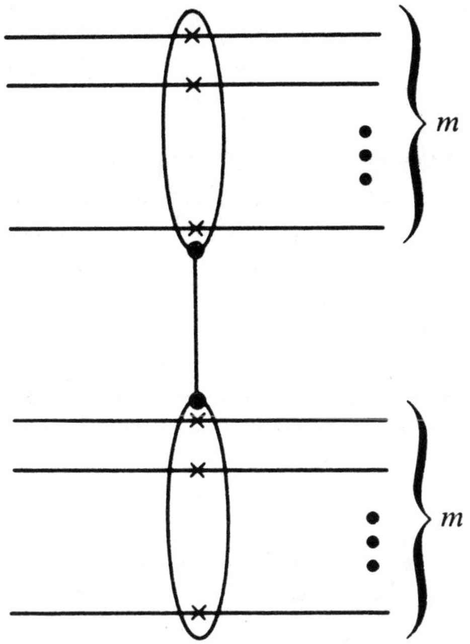

Clearly the constructions described above add the correct number of edges and vertices to a minimal $G(2m-1,0,n)$ or $G(2m,0,n)$ graph for the extension to be a minimal graph satisfying Theorem 4 and include the particular cases covered by Theorems 5, 6 and 7. By arguments similar to those used in the proof of Theorems 6 and 7 it is easy to show that in fact these extensions of minimal command graphs are minimal. Unfortunately, the question of whether every minimal command graph is the result of such an extension of some lower order minimal command graph appears quite difficult.

5. ACKNOWLEDGEMENT

The author wishes to express his appreciation to Mrs. Sharon L. Daniel for developing efficient algorithms for determining the isomorphs of command graphs which made the tabulations given in the Appendix possible.

6. APPENDIX

The following figures tabulate a representative of each equivalence class, under permutation of vertices, of minimal $G(2, 0, n)$ graphs for $1 \leqslant n \leqslant 9$. These representatives were computed by making all possible extensions of minimal $G(2, 0, n-2)$ graphs using Theorem 6, and then computing equivalences by computer processing of the resulting incidence matrices.

$2 - a$

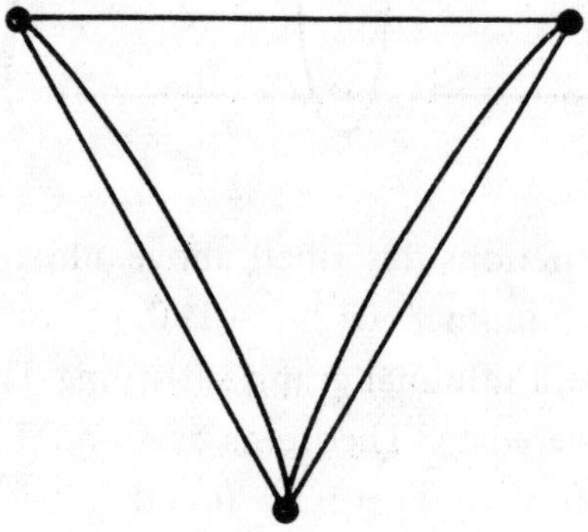

$3 - a$

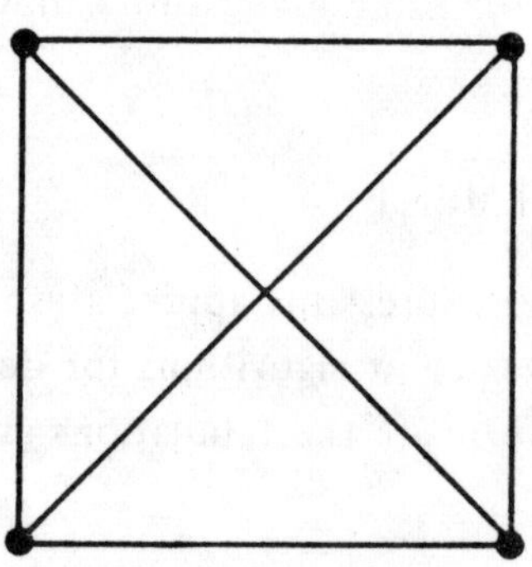

$4 - a$

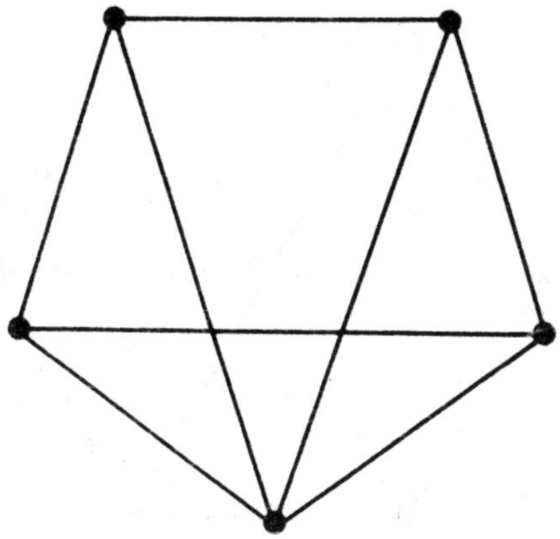

5 – *a*

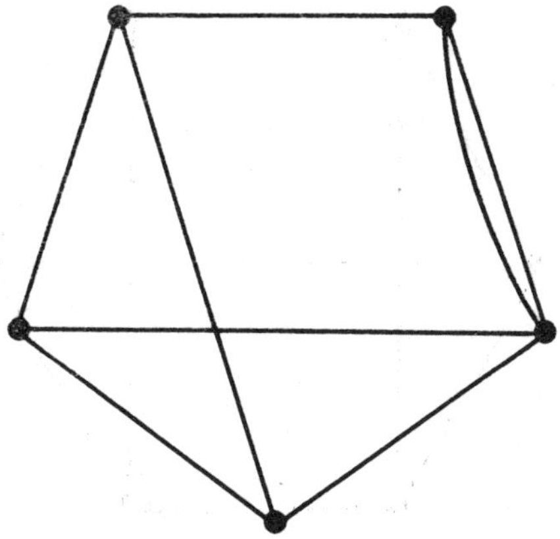

5 – *b*

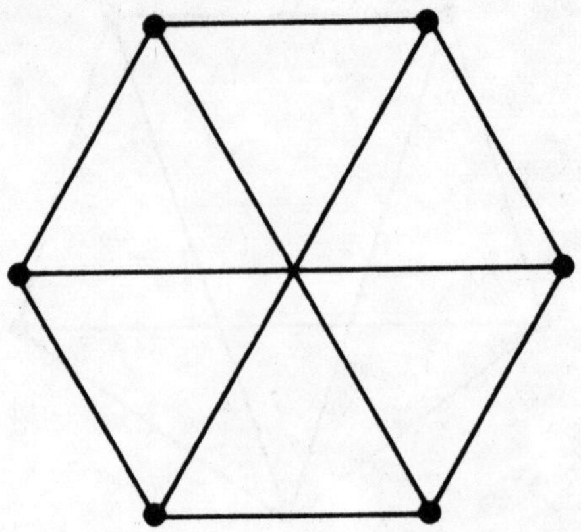

6 – *a*

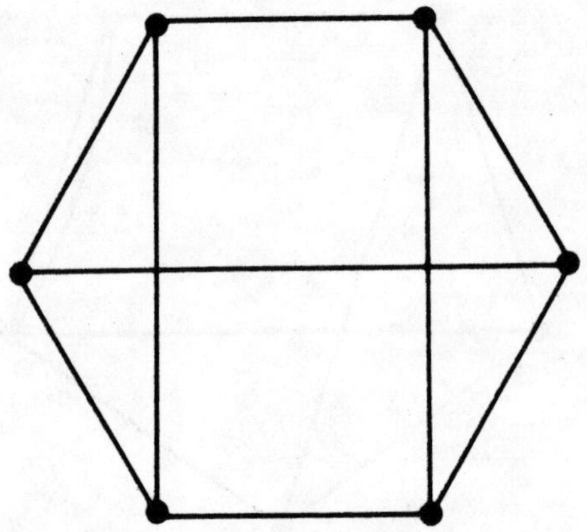

6 – *b*

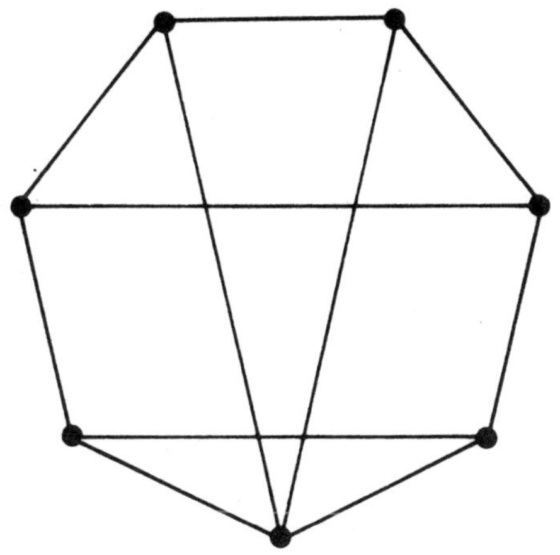

7 − *a*

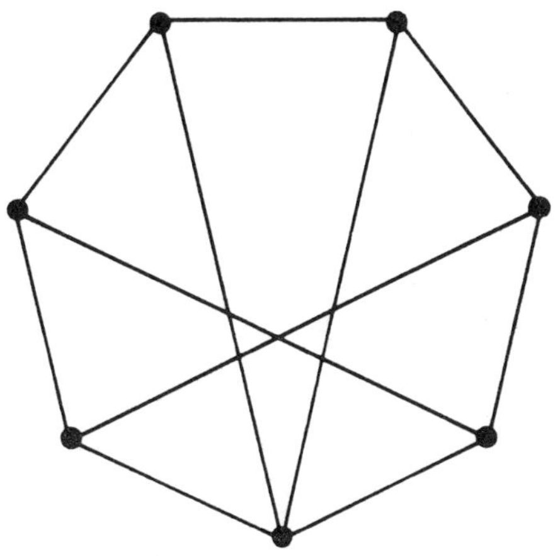

7 − *b*

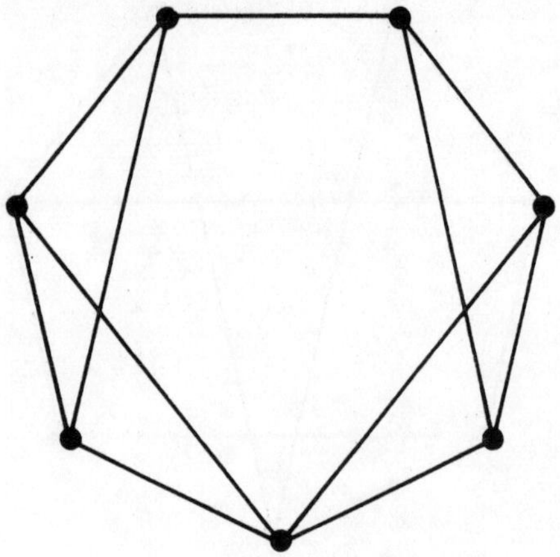

7 – c

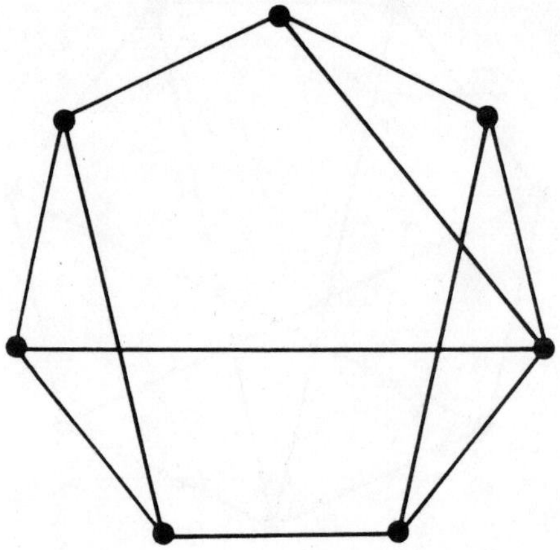

7 – d

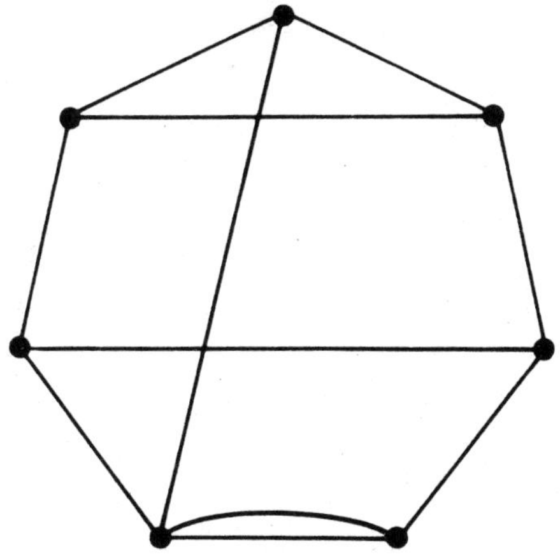

$7-e$

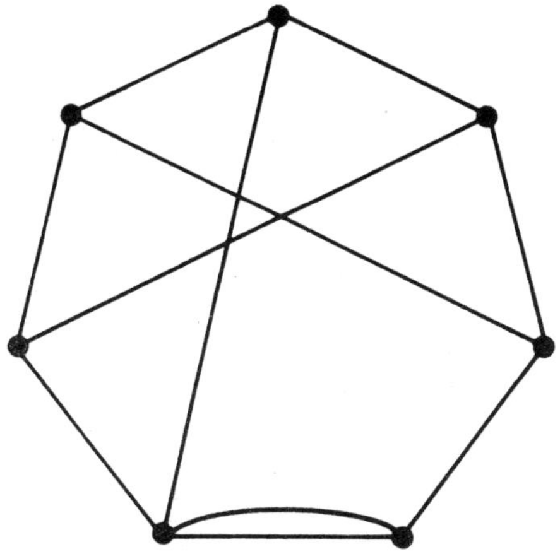

$7-f$

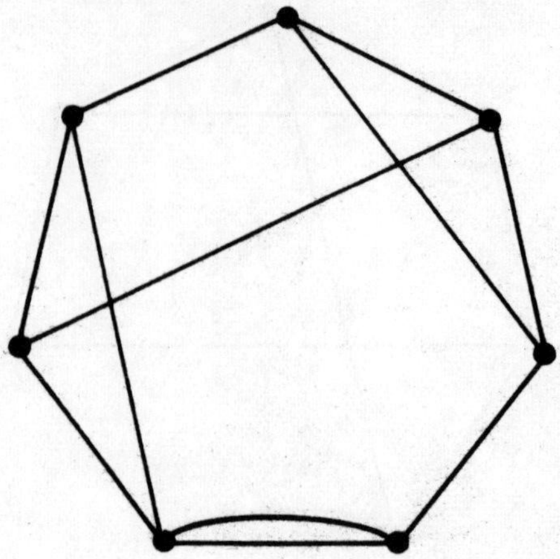

7 – *g*

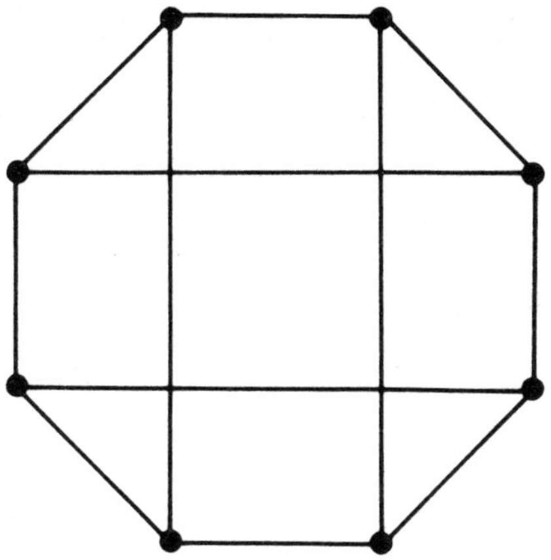

$8-a$

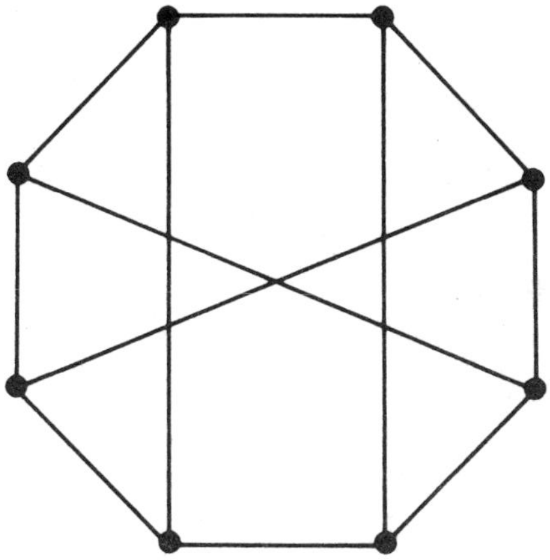

$8-b$

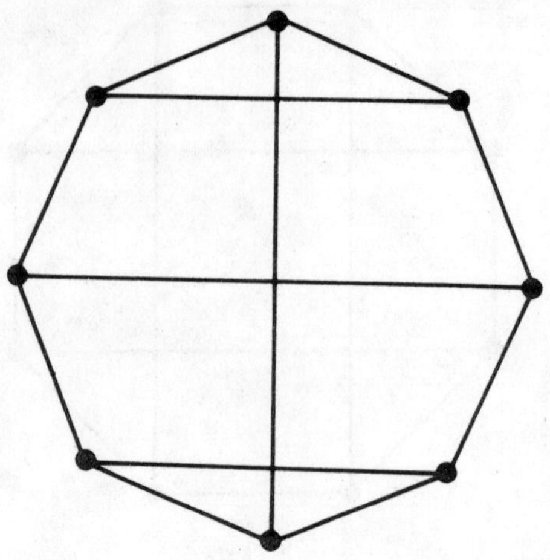

$8-c$

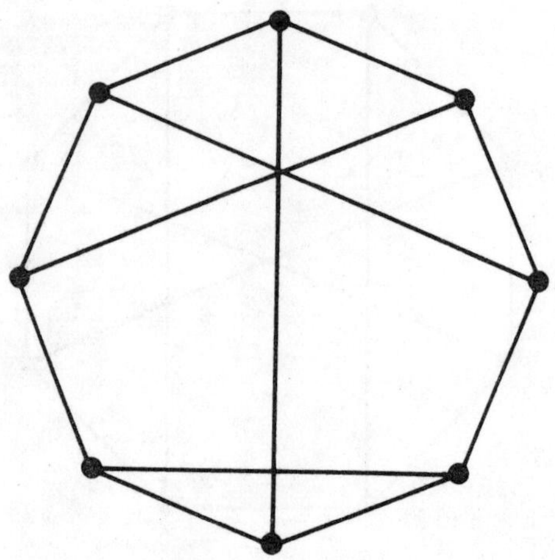

$8-d$

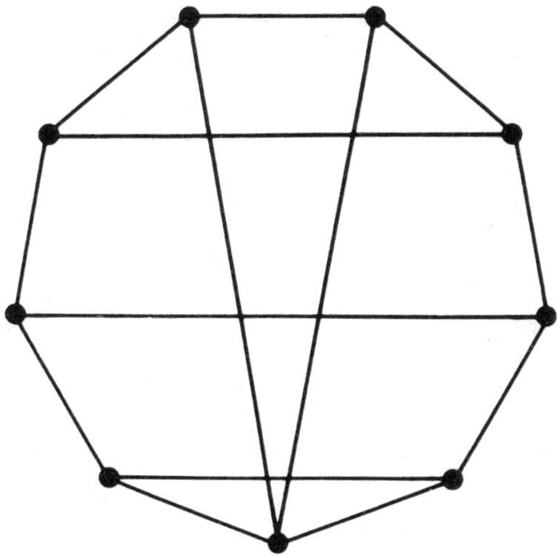

9 – *a*

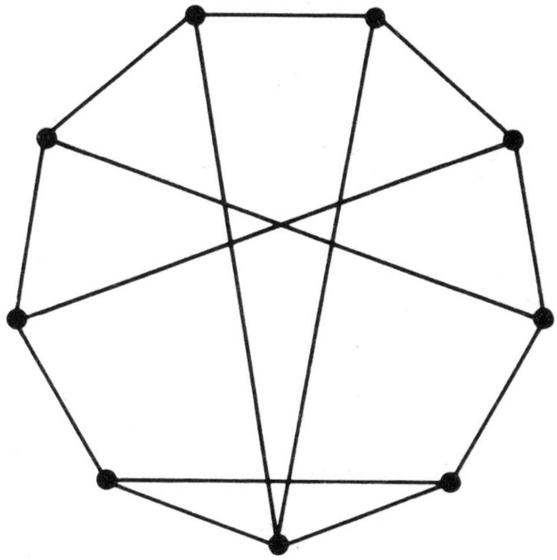

9 – *b*

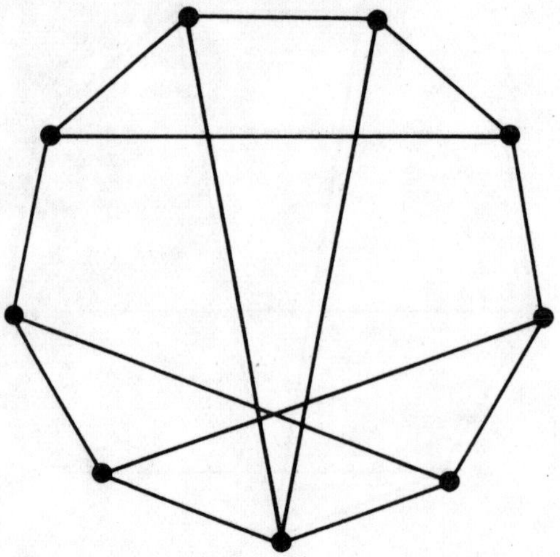

9 – *c*

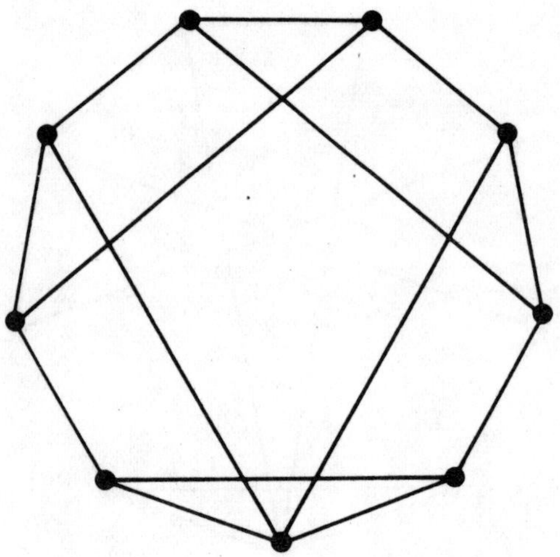

9 – *d*

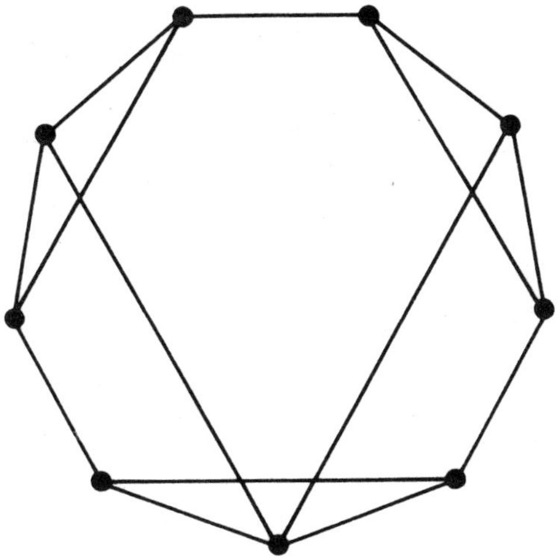

9 − *e*

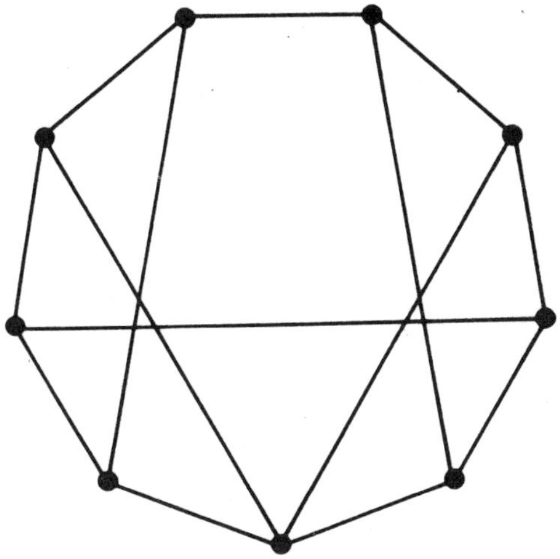

9 − *f*

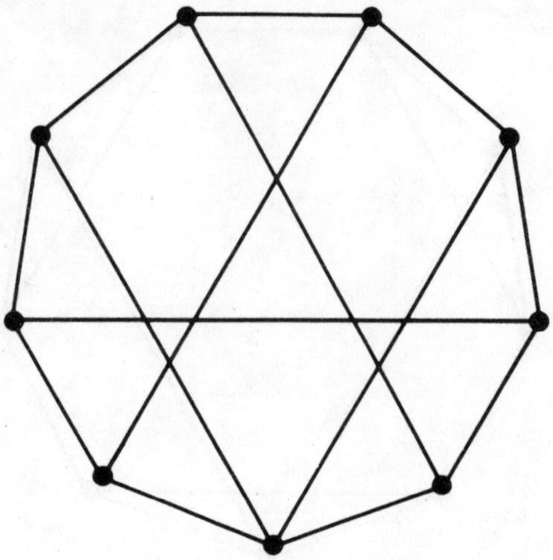

9 – *g*

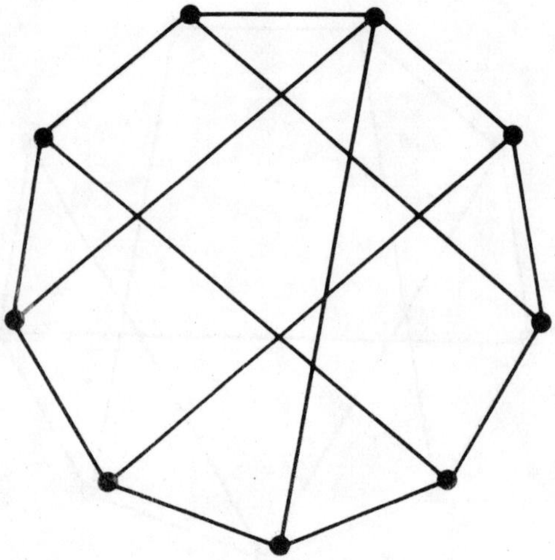

9 – *h*

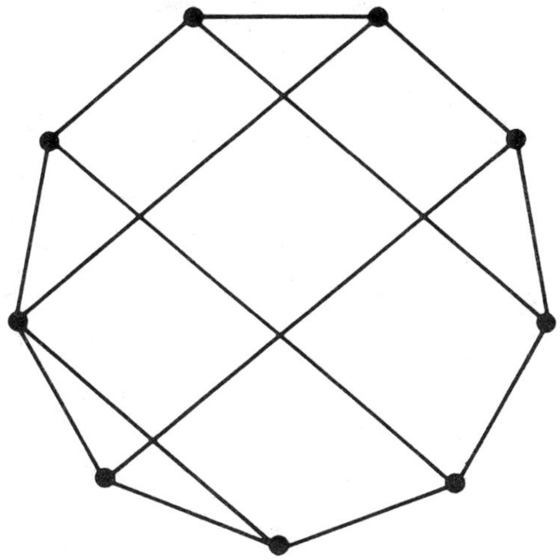

9 – i

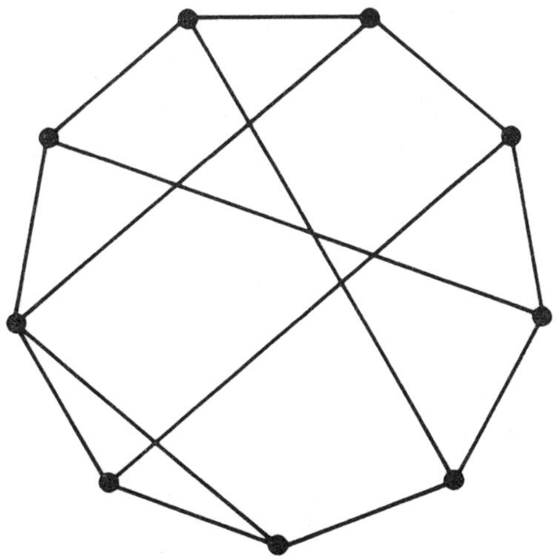

9 – j

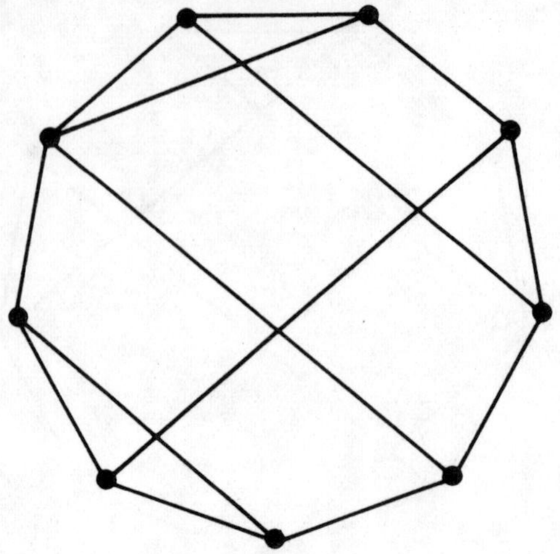

9 – *k*

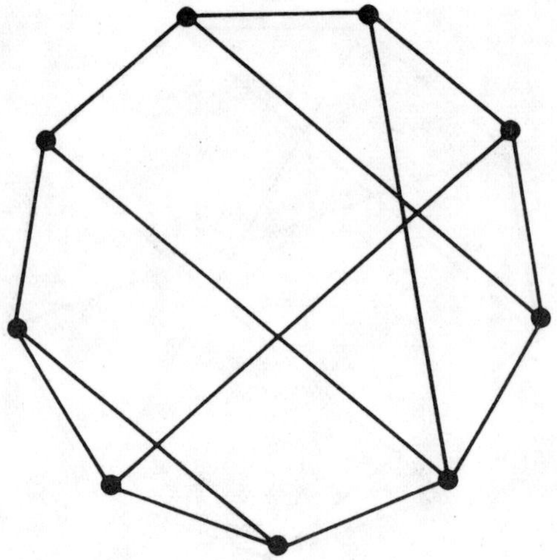

9 – *l*

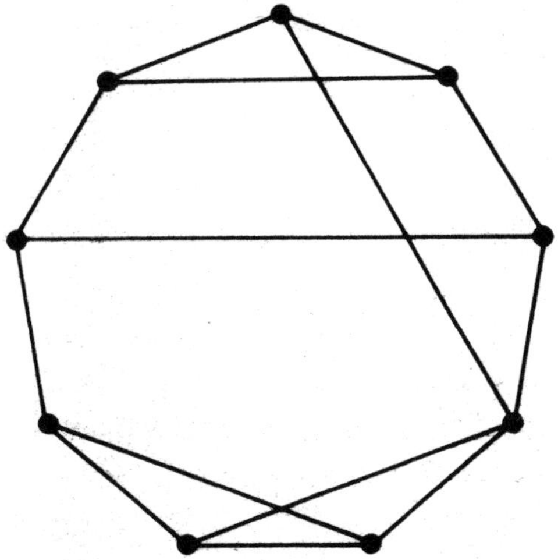

9 – *m*

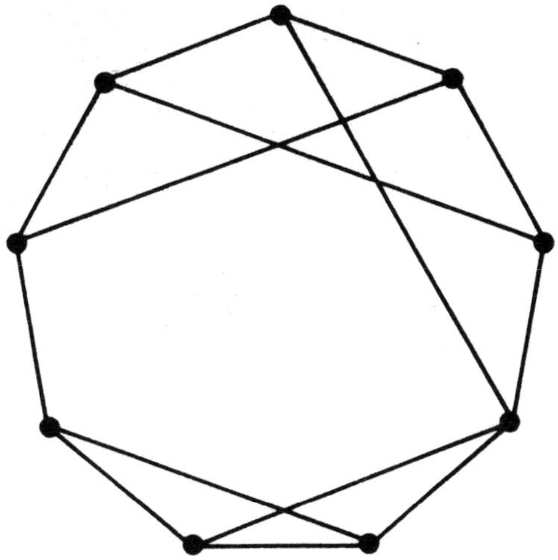

9 – *n*

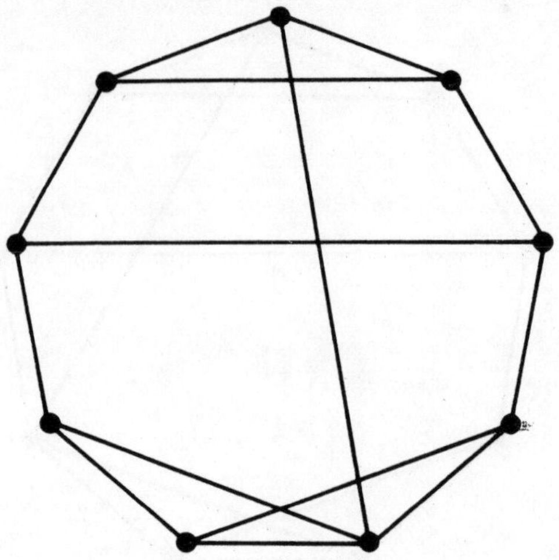

9 − *o*

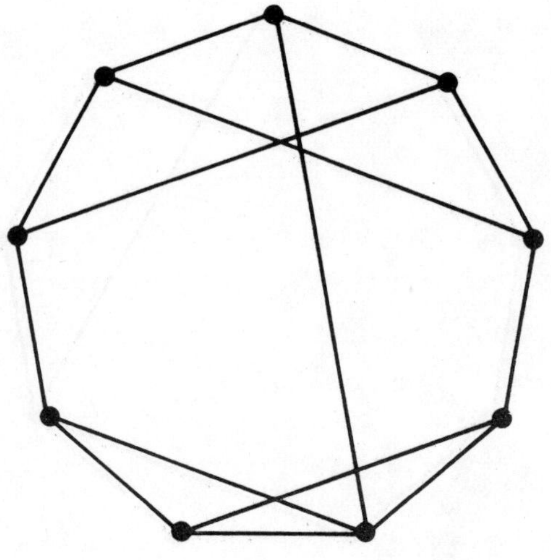

9 − *p*

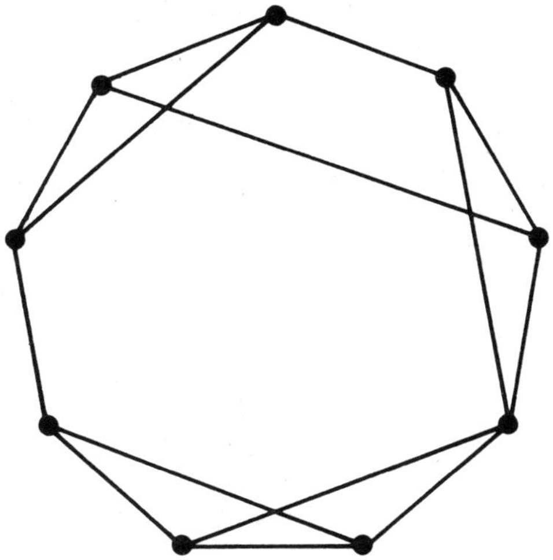

9 – *q*

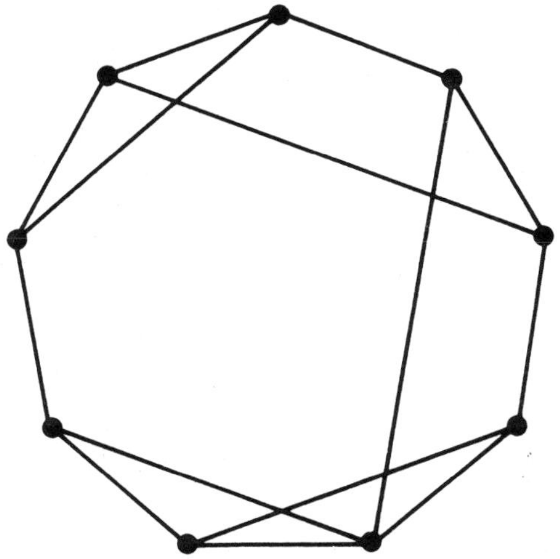

9 – *r*

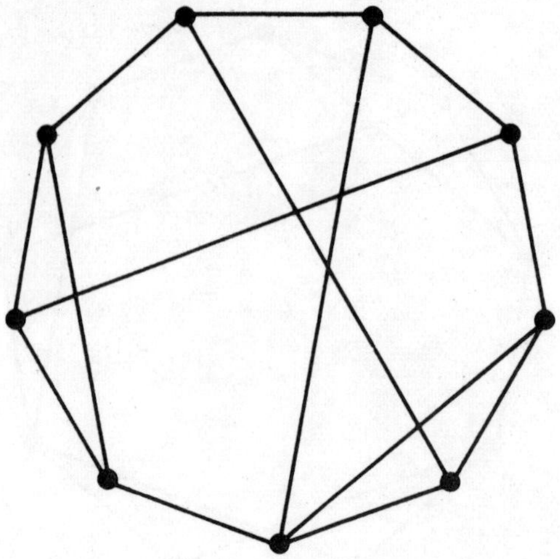

9 – s

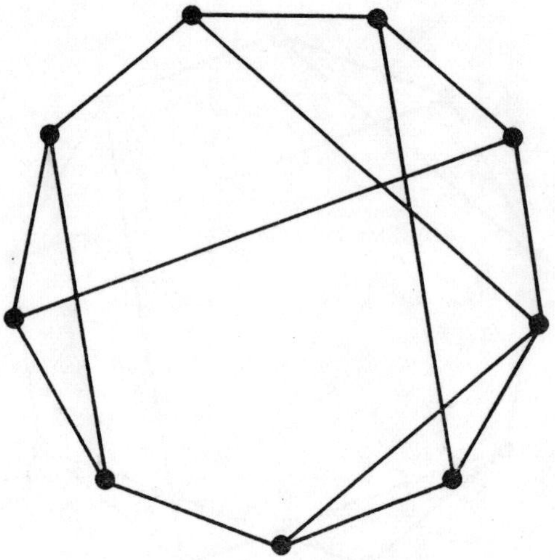

9 – t

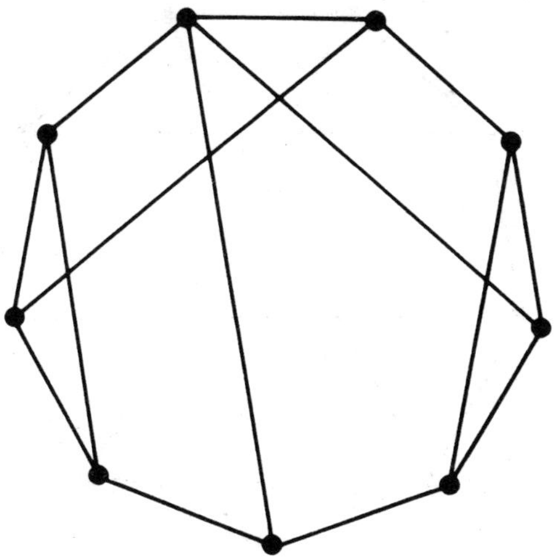

$9-u$

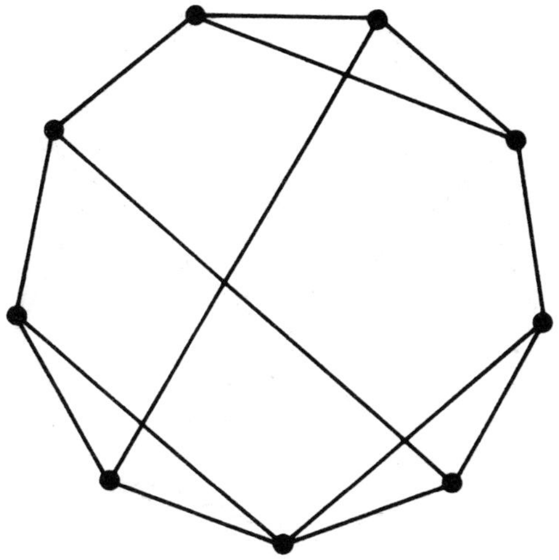

$9-v$

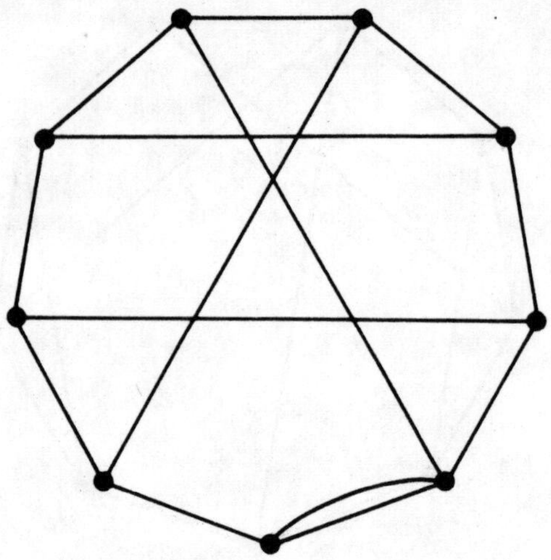

9 – *w*

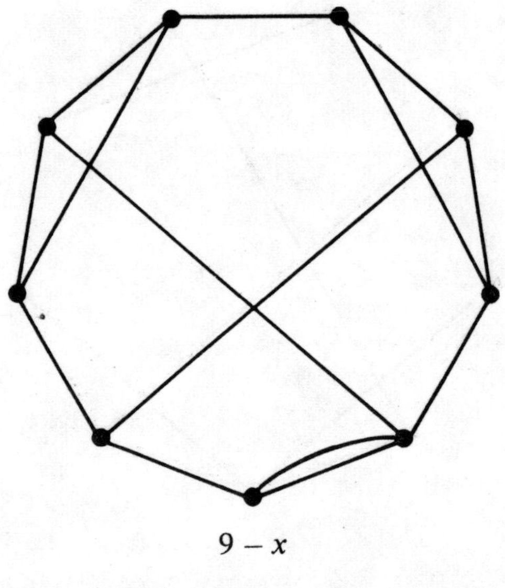

9 – *x*

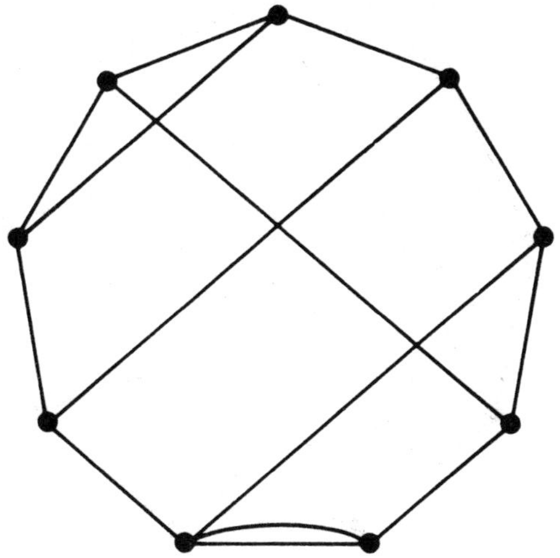

9 − *y*

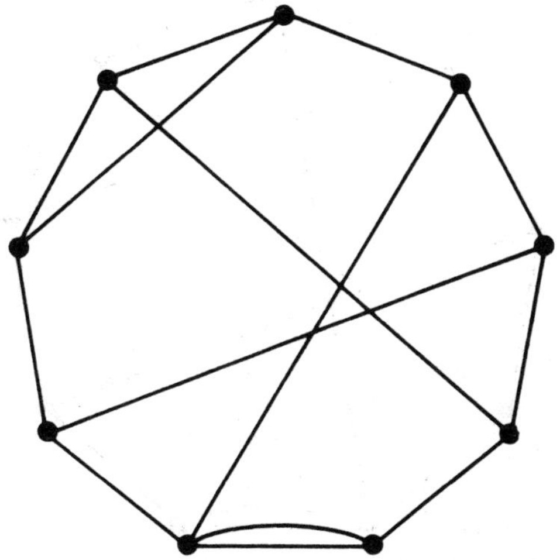

9 − *z*

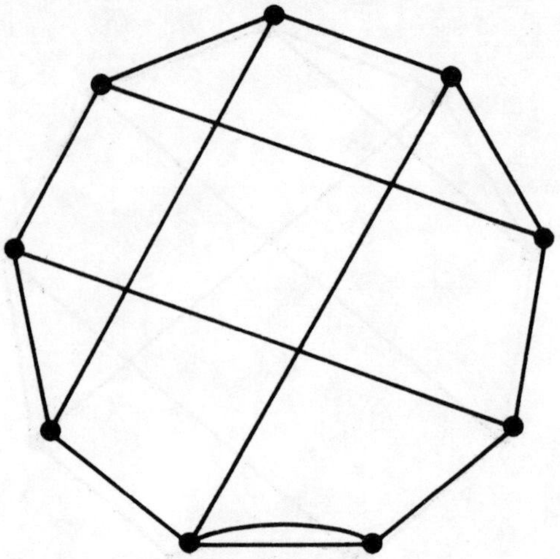

9 – *aa*

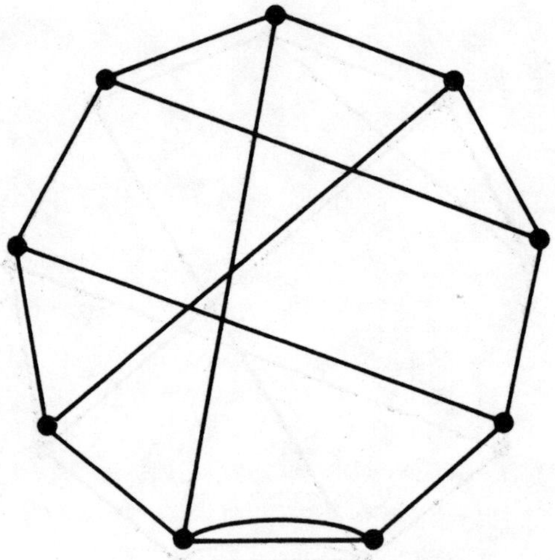

9 – *bb*

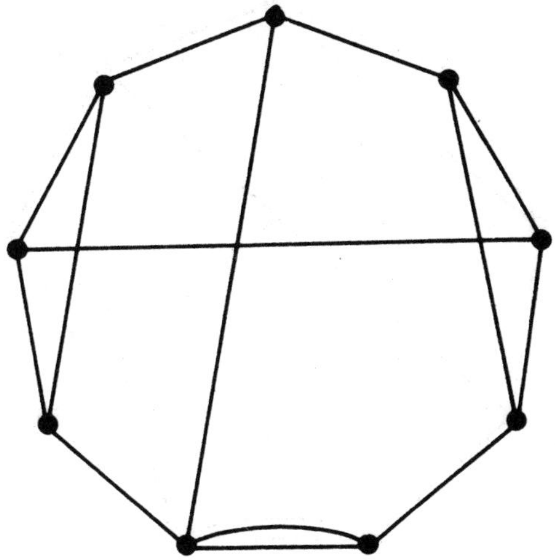

9 – *cc*

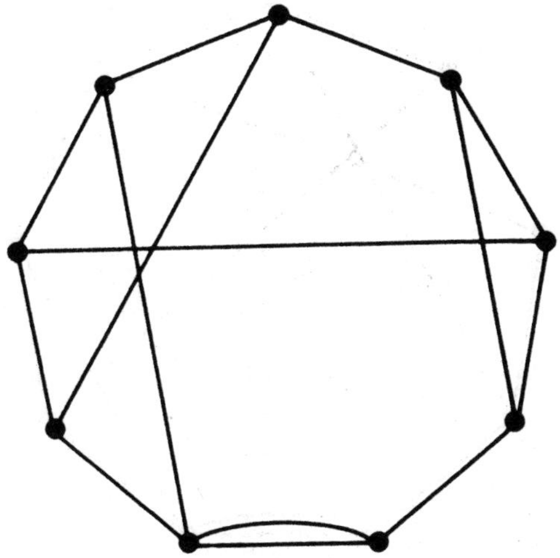

9 – *dd*

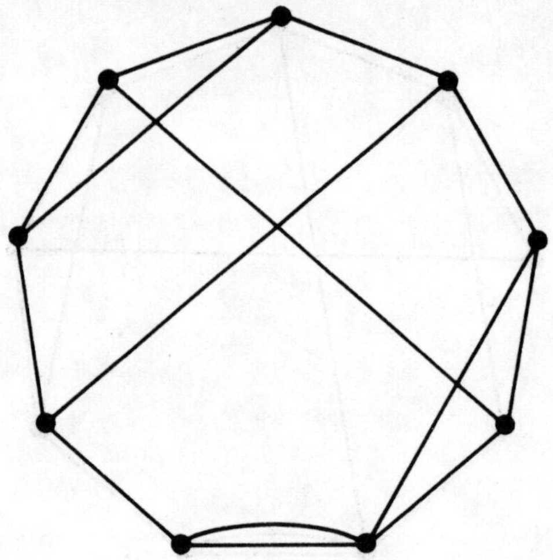

9 – *ee*

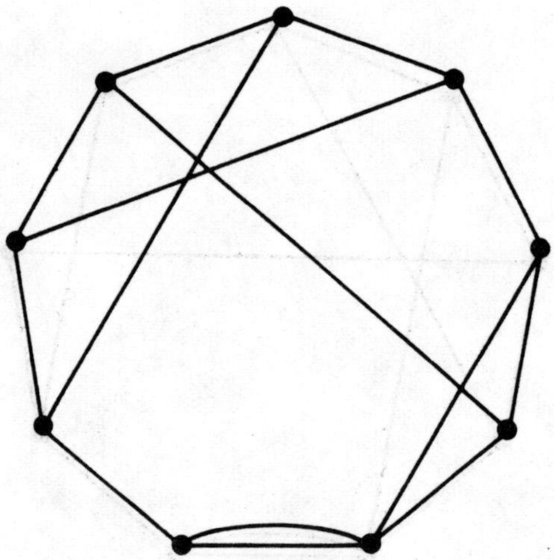

9 – *ff*

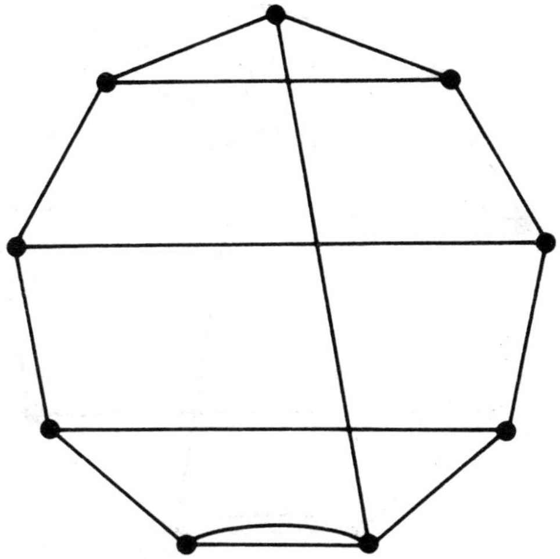

9 – *gg*

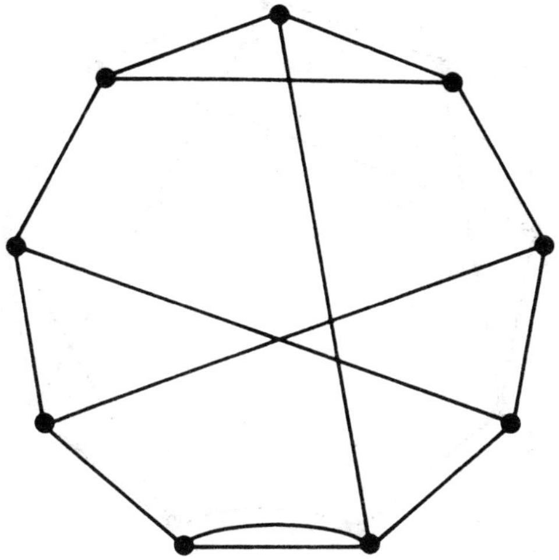

9 – *hh*

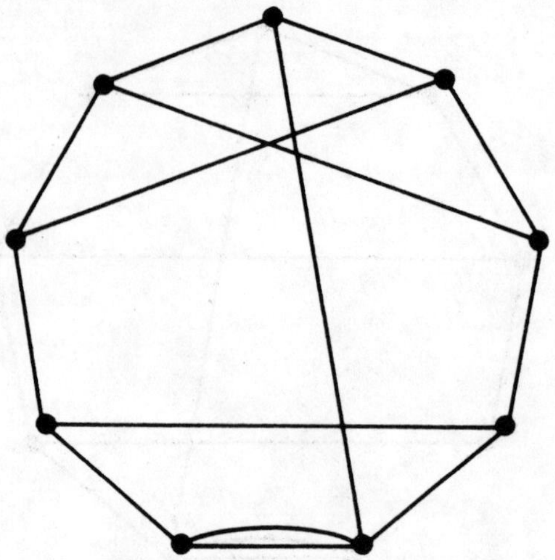

9 – *ii*

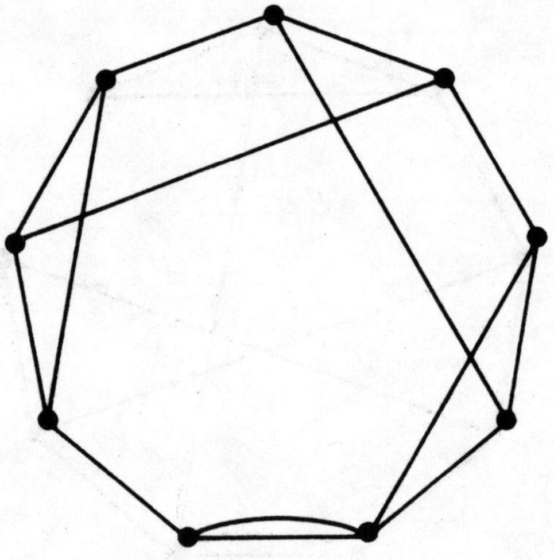

9 – *jj*

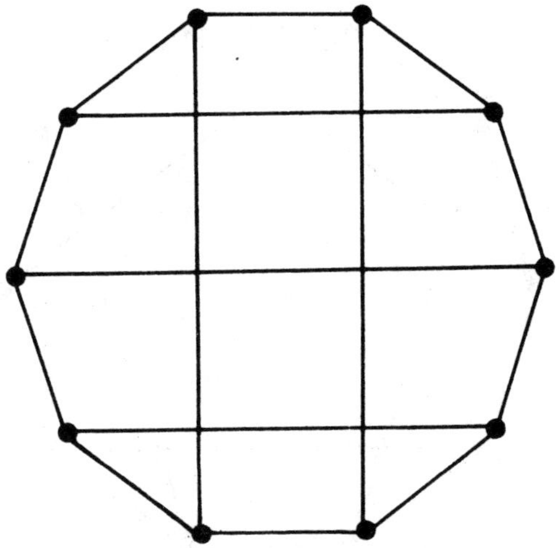

10 – *a*

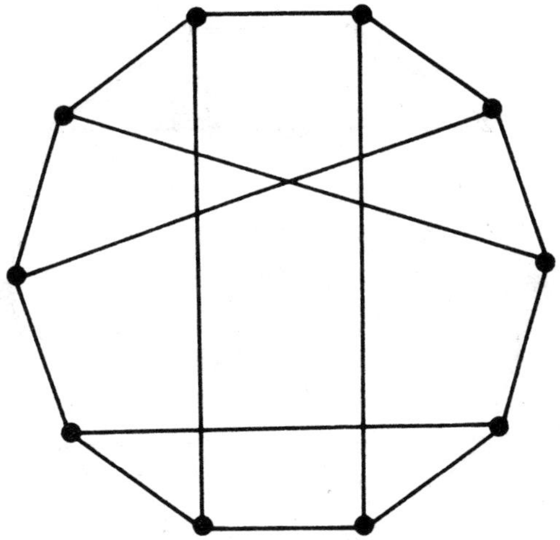

10 – *b*

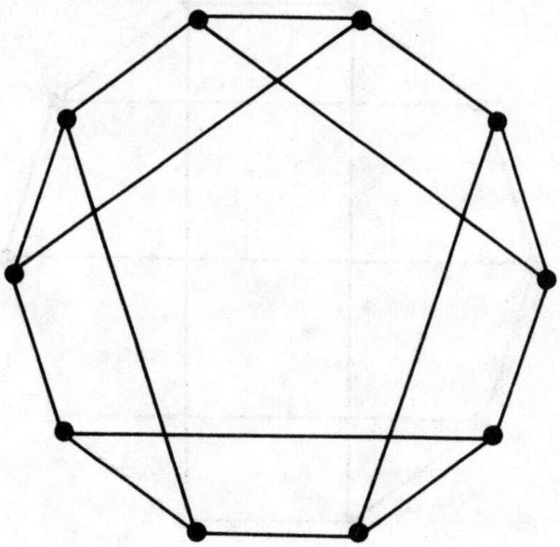

10 – *c*

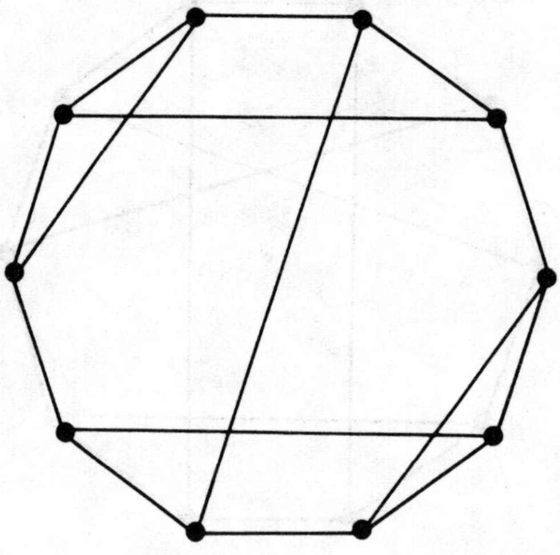

10 – *d*

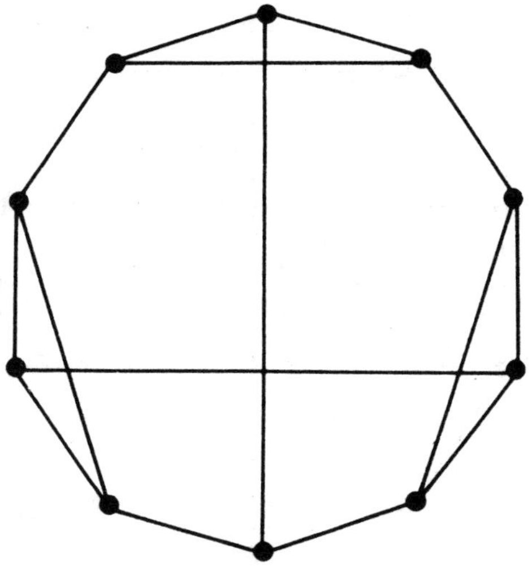

10 − *e*

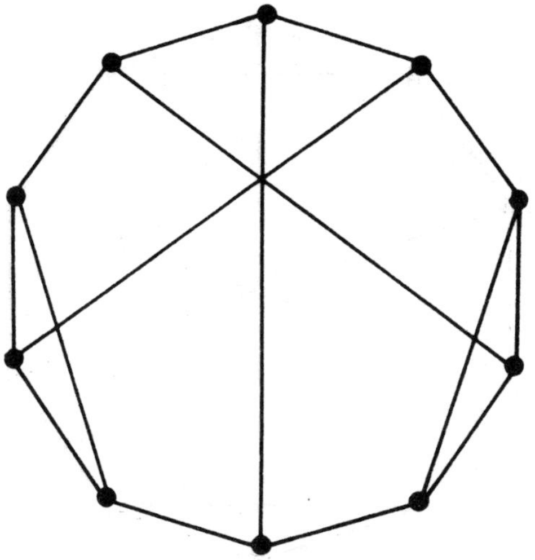

10 − *f*

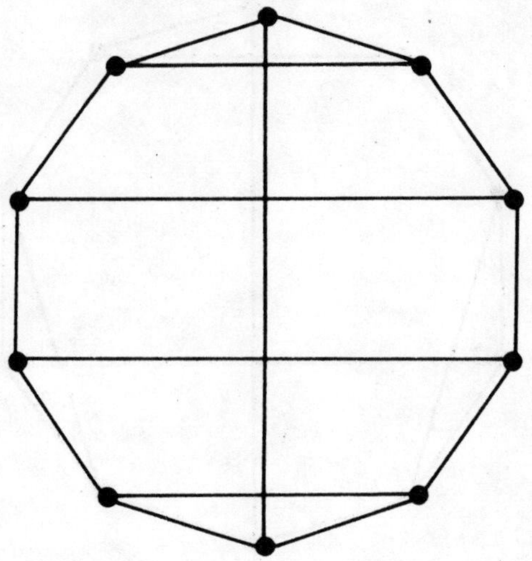

10 – *g*

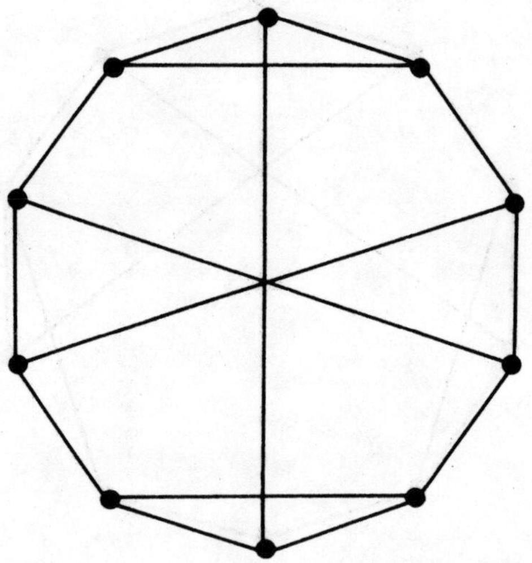

10 – *h*

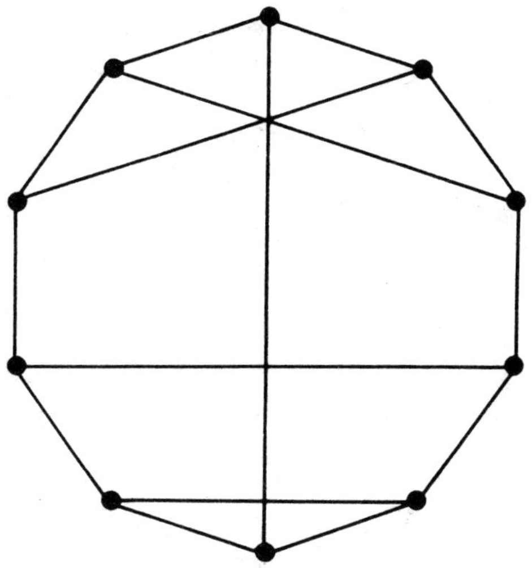

10 − *i*

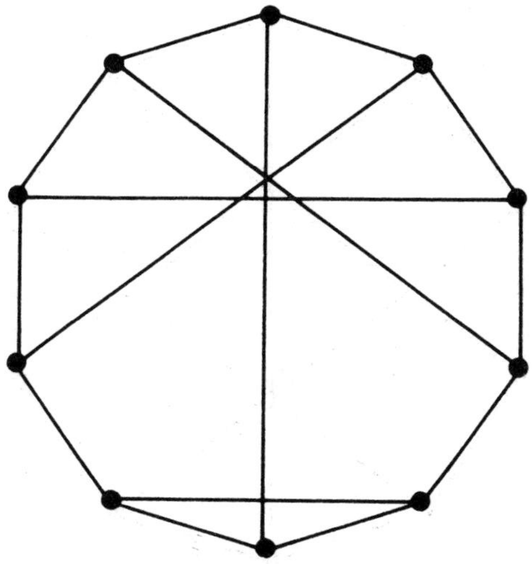

10 − *j*

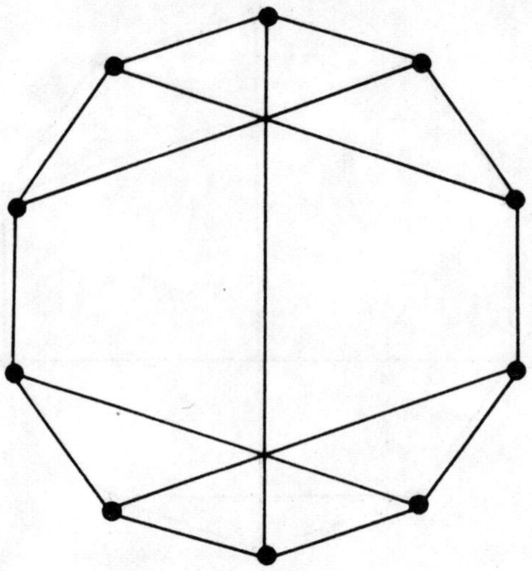

10 – k

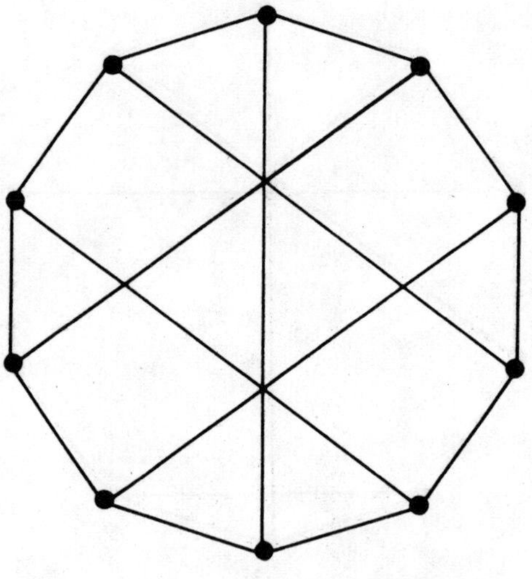

10 – l

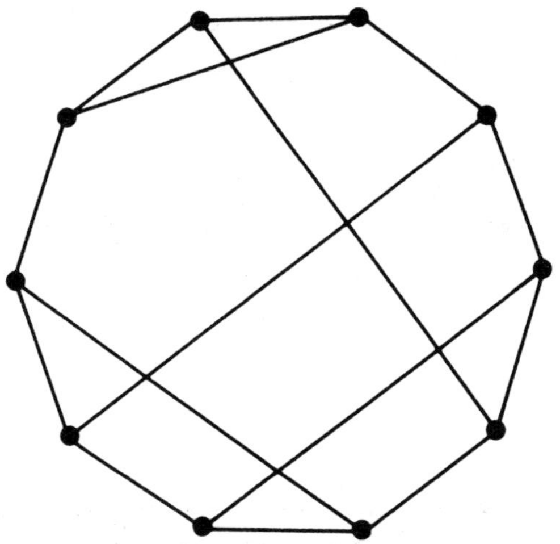

10 − *m*

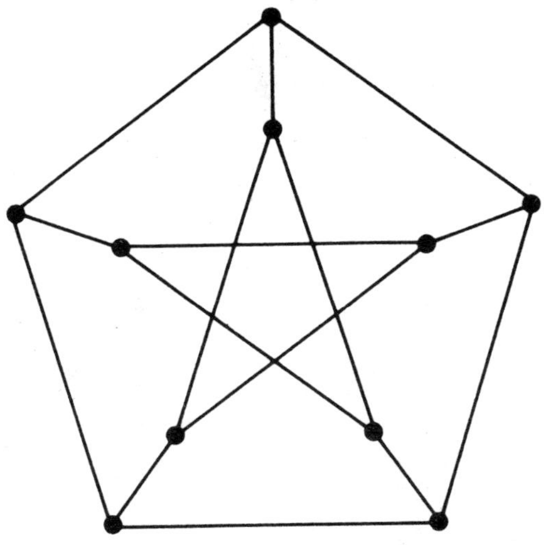

10 − *n*

The following figures tabulate a representative of each equivalence class, under permutation of vertices, of minimal $G(3, 0, n)$ graphs for $1 \leqslant n \leqslant 5$.

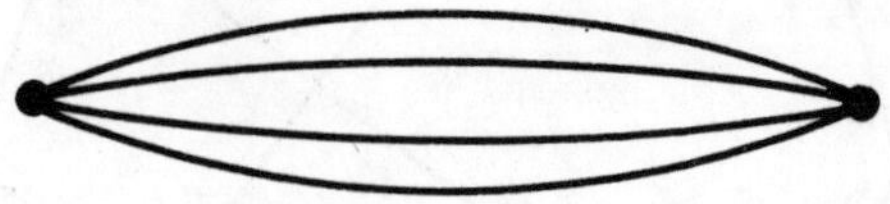

$2 - a$

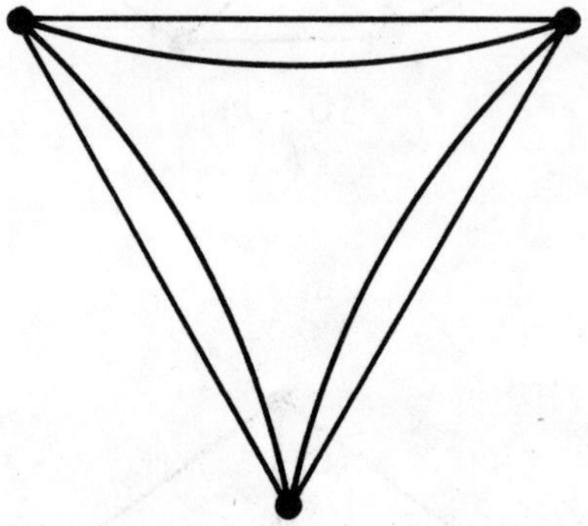

$3 - a$

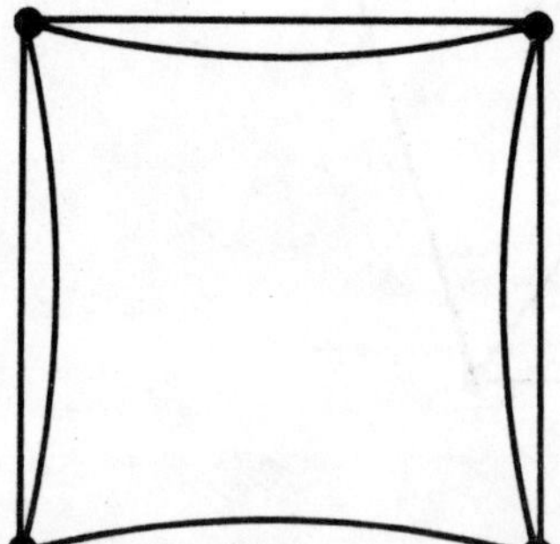

$4 - a$

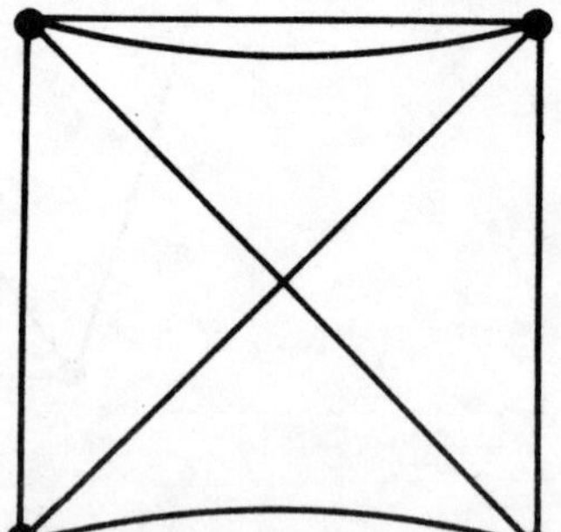

$4 - b$

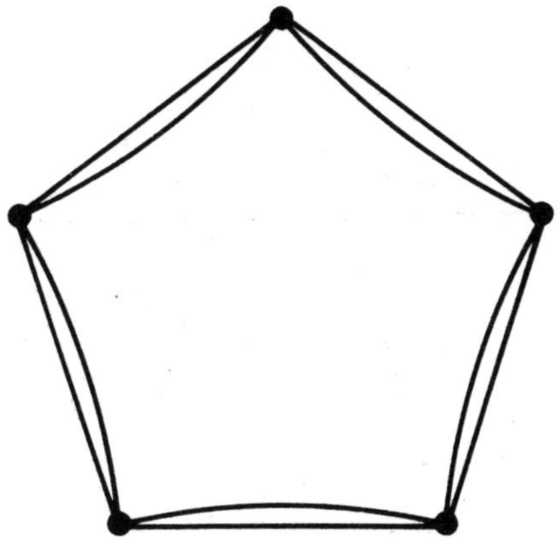

5 – *a*

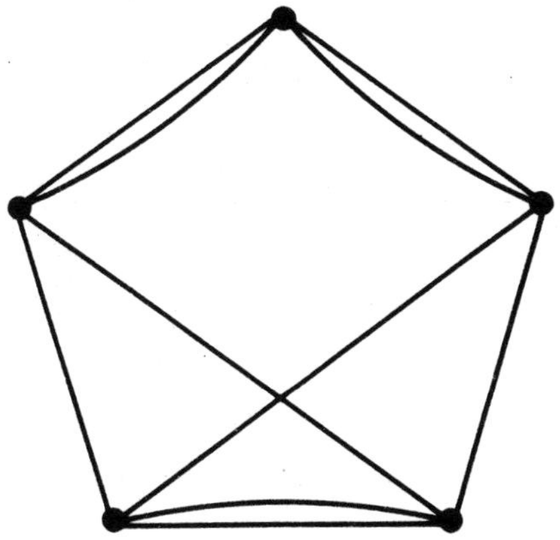

5 – *b*

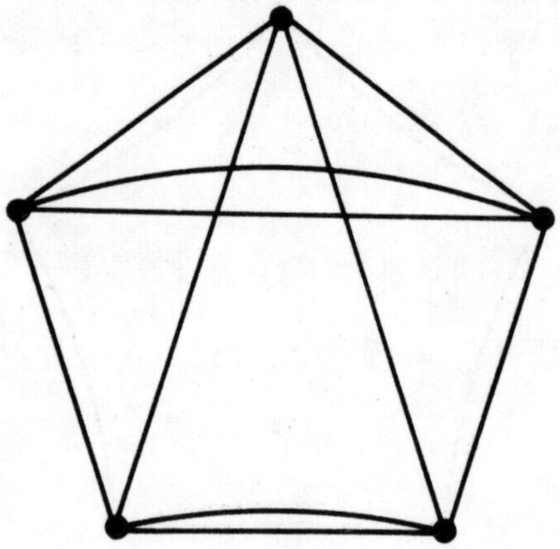

5 – *c*

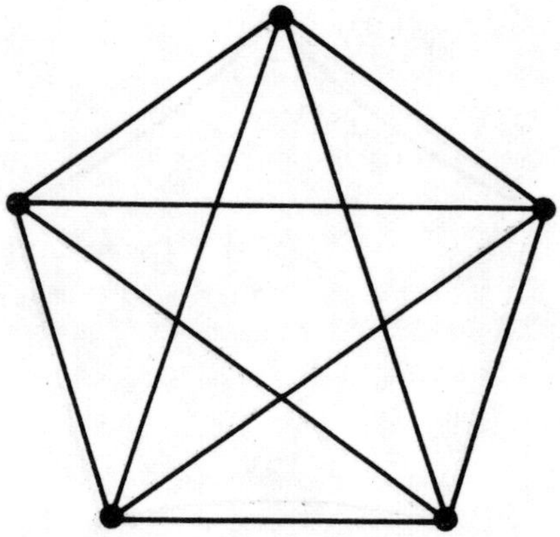

5 – *d*

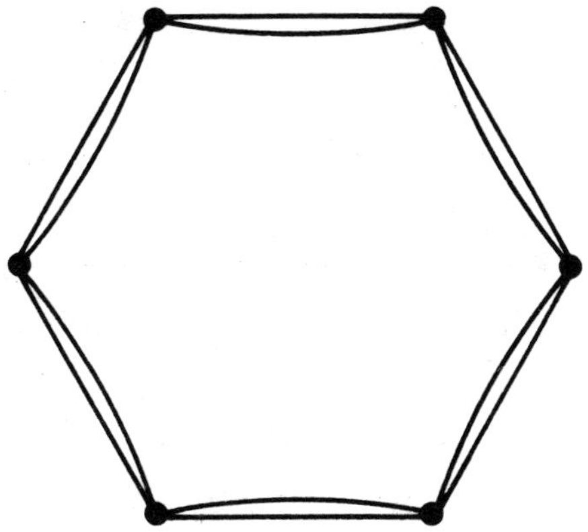

6 – *a*

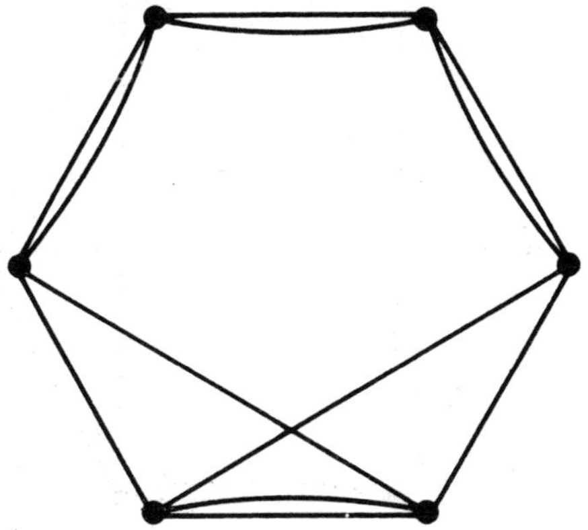

6 – *b*

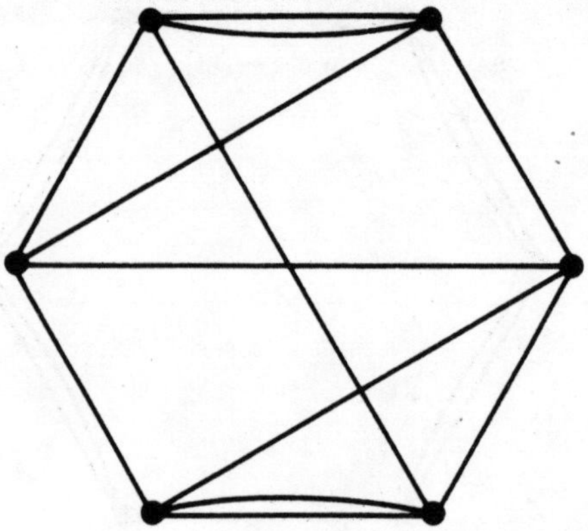

6 – *c*

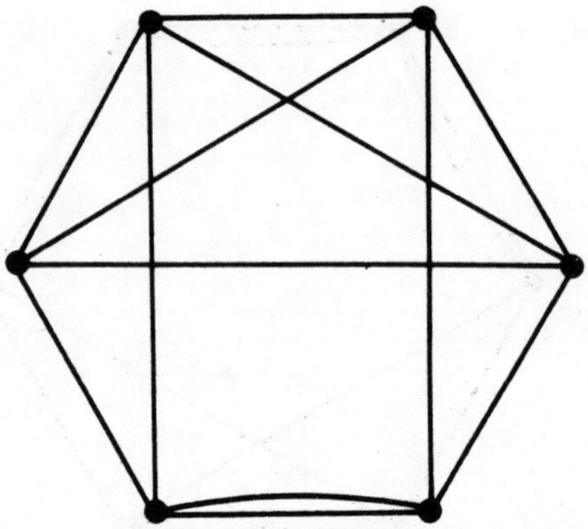

6 – *d*

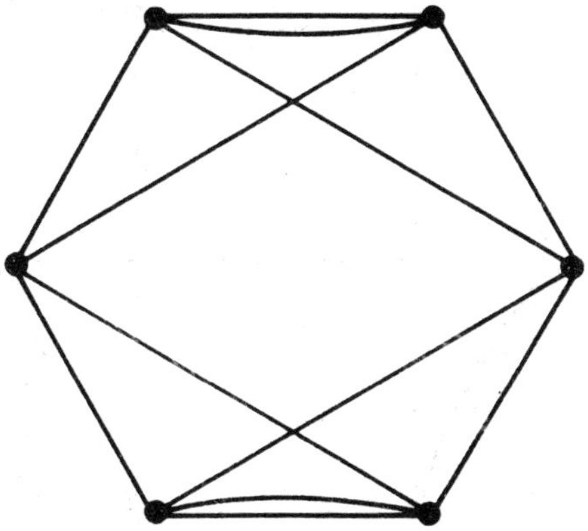

6 – e

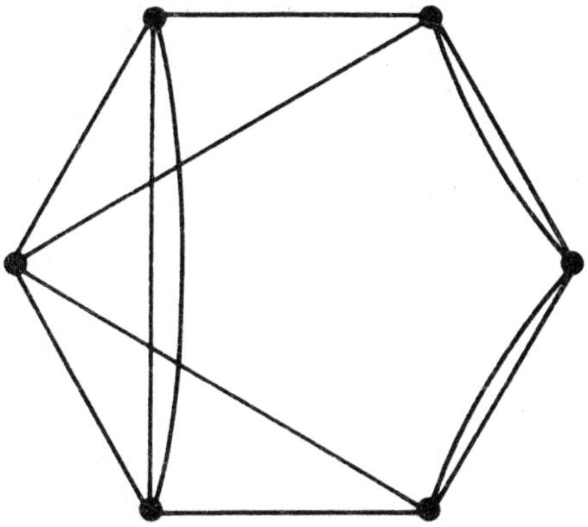

6 – f

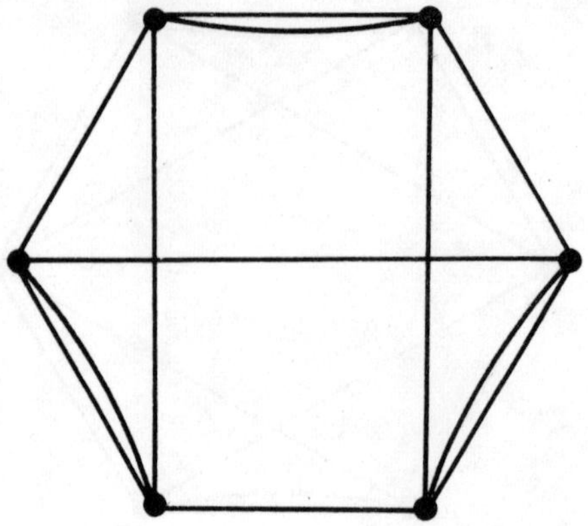

6 – *g*

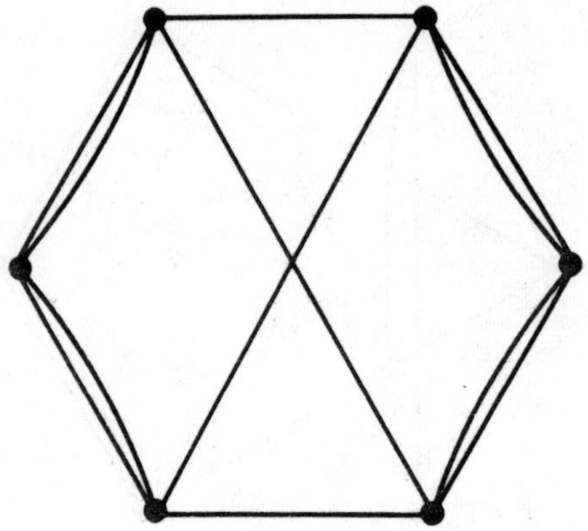

6 – *h*

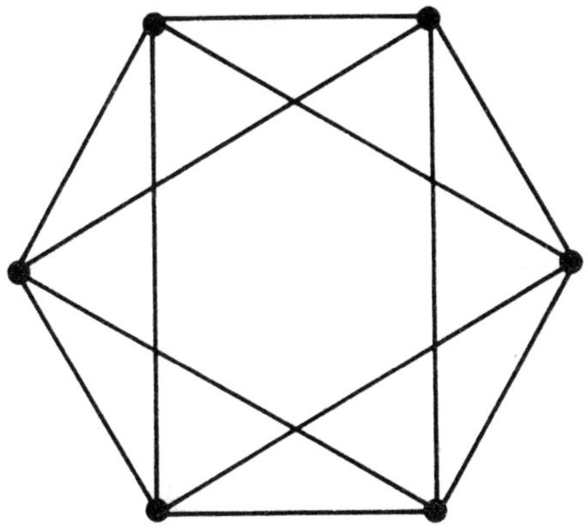

$6-i$

The following figures tabulate a representative of each equivalence class for the minimal $G(4, 0, n)$ graphs, $1 \leqslant n \leqslant 4$.

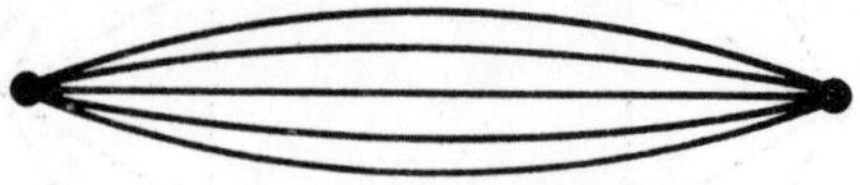

$2 - a$

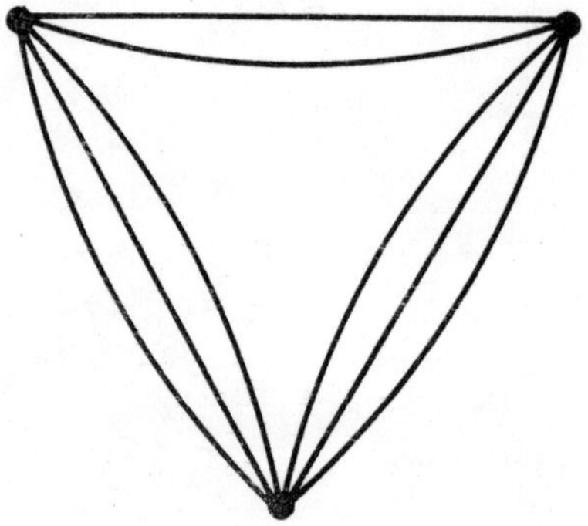

$3 - a$

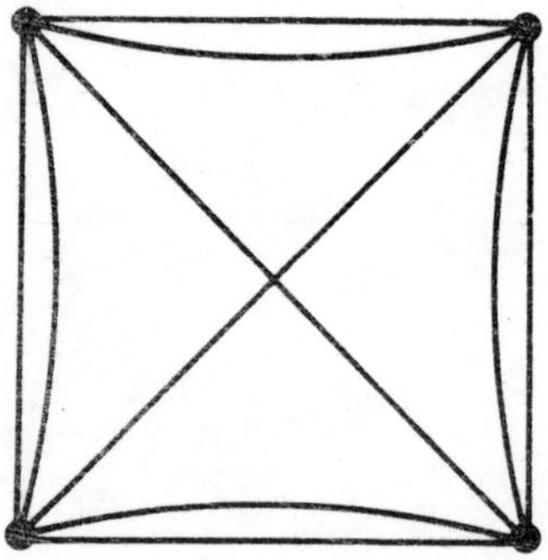

$4 - a$

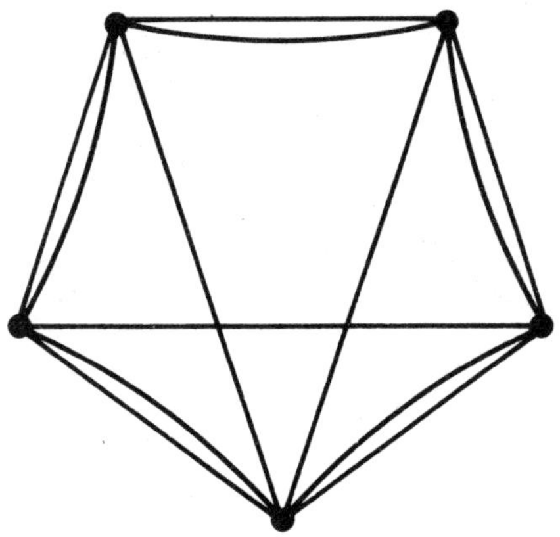

5 − *a*

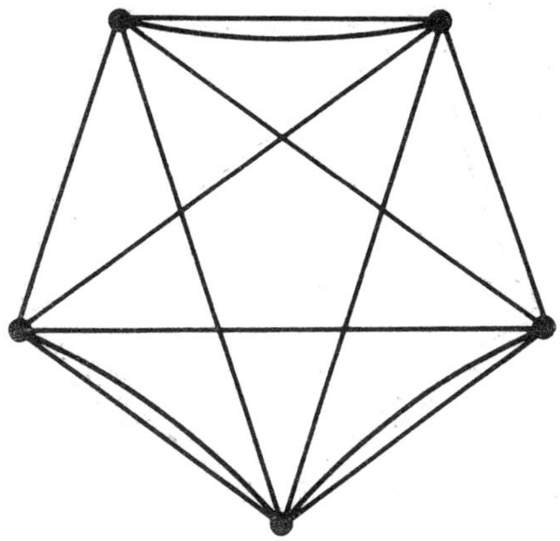

5 − *b*

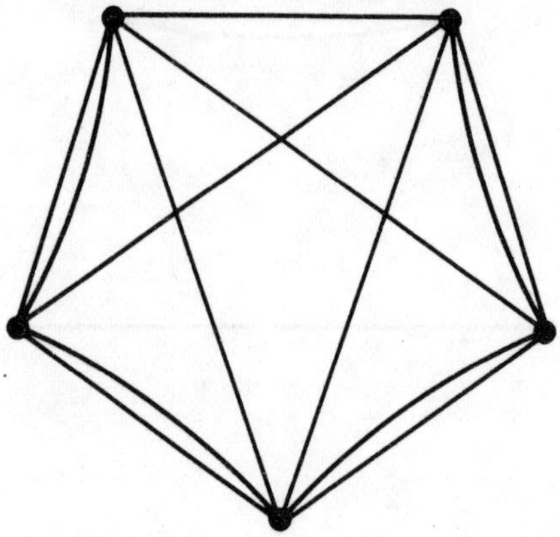

5 – *c*

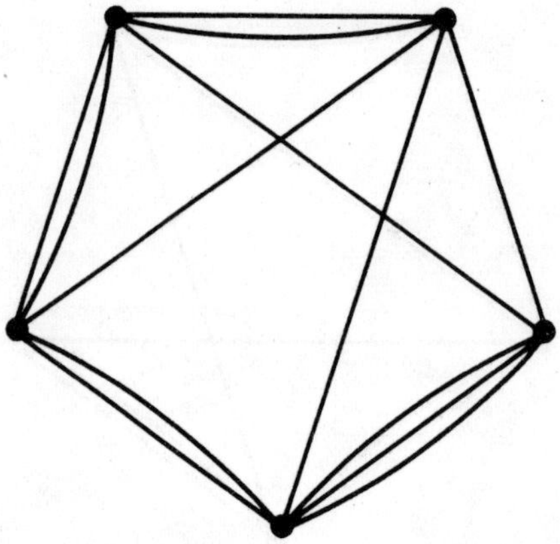

5 – *d*

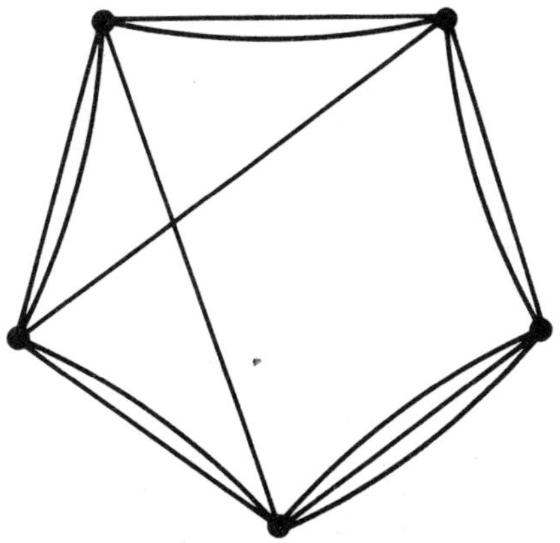

5 − *e*

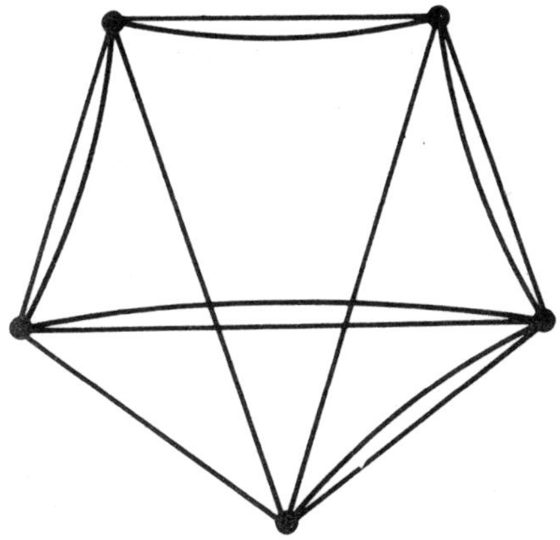

5 − *f*

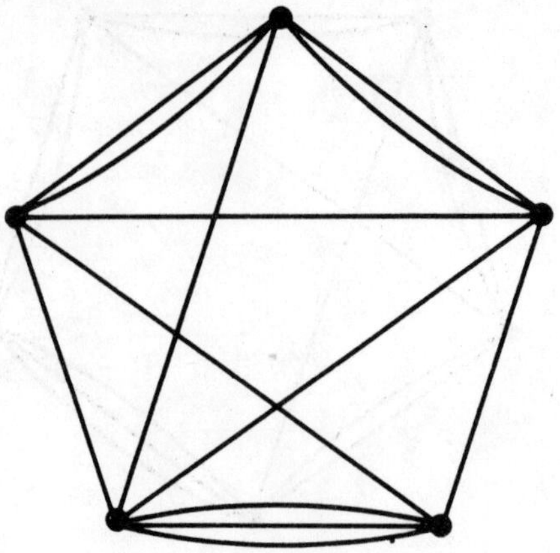

5 – *g*

REFERENCES

[1] G. Chartrand, A graph-theoretic approach to a communications problem, *SIAM Jour. Appl. Math.,* 14 (1966), 778-781.

[2] F.T. Boesch – I.T. Frisch, On the smallest disconnecting set in a graph, *IEEE Trans. on Circuit Theory CT*, 15 (1968), 286-288.

[3] S.L. Hakimi, An algorithm for construction of the least vulnerable communication network on the graph with the maximum connectivity, *IEEE Trans. on Circuit Theory CT,* 16 (1969), 229-230.

[4] F.T. Boesch – R.E. Thomas, On graphs of invulnerable communication nets, *IEEE Trans. on Circuit Theory CT,* 17 (1970), 183-192.

[5] F.T. Boesch – A.P. Felzer, On the minimum m degree vulnerability criterion, *IEEE Trans. on Circuit Theory CT,* 18 (1971), 224-228.

[6] F.T. Boesch – A.P. Felzer, On the invulnerability of the regular complete k-partite graphs, *SIAM Jour. Appl. Math.,* 20 (1971), 176-182.

[7] F.T. Boesch – A.P. Felzer, A general class of invulnerable graphs, *Networks,* 2 (1972), 261-283.

[8] F.T. Boesch, Lower bounds on the vulnerability of a graph, *Networks,* 2 (1972), 329-340.

COLLOQUIA MATHEMATICA SOCIETATIS JÁNOS BOLYAI
10. INFINITE AND FINITE SETS, KESZTHELY (HUNGARY), 1973.

ON GRAPHS UNIQUELY PARTITIONABLE INTO n-DEGENERATE SUBGRAPHS

J.M.S. SIMÕES-PEREIRA

1. INTRODUCTION

One of the present main trends in Graph Theory is the attempt to find always more general settings which may lead to the generalization of known theorems and uncover relationships among results which had previously seemed unrelated. This paper is also an attempt to present some known results on uniquely colorable graphs within the much more general setting given by the theory of k-degenerate graphs, developed by Lick and White [1].

According to Lick and White [1], we say a graph G is n-degenerate if and only if the minimum degree δ of any induced subgraph H of G is at most equal to n. Obviously, if G is n-degenerate, then any induced subgraph of G is also n-degenerate and, for $m > n$, G is also m-degenerate. The minimum value of n for which G is n-degenerate plays an important role in this paper: we shall say that G is strictly n-degenerate to mean that G is n-degenerate but not $(n-1)$-degenerate.

As defined in [1], the point-partition number $\rho_n(G)$ of a graph G is the minimum number of subsets into which the vertex set $V(G)$ must be partitioned so that each subset induces an n-degenerate graph. As a consequence of the definitions we have just given, $\rho_0(G)$ is the chromatic number of G and $\rho_1(G)$ the point arboricity. Since the preceding definition of the point partition numbers generalizes, in a natural way, the definition of these two well-known and largely investigated invariants of Graph Theory, it seems worth trying to extend results known to hold for the chromatic number and/or the point arboricity to the point partition numbers $\rho_n(G)$ where $n \geqslant 2$. We recall that many results concerning the chromatic number have been extended to the point arboricity only recently.

In what follows, no unexpected results will be given. The results we present and their proofs are almost straightforward extensions of properties known to hold for the case in which the point partition number is the chromatic number. Nevertheless, or precisely because of that, they seem to be interesting. They draw our attention to the theory of k-degenerate graphs which, in our opinion, has not yet been investigated as intensively as it deserves to be.

The results to be generalized here concern uniquely m-colorable graphs. The subject has been studied by Cartwright and Harary [2], Harary, Hedetniemi and Robinson [3], and Chartrand and Geller [4] and the basic results of their papers are summarized in Harary [5]. This book is our general reference for graph theoretic concepts not defined here. For brevity, we only recall the definitions and results directly concerned with uniquely colorable graphs.

To begin with, we recall that, denoting by G a labeled graph with chromatic number $\rho_0(G) \leqslant m$, G is said to be uniquely m-colorable if and only if every m-coloring of G induces the same partition of the vertex set $V(G)$. An m-coloring, or a coloring with m colors, means a partition of $V(G)$ into m subsets or color classes, namely, $V_1, \ldots, V_m$, such that no pair of points belonging to the same set V_i, $1 \leqslant i \leqslant m$, are joined by an edge. As an example of a uniquely m-colorable graph, $m = 3$, we have the graph pictured in Fig. 1.

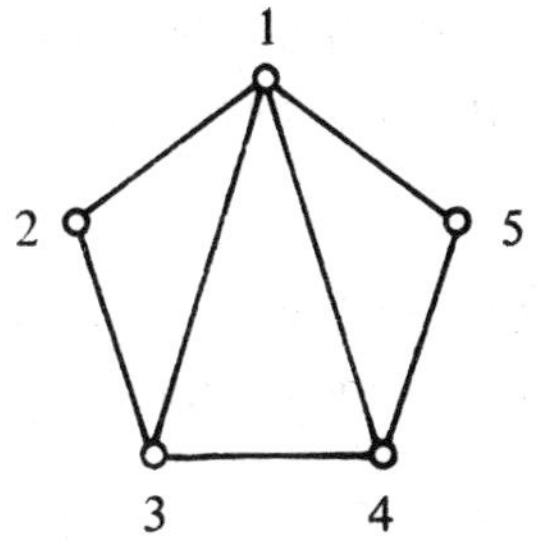

Fig. 1

The following statement is proved in [3].

Theorem A. *If* G *is uniquely* m*-colorable for* $m < |V(G)|$, *then* $\rho_0(G) = m$.

The trivial case in which $m = |V(G)|$ is usually excluded and it is then natural to suppose that a uniquely m-colorable graph G has chromatic number equal to m.

Proofs of the following theorems may be found in [5].

Theorem B. *The minimum degree* δ *of a uniquely* m*-colorable graph is at least* $m-1$.

Theorem C. *Every uniquely* m*-colorable graph is* $(m-1)$*-connected.*

Theorem D. *In any* m*-coloring of a uniquely* m*-colorable graph, the subgraph induced by the union of any* t *color classes,* $2 \leqslant t \leqslant m$, *is* $(t-1)$*-connected.*

Theorems C and D are stronger forms of the following ones which, as pointed out in [5], had been proved first.

Theorem C′. *Every uniquely* m*-colorable graph,* $m \geqslant 2$, *is connected.*

Theorem D′. *In the* m*-coloring of a uniquely* m*-colorable graph, the subgraph induced by the union of any two color classes is connected.*

Finally, we quote the following results from [3].

Theorem E. *Let* G *be uniquely* m*-colorable and let* $V_1, \ldots, V_m$

be its color classes. If G_w *is obtained from* G *by adding a point* w *and lines joining* w *to at least one point in each class* $V_2, \ldots, V_m$, *then* G_w *is uniquely* m*-colorable.*

Theorem F. *Every graph with chromatic number equal to* m *is an induced subgraph of a uniquely* m*-colorable graph.*

Theorem G. *An* m*-critical graph, i.e. a graph* G *such that* $\rho_0(G) = m$ *but every proper subgraph* H *of* G *has* $\rho_0(H) < m$, *has no cut set inducing a uniquely* $(m-1)$*-colorable subgraph.*

The concept of representing set, borrowed from Transversal Theory, will also be used in this paper. For our purposes, it may be defined simply as follows: Given a family of sets $\{A_i : i \in I\}$, we say B is a representing set of this family if and only if, for every $i \in I$, $B \cap A_i \neq \phi$. A general reference for Transversal Theory is Mirsky [6].

2. PRELIMINARIES

We begin with the basic definitions and a few elementary remarks and examples.

We define an (m, n)-partition of a graph G as a partition of the vertex set $V(G)$ into m subsets $V_1, \ldots, V_m$, called partition classes, such that each subset V_i, $1 \leqslant i \leqslant m$, induces an n-degenerate graph. We may also say that G is partitioned into m n-degenerate subgraphs or is (m, n)-partitioned. A labeled graph G is then said to be uniquely m-partitionable into n-degenerate graphs or, more briefly, uniquely (m, n)-partitionable, if and only if $\rho_n(G) \leqslant m$ and G has only one (m, n)-partition. Uniquely m-colorable graphs are obviously the same as uniquely $(m, 0)$-partitionable graphs. The graph of Fig. 1 is then a uniquely $(3, 0)$-partitionable graph.

Another example is the uniquely $(2, 1)$-partitionable graph G represented in Fig. 2. G is the join of two paths of length 3, $P' = \langle\{1, 3, 5, 7\}\rangle$ and $P'' = \langle\{2, 4, 6, 8\}\rangle$. It is not 1-degenerate. Since P' and P'' are 1-degenerate, $\rho_1(G) = 2$. If G has another $(2, 1)$-partition with partition classes denoted H_1, H_2, then both H_1 and H_2 must have odd and

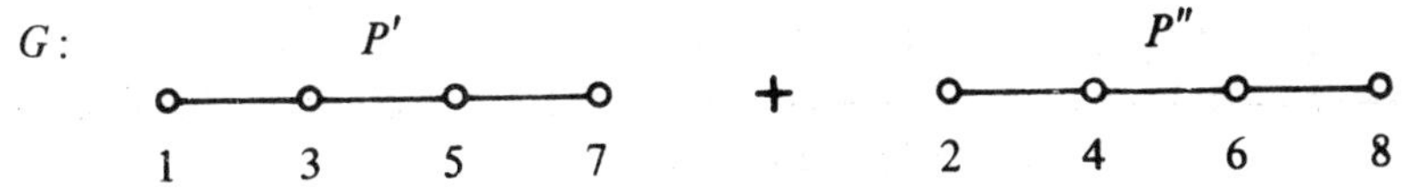

Fig. 2. $G = P' + P''$

even vertices, i.e. H_1 and H_2 are representing sets of the pair $\{V(P'), V(P'')\}$. Without loss of generality, suppose that $|H_1 \cap V(P')| \geqslant 2$. If $|H_1 \cap V(P'')| \geqslant 2$, then $\langle H_1 \rangle$ contains a cycle and it is not 1-degenerate. If $|H_1 \cap V(P'')| < 2$, then $\langle H_2 \rangle$ contains two adjacent vertices in P'' and, since $H_2 \cap V(P') \neq \phi$, $\langle H_2 \rangle$ is not 1-degenerate.

Finally, the join $G = G_1 + G_2$ of the graphs G_1 and G_2, pictured in Fig. 3, is a uniquely (2, 2)-partitionable graph. In fact, if G has another (2, 2)-partition with partition classes denoted H_1, H_2, then H_1 and H_2 are representing sets of the pair of sets $\{V(G_1), V(G_2)\}$. Without loss of generality, let $|H_1 \cap V(G_1)| \geqslant 3$. If $|H_1 \cap V(G_2)| \geqslant 3$, then $\langle H_1 \rangle$ contains a complete bi-partite graph $K_{3,3}$ as subgraph and is not 2-degenerate. If $|H_1 \cap V(G_2)| < 3$, then $\langle H_2 \cap V(G_2) \rangle$ contains a cycle and, since $H_2 \cap V(G_1) \neq \phi$, $\langle H_2 \rangle$ is not 2-degenerate.

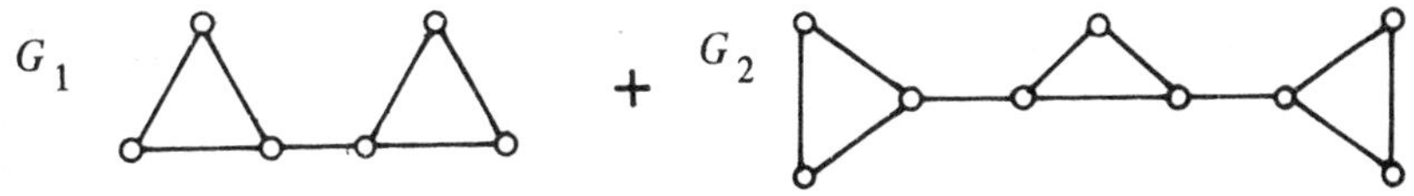

Fig. 3. $G = G_1 + G_2$

As a preliminary observation, we point out that unique (m, n)-partitionability does not imply unique (p, q)-partitionability for $q > n$ or $q < n$. As counterexamples, we may take the graph pictured in Fig. 1 which is uniquely (3, 0)-partitionable but is not uniquely (2, 1)-partitionable, and the join G of the two graphs F' and F'' pictured in Fig. 4. This graph may be proved to be uniquely (2, 1)-partitionable but, is indicated by the labels in the figure, which correspond to color classes, it is not uniquely (4, 0)-partitionable.

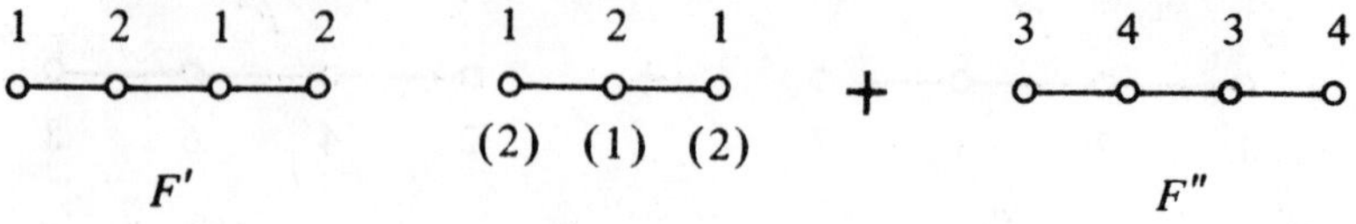

Fig. 4. $G = F' + F''$

Another preliminary observation concerns the graph G pictured in Fig. 1, which has $\rho_1(G) = 2$. Clearly, $V_1 = \{1, 2, 5\}$ and $V_2 = \{3, 4\}$ are classes of a $(2, 1)$-partition where both $\langle V_1 \rangle$ and $\langle V_2 \rangle$ are strictly 1-degenerate, and $V_1' = \{1, 2, 4\}$ and $V_2' = \{3, 5\}$ are classes of another $(2, 1)$-partition where $\langle V_2' \rangle$ is not strictly 1-degenerate. As will be proved (Theorem 2), (m, n)-partitions with classes which do not induce strictly n-degenerate graphs cannot occur in uniquely (m, n)-partitionable graphs.

3. GENERAL PROPERTIES OF UNIQUELY (m, n)-PARTITIONABLE GRAPHS

To begin with, we prove an analogue to Theorem A.

Theorem 1. *If G is uniquely (m, n)-partitionable and $m < |V(G)|$, then $\rho_n(G) = m$.*

Remark. The assumption that $m < |V(G)|$ is made to exclude the trivial case in which $V(G)$ is partitioned into $|V(G)|$ singletons. Trivial graphs are n-degenerate for every $n \geq 0$ and such a partition is obviously unique.

Proof. First note that, if G is (m', n)-partitionable, then it cannot be uniquely (m'', n)-partitionable for $m' < m'' < |V(G)|$. In fact, any partition of one of the m' partition classes with more than one point into two classes yields an $(m' + 1, n)$-partition of G. Obviously, if $m' + 1 < |V(G)|$, then this $(m' + 1, n)$-partition is not unique. By repeating the argument as many times as necessary, G is shown to be (m'', n)-partitionable in several different ways. Hence, if G is uniquely (m, n)-partitionable, then no partition into less than m n-degenerate subgraphs may exist and, by the very definition of $\rho_n(G)$, the theorem is proved.

Theorem 2. *If G is uniquely (m, n)-partitionable, then the m subgraphs induced by the partition classes are strictly n-degenerate.*

Proof. Suppose G is uniquely (m, n)-partitionable and one of the subgraphs induced by the partition classes, say $\langle H_i \rangle$, is $(n - 1)$-degenerate. By definition, $\delta(\langle H_i \rangle) \leqslant n - 1$ and all induced subgraphs of $\langle H_i \rangle$ have also minimal degree $\delta \leqslant n - 1$. Let w be a vertex of G which belongs to a class distinct from H_i, say H_j. Consider a new partition obtained by setting $H_j' = H_j - \{w\}$, $H_i' = H_i \cup \{w\}$, all other classes remaining unchanged. Clearly, $\langle H_j' \rangle$ is n-degenerate. As regards $\langle H_i' \rangle$, $\delta(\langle H_i' \rangle) \leqslant n$. In fact, since the degree of any vertex in $\langle H_i' \rangle$ distinct from w is at most one unit greater than it is in $\langle H_i \rangle$, all induced subgraphs of $\langle H_i' \rangle$ have minimum degree $\delta \leqslant n$. We have then obtained a distinct (m, n)-partition of G which is a contradiction and, therefore, the theorem is proved.

As already pointed out, an (m, n)-partitionable graph may have (m, n)-partitions with the above property and (m, n)-partitions without it. There are also (m, n)-partitionable graphs, $m = \rho_n(G)$, with no (m, n)-partition having such a property, as, for example, the complete graph K_p, when $p/(n + 1)$ is not an integer. In fact, as proved in [1], $\rho_n(K_p) = \{p/(n + 1)\}$, where the brackets denote now the least integer not smaller than $p/(n + 1)$. Hence, when $p/(n + 1)$ is not an integer, at least one class has less than $n + 1$ points and it induces an $(n - 1)$-degenerate subgraph.

The converse of Theorem 2 is not true. A counterexample is the join of three paths of length 2, as pictured in Fig. 5, which admits both $\{1, 2, 3\}$, $\{4, 5, 6\}$, $\{7, 8, 9\}$ and $\{1, 5, 3\}$, $\{4, 8, 6\}$, $\{7, 2, 9\}$ as $(3, 1)$-partitions into classes inducing strictly 1-degenerate subgraphs.

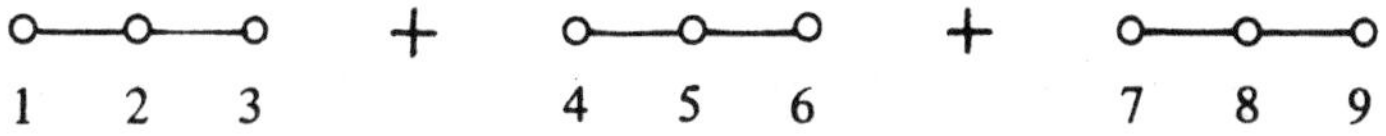

Fig. 5. $\langle\{1, 2, 3\}\rangle + \langle\{4, 5, 6\}\rangle + \langle\{7, 8, 9\}\rangle$

As a corollary of Theorem 2, we obtain:

Corollary 3. *If* G *is uniquely* (m, n)*-partitionable, then, for* $n' < n$, $\rho_{n'}(G) \geqslant m + 1$.

Proof. By the definitions, $\rho_{n'}(G) \geqslant m$. Equality cannot occur because, in such a case, G would have an (m, n')-partition, with $n' < n$, that is to say, a partition into m subsets inducing graphs which are not strictly n-degenerate. Since this contradicts the theorem, the corollary is proved.

Another consequence of Theorem 2 is the following result, which generalizes Theorem B.

Theorem 4. *If* G *is uniquely* (m, n)*-partitionable, then the minimum degree of* G *is at least equal to* $(m - 1)(n + 1)$.

Proof. Let $H_1, \ldots, H_m$ be the partition classes of G, supposed to be uniquely (m, n)-partitionable. By Theorem 2, $\langle H_1 \rangle, \ldots, \langle H_m \rangle$ are strictly n-degenerate. Let $w \in H_j$. Since G is uniquely (m, n)-partitionable, $\langle H_i \cup \{w\} \rangle$, with $i \neq j$, cannot be n-degenerate. We may show that $\langle H_i \cup \{w\} \rangle$ is a strictly $(n + 1)$-degenerate graph. In fact, some subset, say $\widetilde{H}_i$, of $H_i \cup \{w\}$ induces a graph whose minimal degree is $\delta(\langle \widetilde{H}_i \rangle) \geqslant$ $\geqslant n + 1$. But strict inequality cannot occur: it would imply an induced subgraph of $\langle H_i \rangle$ with minimum degree greater than n. Hence $\delta(\langle \widetilde{H}_i \rangle) =$ $= n + 1$. As a consequence, $\langle \widetilde{H}_i \rangle$ is a subgraph of $\langle H_i \cup \{w\} \rangle$ but not of $\langle H_i \rangle$, which implies $w \in \widetilde{H}_i$ and therefore the degree of w in $\langle \widetilde{H}_i \rangle$ is at least equal to $n + 1$.

Since the argument holds for the $m - 1$ classes H_i distinct from H_j and the graphs $\langle H_i \rangle$ are point-disjoint, which in turn implies that the subgraphs $\langle \widetilde{H}_i \rangle$ are line-disjoint, the degree of w in G is at least equal to $(m - 1)(n + 1)$. Since w is arbitrary, the theorem is proved.

Incidentally, we have also proved the following result which will be stated as a theorem for further reference:

Theorem 5. *If* G *is uniquely* (m, n)*-partitionable, then every point of each partition class is adjacent to* r $(\geqslant n + 1)$ *points of each class distinct from that it belongs to and the* r *points, or a subset of them, induce a graph with minimum degree equal to* n.

A result which generalizes Theorem D′ may now be proved

Theorem 6. *In the* (m, n)*-partition of a uniquely* (m, n)*-partitionable graph* G*, the union of any two partition classes induces a connected graph.*

Proof. Let $i \neq j$ and consider the union $H_i \cup H_j$. Suppose that $\langle H_i \cup H_j \rangle$ is not connected and let S_1, S_2 be two subsets into which $H_i \cup H_j$ may be partitioned such that no edge links a point in S_1 to a point in S_2. By Theorem 5, S_1 and S_2 are representing sets of the pair $\{H_i, H_j\}$. Clearly, $\langle H_i \cap S_1 \rangle$ and $\langle H_j \cap S_1 \rangle$ are n-degenerate graphs, as well as $\langle H_i \cap S_2 \rangle$ and $\langle H_j \cap S_2 \rangle$. Since the union of n-degenerate graphs with disjoint point sets is n-degenerate, $\langle (H_i \cap S_1) \cup (H_j \cap S_2) \rangle \equiv \langle H_i \cap S_1 \rangle \cup \langle H_j \cap S_2 \rangle$ and $\langle (H_i \cap S_2) \cup (H_j \cap S_1) \rangle \equiv \langle H_i \cap S_2 \rangle \cup \langle H_j \cap S_1 \rangle$ are n-degenerate. Hence, we may obtain a distinct (m, n)-partition by taking $H_i' = (H_i \cap S_1) \cup (H_j \cap S_2)$ and $H_j' = (H_i \cap S_2) \cup (H_j \cap S_1)$ instead of H_i and H_j, respectively, the other classes remaining unchanged, which is a contradiction. This proves the theorem.

As in immediate consequence, we obtain an analogue to Theorem C′. We state

Corollary 7. *For* $m \geqslant 2$*, any uniquely* (m, n)*-partitionable graph is connected.*

Theorem 6 and Corollary 7 are weak results. Stronger results may also be obtained, which are similar to the Theorems C and D. We have:

Theorem 8. *If* G *is uniquely* (m, n)*-partitionable, then* G *is* $(m-1)$*-connected.*

Proof. Suppose G is not $(m-1)$-connected. Let S be a cutset of G with $|S| \leqslant m-2$. There are at least two partition classes, say H_i, H_j such that $S \cap H_i = S \cap H_j = \phi$. By Theorem 6, any point in H_i may be linked to any point in H_j by a path containing only points of $H_i \cup H_j$. Hence, all points in $H_i \cup H_j$ belong to the same connected component of $\langle V(G) - S \rangle$. We may then choose one point w in another connected component. Let $w \in H_t$, with $t \neq i, j$. Clearly, no edge links w to points in H_i or in H_j. Now taking, for instance, $H_i \cup \{w\}$ instead of H_i and $H_t - \{w\}$ instead of H_t, all other classes remaining unchanged, we obtain a distinct (m, n)-partition of G, which is a contradiction. Hence, the theorem is proved.

As a consequence of this theorem and of the very definitions, we obtain the following corollary, which generalizes Theorem D.

Corollary 9. *In the (m, n)-partition of a uniquely (m, n)-partitionable graph, the subgraph induced by the union of t partition classes, $2 \leq t \leq m$, is $(t-1)$-connected.*

Finally, we prove the following statement, which is a refiniment of Theorem 8.

Theorem 10. *If G is uniquely (m, n)-partitionable and S is a cut-set of G, then S is a representing set of a family of partition classes with at least $m-1$ members.*

Proof. Let $H_1, \ldots, H_m$ be the partition classes and denote by A and B two subsets of $V(G) - S$ such that no line joins a point in A with a point in B. When proving Theorem 8, we have seen that no two classes may exist such that their union is contained in either A or B. Moreover, by Theorem 6, a class, say H_i, contained in A, and another class, say H_j, contained in B, cannot occur. This means that, given any two classes, either at least one of them has points in S or at least one of them has points in A and in B or both hypothesis occur. Since the union $H_i \cup H_j$ of two classes induces a connected subgraph, the hypothesis that one of them has points in A and B implies that $\langle H_i \cup H_j \rangle$ has at least one point in S. Now consider H_1. If $H_1 \cap S = \phi$, then, for $j = 2, \ldots, m$, the reasoning we have just presented, applied to $H_1 \cup H_j$, implies that H_j has at least one point in S. If $H_1 \cap S \neq \phi$, i.e. if H_1 has at least one point in S, then consider H_2 instead of H_1 and repeate this argument. Since m is finite, the theorem is proved.

In view of Theorem 4, we could be tempted to conjecture a result stronger than Theorem 8. An example shows that, in general, no better result may be found. In fact, let G' be the graph pictured in Fig. 6 and

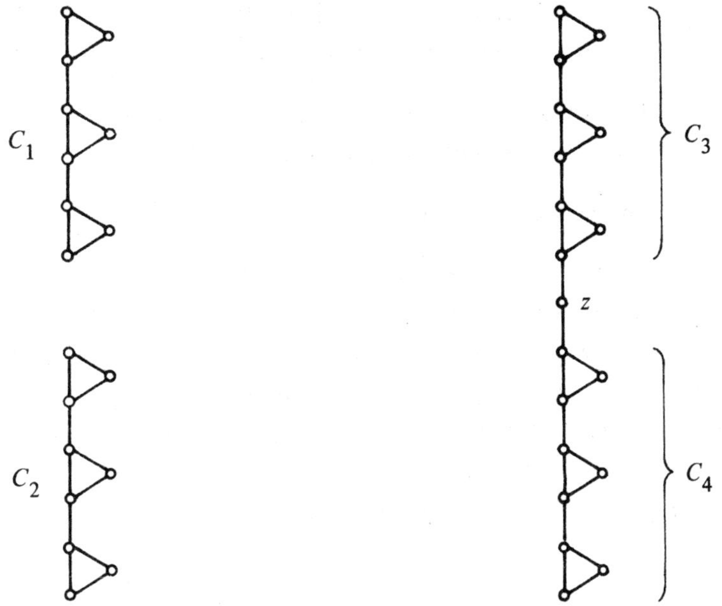

Fig. 6. The graph G'

denote by C_1, C_2, C_3 and C_4, as indicated in the figure, the (isomorphic) connected components of $\langle V(G') - \{z\}\rangle$. For brevity, we write simply V_i instead of $V(C_i)$, for $i = 1, 2, 3, 4$. Now we construct a new graph G obtained from G' as follows: We join all points in V_1 to all points in $V_3 \cup \{z\}$, and all points in V_2 to all points in $V_4 \cup \{z\}$. The graph G is 1-connected, z being a cutpoint, and it has a (2, 2)-partition into the classes $G_1 = V_1 \cup V_2$ and $G_2 = V_3 \cup V_4 \cup \{z\}$. On the other hand, let H_1 and H_2 be the two partition classes of another (2, 2)-partition. Without loss of generality, let $z \in H_1$. By the definition of G, $H_1 \cap G_1 \neq \phi$. Hence, again without loss of generality, suppose that $H_1 \cap V_1 \neq \phi$. Since $z \in H_1$, $\langle H_1 \cap V_1\rangle$ contains no triangle, otherwise $\langle H_1\rangle$ is not 2-degenerate. Hence, $|H_2 \cap V_1| \geqslant 3$. Since $H_1 \cap V_1 \neq \phi$, by the same reason as above, $\langle H_1 \cap V_3\rangle$ contains no triangle. Hence, $|H_2 \cap V_3| \geqslant 3$. This implies a subgraph of $\langle H_2\rangle$ isomorphic to $K_{3,3}$ and, therefore, $\langle H_2\rangle$ is not 2-degenerate. Hence, G is uniquely (2, 2)-partitionable.

4. EMBEDDING PROPERTIES OF UNIQUELY (m, n)-PARTITIONABLE GRAPHS

As regards embedding properties of uniquely (m, n)-partitionable graphs, we generalize now Theorems E, F and G.

To begin with, we prove an analog to Theorem E.

Theorem 11. *Let* G *be uniquely* (m, n)*-partitionable and* $H_1, \ldots, H_m$ *its partition classes. Let* G_w *be obtained from* G *by adding a point* w *to* $V(G)$ *and linking* w *to the points that, in each class* $H_2, \ldots, H_m$, *induce, according to Theorem* 5, *a subgraph of minimum degree equal to* n. *Moreover, any other points in* $H_2, \ldots, H_m$ *may be linked to* w, *and* w *may also be linked to points in* H_1 *but in such a way that* $\langle H_1 \cup \{w\} \rangle$ *is* n*-degenerate. Then* G_w *is uniquely* (m, n)*-partitionable.*

Proof. According to the definitions, we can obtain, from an n-degenerate graph H, another n-degenerate graph, by adding a point w to $V(H)$ and linking w to points in $V(H)$ in such a way that no induced subgraph of minimum degree greater than n arises in $\langle H \cup \{w\} \rangle$. (This condition is satisfied, for instance, when we link w to points in $V(H)$ by no more than n edges). Hence, a possible (m, n)-partition of G_w is given by the subsets $H_1 \cup \{w\}$, $H_2, \ldots, H_m$.

Suppose that $H'_1, \ldots, H'_m$ are classes of another (m, n)-partition of G_w. Suppose, without loss of generality, that they are indexed so that $w \in H'_1$. Clearly, $H'_1 - \{w\}$, $H'_2, \ldots, H'_m$ induce n-degenerate graphs. Hence, they form an (m, n)-partition of G. Since G is uniquely (m, n)-partitionable, these partition classes coincide up to a permutation with $H_1, H_2, \ldots, H_m$. It remains only to show that $H_1 = H'_1 - \{w\}$. This is in fact the case because H_1 is the only H_i which may be contained in H'_1. Otherwise $\langle H'_1 \rangle$ is not n-degenerate. This completes the proof.

As a consequence, a generalization of Theorem F can also be given. We have:

Theorem 12. *Every graph* G *with* $\rho_n(G) = m$ *is an induced sub-*

graph of a uniquely (m, n)*-partitionable graph.*

Proof. Let G be a graph with $\rho_n(G) = m$ and let $G_1, \ldots, G_m$ be a partition of $V(G)$ into subsets inducing n-degenerate graphs. Take a uniquely (m, n)-partitionable graph J and let $J_1, \ldots, J_m$ be the classes of the (unique) (m, n)-partition of J. Take the union $G \cup J$ and link all points of J_r to all points of G_s, for $1 \leqslant s \neq r \leqslant m$. In the graph G^* which we obtain, the subsets $G_1 \cup J_1, \ldots, G_m \cup J_m$ induce n-degenerate graphs. Moreover, G^* may be obtained from J by successively adding the points of G_1, then those of G_2 and so on until all points of G and the edges incident to them are added. Each time a new point is added, the conditions required by the preceding theorem are satisfied. Hence G^* is uniquely (m, n)-partitionable and contains G as an induced subgraph, which completes the proof.

Finally, we may also generalize Theorem G. We recall that, as defined in [1], a graph G with $\rho_n(G) = m$ and such that, for any proper subgraph G', $\rho_n(G') < m$, is said to be m-critical with respect to ρ_n. We write, for brevity, G is (m, n)-critical. We then have:

Theorem 13. *No cutset* S *of an* (m, n)*-critical graph* G *induces a uniquely* $(m-1, n)$*-partitionable subgraph of* G.

Proof. Let S be a cutset of G and $V_1, \ldots, V_q$, $q \geqslant 2$, the point sets of the distinct connected components of $\langle V(G) - S \rangle$. For $i = 1, \ldots, q$, denote by G_i the graph $\langle S \cup V_i \rangle$. Since G is (m, n)-critical, we have $\rho_n(G_1) \leqslant m-1, \ldots, \rho_n(G_q) \leqslant m-1$. We may choose an $(m-1, n)$-partition for each graph G_i. If $\langle S \rangle$ is uniquely $(m-1, n)$-partitionable, then these partitions of the graphs G_i must all partition S in the same way. They then yield an $(m-1, n)$-partition of G, which is a contradiction. Hence, the theorem is proved.

This paper was prepared during a visit to the Technical University of Vienna, Austria, following an invitation by Prof. Wilfried Imrich. The author was partly supported by a scholarship from the Bundesministerium für Wissenschaft und Forschung.

Key words and phrases: Graph, n-degenerate graph, uniquely partitionable, uniquely colorable.

REFERENCES

[1] R. Lick – A.T. White, *k*-degenerate graphs, *Canadian J. Math.*, 22 (1970), 1082-1096.

[2] D. Cartwright – F. Harary, On the coloring of signed graphs, *Elem. Math.*, 23 (1968), 85-89.

[3] F. Harary – S.T. Hedetniemi – R.W. Robinson, Uniquely colorable graphs, *J. Comb. Theory,* 6 (1969), 264-270.

[4] G. Chartrand – D.P. Geller, On uniquely colorable planar graphs, *J. Comb. Theory,* 6, (1969), 271-278.

[5] F. Harary, *Graph Theory,* Addison-Wesley Publishing Company, Reading, 1969.

[6] L. Mirsky, *Transversal Theory,* Academic Press, New York, 1971.

COLLOQUIA MATHEMATICA SOCIETATIS JÁNOS BOLYAI
10. INFINITE AND FINITE SETS, KESZTHELY (HUNGARY), 1973.

ON GRAPHS NOT CONTAINING LARGE SATURATED PLANAR GRAPHS

M. SIMONOVITS

1. INTRODUCTION

In 1967 P. Erdős wrote a paper [1] and dedicated it to H. Grötzsch who was 65 year old that time. In this paper Erdős tried to answer the following question of G. Dirac [1] posed in a conversation:

Problem 1. Let m and n be fixed. How many edges can have a graph G^n of n vertices if it does not contain a saturated planar subgraph of at least m vertices?

Erdős proved the following

Theorem. *There exists a positive constant* c_1 *such that if* G^n *is a graph of* n *vertices not containing saturated planar subgraphs, and* $e(G^n)$ *denotes the number of its edges, then*

$$e(G^n) < \frac{n^2}{4} + c_1 mn .$$

This result is the best possible apart from the value of c_1. To prove his theorem Erdős needed the following other result: Let $C(m, k)$ denote the graph obtained by joining k independent vertices to each vertex of a cycle C^m (having m vertices).

Theorem. *If* G^n *does not contain* $C(m, k)$, *then*

$$e(G^n) \leq \frac{n^2}{4} + c'_k \cdot mn .$$

Again, apart from the value of c'_k this second theorem is sharp, too. The connection between the two theorems is that $C(m, 2)$ is a saturated planar graph of $m + 2$ vertices.

Problem 2. Let us call S^n extremal if (for fixed, given m) S^n has maximum number of edges among the graphs of n vertices not containing saturated planar graphs of m vertices. How can the extremal graphs be characterized?

The main purpose of this paper is to answer Problem 2. In a certain sense our theorem will give a much better answer for Problem 1 than the answer above. The methods used to prove our main theorem work for the extremal problem of $C(m, k)$ as well, giving there quite good description of the extremal graphs. We shall return to the question of $C(m, k)$ in another paper.

Let us call the saturated planar graphs of at least m vertices α-graphs and let us call the 2-connected planar graphs of at least $e = m - 2$ edges β-graphs. We shall need to investigate the following problem:

Problem 3. How many edges can a graph G^n have if it does not contain β-graphs?

Let us call an extremal graph for Problem 1 (for Problem 3) an α-extremal (β-extremal) graph.

The main result of this paper is the following

Theorem 1. (a) *Let* m *be sufficiently large. Then for an arbitrary sufficiently large* N *there exist 2 integers* n_1 *and* n_2 *for which* $N = n_1 + n_2$ *and if* G^{n_1} *is a* β*-extremal graph and* $e(G^{n_2}) = 0$, *then* *the product* $G^{n_1} \times G^{n_2}$ *(obtained by joining each vertex of* G^{n_1} *to each vertex of* G^{n_2} *is an* α*-extremal graph.*

(b) *If* S^N *is an* α*-extremal graph, it can be obtained from a* β*-extremal graph* G^{n_1} *in one of the following two ways:*

(b1) *either* $S^N = G^{n_1} \times G^{n_2}$ *where* $e(G^{n_2}) = 0$ *or*

(b2) *there exist a vertex* $p \in G^{n_1}$ *and a forest* G^{n_2} *such that if for each connected component* T *of* G^{n_2} *(where* $v(T) = 0$ *is also allowed!) we omit* $v(T) - 1$ *edges joining* p *to* T *from the product* $G^{n_1} \times G^{n_2}$, *then we obtain* S^N.

Remark 1. We can see that the extremal graphs of Problem 3 play crucial role in the solution of Problem 1. If we could solve Problem 3, Problem 1 would also be solved.

Conjecture 1. Let $e \neq 6, 9$. The extremal graphs for Problem 2 are connected graphs the 2-connected blocks of which are complete graphs. All these complete graphs but one possible one have exactly $d = \left[\frac{e+5}{3}\right]$ vertices.

Remark 2. Conjecture 1 is valid if $3 \leq e \leq 12$ and $e \neq 6, 9$. For $e = 6$ ($K_4 - 1$ edge) is a better block than K_3 and the β-extremal graphs are connected, each block has $O(1)$ vertices and each but $O(1)$ block is a ($K_4 - 1$ edge). A similar assertion holds for $e = 9$ except that here each block is either a K_4 or a ($K_5 - 2$ edges) if n is sufficiently large.

Remark 3. (a) If Conjecture 1 holds, then both cases (b1) and (b2) do occur, however, in (b2) the forest can have only r edges if $e \equiv r \pmod 3$ and $0 \leq r \leq 2$.

(b) In (K.4) we shall get an extremal graph $S^N = G^{n_1} \times G^{n_2}$ where G^{n_2} is a tree and G^{n_1} is obtained from a β-extremal graph G^{n_1+1} by omitting a vertex p of valence n_1 from it. One could think that this S^N is neither of (b1) nor of (b2) type, however, it is an α-extremal graph of type (b2) which is easily seen if we move a vertex of the tree G^{n_2} from G^{n_2} to G^{n_1}.

Remark 4. Lemma 4 will prove that the graphs in (a) of Theorem 1 do not contain α-graphs. Lemma 5 shows that there exist graphs of type (b2) not containing α-graphs. Moreover, Lemma 5 gives a construction of such graphs from products $B \times G$ where B is a 2-connected graphs not containing β-graphs and G is a forest, and $B \times G$ does not contain α-graphs.

Remark 5. If we knew (b) of Theorem 1. (a) would be trivial: Let S_0^N be an α-extremal graph. S^N is either of type (b1) or of (b2). Anyway, it defines two integers n_1 and n_2 and if we consider any product $S^N = G^{n_1} \times G^{n_2}$ where G^{n_1} is a β-extremal graph and $e(G^{n_2}) = 0$, then $e(S^N) = e(S_0^N)$ and S^N contains no α-graphs (by Lemma 4). This proves (a).

Remark 6. Theorem 1 holds for $m = 4$ but does not hold for $m = 5$. E.g. S^{2n} can be $G^{n_1} \times G^{n_2}$ where $n_1 = n_2 = n$ and both graphs have $n/2$ independent edges.

Definitions, notations. The graphs considered here are non-oriented and have neither loops, nor multiple edges. The notations are mostly the same as in [5]. The number of vertices and edges are denoted by $v(G)$ and $e(G)$ respectively, $d(x)$ denotes the valence of a vertex, $\mathrm{st}\,(x)$ denotes the star of x, i.e. the set of vertices joined to x. If G_1 and G_2 are two graphs, having no common vertices, then $G_1 \times G_2$ denotes the graph obtained from them by joining each vertex of G_1 to each vertex of G_2. If G^n is a graph, then the upper index denotes the number of vertices. C^k and P^k are the cycle and path of k vertices. $K_d(m_1, \dots, m_d)$ denotes the complete d-partite graph with m_i vertices in its i-th class. If E is a set, $|E|$ denotes its cardinality.

$c, c_1, \dots, c_7$ denote positive constants, even if it is not stated explicitely.

2. PRELIMINARY REMARKS ON GRAPHS NOT CONTAINING β-GRAPHS

Lemma 1. *If G is a 2-connected graph not containing β-graphs, then $P^{e^2} \not\subset G$.*

Proof. (Indirectly). Let us suppose that there exists a P^{e^2} joining a to b. Since G is 2-connected, there exist 2 paths P_1 and P_2 joining a to b and having no common vertices but a and b. Either these two paths form a $C^{e'}$ for some $e' \geqslant e$, but this would be a contradiction, or at least one of the two paths has at most $e/2$ common vertices with P^{e^2}. The common vertices divide P^{e^2} at most e arcs at least one of which has at least e edges. This arc and the corresponding arc of the considered P_i form a long cycle: $C^{e'}$ where $e' \geqslant e$. This completes our proof.

Remark. The bound e^2 could be improved but we do not need it.

Lemma 2. *Let G be a 2-connected graph containing no β-graphs. Then the valence of each vertex of G is less than $\frac{e+1}{2}$.*

Proof. Let us suppose that the vertex x is joined to the vertices y_i $(i = 1, 2, \ldots, \mu)$. $G - x$ is connected, thus it contains a spanning tree. Let T be such a spanning tree and T' be a subtree of it containing the y_i's and being minimal under this condition. T' will have no vertices of valence 1 but y_i's. T' and the edges (x, y_i) will determine a 2-connected planar graph of at least $\mu + 1$ vertices, at least μ countries and least $2\mu - 1$ edges.

Q.E.D.

The following lemma will not be needed in the proof, therefore we omit its proof. It is interesting in itself, since it shows that the valence extremal problem is trivial, while the edge-extremal problem is not.

Lemma. *If G be an arbitrary graph each vertex of be has valence $\geqslant m$, the G contains a 2-connected planar subgraph of $\geqslant 3m - 3$ edges.*

This lemma is sharp because if G is the union of complete graphs of $m + 1$ vertices then G does not contain planar 2-connected subgraphs of $3m - 2$ edges.

Let us return to the problem of the structure of β-extremal graphs. By Lemma 1 and Lemma 2 a block of a β-extremal graph cannot contain

pe^2 and vertices of valence $\geqslant \frac{e+1}{2}$. Thus, if G^n is an extremal graph for Problem 3 and $B_1, \ldots, B_t$ are the blocks of it, then $v(B_j) = O(1)$,

$$\sum (v(B_j) - 1) = n - 1$$

and

$$\sum e(B_j) = e(G^n) .$$

Let us call β-extremal blocks those blocks not containing β-graphs for which

$$e(B)/(v(B) - 1)$$

is the maximum possible.

It is trivial, that all but $O(1)$ blocks of a β-extremal graph G^n are β-extremal blocks, since if there were too many non-extremal blocks, some of them could be replaced by β-extremal blocks and the number of edges would increase. Thus we obtain

Lemma 3. (a) *Let G^n be a β-extremal graph. Then all but $O(1)$ blocks of G^n are β-extremal blocks and all the blocks have $O(1)$ vertices.*

(b) *Let us call a graph almost extremal if it does not contain prohibited subgraphs and has $(1 + o(1))$ times as many edges than the extremal graphs. If G^n is β-almost extremal, then all of its blocks have $O(1)$ vertices and all but $o(n)$ blocks of G^n are β-extremal blocks.*

Corollary 1. If G^n is a β-extremal graph, then

$$e(G^n) = c_e n + O(1) .$$

Now we shall prove Lemma 4 and Lemma 5 showing that the conjectured extremal graphs do not contain α-graphs. Then we turn to the proof of Theorem 1. This consists of two parts. In the first part we use only asymptotic estimations, we do not use the fact that S^N is extremal, only that it is nearly extremal. In the second part (from (K)) we apply an operation to S^N under which the number of edges does not increase, *because S^N is extremal.*

3. PROOF OF THEOREM 1

First we show that our conjectured extremal graphs are reasonable candidates: they do not contain β-graphs.

Lemma 4. *Let* G_1 *be a graph not containing any* β*-graph and let* $e(G_2) = 0$. *Then* $G_1 \times G_2$ *contains no* α*-graph.*

Proof. If T is a triangulated shpere of at least m vertices and we omit q independent vertices of it, then the graph will have at least $m - q$ vertices and at least q countries. By Euler's theorem, it will have at least $m - 2$ edges. Therefore, if $G_1 \times G_2$ contains an α-graph, then G_1 contains a planar graph of at least $m - 2 = e$ edges. The only thing to prove is that omitting the q independent vertices we get a 2-connected graph. (This holds for K_3 only if we define K_2 to be 2-connected.) Let $x_1, \ldots, x_q$ be the omitted vertices and let us consider an embedding of T into the plane. If a and b are two vertices of $T - x_1 - \ldots - x_q$, since T is 2-connected (by a well known theorem of Whitney), there exists a cycle $a = z_1, z_2, \ldots, z_k = b, \ldots, z_r = z_1 = a$ in T. Let D_i, $(i = 1, \ldots, q)$ be the domain determined by the triangles incident with x_i. If the cycle $z_1, \ldots, z_r$ contains an x_i, replacing some arcs of this cycle by some arcs of the boundary of D_i we can obtain a new cycle containing a and b but avoiding x_i. Iterating this modification, in finite steps we shall obtain a cycle containing no x_i, but containing a and b. Thus $T - x_1 - x_2 - \ldots - x_q$ is two connected. This completes the proof.

Lemma 5. *Let* F *be a forest and* H_i *be given graphs containing a common vertex* p. *Let us suppose that* H_i *and* H_j *have no common vertices but* p *(for any* $i \neq j$*); let* H *be the union of these graphs. If no* $F \times H_i$ *contains* α*-graphs and* S *is obtained from* $H \times F$ *by omitting edges of it so that* p *be joined to at most one vertex of any subtree* T *of* F, *then* S *neither contains* α*-graphs.*

Proof. Let us suppose indirectly that there exists a triangulation Q in $H \times F$ intersecting at least two H_i. If an $R_i =: H_i \cap Q$ is empty, or if an edge of F is not contained by Q, we omit it. Thus each subtree $F_i \subseteq F$ will be a subgraph of Q. Now, if we omit from the triangulation Q a tree F_i, then the countries (triangles) incident with it will turn into one country D_i which is simply connected. (In the proof of Lemma 4 it was bounded by a simple polygon but now this will not be valid.) Further, if F_i and F_j are two different trees (not joined by an edge) then the omission of F_j will not change D_i, or, in other words, the omission of all the trees F_i will lead to a planar graph in which D_i's are countries, separated by the edges of $H \cap Q$. Now we can use the method of Lemma 4 to show that $H \cap Q$ is connected. Thus each $H_i \cap Q$ must contain p. Embedding Q into the sphere we get a country D_k adjacent to at least to two different R_i's. We may assume that $R_1, \dots, R_t$ are the R_i's adjacent to D_k in their cyclic order. The triangle of Q adjacent to p and H_1 and separating H_1 from H_2 in the cyclic order has one side on the border of H_1, but the other sides cannot join H_1 to H_2, thus there must be a vertex $u \in F$ joined to p such that (p, u) is another side of this triangle. Similarly, there must be an edge (p, u') separating H_t from H_1 and the edges (p, u) and (p, u') must be different, thus $u \neq u'$. Since u and u' are in the same country of $H \cap Q$, u and u' must belong to the same F_i, but only one vertex of F_i can be joined to p. This contradiction completes the proof.

Proof of Theorem 1. We shall consider only the case when $e = m - 2$ is sufficiently large. Let $U^N = D^{n_1} \times D^{n_2}$ where D^{n_1} is β-extremal, $e(D^{n_2}) = 0$ and $n_i = \frac{N}{2} + o(N)$ will be fixed bellow. Since the graphs described in Conjecture 1 do not contain β-graphs, $e(D^{n_1}) = \widetilde{E} \geq dn_1/2 - O(1)$.

(A) Let S^N be an α-extremal graph and let $[N/2] = n$. U^N does not contain α-graphs by Lemma 4, thus

$$(1) \qquad e(S^N) \geq e(U^N) = n_1 n_2 + \widetilde{E}\,.$$

Since $K_3(2, m, m)$ contains an α-graph,

$$(2) \qquad S^N \not\supset K_3(m, m, 2)\,.$$

By Erdős – Stone theorem [2] (2) and (1) imply that

$$(3) \qquad e(S^N) = N^2/4 + o(N^2)\,.$$

By the results of [3, 4, 5] of Erdős and the author

(i) There exists a product $G^{n_1} \times G^{n_2}$ from which S^N can be obtained by omitting $o(n^2)$ edges. Further,

$$n_i = n + o(n) \quad \text{and} \quad e(G^{n_i}) = o(n^2) .$$

(ii) It can also be assumed that the number of edges to be omitted from $G^{n_1} \times G^{n_2}$ to obtain S^N is the minimum possible among all the possible products. Then beside (i) the following assertion will also hold:

For every positive constant c there exists a constant M_c such that the number of vertices of G^{n_i} joined to at least cn vertices of G^{n_i} is at most M_c.

(As a matter of fact, (ii) is stated in [4, 5] only for some extremal graphs, but the proof given there uses only (2) and (3), i.e. remains valid for asymptotically extremal graphs as well.)

We introduce a few notations: G^{n_1} will be called the upper graph and G^{n_2} the lower one and we shall assume that

(4) $$e(G^{n_1}) \geqslant e(G^{n_2}) .$$

If x is a vertex, $d(x)$, $d_1(x)$ and $d_2(x)$ will denote the valence of x, the number of vertices of G^{n_1} and the number of vertices of G^{n_2} joined to x, respectively. L will denote the number of edges we omitted from $G^{n_1} \times G^{n_2}$ to obtain S^N.

(B) Let p denote the minimum number of vertices in G^{n_1} we must omit to ruin all the β-graphs in G^{n_1}. We prove that

(5) $$p = O(1) .$$

(B.1) It is trivial, that if G is a β-graph, then there exists a $G' \subseteq G$ which is a β-graph too and has at most $2m$ countries. Indeed, if we call "hanging chain" a path of a graph each vertex of which is of valence 2 except the endvertices which have valence $\geqslant 3$, then each 2-connected planar graph but the cycle has $\geqslant$ two hanging chains, thus a planar graph G has a hanging chain of less than $v(G)/2$ vertices. Omitting this hanging chain we obtain a 2-connected planar subgraph G' having one less countries and at least $\nu(G)/2$ vertices. Iterating this step either we get a cycle of at least m vertices or a planar 2-connected graph having more than m but less than $2m$ vertices. Since the number of countries is never greater than the number of edges, the proof of (B.1) is complete.

(B.2) To prove (5) we consider $2m$ vertices $x_1, \ldots, x_{2m} \in G^{n_2}$. We may assume that there exists a β-graph in G^{n_1}, hence there exists a β-graph of at most $2m$ countries as well. Let this β-graph be A. At least one edge is missing between A and $\{x_1, \ldots, x_{2m}\}$ since for any G^{2m} $G^{2m} \times A$ contains an α-graph. This implies that at least p edges are missing between $\{x_1, \ldots, x_{2m}\}$ and G^{n_1}. Thus

(6) $$L \geq (n/2m + o(n))p .$$

Clearly,

(7) $$e(S^N) = n_1 n_2 + e(G^{n_1}) + e(G^{n_2}) - L \leq N^2/4 + 2e(G^{n_1}) - L ,$$

further, since the number of vertices of G^{n_1} having valence $\geq \frac{n}{10m}$ is at most M_0 (where M_0 is a constant), thus (by Corollary 1)

(8) $$e(G^{n_1}) \leq M_0 n + (p - M_0) \frac{n}{10m} + O(n) .$$

By (1) $e(S^N) \geq [N^2/4]$. This, (6), (7), and (8) yield (5).

(C) The inequalities of (B) yield some further interesting bounds: By (5) and (8)

(9) $$e(G^{n_i}) = O(n) .$$

By (7) and by $e(S^N) \geq [N^2/4]$ and by (9)

(10) $$L = O(n) .$$

Finally, by the same inequalities $n_1 n_2 \geq N^2/4 - O(n)$. Thus

(11) $$n_i = n + O(\sqrt{n}) .$$

(D) It is useful to assume that if there exists an $x \in S^N$ of valence $\geqslant 2n - n/\log n$, then $x \in G^{n_1}$ (that is, if $x \in G^{n_2}$, then we move it up). By (2) there can be only one such vertex and this change in (ii) of (A) makes so little difference that we could even forget about it in most parts of the proof.

(E) Let V_i be the class of vertices of G^{n_i}

either joined to at most $n/2 + n/\log n$ vertices of $G^{n_{3-i}}$

or of valence $\leqslant n - n/\log n$.

(Later it will turn out that $V_i = \phi$.)

Let $H_i = G^{n_i} - V_i$, $m_i = v(H_i)$. Let L' be the number of missing edges in $H_1 \times H_2$.

We prove that, if E is the maximum number of edges a graph of m_1 vertices can have without containing a β-graph, then

(12) $$e(H_1) \leqslant E + n + o(n) .$$

Let $y_1, \dots, y_p$ be a set of vertices representing each β-graph in H_1. Let $d_i(x)$ denote the number of vertices of G^{n_i} joined to $x \in S^N$. It is enough to show that for $h = \log^2 n$

(13) $$\sum_{j \leqslant h} d_1(y_j) \leqslant n + \binom{p}{2} h .$$

This will imply (12) immediately. To prove (13) it is enough to prove that there cannot be h vertices $x_1, \dots, x_h$ joined to y_j and $y_{j'}$ at the same time. If there are h vertices $x_1, \dots, x_h$ joined to y_j and $y_{j'}$, then, by the definition of V_i and by (11) there exist $\left[\frac{n}{2 \log n}\right]$ vertices in G^{n_2} joined to both y_j and $y_{j'}$. By (10) there exist h vertex $z_1, \dots, z_h$ among them such that only $O(\log^3 n)$ edges are missing between $\{x_i\}$ and $\{z_i\}$. By Erdős – Stone theorem or some other well known results (or Kőváry, T. Sós and Turán [6]) there must be a $K_2(m, m)$ in the graph spanned by the x_i's and z_i's. Thus S^N contains a $K_3(m, m, 2)$. This contradicts (2). Hence (12) is proved.

(F) by (7) and (9)

(14) $$e(S^N) = N^2/4 + O(n).$$

This proves (by a short computation)

(15) $$|V_i| = O(\log n)$$

since

(a) the vertices of first type in V_i are joined to more than $n/3$ vertices of each G^{n_i}: there are at most $O(1)$ such vertices;

(b) If omitting the vertices of valence $\leqslant n - n/\log n$ we get a graph S^M, then the argument above shows that

(16) $$e(\widetilde{S}^M) \leqslant M^2/4 + O(M).$$

Therefore, if S^N contains too many vertices of valence $\leqslant n - n/\log n$, then it must have less than $[N^2/4]$ edges, which is a contradiction.

(The vertices with valence $\leqslant n - n/\log n$ were not considered in (a) even if they were joined to at most $\frac{n}{2} + \frac{n}{\log n}$ vertices of $G^{n_{3-i}}$.)

(G) We would like to prove that for $\delta = E/2n$

(17) $$e(H_1) \geqslant \delta n + o(n).$$

If $V_1 = \phi$, then, by (7) $e(G^{n_1}) + e(G^{n_2}) - L = e(S^N) - n_1 n_2 \geqslant$ $\geqslant e(S^N) - [N^2/4]$; by (1) $e(G^{n_1}) + e(G^{n_2}) - L \geqslant E + o(n)$ proving (17). In the general case let $R_i = e(G^{n_i}) - e(H_i)$, then

(18) $$L \geqslant R_1 + R_2 - o(n),$$

since the vertices of V_i have valence $\leqslant n + o(n)$ and $V_i = O(\log n)$. Thus we obtain

(19) $$\begin{aligned} e(H_1) + e(H_2) &= e(G^{n_1}) + e(G^{n_2}) - R_1 - R_2 \geqslant \\ &\geqslant e(G^{n_1}) + e(G^{n_2}) - L - o(n) \geqslant E - o(n). \end{aligned}$$

Thus

(20) $\max e(H_i) \geqslant \delta n + o(n)$.

(H) Let us assume (17). By (E), more exactly, by (13) H_1 has a subgraph $\bar{H}$ of at least $2n + o(n)$ edges not containing β-graphs. By Lemma 1 and Lemma 2 each block of this graph has only $O(1)$ vertices. It is trivial, that for any M_0 there exists a c_0 such that if each block of a graph of at most $n + o(n)$ vertices has at most M_0 vertices and if the whole graph has at least $2n - o(n)$ edges, then the graph contains $c_0 n$ vertex-independent cycles. Therefore H_1 has at least $c_0 n$ vertex-independent cycles for some positive constant c_0.

Let G' and G'' be two graphs each of which contains m vertex-independent cycles. Then $G' \times G''$ contains an α-graph. Thus between any m vertex-independent cycles of H_1 and of G^{n_2} (if there are such cycles) at least one edge must be missing. $L = O(n)$, hence at most $O(1)$ independent cycles can be in G^{n_2}. Thus

(21) $e(H_2) \leqslant 2n + o(n)$,

since the β graphs can be represented by $O(1)$ vertices, the remaining (short !) cycles again can be represented by $O(1)$ vertices of H_2 and (13) can be applied to these $O(1)$ vertices, while the remaining graph is a tree or a forest.

Now, one can ask, what do we do if (17) does not hold. We avoid this question in the following way: Let us forget the assumption $e(G^{n_1}) \geqslant e(G^{n_2})$ from now on and then, by the symmetry and (20) we may assume (17) without loss of generality. By the way, later we shall see that $V_i = \phi$, and therefore the assumptions $e(G^{n_1}) \geqslant e(G^{n_2})$ and $e(H_1) \geqslant e(H_2)$ are the same, but now we cannot (and need not) use this.

(I) In this paragraph we prove that G^{n_2} contains only $O(1)$ independent edges if e is sufficiently large. Let us divide the blocks of H_1 into 3 groups:

(I.1) If B is a block of H_1 containing a vertex of valence

$\geqslant \frac{e}{3} + 1$, then it contains a planar graph P_0 of at least $\frac{e}{3} + 1$ countries and $\frac{e}{3} + 2$ vertices, see Lemma 2. Now, since $\frac{e}{3} + 2 + 2\left(\frac{e}{3} + 1\right) = $ $= e + 4 > m$, one can easily see that if G contains $\geqslant \frac{e}{3} + 1$ independent edges, then $G \times P_0$ contains a triangulation of the sphere having at least $e + 4$ vertices. Therefore if there exist $c_1 n$ blocks of this type in H_1 for some fixed positive constant c_1, then, by $L = O(n)$ at most $O(1)$, (more exactly, $O_{c_1}(1)$) independent edges can be in G^{n_2}.

(I.2) Let us suppose now that there exist $c_1 n$ blocks of the following type: B does not contain vertices of valence $\geqslant \frac{e}{3} + 1$; the average valence is at least $\frac{e}{3} - 6$; we construct a planar graph P_1 as it is done in the proof of Lemma 2, with at least $\frac{e}{3} - 5$ vertices and at least $\frac{e}{3} - 6$ countries. Finally, let $B - P_1 \supseteq \widetilde{C}^{18}$ where $\widetilde{C}^t$ denotes a cycle of length $\geqslant t$. If we replace P_0 in (I.1) by the planar graph consisting of P_1, $\widetilde{C}^{18}$ and a path joining them, then we obtain again a P_0' such that $G \times P_0'$ contains a triangulation of at least $e + 3$ vertices.

(I.3) Now we suppose that there exist $c_1 n$ blocks of the following type: B does not contain vertices of valence $\geqslant \frac{e}{3} + 1$; the average valence is at least $\frac{e}{3} - 6$; $B - P_1$ (of (I.2)) does not contain $\widetilde{C}^{18}$.

We try to increase P_1 in the following way: if there exists a vertex x in $B - P_1$ joined to at least 2 vertices of a country of P_1, we add this vertex x and two edges to P_1, thus obtaining a P_2 having one more vertex and one more country than P_1. If we can, we increase P_2 in the same way, obtaining P_3, and so on. In at most 7 steps the procedure must stop, otherwise we get an α-graph in the way described in (I.1). If the last graph obtained is P_j, since there is no $\widetilde{C}^{18}$ outside, in $B - P_j$,

(22) $$e(B - P_j) \leqslant 9v(B - P_j)$$

(by the well known theorem of Erdős and Gallai [7]). Further, the number of edges joining a vertex $x \in B - P_j$ to P_j is at most $\frac{e}{6} + 1$, otherwise P_j could be increased. Hence, if $\tilde{d}(x) = |\operatorname{st}(x) \cap B|$, then

(23) $$\sum_{x \in B - P_j}' \tilde{d}(x) \leqslant \left(\frac{e}{6} + 20\right) \nu(B - P_j) .$$

Since the average of $\tilde{d}(x)$ is $\geqslant \frac{e}{3} - 6$ and $\max \tilde{d}(x) \leqslant \frac{e}{3} + 1$, there cannot be many vertices in $B - P_j$. An easy calculation shows that

(24) $$\nu(B - P_j) = O(1)$$

independently of e. Indeed, $\nu(P_j) = \frac{e}{3} + O(1)$, therefore

(25) $$\sum_{x \in P_j} \tilde{d}(x) \leqslant \left(\frac{e}{3} - 6\right) \nu(P_j) + O(e) .$$

By (23), if e is sufficeintly large, then

(26) $$\sum_{x \in B - P_j} \tilde{d}(x) \leqslant \left(\frac{e}{3} - 6\right) \nu(B - P_j) - \frac{e}{7} \cdot \nu(B - P_j) .$$

Thus, since the average valence is at least $\frac{e}{3} - 6$, (24) must hold. Now, if e is large, we can find two vertices of valence $\geqslant \frac{e}{3} - 6$. Let us denote the 2 vertices by a and b and the graph spanned by the set of vertices joined to both a and b by T. By (24) $\nu(T) = \frac{e}{3} + O(1)$. If this T contained a path of $r = \frac{e}{3} + 1 - \nu(T) = O(1)$ vertices, then a, b, and this path and the vertices of T would determine a planar graph of $\geqslant \frac{e}{3} + 1$ countries and $\geqslant e/3 + 2$ vertices and the method of (I.1) could be applied. But if there were no such path, then the number of edges in T would be $O(r\nu(T)) = O(\nu(T))$, while by $\nu(B - T) = O(1)$, $e(B) = e(T) + O(e) = O(e)$ would hold. This is a contradiction if e is large enough.

(J) We have seen in (I) that G^{n_2} contains only $O(1)$ independent edges. We shall consider the following two cases separately:

(27) $$e(H_1) \geqslant E + \frac{n}{\log n}$$

may hold or may not. First we assume that (27) holds.

By (10) all but $O(n/\log n)$ vertices of H_2 have less than $\log n$ non-neighbours. (If $x \in G^{n1}$ and $y \in G^{n2}$ and (x, y) does not belong to S^N, then x and y will be called non-neighbours, the set of the non-neighbours of $x \in S^N$ will be denoted by $N^-(x)$ and the cardinality of the non-neighbours of an $x \in H_1$ will be denoted by $d^-(x)$.) We shall prove that if W_1 is the set of vertices of H_1 having in H_1 valence $\geq n/\log^3 n$, then

$$(28) \qquad \sum_{x \in W_1} d^-(x) \geq n - o(n).$$

Let us assume the contrary. Then we can find $2m$ vertices $x_1, \dots, x_{2m} \in \in H_2$ such that $d^-(x_i) < \log n$ and $N^-(x_i) \cap W_1 = \phi$, $\bigcup_{1 \leq i \leq 2m} N^-(x_i)$ represents all the β-graphs. Hence

$$m \log n \frac{n}{\log^3 n} = O(n/\log^2 n)$$

edges may represent all the β-graphs in H_1. This contradicts (27). Thus we proved (28) which means that the vertices of W_1 contribute to L' at least $n - o(n)$. At the same time, (13) gives here that W_1 contributes to $e(H_1)$ by at most $n + o(n)$. This means that "in average the vertices of W_1 are similar" to the vertices of V_1, this is, why we turn to the investigation of $H_1 - W_1$). Let $H_1 - W_1 = H'$.

(K) If (27) holds, let H' be the graph, defined above, if (27) does not hold, let $H' =: H_1$. Let L'' be the number of missing edges between H' and H_2. The method of (G) now gives that

$$(29) \qquad e(H') + e(H_2) \geq E - o(n).$$

Since H_2 contains only $O(1)$ independent edges, we may represent its edges by $O(1)$ vertices. Let us denote a representing set of $O(1)$ vertices by W_2 and let $H'' = H_2 - W_2$. The vertices of H'' will be divided into 4 groups as follows:

A_1 is the set of vertices joined to H' completely and to exactly one vertex of W_2.

A_2 is the set of vertices joined to each but exactly one vertex of H' and to exactly one vertex of W_2.

A_3 is the set of vertices joined to H' completely but not joined to any vertex of W_2.

A_4 contains the remaining vertices.

To avoid the lengthy explanations let us use the word "valence" for $d(x)$ and the word "degree" for the valence in $(H' \times H_2) \cap S^N$.

The *degree* of a vertex in A_1 is always the same and just by 1 greater than the *degree* of a vertex of $A_2 \cup A_3$. The method used to prove (13) gives that only $O(1)$ vertices can have higher *degree* than the vertices in A_1 and their *degree* is higher by at most $O(1)$.

(K.1) Let $M = 4|W_2|e$. First we consider the case when $|A_1| > M$. Some vertices will be called symmetric if they are independent and joined to exactly the same vertices of $H' \cup H_2$. The vertices of A_1 are symmetric if they are joined to the same vertex of W_2. There exist $4e$ symmetric vertices in A_1. Let us denote them by $a_1, \ldots, a_{4e}$. Now we symmetrize: This means that we omit all the edges of S^N represented by W_1, V_1 and G^{n_2}; then we join each vertex of G^{n_2}, W_1 and V_1 to the vertex $p \in W_2$ which was joined to the a_i's, (except, of course this p) and join all these vertices to H' completely (including p too). The obtained graph Z^N cannot contain α-graphs: let us suppose indirectly, that it does. Then the subgraph H^+ spanned by H' and p (which remained untouched during the symmetrization) must contain a β-graph, since Z^N is the product of H^+ and a graph having no edges. H^+ must also contain a β-graph of at most $2e$ countries. Arbitrary $2m$ vertices downstairs (i.e. in $Z^N - H' - p$) will form an α-graph with this β-graph of at most $2m$ countries. Thus we may select the $2m$ vertices from the a_i's as well. But then the subgraph defined by this β-graph and the corresponding a_i' must be an α-graph in S^N, which is a contradiction.

Since the number of edges decreased by at most $o(n)$ edges when we symmetrized the vertices of $V_1 \cup W_1 \cup V_2 \cup (W_2 - p)$ and by $O(1)$ in the case of the vertices of A_4 and it increased by 1 in the case of each vertex of A_2, A_3, and since Z^N does not contain α-graphs (and consequently $e(Z^N) \leqslant e(S^N)$) therefore $|A_2 \cup A_3| = o(n)$. This proves that

(30) $$\sum_{x \in H''} |N^-(x) \cap H'| = o(n) .$$

We prove that

(31) $$\sum_{x \in H_2} |N^-(x) \cap H'| = o(n)$$

too. This is trivial if there are at least 2 vertices p' and p'' which can be taken for p in the proof above, since what we have proved above is that the number of missing edges between H' and $H_2 - p$ is $o(n)$. Now we know that it is $o(n)$ between $H_2 - p'$ and H' and it is $o(n)$ between $H_2 - p''$ and H', which implies (31). If there were only one suitable p, then at least $n - o(n)$ vertices of H'' would be joined to it, thus p would be joined to at least $n - o(n)$ vertices of G^{n1} and by (D) $p \in G^{n1}$ which would be a contradiction.

(K.2) We show that

(32) $$W_1 = \phi .$$

Since the symmetrization did not essentially increase the number of edges, and the number of missing edges incident to W_1 is at least $n - o(n)$ (if (32) does not hold), the number of edges of H_1 represented by W_1 must also be $n - o(n)$. (It cannot be larger by (13).) This means that the number of missing edges between W_1 and H_2 cannot be larger, than $n + o(n)$: it is exactly $n + o(n)$. We have seen in (J) that almost all the vertices of H'' have a non-neighbour in W_1. Thus one can find $c_2 n$ vertices in H'' which are joined to exactly one vertex of W_1 by a missing edge and this vertex of W_1 is the same p for all the $c_2 n$ vertices. There must be $2e$ vertices among these $c_2 n$ ones having $O(1)$ non-neighbours and these non-neighbours will represent all the β-graphs of H_1. Since W_1 represents $n + o(n)$ edges and since H' has only by $o(n)$ edges less than the maximum, thus

(33) $$e(H_1) = e(H') + n + o(n) = E + n + o(n) .$$

Thus at least $n - o(n)$ edges are needed to represent the β-graphs of H_1. Since $d_1(x)$ is small for all the considered non-neighbours but p,

(34) $\quad d_1(p) = n - o(n)$.

Therefore

(35) $\quad d_2(p) = n - o(n)$.

This contradicts to the fact that $c_2 n$ vertices of H'' are not joined to p. Thus (32) is proved. This means that H_1 does not contain β-graphs.

(K.3) We prove that V_i is also empty. If there were an $x \in V_1$ joined to $\frac{n}{2} + o(n)$ vertices of each H_i, then H^+ spanned by H_1 and x would contain β-graphs by $e(H^+) = E + \frac{n}{2} + o(n)$ and the non-neighbours of arbitrary $2e$ vertices of H'' would represent the β-graphs of H^+. Thus at least $n - o(n)$ edges are missing between H'' and H^+. On the other hand, by (30) only $\frac{n}{2} + o(n)$ edges are missing between H'' and H^+. This contradiction proves that all the vertices of V_1 have valence $\leqslant n - n/\log n$. The same holds for V_2. Since during the symmetrization the number of edges represented by H_2 diminished by at most $O(1)$, if $V_1 \cup V_2$ were not empty, Z^N would have more edges than S^N (the valence of a vertex of V_i increases by at least $n/2 \log n$). This proves that $V_i = \phi$ and $L = o(n)$.

(K.4) We prove that H_2 contains no cycles. Of course it cannot contain a cycle longer than e because if it contained, at least $n - o(n)$ missing edges could be found between H_1 and H_2. Let us assume that H_2 contains a short cycle. Since $e(H_1)/v(H_1) \geqslant e/3 + o(e)$, H_1 must contain $c_3 n$ blocks for which the average valence is at least $\frac{e}{3} + o(e)$. The method used to prove Lemma 2 gives that these blocks contain planar graphs of $\geqslant \frac{e}{3} + o(e)$ countries and vertices. It can also be assumed that there are $c_4 n$ such graphs no two of which have common vertices. Putting one of then inside, the other outside the cycle of H_2 and putting some suitable vertices of H_2 into the inner countries of the two planar graphs we obtain a triangulation of $\geqslant \frac{4}{3}e + o(e)$ vertices, unless at least one edge is missing between the considered vertices of H_2 and the considered vertices of H_1. Since S^N contains no α-graph, at least $c_5 n$ edges must be missing between H_1 and H_2 (if e is large enough). This proves that no cycle is in H_2. Hence during the symmetrization the number of edges remains constant downstairs. The global number of edges will increase unless $H_2 - p$ is joined completely to H_1. We apply again the trick used to prove (31): there exist at least 2 vertices which can be taken for p, thus H_2 is completely joined to H_1. Thus we get the $S^N = G^{n_1} \times G^{n_2}$ of Remark 3/b: G^{n_2} is a tree and $H^+ - p = G^{n_1}$, (see (K.1)). H^+ contains no β-graphs. It is easily seen that H^+ is β-extremal graph.

(L) Let $|A_1| \leqslant M$. We prove that $|A_2| = o(n)$. Let $p \in W_2$ be joined to ν vertices of A_2. If H^+ is the graph spanned by p and H, then

(36) $$e(H^+) = E + n + o(n)\,.$$

To prove (36) let us replace each vertex of $S^N - H'$ by independent vertices completely joined to H'. Since $|A_1| = O(1)$, the number of edges will decrease by at most $o(n)$. If $\bar{U}^N$ is the graph obtained from an extremal graph of Problem 3 of $\nu(H_1)$ vertices and $N - \nu(H_1)$ independent vertices by joining the later ones to the extremal graph completely, then $e(\bar{U}^N) \leqslant e(S^N)$. This proves that H_1 must be almost extremal graph, it has by at most $o(n)$ less edges than the maximum is. Thus we obtained that (36), holds and the β-graphs of H^+ cannot be represented by less than $n - o(n)$ edges. Let $\widetilde{A} =: \mathrm{st}\,(p) \cap A_3$ and $\widetilde{B} = \{N^-(x) \colon x \in \widetilde{A}\}$. $(|N^-(x)| = 1$ and $|\widetilde{A}| = \nu$. Each $s \in \widetilde{B}$ but $\leqslant 2e$ has valence $\geqslant \dfrac{n}{2e}$ in H', otherwise $n/2$ edges could represent the β-graphs of $e(H^+)$. Hence each $s \in \widetilde{B}$ but at most $2e$ is chosen from a set of cardinality $4e^2$, i.e. at least $r = \left[\dfrac{\nu - 2e}{4e^2}\right]$ vertices have the same non-neighbour $N^-(x) = q$. This q represents all the β-graphs of H^+, unless $r < 2e$. But if q represents all the β-graphs of H^+, then $d_1(q) = n - o(n)$, thus $d_2(q) = n - o(n)$, too. Since it is not joined to $r \geqslant c_5 \nu$ vertices of H'', $\nu = o(n)$. Since each $x \in A_2$ is joined to one of $O(1)$ vertices of W_2, $|A_2| = o(n)$.

We obtained that almost all the vertices of H'' belong to A_3. (For this we need that $|A_4| = o(n)$, but this follows from the fact that $\bar{U}^N$

has at most as many edges as S^N!) Since $|A_3| = n - o(n)$, H_2 has $o(n)$ edges. Again, the comparison of $\bar{U}^N$ and S^N gives that at most $o(n)$ edges are missing between H_2 and H'. The argument of (K.2) remains valid for this case as well, thus $W_1 = \phi$ again. Then we can repeat the argument of (K.3) just replacing Z^N by $\bar{U}^N$. Thus we obtain that $V_i = \phi$. Then we can apply (K.4) without any change obtaining that H_2 contains no cycle. Hence H_2 is a forest.

(L.1) Let us suppose that B_1 and B_2 are two blocks of H_1 containing 2-connected planar subgraphs of $\frac{e}{3} + o(e)$ countries and $\frac{e}{3} + o(e)$ vertices (we do know that there exist $c_6 n$ such blocks in H_1 since it is almost extremal for Problem 3). Let us suppose that p is a common vertex of B_1 and B_2. One can easily show that if there exists a path in H_2 the endpoints of which are joined to each vertex of B_1 and B_2, and each other vertex of the path is joined to each vertex of $B_i - p$ $(i = 1, 2)$, then S^N contains an α-graph. Therefore, if T is a tree in H_2 each vertex of which is joined to $B_1 - p$ and $B_2 - p$ completely and two vertices of T are joined to p as well, then S^N contains an α-graph.

(L.2) $H_2 = G^{n_2}$ contains $2e$ independent vertices joined to $H_1 = G^{n_1} = H'$ completely. Therefore if we omit all the edges of H_2 and join its vertices to H_1 completely, then the obtained graph Z^N will not contain α-graphs, since S^N had not contained. Since H_1 is an almost extremal graph for Problem 3, it contains two blocks described in (L.1). Hence, if T is an arbitrary tree in H_2, at least $\nu(T) - 1$ edges are missing between T and H_1: omitting the edges of T and joining completely its vertices to H_1 we increased the number of edges unless the number of missing edges between T and H_1 was exactly $\nu(T) - 1$. On the other hand, since S^N is an extremal graph, we did not increase the number of edges. Therefore H_1 must be an extremal graph for Problem 3 and each tree T in H_2 represents $\nu(T) - 1$ missing edges. Thus each extremal block B_i in H_1 must contain the same vertex p unless $\nu(T) - 1 = 0$ for any tree in H_2, i.e. H_2 has no edges. In this later case the argument above works: each vertex represents 0 missing edges and H_1 is an extremal graph. This completes our proof.

Remark. The proof being rather complicated we did not try to make it even more complicated by proving an effective bound for e for which Theorem 1 holds. However, it is not too difficult to see that (I) was the most critical point from this view-point. If we could prove Conjecture 1, then this difficulty would disappear at once. This is the other reason, why it is not worth looking for such a bound. Anyway, if we wanted to get a good bound, one could easily replace (21) by $e(H_2) \leqslant 3n/2 + o(n)$. Further, the cycle $\widetilde{C}^{18}$ could be replaced by a smaller planar graph: a planar graph of 5 countries and 6 vertices would do.

REFERENCES

[1] P. Erdős, Über die in Graphen enthaltenen saturierten planaren Graphen, *Math. Nachrichten,* 40 (H. 1-3), (1969).

[2] P. Erdős – A.H. Stone, On the structure of linear graph, *Bull. Amer. Math. Soc.,* 52 (1946), 1089-1091.

[3] P. Erdős, Some recent results on extremal problems in graph theory (Results), *Theory of graphs Intern. Symp., Rome* (1966), 118-123.

[4] P. Erdős, On some new inequalities concerning extremal properties of graphs, *Theory of Graphs Proc. Coll. Tihany,* (1966), 77-81.

[5] M. Simonovits, A method for solving extremal problems in graph theory. Stability problems, *Theory of Graphs, Proc. Coll. Tihany,* (1966), 279-319.

[6] T. Kővári – V.T. Sós – P. Turán, On a problem of Zarankievicz, *Coll. Math.,* 3 (1954), 50-57.

[7] P. Erdős – T. Gallai, On maximal path and circuits of graphs, *Acta Math. Acad. Sci. Hungar.*, 10 (1959), 337-356.

COLLOQUIA MATHEMATICA SOCIETATIS JÁNOS BOLYAI
10. INFINITE AND FINITE SETS, KESZTHELY (HUNGARY), 1973.

A NEW COMBINATORIAL PARAMETER

F. STERBOUL

INTRODUCTION

Several parameters can be associated with a hypergraph H: for instance $n(H)$, the number of its vertices; $m(H)$, the number of its edges, or more sophisticated ones such as $\alpha(H)$, the weak stability number; $\chi(H)$, weak chromatic number, and many others. . . (We refer the reader to [1] for the elementary definitions of hypergraph theory).

Recently, C. Berge defined a new parameter: $\bar{\gamma}(H)$, called the chromatic number. The study of a new parameter may be carried out in many directions. The aim of this paper is to give some properties of $\bar{\gamma}(H)$, emphasizing the following general idea: "$\bar{\gamma}$ is closely related to α, the weak stability number".

§I contains, after the fundamental definitions, an inequality relating $\bar{\gamma}$ to α, and a result on the cases of equality in this inequality.

§II is devoted to the study, with respect to $\bar{\gamma}$, of a special class of hypergraphs, namely the direct products. This class of hypergraphs is indeed particularly worthy of interest, as regards the parameter α; and, since Z a r a n k i e w i c z [11] (1951), many authors have tackled this problem. (An extensive bibliography may be found in [6], see also the recent papers [2] and [9]).

In §III, the following extremal problem is studied: given the integers n, h, c, what is the minimum number of edges of a h-graph H (i.e. a uniform hypergraph of rank h) with n vertices and $\bar{\gamma}(H) \leq c$? $G(n, h, c)$ will denote this minimum number of edges. We observe that the relation between the function G and the parameter $\bar{\gamma}$ is identical with the relation between the function T (of T u r á n) and the parameter α. (On T u r á n's function, see the references [3], [4], [7], [8], [10]).

§I

Let $H = (X, \mathscr{E})$ be a hypergraph whose set of vertices is X and whose set of edges is $\mathscr{E}$: $\mathscr{E} \subset \mathscr{P}(X)$.

Definition 1. A set $S \subset X$ is weakly stable if no edge of H is included in S: $\forall E \in \mathscr{E},\ E \not\subset S$.

Definition 2. The weak stability number of H, called $\alpha(H)$, is the maximum number of elements of a weakly stable set.

$$\alpha(H) = \max \{|S|: S \text{ weakly stable}\} .$$

Definition 3. A colouring of the set X of vertices of H is a mapping f from X *onto* a set C of "colours". (f, C) denotes such a colouring.

Definition 4. The edge E of H is strongly coloured by the colouring (f, C) if for any two distinct vertices x and y in E: $f(x) \neq f(y)$.

Definition 5. $\bar{\gamma}(H)$ denotes the smallest integer k fulfilling the following condition: for every colouring (f, C) with $|C| = k$, there exists a strongly coloured edge of H.

A fundamental relation between $\bar{\gamma}$ and α is the following:

Lemma 6. *For every hypergraph* H: $\bar{\gamma}(H) \leq \alpha(H) + 1$.

Proof. Let (f, C) be a colouring of the vertices of H, with $|C| = \alpha(H) + 1$. Let $C = \{c_1, \ldots, c_k\}$ with $k = \alpha(H) + 1$ and $X_i = \{x : x \in X, f(x) = c_i\}$. Then $(X_1, \ldots, X_k)$ is a partition of the set X of vertices of H. For every index i, $(1 \leqslant i \leqslant k)$, let us choose a vertex $x_i \in X_i$. The set $\{x_1, \ldots, x_k\}$ cannot be stable, so it contains an edge E of H, and $|E \cap X_i| \leqslant 1$ for every i, $(1 \leqslant i \leqslant k)$. Hence, for every colouring of X with $\alpha(H) + 1$ colours, there exists a strongly coloured edge of H.

Theorem 7. *For every triple* (n, c, h) *of integers with* $n \geqslant c \geqslant h \geqslant 2$, *there exists a hypergraph* H *such that*

(1) H *is a* h*-graph*

(2) H *has* n *vertices*

(3) $\alpha(H) = c - 1$

(4) $\bar{\gamma}(H) = c$

Proof. H is constructed in the following way:

Let X be a set of cardinality n. Let $(X_1, \ldots, X_{c-1})$ be a partition of X into $c - 1$ non-empty sets. Let A be a subset of X, with $|A| = c$. By the "pigeon-hole principle", there exists an index i_0 such that $|A \cap X_{i_0}| \geqslant 2$. Let $E_1, \ldots, E_p$ be such that

$$E_j \subset A, \quad |E_j| = h, \quad |E_j \cap X_{i_0}| \geqslant 2 \qquad (1 \leqslant j \leqslant p),$$

$$\text{and} \quad \bigcup_{j=1}^{p} E_j = A\,.$$

We take $E_1, \ldots, E_p$ as the edges of H, and repeat this operation for each of the $\binom{n}{c}$ subsets A of X, with $|A| = c$. The hypergraph H obtained in this way is a h-graph; the union of its edges is X; we have $\alpha(H) \leqslant c - 1$, because every subset A, with $|A| = c$, contains at least one edge; and $\bar{\gamma}(H) \geqslant c$, because the partition $(X_1, \ldots, X_{c-1})$ is equivalent to a colouring of X with $c - 1$ colours in which no edge is strongly coloured. From Lemma 6, we have then: $\bar{\gamma}(H) = \alpha(H) + 1 = c$.

§II

Definition 8. Let $H_1 = (X_1, \mathscr{E}_1)$ and $H_2 = (X_2, \mathscr{E}_2)$ be two hypergraphs. The direct product $H_1 \times H_2$ is the hypergraph whose set of vertices is the cartesian product $X_1 \times X_2$, and whose set of edges is

$$\mathscr{E} = \{A \colon A \subset X_1 \times X_2,\ A = E_1 \times E_2,\ E_1 \in \mathscr{E}_1,\ E_2 \in \mathscr{E}_2\}.$$

Definition 9. K_n^h denotes the complete h-graph with n vertices: $K_n^h = (X, \mathscr{E})$, with $|X| = n$ and $\mathscr{E} = \mathscr{P}_h(X)$. ($\mathscr{P}_h(X)$ is the set of all the subsets, of cardinality h, of X).

The notation K_n is equivalent to K_n^2, the complete graph with n vertices.

The problem, raised by Zarankiewicz [11], of computing the value of $\alpha(K_{n_1}^{p_1} \times K_{n_2}^{p_2})$, and, in general of $\alpha(H_1 \times H_2)$, is far from being solved in spite of the many papers on this subject (see the Introduction). The following results show that finding the value of $\bar{\gamma}$ for products of hypergraphs may be less difficult, at least for the direct product of two graphs.

Theorem 10. *Let G_1 be a graph with n_1 vertices and whose stability number is α_1, and let K_{n_2} be the complete graph with n_2 vertices; then, $\bar{\gamma}(G_1 \times K_{n_2}) = n_1 + \alpha_1(n_2 - 1) + 1$.*

Proof. First, we prove that $\bar{\gamma}(G_1 \times K_{n_2}) \geq n_1 + \alpha_1(n_2 - 1) + 1$. Let X_1 (resp. X_2) be the set of vertices of G_1 (resp. K_{n_2}). Let S be a maximum stable set of G_1: $|S| = \alpha_1$. We construct a colouring of the elements of $X_1 \times X_2$ with $n_1 + \alpha_1(n_2 - 1)$ colours in the following way: $\alpha_1 n_2$ distinct colours are assigned to the elements of $S \times X_2$; and $n_1 - \alpha_1$ other colours are assigned to the elements of $(X_1 - S) \times X_2$, so that the subsets $\{x\} \times X_2$, for every $x \in X_1 - S$, be unicoloured. No edge of $G_1 \times K_{n_2}$ is strongly coloured by this colouring; hence we get the required inequality.

Then, we prove that $\bar{\gamma}(G_1 \times K_{n_2}) \leqslant n_1 + \alpha_1(n_2 - 1) + 1$, by induction with regard to n_2.

Case $n_2 = 2$. Let us consider an arbitrary colouring of the set $X_1 \times X_2 = \{x_1, \ldots, x_{n_1}\} \times \{y_1, y_2\}$ such that no edge of $G_1 \times K_{n_2}$ is strongly coloured. Let k be the number of colours which are present in $X_1 \times \{y_1\}$ and are *not* in $X_1 \times \{y_2\}$. If $k \leqslant \alpha_1$, the total number of colours is less or equal to $n_1 + \alpha_1$, and we get the required inequality. If $k > \alpha_1$, let $(x_1, y_1), \ldots, (x_k, y_1)$ be k vertices coloured with the k distinct colours that are not present in $X_1 \times \{y_2\}$. (In order to avoid heavy notation, we give a new indexing to the elements of X_1). The set $\{x_1, \ldots, x_k\}$ cannot be stable and spans a subgraph G' of G_1. Let $C_1, \ldots, C_p$ be the connected components of G'. We have $p \leqslant \alpha_1$: indeed, the set obtained by choosing one vertex in each component C_i is stable. Moreover, each one of the sets $C_i \times \{y_2\}$, $(1 \leqslant i \leqslant p)$ is unicoloured: indeed, if x and y are two adjacent vertices of C_i, the edge $\{x, y\} \times \{y_1, y_2\}$ of $G_1 \times K_2$ is not strongly coloured and the vertices (x, y_2) and (y, y_2) have the same colour. Thus, the number of colours present in $X_1 \times \{y_2\}$ is at most $n_1 - k + p \leqslant n_1 - k + \alpha_1$ and the total number of colours is at most equal to $n_1 + \alpha_1$.

Case $n_2 > 2$. Let us assume (by induction) that the formula of Theorem 10 is true up to the value $n_2 - 1$. Let us consider an arbitrary colouring of the set $X_1 \times X_2 = \{x_1, \ldots, x_{n_1}\} \times \{y_1, \ldots, y_{n_2}\}$ such that no edge of $G_1 \times K_{n_2}$ is strongly coloured. Let k be the number of colours which are present in $X_1 \times \{y_1\}$ and are *not* in $X_1 \times (X_2 - \{y_1\})$. If $k \leqslant \alpha_1$, by the induction hypothesis, the number of colours which are present in $X_1 \times (X_2 - \{y_1\})$ is less or equal to $n_1 + \alpha_1(n_2 - 2)$, so that the total number of colours is less or equal to $n_1 + \alpha_1(n_2 - 1)$, which gives the required inequality. If $k > \alpha_1$, as in the case $n_2 = 2$, let $(x_1, y_1), \ldots, (x_k, y_1)$ be k vertices coloured with the k distinct colours which are not present in $X_1 \times (X_2 - \{y_1\})$ and let $C_1, \ldots, C_p$ be the connected components of the subgraph G' of G_1 spanned by the set $\{x_1, \ldots, x_k\}$. As in the case $n_2 = 2$, we have $p \leqslant \alpha_1$ and each one of the sets $C_i \times \{y_j\}$, $(1 \leqslant i \leqslant p,\ 2 \leqslant j \leqslant n_2)$ is unicoloured. Let then

H_1 be the graph obtained from G_1 by contracting (see [1], p. 28) each component C_i into one single vertex d_i $(1 \leq i \leq p)$. The set of vertices of the graph H_1 is $\{d_1, \ldots, d_p, x_{k+1}, \ldots, x_{n_1}\}$. The initial colouring of the vertices of $G_1 \times K_{n_2}$ induces a colouring of the vertices of $H_1 \times K_{n_2-1}$ in the following way: the set of vertices of $H_1 \times K_{n_2-1}$ is $\{d_1, \ldots, d_p, x_{k+1}, \ldots, x_{n_1}\} \times (X_2 - \{y_1\})$; the vertex (d_i, y_j) is given the common colour of the set $C_i \times \{y_j\}$, $(1 \leq i \leq p,\ 2 \leq j \leq n_2)$, and the vertex (x_i, y_j), $(k+1 \leq i \leq n_1,\ 2 \leq j \leq n_2)$ is given the same colour it had in the former colouring. Thus, we obtain a colouring by which no edge of $H_1 \times K_{n_2-1}$ is strongly coloured; by the induction hypothesis, the number of colours in this colouring is less or equal to $n_1 - k + p + \alpha(H_1)(n_2 - 2) \leq n_1 - k + \alpha_1(n_2 - 1)$. Hence, the number of colours in the initial colouring of $G_1 \times K_{n_2}$ is less or equal to $n_1 + \alpha_1(n_2 - 1)$, which gives the required inequality.

Let us now consider two graphs G_1 and G_2. X_1 (resp. X_2) denotes the set of vertices, n_1 (resp. n_2) the number of vertices, α_1 (resp. α_2) the stability number, θ_1 (resp. θ_2) the clique-partition index (i.e. the minimum number of classes of a partition of the vertex-set such that each class is a clique of the graph) of G_1 (resp. G_2).

We give some bounds for $\bar{\gamma}(G_1 \times G_2)$. We first observe that, since $G_1 \times G_2$ is a partial hypergraph of $G_1 \times K_{n_2}$, we have

$$\bar{\gamma}(G_1 \times G_2) \geq n_1 + \alpha_1(n_2 - 1) + 1 .$$

Theorem 11. $\bar{\gamma}(G_1 \times G_2) \geq (n_1 - \alpha_1)\alpha_2 + (n_2 - \alpha_2)\alpha_1 + 2.$

Proof. Let S_1 (resp. S_2) be a maximum stable set of G_1 (resp. G_2): $|S_1| = \alpha_1$ and $|S_2| = \alpha_2$. Let us consider the following colouring of $X_1 \times X_2$, the set of vertices of $G_1 \times G_2$: $\alpha_1(n_2 - \alpha_2) + \alpha_2(n_1 - \alpha_1)$ distinct colours are assigned to the elements of the set $\{S_1 \times (X_2 - S_2)\} \cup \cup \{(X_1 - S_2) \times S_2\}$, and one single other colour is assigned to the elements of the set $(S_1 \times S_2) \cup \{(X_1 - S_1) \times (X_2 - S_2)\}$. It suffices to prove that no edge of $G_1 \times G_2$ is strongly coloured by this colouring. Let $E \times F$ be an edge of $G_1 \times G_2$, with $E = \{x_1, y_1\}$ and $F = \{x_2, y_2\}$. Since E is an edge of G_1, at least one of its extremities belongs to $X_1 - S_1$: for example $x_1 \in X_1 - S_1$, and likewise $x_2 \in X_2 - S_2$.

First case. $y_1 \in X_1 - S_1$ (or likewise $y_2 \in X_2 - S_2$). Then, the vertices (x_1, x_2) and (y_1, x_2) have the same colour, and the edge $E \times F$ is not strongly coloured.

Second case. $y_1 \in S_1$ and $y_2 \in S_2$. Then, the vertices (x_1, x_2) and (y_1, y_2) have the same colour and the edge $E \times F$ is not strongly coloured.

Remark. This last *inequality* as well as the preceding one, can be easily generalized to the case of the product of any two hypergraph.

Theorem 12. $\bar{\gamma}(G_1 \times G_2) \leqslant (n_1 - \alpha_1)\theta_2 + \alpha_1 n_2 + 1.$

Proof. Let us consider a colouring of the vertices of $G_1 \times G_2$ such that no edge is strongly coloured. We must give an upper bound for the number of colours in this colouring. Let $(C_1, \ldots, C_{\theta_2})$ be a partition of the vertex-set X_2 of G_2, each C_i being a clique, and $|C_i| = q_i$, $(1 \leqslant i \leqslant \theta_2)$. The sub-hypergraph of $G_1 \times G_2$ spanned by the set $X_1 \times C_i$ is isomorphic to $G_1 \times K_{q_i}$. The colouring of $G_1 \times G_2$ induces a colouring of $G_1 \times K_{q_i}$ by which no edge is strongly coloured. Hence, according to Theorem 10, the number of colours present in $X_1 \times C_i$ is less or equal to $n_1 + \alpha_1(q_i - 1)$. Thus, the total number of colours is less or equal to $\sum_{i=1}^{\theta_2} (n_1 + \alpha_1(q_i - 1)) = (n_1 - \alpha_1)\theta_2 + \alpha_1 n_2$.

The following results deal with the direct product of two complete hypergraphs. Proposition 14 and Lemma 15 are necessary for the proof of the main result, Theorem 16, which is quite analogous to the following Theorem of Čulik [5]

Theorem 13 (Čulik). *Let* $p_1 \geqslant 2$, $p_2 \geqslant 2$ *and* $n_2 \geqslant (p_2 - 1)\binom{n_1}{p_1}$, *then*

$$\alpha(K_{n_1}^{p_1} \times K_{n_2}^{p_2}) = (p_2 - 1)\binom{n_1}{p_1} + n_2(p_1 - 1)\,.$$

Proposition 14. *Let* $p_1 \geqslant 2$, $p_2 \geqslant 3$, *and* $n_2 \geqslant (p_2 - 2)\binom{n_1}{p_1}$, $n_1 > p_1$, *then*

$$\bar{\gamma}(K_{n_1}^{p_1} \times K_{n_2}^{p_2}) \geqslant 2 + (p_2 - 2)\binom{n_1}{p_1} + n_2(p_1 - 1)\,.$$

Proof. Let $n_2 = (p_2 - 2)\binom{n_1}{p_1} + q$, $q \geqslant 0$. The set of vertices of $K_{n_1}^{p_1} \times K_{n_2}^{p_2}$ is $X_1 \times X_2$, with $X_1 = \{x_1, \dots, x_{n_1}\}$ and $X_2 = \left\{y_{s,t};\ (1 \leqslant s \leqslant p_2 - 2),\ \left(1 \leqslant t \leqslant \binom{n_1}{p_1}\right)\right\} \cup \{z_1, \dots, z_q\}$. Moreover, let $\left\{I_t;\ \left(1 \leqslant t \leqslant \binom{n_1}{p_1}\right)\right\}$ be the set of all the subsets of cardinality p_1 of the segment $\{1, \dots, n_1\}$. Let C be a set of colours

$$C = \left\{c_{s,t,u}\ ;\ (1 \leqslant s \leqslant p_2 - 2),\ \left(1 \leqslant t \leqslant \binom{n_1}{p_1}\right),\ (1 \leqslant u \leqslant p_1)\right\} \cup$$

$$\cup \{d_{v,w};\ (1 \leqslant v \leqslant q),\ (1 \leqslant w \leqslant p_1 - 1)\} \cup \{\beta\}$$

so that $|C| = 1 + (p_2 - 2)\binom{n_1}{p_1} + n_2(p_1 - 1)$. We are going to construct a colouring of the elements of $X_1 \times X_2$, ($f(x, y)$ will denote the element of C which is the colour assigned to the vertex $(x, y) \in X_1 \times X_2$):

(1) let $f(x_i, y_{s,t}) = \beta$ if $i \notin I_t$

(2) s and t being fixed, each one of the p_1 vertices $(x_i, y_{s,t})$ (i ranging over I_t) is given one of the p_1 colours $c_{s,t,u}$ $(1 \leqslant u \leqslant p_1)$, so that the image by f of the set $\{(x_i, y_{s,t});\ (i \in I_t)\}$ is the set $\{c_{s,t,u};\ (1 \leqslant u \leqslant p_1)\}$.

(3) v being fixed, each one of the n_1 vertices $(x\ , z_v)$, $(1 \leqslant i \leqslant n_1)$ is arbitrarily given one of the $p_1 - 1$ colours $d_{v,w}$ $(1 \leqslant w \leqslant p_1 - 1)$, so that the image by f of the set $\{(x_i, z_v);\ (1 \leqslant i \leqslant n_1)\}$ is the set $\{d_{v,w};\ (1 \leqslant w \leqslant p_1 - 1)\}$.

It is sufficient to prove that no edge of $K_{n_1}^{p_1} \times K_{n_2}^{p_2}$ is strongly coloured by this colouring. Let $E \times F$ be an edge of this hypergraph: $E \subset X_1$, $F \subset X_2$, $|E| = p_1$, $|F| = p_2$. If $F \cap \{z_1, \ldots, z_q\} \neq \phi$, the edge $E \times F$ cannot be coloured with $p_1 p_2$ colours, hence it is not strongly coloured. If $F \cap \{z_1, \ldots, z_q\} = \phi$, the colour β is present at least twice in the edge $E \times F$, hence it is not strongly coloured.

Lemma 15. *Let us consider a colouring of* $X_1 \times X_2$, *the set of vertices of the hypergraph* $K_{n_1}^{p_1} \times K_{n_2}^{p_2}$, *with* $n_1 = p_1 + 1$, $n_2 = (p_1 + 1)(p_2 - 2) + 1$, $p_1 \geq 2$, $p_2 \geq 3$. *Let us assume that, for every* $x \in X_1$, *the colours which are present in* $\{x\} \times X_2$ *are not present in* $(X_1 - \{x\}) \times X_2$, *and that the number of colours present in* $\{x\} \times X_2$ *is greater or equal to* $1 + p_1(p_2 - 2)$. *Then, there exists a strongly coloured edge of* $K_{n_1}^{p_1} \times K_{n_2}^{p_2}$.

Proof. The lemma is proved by induction on p_1.

Case 1. $p_1 = 2$. Let $X_1 = \{x_1, x_2, x_3\}$ and $X_2 = \{y_j;\ (1 \leq j \leq n_2)\}$.

Case 1.1. Let us assume in this case that there exists $x \in X_1$ such that the number of colours present in $\{x\} \times X_2$ is strictly greater than $1 + 2(p_2 - 2)$. We take, for example $x = x_1$. Let then $L_1 = \{j_k;\ (1 \leq k \leq 2 + 2(p_2 - 2))\}$ be a set of indexes such that the colours of the vertices of $\{x_1\} \times \{y_j;\ (j \in L_1)\}$ are pairwise distinct. Let L_2 be a set of indexes such that $|L_2| = 1 + 2(p_2 - 2)$ and the colours of the vertices of $\{x_2\} \times \{y_j;\ (j \in L_2)\}$ are pairwise distinct. Then, $|L_1 \cap L_2| = |L_1| + |L_2| - |L_1 \cup L_2|$, hence $|L_1 \cap L_2| \geq 3 + 4(p_2 - 2) - n_2 \geq p_2$. Then the set $\{x_1, x_2\} \times \{y_j;\ (j \in L_1 \cap L_2)\}$ contains a strongly coloured edge.

Case 1.2. Let us assume in this case that, for every $x \in X_1$, the number of colours present in $\{x\} \times X_2$ is equal to $1 + 2(p_2 - 2)$. For $i = 1, 2$ or 3, let J_i be a set of indices, with $|J_i| = 1 + 2(p_2 - 2)$, such that the colours of the vertices of $\{x_i\} \times \{y_j;\ (j \in J_i)\}$ are pairwise distinct; and let $K_i = \{1, 2, \ldots, n_2\} - J_i$. Let us assume (ad absurdum) that no edge is strongly coloured. The following points are successively proved

$|J_1 \cap J_2| \leqslant p_2 - 1$. Otherwise, the set $\{x_1, x_2\} \times \{y_j;\ (j \in J_1 \cap J_2)\}$ would contain a strongly coloured edge.

$K_1 \subset J_2$. Indeed, $1 + 2(p_2 - 2) = |J_2| = |J_1 \cap J_2| + |K_1 \cap J_2|$, hence $|K_1 \cap J_2| \geqslant p_2 - 2 = |K_1|$ and $K_1 \subset J_2$.

$|J_1 \cap J_2| = p_2 - 1$. Otherwise, $|J_1 \cap J_2| \leqslant p_2 - 2$, and we would have $p_2 - 2 = |K_1| = |J_2| - |J_1 \cap J_2| \geqslant p_2 - 1$.

$|J_1 \cap J_2 \cap J_3| = 1$. Indeed, $K_1 \subset J_2$ and similarly $K_3 \subset J_2$ imply $K_1 \cup K_3 \subset J_2$, hence $J_1 \cap J_3 \supset K_2$, hence $|J_1 \cap J_2 \cap J_3| = |J_1 \cap J_3| - |K_2| = 1$.

Let then $\{j_0\} = J_1 \cap J_2 \cap J_3$, and let j_2 (resp. j_3) be an element of K_2 (resp. K_3). By the definition of J_2, there exists one single vertex j_4 such that $j_4 \in J_2$ and the colour of the vertex (x_2, y_{j_4}) is the same as the colour of the vertex (x_2, y_{j_2}). We prove now

$j_4 \in K_1 \cup \{j_0\}$. Otherwise, the set $\{x_2, x_3\} \times \{y_j;\ (j \in K_1 \cup \{j_0\} \cup \cup \{j_2\})\}$ would contain a strongly coloured edge.

$j_4 = j_0$. Indeed, $j_4 \in K_1 \cup \{j_0\}$ and similarly $j_4 \in K_3 \cup \{j_0\}$ and, since $K_3 \subset J_1$, we have $K_1 \cap K_3 = \phi$.

Similarly, j_0 is the only index j of J_3 such that the colour of (x_3, y_j) is the same as the colour of (x_3, y_{j_3}). This is absurd, because the set $\{x_2, x_3\} \times \{y_j;\ (j \in K_1 \cup \{j_2\} \cup \{j_3\})\}$ would contain a strongly coloured edge.

Case 2. $p_1 > 2$. Let us assume (by induction) that the lemma is true up to the value $p_1 - 1$. We consider a colouring satisfying the conditions of the lemma. Let $X_1 = \{x_1, \ldots, x_{n_1}\}$ and $X_2 = \{y_1, \ldots, y_{n_2}\}$. Let J be a set of indexes such that $|J| = 1 + p_1(p_2 - 2)$ and the colours of the vertices of $\{x_1\} \times \{y_j;\ (j \in J)\}$ are pairwise distinct. For every index i, $(2 \leqslant i \leqslant n_1)$ the number of colours present in $\{x_i\} \times \{y_j;\ (j \in J)\}$ is greater or equal to $1 + (p_1 - 1)(p_2 - 1)$. According to the induction hypothesis, the set $(X_1 - \{x_1\}) \times \{y_j;\ (j \in J)\}$ contains the strongly coloured subset: $\{x_i;\ (i \in I)\} \times \{y_j;\ (j \in K)\}$, with $I \subset \{2, \ldots, n_1\}$, $K \subset J$, $|I| = p_1 - 1$,

$|K| = p_2$. Hence, the set $(\{x_i;\ (i \in I)\} \cup \{x_1\}) \times \{y_j;\ (j \in K)\}$ is a strongly coloured edge of $K_{n_1}^{p_1} \times K_{n_2}^{p_2}$.

Theorem 16. *Let* $p_1 \geqslant 2$, $p_2 \geqslant 3$, *and* $n_2 \geqslant (p_2 - 2)\binom{n_1}{p_1}$, *then*

$$\bar{\gamma}(K_{n_1}^{p_1} \times K_{n_2}^{p_2}) = 2 + (p_2 - 2)\binom{n_1}{p_1} + n_2(p_1 - 1)\,.$$

Proof. Let $X_1 \times X_2$ be the set of vertices, with $X_1 = \{x_1, \ldots, x_{n_1}\}$ and $X_2 = \{y_1, \ldots, y_{n_2}\}$.

Case 1. $n_1 = p_1$. We must prove that $\bar{\gamma} = n_2(p_1 - 1) + p_2$. First, we prove that $\bar{\gamma} \geqslant n_2(p_1 - 1) + p_2$ by constructing a colouring, with $n_2(p_1 - 1) + p_2 - 1$ colours, by which no edge is strongly coloured: $(p_1 - 1)n_2$ distinct colours are assigned to the vertices of $(X_1 - \{x_1\}) \times \times X_2$, and $p_2 - 1$ other colours are arbitrarily assigned to the vertices of $\{x_1\} \times X_2$.

Then, we prove that $\bar{\gamma} \leqslant n_2(p_1 - 1) + p_2$. Indeed, let us consider an arbitrary colouring of $X_1 \times X_2$ with $n_2(p_1 - 1) + p_2$ colours, it suffices to show that there exists a strongly coloured edge. For every index j, $(2 \leqslant j \leqslant n_2)$, let l_j be the number of colours which are present in $X_1 \times \{y_j\}$ and are *not* in $X_1 \times \{y_1, \ldots, y_{j-1}\}$; and let l_1 be the number of colours present in $X_1 \times \{y_1\}$. The total number of colours is $\sum_{j=1}^{n_2} l_j = n_2(p_1 - 1) + p_2$. Let $k = |\{j\colon l_j = n_1 = p_1\}|$. Then, $n_2(p_1 - 1) + p_2 \leqslant kn_1 + (n_2 - k)(n_1 - 1)$, hence $k \geqslant p_2$. Let, then, $J \subset \{j\colon l_j = n_1 = p_1\}$, with $|J| = p_2$. The set $X_1 \times \{y_j;\ (j \in J)\}$ is a strongly coloured edge.

For $n_1 > p_1$, the theorem will be proved by induction on n_1. Therefore, the theorem is assumed to be true up to the value $n_1 - 1$. It is known (Proposition 14) that $\bar{\gamma} \geqslant A = 2 + (p_2 - 2)\binom{n_1}{p_1} + n_2(p_1 - 1)$, hence it suffices to prove that $\bar{\gamma} \leqslant A$, and this will be done by considering an arbitrary colouring with A colours and proving the existence of a strongly coloured edge.

Case 2. $n_1 > p_1$ and $n_2 = (p_2 - 2)\binom{n_1}{p_1} + \alpha$, with $\alpha = 0$ or 1.

For every index i, $(1 \leq i \leq n_1)$, let k_i be the number of colours which are present in $\{x_i\} \times X_2$ and are *not* in $(X_1 - \{x_i\}) \times X_2$. The total number of colours, A, is greater or equal to $\left(\sum_{i=1}^{n_1} k_i\right) + \epsilon$ with $\epsilon = 1$ in general, and $\epsilon = 0$ in the special case when, for every index i, every colour which is present in $\{x_i\} \times X_2$ is not present in $(X_1 - \{x_i\}) \times X_2$.

Case 2.1. In this case, we assume that there exists an index i_0 such that $k_{i_0} \leq (p_2 - 2)\binom{n_1 - 1}{p_1 - 1}$. Then, at least $2 + (p_2 - 2)\binom{n_1 - 1}{p_1} +$ $+ n_2(p_1 - 1)$ colours are present in $(X_1 - \{x_{i_0}\}) \times X_2$, and we have $n_2 \geq (p_2 - 2)\binom{n_1 - 1}{p_1}$. Hence, according to the induction hypothesis, there exists a strongly coloured edge (included in $(X_1 - \{x_{i_0}\}) \times X_2$).

Case 2.2. In this case, we assume that for every index i, $(1 \leq i \leq n_1)$, we have $k_i \geq (p_2 - 2)\binom{n_1 - 1}{p_1 - 1} + 1$. Then,

$$A = 2 + (p_2 - 2)\binom{n_1}{p_1} + n_2(p_1 - 1) \geq$$

$$\geq \sum_{i=1}^{n_1} k_i + \epsilon \geq \epsilon + n_1(p_2 - 2)\binom{n_1 - 1}{p_1 - 1} + n_1$$

hence $n_1 \leq 2 - \epsilon + \alpha(p_1 - 1)$. This inequality, together with $n_1 > p_1$, implies $\epsilon = 0$, $\alpha = 1$ and $n_1 = p_1 + 1$, so that we have $n_2 = (p_1 + 1)(p_2 - 2) + 1$, as well as the other hypotheses of Lemma 15: by Lemma 15, there exists a strongly coloured edge.

Case 3. $n_2 \geq (p_2 - 2)\binom{n_1}{p_1} + 2$, and $n_1 > p_1$.

As in case 2, we consider a colouring of $X_1 \times X_2$ with A colours. We will prove the existence of a strongly coloured edge by induction with regard to n_2, the values n_1, p_1, p_2 being fixed. Therefore, we assume that the theorem is true up to the value $n_2 - 1$.

For every index j, $(1 \leqslant j \leqslant n_2)$, let s_j be the number of colours which are present in $X_1 \times \{y_j\}$ and are *not* in $X_1 \times (X_2 - \{y_j\})$.

Case 3.1. In this case, we assume that there exists an index j_0 such that $s_{j_0} \leqslant p_1 - 1$. Then, the number of colours present in $X_1 \times$ $\times (X_2 - \{y_{j_0}\})$ is greater or equal to $2 + (p_2 - 2)\binom{n_1}{p_1} + (n_2 - 1)(p_1 - 1)$. According to the induction hypothesis, there exists a strongly coloured edge (included in $X_1 \times (X_2 - \{y_{j_0}\})$).

Case 3.2. In this case, we assume that, for every index j, $s_j \geqslant p_1$. The total number of colours is greater or equal to $\left(\sum_{j=1}^{n_2} s_j\right) + \epsilon$ with $\epsilon = 1$ in general, and $\epsilon = 0$ in the special case when, for every index j, every colour which is present in $X_1 \times \{y_j\}$ is not in $X_1 \times (X_2 - \{y_j\})$. Hence,

$$A = 2 + (p_2 - 2)\binom{n_1}{p_1} + n_2(p_1 - 1) \geqslant \sum_{j=1}^{n_2} s_j + \epsilon \geqslant n_2 p_1 + \epsilon .$$

Hence, $n_2 \leqslant 2 - \epsilon + (p_2 - 2)\binom{n_1}{p_1}$, which implies $\epsilon = 0$ and $n_2 =$ $= 2 + (p_2 - 2)\binom{n_1}{p_1}$.

Then, for every index j, $(1 \leqslant j \leqslant n_2)$, let t_j be the number of p_1-tuples $I = (i_1, \ldots, i_{p_1})$ such that the colours of the vertices of $\{x_i;$ $(i \in I)\} \times \{y_j\}$ are pairwise distinct. The inequality $s_j \geqslant p_1$ implies $t_j \geqslant 2$ (A better lower bound could be given, but is not needed here.)

If $\sum_{j=1}^{n_2} t_j > (p_2 - 1)\binom{n_1}{p_1}$ (inequality (D)), the "pigeon-hole principle" implies that there exist a p_1-tuple $I = (i_1, \ldots, i_{p_1})$ and a p_2-tuple

$J = (j_1, \ldots, j_{p_2})$ such that the edge $\{x_i;\ (i \in I)\} \times \{y_j;\ (j \in J)\}$ is strongly coloured. Let us prove inequality (D)

$$\sum_{j=1}^{n_2} t_j \geqslant 2n_2 = 2\left(2 + (p_2 - 2)\binom{n_1}{p_1}\right) \geqslant 4 + (p_2 - 1)\binom{n_1}{p_1}.$$

Q.E.D.

§III

In this § we consider h-graphs (i.e. hypergraphs $H = (X, \mathscr{E})$ with $\mathscr{E} \subset \mathscr{P}_h(X)$), but we do not impose upon H the classical condition $\bigcup_{E \in \mathscr{E}} E = X$.

Definition 17. Let $n \geqslant c \geqslant h \geqslant 2$. $T(n, h, c)$ denotes the smallest of those integers m for which there exists a h-graph $H = (X, \mathscr{E})$ such that $|X| = n$, $|\mathscr{E}| = m$, and $\alpha(H) \leqslant c - 1$.

The problem of computing the value of $T(n, h, c)$ was solved by Turán [10] in the case $h = 2$. For the case $h > 2$, see the references [3], [4], [7], [8].

Definition 18. Let $n \geqslant c \geqslant h \geqslant 2$. $G(n, h, c)$ denotes the smallest of those integers m for which there exists a h-graph $H = (X, \mathscr{E})$ such that $|X| = n$, $|\mathscr{E}| = m$, and $\bar{\gamma}(H) \leqslant c$.

The two following results connect the function G to the function T.

Proposition 19. $G(n, h, c) \leqslant T(n, h, c)$.

Proof. Immediate from Lemma 6.

Proposition 20. $T(n, h - 1, c - 1) \leqslant hG(n, h, c)$.

Proof. Let $H = (X, \mathscr{E})$ be a h-graph such that $|X| = n$, $|\mathscr{E}| = G(n, h, c)$ and $\bar{\gamma}(H) \leqslant c$. Let $H_1 = (X, \mathscr{E}_1)$ be the $(h - 1)$-section of H: $A \in \mathscr{E}_1 \Leftrightarrow |A| = h - 1$ and $\exists E \in \mathscr{E},\ A \subset E$.

Let us prove that $\alpha(H_1) \leqslant c-2$. Indeed, let B be a subset of X, with $|B| = c-1$. Let us assign to the elements of B, $c-1$ distinct colours, and to the elements of $X - B$ one single other colour: we obtain a colouring of X with c colours. Since $\bar{\gamma}(H) \leqslant c$, there exists a strongly coloured edge E of H; hence, $|E \cap (X-B)| \leqslant 1$, and $|E \cap B| \geqslant h-1$. Hence, $E \cap B$ contains an edge of H_1, and B is not a weakly stable set of H_1.

Moreover, $|\mathscr{E}_1| \leqslant h|\mathscr{E}| = hG(n, h, c)$, and this proves the Proposition.

The three following propositions state functional relations satisfied by both function G and T.

Proposition 21. $G(n, h, c) \leqslant G(n-1, h-1, c-1) + G(n-1, h, c)$.

Proof. (The following construction is identical with the one used by Chvátal [3] for the function T). Let $H_1 = (X, \mathscr{E}_1)$ be a $(h-1)$-graph with $|X| = n-1$, $|\mathscr{E}_1| = G(n-1, h-1, c-1)$ and $\bar{\gamma}(H_1) \leqslant c-1$. Let $H_2 = (X, \mathscr{E}_2)$ be a h-graph with $|\mathscr{E}_2| = G(n-1, h, c)$ and $\bar{\gamma}(H_2) \leqslant c$. Let $x_0 \notin X$ and $Y = X \cup \{x_0\}$. Let $H = (Y, \mathscr{E})$ be the h-graph defined by $\mathscr{E} = \mathscr{E}_2 \cup \{E\colon E = A \cup \{x_0\},\ A \in \mathscr{E}_1\}$. If suffices to prove $\bar{\gamma}(H) \leqslant c$. Hence, let us consider an arbitrary colouring of Y with c colours.

Case 1. In this case, we assume that the colour assigned to x_0 is not present in X. Hence, X is coloured with $c-1$ colours, and there exists a strongly coloured edge A of H_1. Then, $E = A \cup \{x_0\}$ is a strongly coloured edge of H.

Case 2. In this case, we assume that the colour assigned to x_0 is also present in X. Hence, X is coloured with c colours, and there exists a strongly coloured edge E of H_2; but E is also a strongly coloured edge of H.

Proposition 22. $G(n, h, c) \leqslant G(n, h, c-1)$.

Proof. Immediate from the definitions.

Proposition 23. $G(n, h, c) \geqslant G(n, h-1, c)$.

Proof. Let $H_1 = (X, \mathscr{E}_1)$ be a h-graph with $|X| = n$, $|\mathscr{E}_1| = G(n, h, c)$ and $\bar{\gamma}(H_1) \leqslant c$. If $\mathscr{E}_1 = \{E_i;\ (i \in I)\}$, let $H = (X, \mathscr{E})$ be a $(h-1)$-graph constructed by arbitrarily choosing in each edge E_i of H_1 a subset A_i, with $|A_i| = h - 1$, and by taking $\mathscr{E} = \{A_i;\ (i \in I)\}$. Then, $|\mathscr{E}| = |\mathscr{E}_1| = G(n, h, c)$ and $\bar{\gamma}(H) \leqslant c$. This proves the proposition.

The following lemma is due to C. Berge.

Lemma 24. *Let H be a 2-graph (i.e. a graph) with exactly p connected components. Then,* $\bar{\gamma}(H) = p + 1$.

Proof.

(1) $\bar{\gamma}(H) \geqslant p + 1$. Indeed, let $H_i = (X_i, \mathscr{E}_i)$, $(1 \leqslant i \leqslant p)$, be the connected components of the graph $H = (X, \mathscr{E})$. $(X_1, \ldots, X_p)$ is a partition of X. Let $C = \{c_1, \ldots, c_p\}$ be a set of p distinct colours, and let (f, C) be the colouring of X defined by $f(x) = c_i$ if $x \in X_i$. No edge of H is strongly coloured by this colouring, so that $\bar{\gamma}(H) \geqslant p + 1$.

(2) $\bar{\gamma}(H) \leqslant p + 1$. Indeed, let (f, C) be an arbitrary colouring of X, with $|C| = p + 1$. By the "pigeon-hole principle", there exists an index i_0 and two vertices a and b, such that $a \in X_{i_0}$, $b \in X_{i_0}$, and $f(a) \neq f(b)$. Let $[a = a_1, a_2, \ldots, a_q = b]$ be an elementary chain connecting a to b in the component H_{i_0}. There exists an index k, $(1 \leqslant k \leqslant q - 1)$ such that $f(a_k) \neq f(a_{k+1})$. Hence, the edge $[a_k, a_{k+1}]$ is strongly coloured.

Proposition 25. $G(n, 2, c) = n - c + 1$.

Proof.

(1) $G(n, 2, c) \geqslant n - c + 1$. Indeed, let $H = (X, \mathscr{E})$ be a graph such that $|X| = n$, $\bar{\gamma}(H) \leqslant c$, $|\mathscr{E}| = G(n, 2, c)$ and let $p(H)$ be the number of its connected components. According to Lemma 24, $p(H) \leqslant c - 1$. The cyclomatic number $|\mathscr{E}| - |X| + p(H)$ is non-negative, hence $|\mathscr{E}| \geqslant n - p(H) \geqslant n - c + 1$.

(2) $G(n, 2, c) \leqslant n - c + 1$. Indeed, let $H = (X, \mathscr{E})$ be a graph such that $|X| = n$, $p(H) = c - 1$, and whose $c - 1$ connected components

are trees. Then, $\bar{\gamma}(H) = p(H) + 1 = c$, and $|\mathscr{E}| = n - p(H) = n - c + 1$, and this proves the proposition.

Proposition 26. $G(n, h, c) \leqslant \binom{n - c + h - 1}{h - 1}$.

Proof. By induction, using the inequality of Proposition 21, Proposition 25, and the immediate formula: $G(n, h, n) = 1$.

Proposition 26 could also be deduced from the following

Proposition 27. $G(n, h, c) \leqslant T(n - 1, h - 1, c - 1)$.

Proof. Let $H_1 = (X, \mathscr{E}_1)$ be a $(h - 1)$-graph such that $|X| = n - 1$, $|\mathscr{E}_1| = T(n - 1, h - 1, c - 1)$ and $\alpha(H_1) \leqslant c - 2$. Let $x_0 \notin X$ and $Y = X \cup \{x_0\}$. Let $H = (Y, \mathscr{E})$ be the h-graph defined by $\mathscr{E} = \{E\colon E = A \cup \{x_0\}, A \in \mathscr{E}_1\}$. It suffices to prove $\bar{\gamma}(H) \leqslant c$. Let (f, C) be a colouring of Y with c colours. Let B be a subset of X such that:

(1) $|B| = c - 1$

(2) $x \in B \Rightarrow f(x) \neq f(x_0)$

(3) $x \in B,\ y \in B,\ x \neq y \Rightarrow f(x) \neq f(y)$.

Then, condition (1) implies that B contains an edge A of H_1 and conditions (2) and (3) imply that the edge $E = A \cup \{x_0\}$ of H is strongly coloured.

REFERENCES

[1] C. Berge, *Graphes et hypergraphes,* Dunod, Paris, 1970.

[2] C. Berge – M. Simonovits, *The coloring numbers of the direct product of two hypergraphs,* in Hypergraph Seminar, Lecture Notes, Springer Verlag, Berlin. (*to be published*)

[3] V. Chvátal, Hypergraphs and Ramseyian theorems, *Ph. D. Thesis,* University of Waterloo, 1970.

[4] V. Chvátal, Hypergraphs and Ramseyian theorems, *Proc. American Math. Soc.*, 27, 3 (1971), 434-440.

[5] K. Čulik, Teilweise Lösung eines verallgemeinerten Problems von K. Zarankiewicz, *Ann. Polon. Math.*, 3 (1956), 165-168.

[6] R.K. Guy, *A many-facetted problem of Zarankiewicz,* in The many facets of graph theory, Lecture Notes 110, Springer Verlag, Berlin, 1969, 129-148.

[7] G. Katona – T. Nemetz – M. Simonovits, On a graph-problem of Turán (in Hungarian), *Mat. Lapok,* 15 (1964), 228-238.

[8] J. Schönheim, On coverings, *Pacific J. Math.*, 14 (1964), 1405-1411.

[9] F. Sterboul, *On the chromatic number of the direct product of two hypergraphs,* in Hypergraph Seminar, Lecture Notes, Springer Verlag, Berlin. (to be published)

[10] P. Turán, An extremal problem in graph theory (in Hungarian), *Mat. Fiz. Lapok,* 48 (1941), 436-452. See also: On the theory of graphs, *Colloq. Math.*, 3 (1954), 19-30.

[11] K. Zarankiewicz, Problem 101, *Colloq. Math.*, 2 (1951), 301.

COLLOQUIA MATHEMATICA SOCIETATIS JÁNOS BOLYAI
10. INFINITE AND FINITE SETS, KESZTHELY (HUNGARY), 1973.

LOCALLY MAXIMAL SUM-FREE SETS

A.P. STREET – E.G. WHITEHEAD, JR.*

1. INTRODUCTION

Definition. Let S be a subset of a finite group G, written additively. S is *sum-free* if $S \cap (S + S) = \phi$.

Definition. A sum-free set S is *maximal* if for every sum-free $T \subseteq G$, $|T| \leqslant |S|$. Let $\lambda(G)$ be the cardinality of a maximal sum-free set in G.

From results of P. Erdős [1], it follows that

$$2|G|/7 \leqslant \lambda(G) \leqslant |G|/2$$

where G is a finite abelian group.

We now generalize the concept of a maximal sum-free set.

Definition. A sum-free set S is *locally maximal* if for every sum-free T where $S \subseteq T \subseteq G$, $S = T$ holds. Let $\Lambda(G)$ be the set of cardinalities of all locally maximal sum-free sets in G.

*Supported in part by the U.S. Atomic Energy Commission, Contract No. AT(11-1)-3077 and in part by the U.S. Air Force, Contract No. AF-AFOSR-2008.

Clearly $\lambda(G) = \max \Lambda(G)$. Let $\mu(G) = \min \Lambda(G)$. A. A. Mullin [3] considered locally maximal sum-free sets under the name of "maximal mutant sets." We shall obtain lower bounds on $\mu(G)$ for the elementary abelian 2-groups.

2. GROUP RAMSEY THEORY

This section highlights a recent paper by Street and Whitehead [5], "Group Ramsey Theory". We mention results from that paper which are needed in the next two sections of this paper.

Every partition of G^* (the nonidentity elements of group G) into k sum-free sets can be extended to at least one covering of G^* by k locally maximal sum-free sets. This extensions is done by adjoining elements to each partition set until a locally maximal set is reached. By using the Greenwood and Gleason [2] subgraph technique, one can show that every set in such a covering, has cardinality less than the Ramsey number $N_{k-1} \equiv N(\underbrace{3, \ldots 3}_{k-1 \text{ threes}}; 2)$.

Definition. Let S be a locally maximal set in G. If $S \cup (S+S) \supseteq G^*$, then S *fills* G. If every locally maximal sum-free set in G fills G, then G is a *filled group.*

Theorem 1. *A finite abelian group is a filled group if and only if it is*

(1) *an elementary abelian* 2*-group,*

(2) Z_3, *or*

(3) Z_5.

3. ABELIAN GROUPS OF ORDER 16

Theorem 2. *Let* G *be an abelian group of order* 16. G^* *can be partitioned into* 3 *sum-free sets if and only if* G *is*

(1) $(Z_2)^4$ *or*

(2) $(Z_4)^2$.

Proof. Greenwood and Gleason [2] partition $(Z_2)^4$ into 3 sum-free sets. Whitehead [6] partitions $(Z_4)^2$ into 3 sum-free sets. Therefore the "only if" is all that is left to prove.

To complete the proof, we show that $(Z_{16})^*$, $(Z_2 \times Z_8)^*$, and $(Z_2 \times Z_2 \times Z_4)^*$ cannot be covered by 3 locally maximal sum-free sets of order 5. This is equivalent to showing that these groups cannot be partitioned into 3 sum-free sets by the results in the previous section and the fact that $N_2 = 6$ [2].

Z_{16}: Computer computations showed that *all* locally maximal sum-free 5-sets in Z_{16} contain 8, the unique element of order 2; thus $(Z_{16})^*$ cannot be so partitioned. (Schür [4] proved that $\{1, \ldots, 14\}$ cannot be partitioned into three sum-free sets. From Schür's result it follows that $(Z_{16})^*$ cannot be so partitioned.)

$Z_2 \times Z_8$: Computer computations showed that there are *no* locally maximal sum-free 5-sets in $Z_2 \times Z_8$; thus $(Z_2 \times Z_8)^*$ cannot be so partitioned.

$Z_2 \times Z_2 \times Z_4$: Computer computations showed that *all* locally maximal sum-free 5-sets in $Z_2 \times Z_2 \times Z_4$ contain exactly 2 elements of order 4. Thus $(Z_2 \times Z_2 \times Z_4)^*$ cannot be so partitioned, since this group contains 8 elements of order 4.

Thus we have proven the theorem. Actually these computer computations also gave the following results:

$$\Lambda(Z_{16}) = \{4, 5, 6, 8\},$$

$$\Lambda((Z_4)^2) = \{3, 4, 5, 8\},$$

$$\Lambda(Z_2 \times Z_8) = \{4, 6, 8\},$$

$$\Lambda((Z_2)^2 \times Z_4) = \{4, 5, 8\},$$

$$\Lambda((Z_2)^4) = \{5, 8\},$$

4. ELEMENTARY ABELIAN 2-GROUPS

We write our elementary abelian 2-groups as $(Z_2)^n$ for $n \geqslant 1$. By the results stated in Section 2, we know that $(Z_2)^n$ is a *filled group* for all $n \geqslant 1$. Below we use this result to obtain a lower bound for $\mu((Z_2)^n) \equiv$ $\equiv \min \Lambda((Z_2)^n)$.

Theorem. *$\mu((Z_2)^n) \geqslant m$ where m is the smallest integer such that*

$$m + \binom{m}{2} \geqslant |((Z_2)^n)^*| .$$

Proof. Clear, since $(Z_2)^n$ is filled, abelian and $g + g = 0$ for all $g \in (Z_2)^n$.

Remarks. Using this theorem to calculate lower bounds for $\mu((Z_2)^n)$ for $n = 3, 4, 5, 6,$ we obtain

$$4 \leqslant \mu((Z_2)^3) ,$$
$$5 \leqslant \mu((Z_2)^4) ,$$
$$8 \leqslant \mu((Z_2)^5) ,$$
$$11 \leqslant \mu((Z_2)^6) .$$

Compare the above lower bounds with the following computer results

$$\Lambda((Z_2)^3) = \{4\} ,$$
$$\Lambda((Z_2)^4) = \{5, 8\} ,$$
$$\Lambda((Z_2)^5) \supseteq \{9, 10, 16\} ,$$
$$\Lambda((Z_2)^6) \supseteq \{13, 17, 18, 20, 32\} .$$

Thus $\mu((Z_2)^n)$ is known exactly for $n \leqslant 4$.

Finally, *if* we could show that $16 \notin \Lambda((Z_2)^6)$, *then* we could prove the following conjecture.

Conjecture. $((Z_2)^6)^*$ cannot be partitioned into 4 sum-free sets.

Acknowledgements. The computer computations were done on the A.E.C. CDC 6600 computer at the Courant Institute of Mathematical Sciences.

REFERENCES

[1] P. Erdős, Extremal Problems in Number Theory, *Proc. Sympos. Pure Math.*, 8, Amer. Math. Soc., (1965), 181-9.

[2] R.E. Greenwood – A.M. Gleason, Combinatorial Relations and Chromatic Graphs, *Canadian J. Math.*, 7 (1955), 1-7.

[3] A.A. Mullin, On Mutant Sets, *Bull. Math. Biophysics*, 24 (1962), 209-15.

[4] I. Schür, Über die Kongruenz $x^m + y^m \equiv z^m \pmod{p}$, *Jahresbericht der Deutschen Mathematiker Vereinigung*, 25 (1916), 114-7.

[5] A.P. Street – E.G. Whitehead, Jr., Group Ramsey Theory, *J. Combinatorial Theory*, (A) 17 (1974), 219-226.

[6] E.G. Whitehead, Jr., Algebraic Structure of Chromatic Graphs Associated with the Ramsey Number $N(3, 3, 3; 2)$, *Discrete Math.*, 1 (1971), 113-4.

COLLOQUIA MATHEMATICA SOCIETATIS JÁNOS BOLYAI
10. INFINITE AND FINITE SETS, KESZTHELY (HUNGARY), 1973.

ON LINE CRITICAL GRAPHS

L. SURÁNYI

§0. INTRODUCTION

Throughout this paper graph always means a finite graph without loops and multiple edges. G always denotes a graph, its vertex- resp. edge-set will be denoted by $V(G)$ resp. $E(G)$. $\rho(G, x)$ denotes the degree of the vertex x in G. We use the usual set-theoretical notations. A subset H of $V(G)$ is called *independent* iff H spans no edge in G. We put

$$\mathscr{F}(l, G) = \{H \subset V(G).\ H \text{ is independent in } G,\ |H| = l\}.$$

$$(\mathscr{F}(0, G) = \{\phi\}.)$$

G is said to be *α-critical* (*α-crit.*) iff $\mathscr{F}(\alpha + 1, G) = \phi$ but $\mathscr{F}(\alpha + 1, G - e) \neq \phi$ for all $e \in E(G)$.

G is said to be *strongly α-critical* (*str. α-crit.*) iff G is α-crit. and $\rho(G, x) \geqslant 1$ for each $x \in V(G)$.

The concept of α-saturated graphs – which are the complements of α-critical graphs – was introduced in [17] and the concept of "point and edge critical graphs" – which are the graphs being str. α-crit. for some

α – was introduced in [5]. These graphs are now usually called line-critical. Recently a lot of papers study the properties of α-crit. graphs (see [1], [8], [11], [13], [14], [18], [19] and partially [2], [5], [6], [10], [20]). In [5] Erdős and Gallai proved the following

Theorem 1. [5, Theorem 3.10 and 4.8]. *If* G *is str.* α*-crit., then* $|V(G)| \geqslant 2\alpha$. *Equality holds iff* G *is a* 1*-factor, i.e.* $\rho(G, x) = 1$ *for each* $x \in V(G)$.

(Of course they used an other terminology equivalent to ours.) Hajnal [8, Lemma] and Folkmann [7, Lemma 2.4] gave new proof of the first part of this theorem. Hajnal also generalized it proving

Theorem 2. [8, Theorem 1]. *If* G *is str.* α*-crit. then* $\rho(G, x) \leqslant \delta(G) + 1$ *for each* $x \in V(G)$. *Here* $\delta(G)$ *is defined only for str.* α*-crit. graphs, and* $\delta(G) = |V(G)| - 2\alpha$ *for such a graph* G.

This is clearly a generalization of Theorem 1:

$$1 \leqslant \rho(G, x) \leqslant \delta(G) + 1 \quad \text{implies} \quad \delta(G) \geqslant 0 \quad \text{and}$$

$$\rho(G, x) = 1 \quad \text{for} \quad \delta(G) = 0 .$$

Now in §2 we prove the following generalization of Theorem 2.

Theorem 2′. *If* G *is str.* α*-crit.,* $L \in \mathscr{F}(l, G)$, $I = \{x\colon$ *there is a* $y \in L$ *with* $xy \in E(G)\}$, *then*

$$\rho(G, y) \leqslant |I| - l + 1 \quad \textit{for all} \quad y \in L .$$

This can be proved on a similar way as Hajnal proved Theorem 2, however our proof goes a different way using a lemma also having some interest in itself (Lemma 2.2) and which is also useful in §3-4, Theorem 2′ enables us to prove in §3 a sharpening of Theorem 2 originally conjectured by Gallai.

Theorem 4. *If* G *is str.* α*-crit,* $\delta(G) \geqslant 2$, *then* $\{x\colon \rho(G, x) \geqslant \delta(G) + 1\} = \{x\colon \rho(G, x) = \delta(G) + 1\}$ *has* $\leqslant \delta(G) + 2$ *points. So* $|E(G)| \leqslant (\alpha - 1)\delta(G) + \binom{\delta(G) + 2}{2}$.

This also settles an other conjecture of Gallai (see 3.14). The upper bound for $|E(G)|$ is a sharpening of [6, Theorem 2].

While investigating the str. α-crit. graphs we may restrict ourselves to the case, when G is connected (see 2.7). Here the following problem arises: characterize all the connected α-crit. graphs with fixed $\delta(G)$. The case $\delta(G) \leqslant 1$ is easily settled using Theorem 2 (see 2.8 or e.g. [1]). However the case $\delta(G) \geqslant 2$ is more complicated. Andrásfai [1, Theorem (1.2)] solved this problem showing, that the following is true.

Theorem 3. [1, Theorem (1.2)]. *G is a connected α-crit. with $\delta(G) = 2$ iff $G \in \mathscr{A}(K_4)$.*

Here K_p is the complete p-graph, the class $\mathscr{A}(H)$ is defined as the class of all graphs coming into being by replacing the edges of H with mutually disjoint and disconnected even paths. (For exact definition see 3.4.) To prove Theorem 3 Andrásfai used Petersen's theorem on 3-regular graphs. In §3 we give a new proof of Theorem 3 using Theorem 2′ instead of this. Our main result however is contained in §4, where we prove an analogous result for $\delta(G) = 3$.

Theorem 5. *G is connected α-crit. with $\delta(G) = 3$ iff $G \in \in \bigcup_{H \in \mathscr{H}_\alpha} \mathscr{A}(H)$, where $\mathscr{H} = \{H$: H is connected α-crit. with $\delta(H) = 3$ and $\alpha \leqslant 27\}$ so $|\mathscr{H}| < \infty$.*

This means, that the class of all connected α-crit. graphs with $\delta(G) = = 3$ is "finitely characterizable" using only one simple operation.

Analogous conjectures are stated for $\delta(G) \geqslant 4$.

In §5 we prove some consequences of Theorem 5 to the arrow-symbol introduced in [4] and investigated in [4], [12], [14], [15], [16].

In §3 we also state some other conjectures arising from Theorem 3, and in §5 some other ones arising from Theorem 6.

All non-trivial results (lemmas and theorems) seem to be new if the opposite is not explicitely stated.

§1. NOTATIONS, DEFINITIONS

Throughout this paper we put

$$E(G, A, B) = \{xy \in E(G) \colon x \in A, y \in B\},$$

$$E(G, x) = \{y \colon xy \in E(G)\}, \quad E(G, A) = \bigcup_{x \in A} E(G, x).$$

(Clearly $\rho(G, x) = |E(G, x)|$.) If $H \subset V(G)$, then H always denotes the subgraph spanned by H in G. If $x \in V(G)$, $A \subset V(G)$, then $G - x$, $G - A$ denotes the subgraph $V(G) - \{x\}$, $V(G) - A$ of G. If $e \in E(G)$, then $G - e$ denotes the subgraph of G with $V(G - e) = V(G)$, $E(G - e) = E(G) - \{e\}$, and for $x, y \in V(G)$, $G \cup \{xy\}$ denotes the graph defined by $V(G \cup \{xy\}) = V(G)$, $E(G \cup \{xy\}) = E(G) \cup \{xy\}$. We put $S_i(G) = \{x \in V(G) \colon \rho(G, x) = i\}$ and $S_i'(G) = \{x \in V(G) \colon \rho(G, x) \geqslant i\}$. Also we put for $s = 0, 1$,

$$\begin{aligned}\mathfrak{C}_s^0(G) = \{C \subset V(G) \colon\ & C = \{x_1, x_2, \dots, x_k\};\ k \equiv s \bmod 2;\\ & x_1 x_k \in E(G),\ x_i x_{i+1} \in E(G);\\ & \text{for } 1 \leqslant i \leqslant k-1;\ k \geqslant 3\} =\\ = \{C \subset V(G) \colon\ & C \text{ is a cycle of length } k \text{ in } G;\\ & k \equiv s \bmod 2;\ k \geqslant 3\}.\end{aligned}$$

$$\begin{aligned}\mathfrak{C}_s^1(G) = \{C \subset V(G) \colon\ & C = \{x_1, x_2, \dots, x_k\};\ k \equiv s \bmod 2;\\ & x_i x_j \in E(G) \text{ iff } i = 1, j = k \text{ or } j = i + 1;\ k \geqslant 3\} =\\ = \{C \subset V(G) \colon\ & C \text{ is a cycle of length } k \text{ without}\\ & \text{diagonals in } G;\ k \equiv s \bmod 2,\ k \geqslant 3\}.\end{aligned}$$

$$\begin{aligned}\mathscr{P}_s^0(G) = \{P \subset V(G) \colon\ & P = \{x_1, \dots, x_k\};\ k \equiv s \bmod 2;\\ & x_i x_{i+1} \in E(G) \text{ for } 1 \leqslant i \leqslant k-1\} =\\ = \{P \subset V(G) \colon\ & P \text{ is a path of length } k \text{ in } G;\\ & k \equiv s \bmod 2;\ k \geqslant 3\}.\end{aligned}$$

$$\mathscr{P}_s^1(G) = \{P \subset V(G):\ P = \{x_1, \dots, x_k\},\ k \equiv s \bmod 2;$$

$$x_i x_j \in E(G) \text{ iff } j = i + 1\} =$$

$$= \{P \subset V(G):\ P \text{ is a path of length } k \text{ in } G$$

$$\text{without diagonals}\}.$$

$$\mathscr{R}^0(G) = \{R \subset V(G):\ R \text{ has a 1-factor in } G\},$$

$$\mathscr{R}^1(G) = \{R \subset V(G):\ \rho(R, x) = 1 \text{ for all } x \in R\}.$$

Clearly $\mathfrak{C}_s^0(G) \supset \mathfrak{C}_s^1(G)$, $\mathscr{P}_s^0(G) \supset \mathscr{P}_s^1(G)$, $\mathscr{R}^0(G) \supset \mathscr{R}^1(G)$. K_j denotes the complete graph with j vertices. We put $[S]^k = \{T \subset S:\ |T| = k\}$. $\bigcup_{i=1}^{l}{}^{*} A_i$ is the union of the pairwise disjoint sets A_i.

Γ is a subgraph on G means, that Γ is a subgraph of G and $V(\Gamma) = V(G)$.

§2. GENERAL PROPERTIES OF STR. α-CRIT. GRAPHS AND THE CASE $\delta(G) \leqslant 1$

2.1. *Let* G *be* α*-crit.,* $y \in V(G)$, $D \subset E(G, y)$, *then*

(a) $\mathscr{F}(\alpha, G - y - D) = \phi$ *iff* $D = E(G, y)$.

(b) *Suppose* $x \in E(G, y)$. *Then* $\rho(G, x) = 1$, $(\rho(G, x) \geqslant 2)$ *implies* $\rho(G, y) = 1$, $(\rho(G, y) \geqslant 2)$.

(c) $V(G) = S_0(G) \cup^* S_1(G) \cup^* S_2'(G)$ *and* $E(G, S_0(G), S_1(G)) = E(G, S_0(G) \cup S_1(G), S_2'(G)) = \phi$.

Proof.

(a) If F is independent in $G - y - E(G, y)$ then $F \cup \{y\}$ is independent in G. This proves the if part. If $xy \in E(G)$, then there is an F in $\mathscr{F}(\alpha + 1, G - xy)$ with $x, y \in F$. So $F - \{y\} \in \mathscr{F}(\alpha, G - y - (E(G, y) - \{x\}))$, proving the only if part.

(b) and (c) are trivial consequences of (a).

The following lemma will be essential throughout this paper.

2.2. *Suppose* G *is str.* α*-crit.,* $x \in V(G)$ *and* $\rho(G, x) > 1$. *Then there exists a str.* α*-crit. subgraph* Γ *on* $G - x$.

Remarks.

(1) There is no problem to find an α-crit. subgraph Γ' on $G - x$.

(2) This lemma will enable us in certain cases to use induction on $\delta(G)$.

(3) This lemma is a weaker from of Wade's conjecture stating $\mathscr{F}(\alpha, G - x - y) \neq \phi$ for all $x, y \in V(G) - S_1(G)$.

Proof. Suppose G is str. α-crit., $x \in V(G)$, $\rho(G, x) > 1$. We consider 2 vertex sequences: $\langle x_0, \ldots, x_k \rangle$ and $\langle y_0, \ldots, y_{k-1} \rangle$ ($k \geqslant 0$, the second sequence is empty when $k = 0$), and set-sequences $\langle X_0, \ldots, X_k \rangle$, $\langle Y_0, \ldots, Y_{k-1} \rangle$, $\langle D_0, \ldots, D_k \rangle$ and subgraph-sequences $\langle G_0, \ldots, G_k \rangle$, $\langle G'_0, \ldots, G'_k \rangle$ fulfilling (a), (b) and also (c)-(g) for $0 \leqslant i \leqslant k - 1$. We also suppose, that k is the largest integer, for which we can find such sequences. (Clearly $2k + 1 \leqslant |V(G)|$.)

(a) $x_0 = x$, $X_0 = \{x_0\}$, $G_0 = G$, $G'_0 = G - x_0$, $D_0 = \bigcap \mathscr{F}(\alpha, G'_0)$;

(b) $X_k \cap Y_{k-1} = \phi$, $|X_k| = k + 1$, $|Y_{k-1}| = k$;

(c) $X_{i+1} = \{x_0, \ldots, x_{i+1}\}$, $Y_i = \{y_0, \ldots, y_i\}$, $G_{i+1} = G - X_i - Y_i$, $G'_{i+1} = G - X_{i+1} - Y_i$;

(d) $D_{i+1} = \bigcap \mathscr{F}(\alpha - 1 - i, G'_{i+1})$ if $i < \alpha - 1$ and $\mathscr{F}(\alpha - i - 1, G'_{i+1}) \neq \phi$, $D_{i+1} = V(G)$ otherwise;

(e) $y_i x_{i+1} \in E(G)$;

(f) $y_i \in D_i$;

(g) If $E(G, \{x_i\}, D_i) \neq \phi$ then $x_i y_i \in E(G)$.

Claim. $k \leqslant \alpha - 1$ and $D_k = \phi$. Once we proved this, we are done. If $k \geqslant 1$, then (f) for $i = k - 1$ gives $D_{k-1} = \phi$. So (a) and (d) give either $\mathscr{F}(\alpha - k + 1, G'_{k-1}) = \phi$ or $D_{k-1} = \bigcap \mathscr{F}(\alpha - k + 1, G'_{k-1})$. In both cases (and also if $k = 0$) we get $\phi = \mathscr{F}(\alpha - k + 1, G'_k) = = \mathscr{F}(\alpha - k + 1, G_k)$. So we can choose a subgraph Γ_k of G'_k which is $\alpha - k$-crit. Clearly Γ_k has no isolated vertex y, since this would mean $y \in \bigcap \mathscr{F}(\alpha - k, \Gamma_k) \subset \bigcap \mathscr{F}(\alpha - k, G'_k) = D_k = \phi$ which is absurd. So Γ_k is str. $\alpha - k$-crit. Define Γ as follows: $V(G - x_0) = V(\Gamma)$, $\{x_{i+1}y_i : 0 \leqslant i \leqslant k-1\} \cup^* E(\Gamma_k) = E(\Gamma)$. Clearly Γ is a str. α-crit. subgraph on $G - X$.

So we have to prove our claim. To do this put $m = \min(k-1, \alpha - 1)$ if $D_k = \phi$ and put $m = \min(k, \alpha)$ if $D_k \neq \phi$. In the latter case let y_k be a member of D_k with property (g), i.e. $x_k y_k \in E(G)$ if $E(G, \{x_k\}, D_k) \neq \phi$. Put $Y_k = Y_{k-1} \cup^* \{y_k\}$ in case $D_k \neq \phi$.

(α) $\mathscr{F}(\alpha - 1, G'_i) \neq \phi$ for $0 \leqslant i \leqslant \min(k, \alpha)$. But this is true for $i = 0$. Once this is true for some $i < \min(k, \alpha)$, then $\mathscr{F}(\alpha - 1, G'_i)$ has a member F. $i < k$, so y_i, x_{i+1} are defined. (e) implies $|\{x_{i+1}, y_i\} \cap F| \leqslant 1$. $F - \{x_{i+1}, y_i\}$ is either in $\mathscr{F}(\alpha - i, G'_{i+1})$, or in $\mathscr{F}(\alpha - i - 1, G'_{i+1})$ (note, that $|F| \geqslant 1$, since $i < n$). So in both cases $\mathscr{F}(\alpha - i - 1, G'_{i+1}) \neq \phi$ follows.

(β) Suppose $0 \leqslant i \leqslant j \leqslant m$. Then $y_j \in D_i$. (y_j is defined for $i \leqslant j \leqslant m$.) This is clear for $j = i$. Let j be any integer with $i \leqslant j \leqslant m$ and suppose, (β), is true for all $i \leqslant j' \leqslant j$. We want to prove it for $j + 1$. $\mathscr{F}(\alpha - i, G'_i) \neq \phi$ according to (α). Let F be any member of $\mathscr{F}(\alpha - i, G_i)$. We have to show, that $y_{j+1} \in D_i$. But the induction hypothesis gives $\{y_i, \ldots, y_j\} \subset F$, and so $\phi = F \cap \{x_{i+1}, \ldots, x_{j+1}\}$. Thus $j - i \leqslant \alpha - i$ gives $F - \{y_i, \ldots, y_j\} \in \mathscr{F}(\alpha - j - 1, G'_{j+1})$. But $j \leqslant \alpha - 1$, so $D_{j+1} = = \bigcap \mathscr{F}(\alpha - j + 1, G'_{j+1})$. (See (d)). $j + 1 \leqslant m$ means, that y_{j+1} is defined and is in D_{j+1}, so $y_{j+1} \in F - \{y_i, \ldots, y_j\}$ that was to be proved.

(γ) *$Y_m \cup^* \{x_0\}$ is independent in G.* First let F be in $\mathscr{F}(\alpha, G'_0) \neq \neq \phi$. Then use ($\beta$), for $i = 0$, to get $Y_m \subset F$. (Note, that $\mathscr{F}(\alpha, G'_0) \neq \phi$ by (α) and $0 < \alpha$, so $D_0 = \bigcap \mathscr{F}(\alpha, G'_0) \subset F$).) So Y_m is independent in G. Next use (β), for $i = 0$, $0 \leqslant j \leqslant m$, to get $\mathscr{F}(\alpha, G - x_0 - y_j) = \phi$. ($y_j \in D_0 = \bigcap \mathscr{F}(\alpha, G'_0)$.) So $x_0 y_j \in E(G)$ would mean $\rho(G, x_0) = 1$, but (a) gives $x_0 = x$ so $\rho(G, x_0) > 1$.

So $Y_m \cup^* \{x_0\} \in \mathscr{F}(m+2, G)$ which implies $m \leqslant \alpha - 2$. So the definition of m gives $k \leqslant \alpha - 1$, which is the first part of our claim.

We prove $D_k = \phi$ indirectly, so we assume $D_k \neq \phi$. Then y_k is defined (and $m = \min(k, \alpha) = k$). We prove that $E(G, \{y_k\}, V(G'_k) - \{y_k\}) = \phi$. If this is not true then we have an $x_{k+1} \in V(G'_{k+1}) - \{y_k\}$ with $x_{k+1}y_k \in E(G)$. Then defining $X_{k+1} = X_k \cup^* \{x_{k+1}\}$, $G_k = G - X_k - Y_k$, $G'_k = G - X_{k+1} - Y_k$, $D_{k+1} = \bigcap \mathscr{F}(\alpha - k - 1, G'_{k+1})$ (note, that $k \leqslant \alpha - 1$, (α) and $D_k \neq \phi$ implies $\mathscr{F}(\alpha - k - 1, G'_{k+1}) \neq \phi$), we have the vertex-sequences $\langle x_0, \ldots, x_{k+1}\rangle$, $\langle y_0, \ldots, y_k\rangle$ the set-sequences $\langle X_0, \ldots, X_{k+1}\rangle$, $\langle Y_0, \ldots, Y_k\rangle$, $\langle D_0, \ldots, D_{k+1}\rangle$ and the subgraph sequences $\langle G_0, \ldots, G_{k+1}\rangle$, $\langle G'_0, \ldots, G'_{k+1}\rangle$ with the properties (a), (b) and with the properties (c)-(g) for $0 \leqslant i \leqslant k$. But this contradicts to the maximality of k. So we get $E(G, \{y_k\}, V(G'_k) - \{y_k\}) = \phi$. But this means $E(G, y_k) \subseteq X_k \cup^* Y_{k-1}$ so $E(G, y_k) \subseteq X_k - \{x_0\}$ by (γ). So $k \geqslant 0$ follows. Now we want to prove $E(G, y_k) \cap X_{k-1} - \{x_0\} \neq \phi$. But the opposite would mean $E(G, y_k) = \{x_k\}$, so $\rho(G, y_k) = 1$ and $\rho(G, x_k) \geqslant 2$ a contradiction by 2.1 (b). So there is an $1 \leqslant i \leqslant k - 1$ with $y_k x_j \in E(G)$. (β) and $m = k$ means $y_k \in D_i$. So $E(G, \{x_i\}, D_i) \neq \phi$ which implies $x_i y_j \in E(G)$ by (g). (Note $i \leqslant k - 1$.) So there is an F in $\mathscr{F}(\alpha, G - x_i y_i)$ with $x_i, y_i \in F$. Clearly $y_{i-1}, y_k \notin F$. Put $F - \{x_i\} = F_1$. So we have $F_1 \cap (X_{i-1} \cup^* Y_{i-1}) \subset X_{i-1} \cup^* Y_{i-2}$. But then $|F_1 \cap X_{i-1} \cup^* Y_{i-1}| \leqslant i$, since $(X_{i-1} \cup^* Y_{i-2}) - \{x_0\}$ has $i - 1$ independent edges: $y_0 x_1, \ldots, y_{i-2} x_{i-1}$. (See (e)). So $F_1 \cap V(G_i) = F_1 \cap V(G'_i)$ has at least $\alpha - i$ vertices. Now $i \leqslant k - 1 < \alpha - 1$ implies $\bigcap \mathscr{F}(\alpha - i, G'_i) = D_i$ by (α) and (d), and $D_i \neq \phi$ by $y_i \in D_i$. So $\mathscr{F}(\alpha - i + 1, G'_i) = \phi$, i.e. $F_1 \cap V(G'_i) \in \mathscr{F}(\alpha - i, G'_i)$. But then $D_i \subset F_1 \cap V(G'_i)$. Now (β) implies $y_k \in D_i$, (put $j = k$ and note $k = m$). So $y_k \in F_1 \cap V(G'_i) \subset F$ follows contradicting to $y_k \notin F$. So we completed our proof.

To prove Theorem 2′ we need one more lemma.

2.3. *G is α-crit., $x, x' \in V(G)$, $xx' \in E(G)$, F an α-crit. subgraph on $G - x'$. Then $\rho(\Gamma, x) = \rho(G, x) - 1$.*

Proof. $\mathscr{F}(\alpha, \Gamma - x - E(\Gamma, x)) = \phi$ by 2.1 (a), so

$\mathcal{F}(\alpha, G - x' - E(\Gamma, x) - x) = \phi$. $E(\Gamma, x) \cup \{x'\} \subset E(G, x)$, and 2.1 (a) gives $E(\Gamma, x) \cup \{x'\} = E(G, x)$.

Now we turn to prove the following

2.4. Theorem 2′. *G is str. α-crit.; $\phi \neq L \in \mathcal{F}(l, G)$ and $I \supset E(G, L)$, $|I| = l + k$. Then $\rho(G, x) \leqslant k + 1$ for all $x \in L$.*

Remarks.

(1) The original proof of Theorem 2 given by Hajnal in [8] would also give this result without essential change.

(2) Theorem 2′ is the generalization of the following theorems:

2.5. Theorem 2. (Hajnal [8]). *G is str. α-crit, then $\rho(G, x) \leqslant \delta(G) + 1$ for all $x \in V(G)$.*

2.6. Theorem 1. (Erdős – Gallai [2]). *G is str. α-crit,, then $\delta(G) \geqslant 0$. $\delta(G) = 0$ iff G is a 1-factor.*

Theorem 2′ ⇒ Theorem 2. Let $x \in V(G)$, G str. α-crit. Clearly we have an F in $\mathcal{F}(\alpha, G)$ with $x \in F$. Choose $L = F$, $I = V(G) - L$. Clearly $|L| = \alpha$, $|I| = \alpha + \delta(G)$ and $I = E(G, L)$. So Theorem 2′ gives $\rho(G, x) \leqslant \delta(G) + 1$.

Proof of Theorem 2′. Goes by induction. Theorem 2′ is trivial for $l = 1$, k arbitrary. Now let l be arbitrary. $k \leqslant -l$. If G is str. α-crit., $|L| \geqslant 1$, then $|E(G, L)| \geqslant 1$. So Theorem 1′ holds vacuously for $k \leqslant -l$.

Now suppose $l_0 \geqslant 2$, $k_0 \geqslant 1 - l_0$ and suppose, that Theorem 2′ is true for $l = l_0 - 1$, $k = k_0$ and for $l = l_0$, $k = k_0 - 1$. We want to prove it for $l = l_0$, $k = k_0$.

So let G be str. α-crit., $L \in \mathcal{F}(l_0, G)$, $I = E(G, L)$, $|I| = l_0 + k_0$, $x \in L$.

(a) First suppose $\rho(G, x) = 1$, $y \in E(G, x)$. Then $y \in I$, $\rho(G, y) = 1$.

Put $L' = L - \{x\}$, $I' = I - \{y\}$. Clearly $L' \in \mathcal{F}(l_0 - 1, G)$, $I' = E(G, L')$ and $|I'| = (l_0 - 1) + k_0$.

So the induction hypothesis gives $\rho(G, z) \leqslant k_0 + 1$ for any z in L'. But $L' \neq \phi$, so there is a $z \in L'$ and $1 \leqslant \rho(G, z) \leqslant k_0 + 1$. So we have $1 = \rho(G, x) \leqslant k_0 + 1$ as we wanted.

(b) Next suppose $\rho(G, x) \geqslant 2$, $y \in E(G, x)$. Then $\rho(G, y) \geqslant 2$ by 2.1 (b). So 2.2 gives a str. α-crit. Γ on $G - y$. $L \in \mathscr{F}(l_0, \Gamma)$ and $I' = I - \{y\} \supset E(\Gamma, L)$. $|I'| = l_0 + k_0 - 1$, so the induction hypothesis for l_0, $k_0 - 1$ gives $\rho(\Gamma, x) \leqslant k_0$. 2.3 shows $\rho(G, x) = \rho(\Gamma, x) + 1 \leqslant$ $\leqslant k_0 + 1$, that was to be proved.

(a) and (b) proves Theorem 2′ for l_0, k_0.

The following consequence of Theorem 1 shows, that while investigating the structure of str. α-crit. graphs we may restrict ourselves to the case, when the graph is connected.

2.7. *Let G be str. α-crit. Then $V(G) = S_1(G) \cup^* \bigcup\limits_{i=1}^{l}{}^* V(G_i)$ such that $S_1(G)$ is a 1-factor, and is not connected to the rest of $V(G)$, the G_i's are the connected components of $G - S_1(G)$, G_i is str. α_i-crit. for some α_i, $\sum\limits_{1}^{l} \alpha_i = \alpha - \dfrac{|S_1(G)|}{2}$, $1 \leqslant \delta(G_i) \leqslant \delta(G)$, $\sum\limits_{1}^{l} \delta(G_i) = \delta(G)$. If $1 < l$, then $\delta(G_i) < \delta(G)$ for all $1 \leqslant i \leqslant l$.*

Proof. A consequence of 2.1 (c) and Theorem 1.

So from now up to the end of §4 we investigate the structure of connected α-crit. graphs with fixed $\delta(G)$. The case $\delta(G) \leqslant 0$ is handled in Theorem 1. So now we turn to the case $\delta(G) = 1$, which is an easy consequence of Theorem 2 and we include it only for the sake of completeness (see e.g [1, page 10]).

2.8. *G is connected α-crit. with $\delta(G) = 1$ iff G is an odd cycle without diagonals.*

Proof. Theorem 2 implies $1 \leqslant \rho(G, x) \leqslant 2$ for all $x \in V(G)$ if G is str. α-crit. with $\delta(G) = 1$. 2.1 (c) also implies $\rho(G, x) \geqslant 2$ if G is connected. So the connected components of G are cycles without diagonals, proving the only if part. The if part is trivial.

§3. THE CASE $\delta(G) = 2$

The case $\delta(G) \leqslant 1$ depended on Theorem 2. It will be essential in the case $\delta(G) \geqslant 2$, too, however lemma 2.2 will be more important, and we also need some new definitions and observations. (We will be able to prove results in cases $\delta(G) = 2, 3$.)

3.1. Definition. Suppose G is a graph, $xy \in E(G)$ and $E(G, x) \cap \cap E(G, y) = \phi$. Then we define the graph $G(x, y)$ by

$$V(G(x, y)) = V(G) - \{x, y\},$$

$$E(G(x, y)) = E(G - x - y) \cup \{zz' : z \in E(G, x) - \{y\}, \quad z' \in E(G, y) - \{x\}\}.$$

Remark. We use this definition essentially only for the case, when $|E(G, x)| = 2$. However in this case our operation together with the one defined in 3.3 is a special case of a more general operation first used by Hajós [9] and Dirac [3] for chromatical problems, and introduced in the study of α-crit. graphs – as far as I know – by Plummer [13] and Andrásfai [1]. So generalizations of 3.2 (b) and 3.5 are proved in [13, Theorem 3] and more generally in [18, Satz 61 and 71] (see also [1, Theorem of Gallai 2.5].) So we include the proof of 3.2 (b) and 3.5 only for the sake of self-containedness. We mention, that none of the notations used in the above mentioned papers are adequate for our purposes, so this is one point, why we introduced a new one. However in the settling of Conjecture 1 (see §4) it would be perhaps essential to get more information for the case when $|E(G, x)|$, $|E(G, y)| \geqslant 3$ in 3.1, and this is the other aim of this definition.

3.2. *Suppose* $\mathscr{F}(\alpha + 1, G) = \phi$, $xy \in E(G)$, $E(G, x) \cup E(G, y) = \phi$. *Then*

(a) $\mathscr{F}(\alpha, G(x, y)) = \phi$.

If G *is also* α*-crit.,* $E(G, x) = \{y, x'\}$, $(yx' \notin E(G))$, *then*

(b) $G(x, y)$ *is* $\alpha - 1$*-crit., and* $E(G, \{x'\}, E(G, y) - \{x\}) = \phi$.

Proof. To prove (a) let G be a graph with $\mathscr{F}(\alpha+1, G) = \phi$ and suppose $xy \in E(G)$, $E(G, x) \cap E(G, y) = \phi$ for some x and y. Then $G(x, y)$ is defined. Assume indirectly $F \in \mathscr{F}(\alpha, G(x, y))$. $\mathscr{F}(\alpha+1, G) = = \phi$ so $E(G, F \cup \{x\}, F \cup \{x\}), E(G, F \cup \{y\}, F \cup \{y\}) \neq \phi$. But $F \in \in \mathscr{F}(\alpha, G - x - y)$ so $e \in E(G, F \cup \{x\}, F \cup \{x\})$ implies $e = xu$, $u \in F$. Similarly $yu' \in E(G)$ for some $u' \in F$. But then $uu' \in E(G(x, y))$ which is absurd.

To prove (b) we suppose G to be α-crit., and $E(G, x) = \{y, x'\}$. Clearly $x'y \notin E(G)$ by $E(G, x) \cap E(G, y) = \phi$. 2.1 (b) and $\rho(G, x) = 2$ implies the existence of an y' in $E(G, y) - \{x\}$. $y' \neq x'$. To prove that $G(x, y)$ is $\alpha - 1$-crit. we have to prove $\mathscr{F}(\alpha, G(x, y) - e) \neq \phi$ for all $e \in E(G(x, y))$. ($\mathscr{F}(\alpha, G(x, y)) = \phi$ by (a)). Put $E_1 = E(G - x - y)$ and $E_2 = \{x'z\colon z \in E(G, y) - \{x\}\}$. Clearly $E(G(x, y)) = E_1 \cup E_2$.

Suppose $uv \in E_1$. Then $uv \in E(G)$ and we have an F in $\mathscr{F}(\alpha+1, G - uv)$. If $F \cap \{x, y\} = \phi$, there $F - \{x'\}$ is independent in $G - x - y - x' - uv$. But $G(x, y) - x' = G - x - y - x'$ by the definition of $G(x, y)$, so $F - \{x'\}$ is independent in $G(x, y) - x' - uv$ so also in $G(x, y) - uv$. $|F - \{x'\}| \geqslant \alpha$ proves $\mathscr{F}(\alpha, G(x, y) - uv) \neq \phi$ in this case.

Next suppose $F \cap \{x, y\} \neq \phi$. Then $F \cap \{x, y\} = \{x\}$ or $\{y\}$. Then $x' \notin F$ resp. $(E(G, y) - \{x\}) \cap F = \phi$, so $F - \{x, y\}$ spans the same subgraph in $G - x - y - uv$ and in $G(x, y) - uv$ in both cases. But $F - \{x, y\}$ is independent in $G - x - y - uv$, so it is independent in $G(x, y) - uv$. $|F - \{x, y\}| = \alpha$ proves $\mathscr{F}(\alpha, G(x, y) - uv) \neq \phi$ in this case.

Now suppose $uv \in E_2$, i.e. $u = x'$, $v \in E(G, y) - \{x\}$. Then we have an F in $\mathscr{F}(\alpha+1, G - yv)$, with $y, v \in F$. We claim, that $x' \in F$. Suppose $x' \notin F$. But $y \notin F$, so then $F - \{y\} \in \mathscr{F}(\alpha, G - x - y - x') = = \mathscr{F}(\alpha, G(x, y) - x')$, a contradiction. So $x' \in F$. $y \in F$ also means $E(G, y) \cap F = \{v\}$. So $E(G(x, y), F - \{y\}, F - \{y\}) = \{x'v\} \cup \cup E(G, F - \{y\}, F - \{y\})$. But $E(G, F - \{y\}, F - \{y\}) = \phi$ so $F - \{y\} \in \in \mathscr{F}(\alpha, G(x, y) - x'v)$ which completes the proof of the first part of (b).

Now we have to prove that $E(G, \{x'\}, E(G, x) - \{x\}) = \phi$. But assume the contrary. Then there is an $u \in E(G, y) - \{x'\}$ with $x'u \in E(G)$. So there is an F in $\mathscr{F}(\alpha+1, G-x'u)$. $x', u \in F$ so $x, y \notin F$. But this gives $F-\{x'\} \in \mathscr{F}(\alpha, G-x-y-x') = \quad (\alpha, G(x,y)-x') \subset \mathscr{F}(\alpha, G(x,y)) = \phi$, which is absurd.

The following definition is in some sense the "inverse" of 3.1 (see the remark in 3.1).

3.3. Definition. Suppose G is a graph, $x \in V(G)$, $A \cup^* B = E(G,x)$, $A \neq \phi$, $B \neq \phi$, $x', y' \notin V(G)$. Then we define $G[x, A, B, x', y']$ by

$$V(G[x,A,B,x',y']) = V(G) \cup^* \{x', y'\},$$

$$E(G[x,A,B,x',y']) =$$

$$= (E(G) - \{xz\colon\ z \in B\}) \cup^* \{x'x, x'y'\} \cup^* \{y'z\colon\ z \in B\}.$$

Furthermore we put

$$G[x, E(G,x)-\{y\}, \{y\}, x', y'] = G[x, y, x', y']$$

in case $\rho(G,x) \geqslant 2$, and $xy \in E(G)$.

3.4. Let G be a graph. We put $\mathscr{A}_0(G) = \{G\}$ and for $i \geqslant 0$

$$\mathscr{A}_{i+1}(G) = \{H[x,y,x',y']\colon\ x, y \in V(H);\ \rho(H,x), \rho(H,y) \geqslant 2;$$
$$x', y' \notin V(H),\ xy \in E(H),\ H \in \mathscr{A}_i(G)\},$$

and finally $\mathscr{A}(G) = \bigcup_{i=0}^{\infty} \mathscr{A}_i(G)$.

So $H \in \mathscr{A}(G)$ iff H comes into being from G by replacing the edges of G by even pairwise disjoint and disconnected paths. So e.g.

$$\mathscr{A}(K_3) = \{C\colon\ C \text{ is an odd cycle without diagonals}\}.$$

3.5. *Suppose* G *is* $\alpha-1$*-crit.,* $x \in V(G)$, $x', y' \in V(G)$; $A \neq \phi$, $B \neq \phi$, $A \cup^* B = E(G,x)$. *Then* $G[x,A,B,x',y']$ *is* α*-crit. (So e.g. the members of* $\mathscr{A}_j(G)$ *are* $\alpha+j-1$*-crit.)*

Proof. Throughout this proof we put $G[x,A,B,x',y'] = G_0$, and we suppose, that G, x, A, B fulfil the above conditions.

(a) Suppose $\mathscr{F}(\alpha+1, G_0) \neq \phi$ and $F \in \mathscr{F}(\alpha+1, G_0)$. $G_0 - x - x' - y' = G - x$ and $\phi = \mathscr{F}(\alpha, G) \supset \mathscr{F}(\alpha, G-x)$, so $\mathscr{F}(\alpha, G-x) = \mathscr{F}(\alpha, G_0 - x - x' - y') = \phi$. $F - \{x, x', y'\}$ is independent in G_0, so in $G_0 - x - x' - y' = G - x$, so $|F - \{x, x', y'\}| \leqslant \alpha - 1$. This implies $|F \cap \{x, x', y'\}| \geqslant 2$ i.e. $F \cap \{x, x', y'\} = \{x, y'\}$. So $(A \cup^* B) \cap F = \phi$ follows, and we get $F - \{x, x', y'\} = F - \{x, y'\} \subset$ $\subset \mathscr{F}(\alpha - 1, G - x - A - B)$. But 2.1 (a) and $E(G,x) = A \cup^* B$ shows $\mathscr{F}(\alpha+1, G - x - A - B) = \phi$ if G is $\alpha - 1$-crit, which is a contradiction.

(b) Suppose $e \in E(G_0)$. We have to show, that $\mathscr{F}(\alpha+1, G_0 - e) \neq$ $\neq \phi$. We distinguish several cases:

(α) $e = x'y'$. $A \neq \phi$, so there is an $u \in A$. We find an F in $\mathscr{F}(\alpha, G - xu) \neq \phi$. Clearly $B \cap F \neq \phi$. So $F - \{x\} \cup \{x', y'\}$ is clearly in $\mathscr{F}(\alpha+1, G_0 - x'y')$.

(β) $e \in E(G) - \{xz\colon z \in B\}$. Then we find an F in $\mathscr{F}(\alpha, G-e)$. Either $x \notin F$ or $F \cap B = \phi$. (Note that x and B are completely connected in G). Put $F_0 = F \cup \{x'\}$ in case of $x \notin F$ and $F_0 = F \cup \{y'\}$ in case of $F \cap B = \phi$. Clearly $F_0 \in \mathscr{F}(\alpha+1, G_0 - e)$.

(γ) The case $e = xx'$ is symmetrical to the case (α) and

(δ) the case $e \in E(G) - \{xz\colon z \in A\}$ is symmetrical to the case (β).

So (α)-(δ) together give the proof of (b).

Now in case of $\delta(G) = 2$ the following observation will be of particular interest:

3.6. *Suppose* G *is str.* α*-crit.,* $x \in V(G)$, $\rho(G,x) = 2$. *Then* $\sum_{y \in E(G,x)} \rho(G,y) \leqslant \delta(G) + 3$, *so we have* $\rho(G,y) \leqslant \left[\frac{\delta(G)+3}{2}\right]$ *for some* $y \in E(G,x)$.

Proof. 3.6 is vacuous for $\delta(G) \leqslant 0$ by Theorem 1. So we may suppose $\delta(G) \geqslant 1$, G is str. α-crit., $\rho(G,x) = 2$, $E(G,x) = \{y, x'\}$. If $x' \in E(G,y)$ then $\phi = \mathscr{F}(\alpha, G - x - E(G,x)) = \mathscr{F}(\alpha, G - y - x' - x)$ by

2.1 (a), so 2.1 (a) gives $E(G, y) = \{x', x\}$ and analogously $\phi = \mathscr{F}(\alpha, G - x - E(G, x)) = \mathscr{F}(\alpha, G - x' - y - x)$ gives $E(G, x') = \{y, x\}$. So $\sum_{z \in E(G,x)} \rho(G, z) = \rho(G, y) + \rho(G, x') = 4 \leqslant \delta(G) + 3$, yielding the conclusion of 3.6 in this case. If $x' \notin E(G, y)$, then $G(x, y)$ is $\alpha - 1$-crit. by 3.2 (b). Clearly $G(x, y)$ is str. $\alpha - 1$-crit. since G is str. α-crit. and also $\delta(G(x, y)) = \delta(G)$. Now $\rho(G(x, y), x') \leqslant \delta(G(x, y)) + 1 = \delta(G) + 1$ by Theorem 2. To prove 3.6 we only need to prove $\rho(G(x, y), x') + 2 = \rho(G, x') + \rho(G, y)$. Put $E_1 = E(G, x') - \{x\}$, $E_2 = E(G, y) - \{x\}$. $E_1 \cap E_2 = \phi$ by the second part of 3.2 (b). So $E(G(x, y), x') = E_1 \cup^* E_2$ and $E(G, x') = E_1 \cup^* \{x\}$, $E(G, y) = E_2 \cup^* \{x\}$ proving $\rho(G(x, y), x') + 2 = \rho(G, x') + \rho(G, y)$ and so 3.6.

Now we prove a consequences of Theorem 2′. This will be the main tool in the proof of Theorem 3 and Theorem 4.

3.7. *Suppose G is str. α-crit.*

(a) *If $x \in V(G)$, $\rho(G, x) = \delta(G) + 1$, Γ is an α-crit. subgraph on $G - x$, then $\rho(G, y) \leqslant \rho(\Gamma, y) + 1$ for all $y \in V(G) - \{x\}$.*

(b) *If $x, y, z \in V(G)$, $yz \in E(G)$, $\mathscr{F}(\alpha, G - \{x, y, z\}) = \phi$ then either* $\min\{\rho(G, y), \rho(G, z)\} \leqslant 2$ *or* $\rho(G, x) \leqslant \delta(G) - 1$.

Proof. Suppose $x, y \in V(G)$, $A \subset E(G, y)$, $\mathscr{F}(\alpha, G - x - y - A) = \phi$. We distinguish 2 cases.

Case 1. $\mathscr{F}(\alpha, G - A - y) = \phi$. Then $E(G, y) = A$ by 2.1.

Case 2. $\mathscr{F}(\alpha, G - A - y) \neq \phi$. Put $A \cup \{y\} = A_0$, $\bigcap \mathscr{F}(\alpha, G - A_0) = L$, $(\bigcup \mathscr{F}(\alpha, G - A_0)) - L = B$, $|L| = l$, $I_0 = G - A_0 - (\bigcup \mathscr{F}(\alpha, G - A_0)$. Clearly $A_0 \cup^* B \cup^* I_0 \cup^* L = G$, $x \in L$.

(α) L is independent and $E(G, L, B) = \phi$. L is independent, since $\mathscr{F}(\alpha, G - A_0) \neq \phi$. Let $F \in \mathscr{F}(\alpha, G - A_0)$ arbitrary then $L \subset F$, so $u \in L$, $uv \in E(G)$ implies $u \in F$, $v \notin F$. So $y \notin \dot{\bigcup} \mathscr{F}(\alpha, G - A_0) = B$ follows.

(β) $\mathscr{F}(\alpha - l + 1, B) = \phi$, $|B| \geqslant 2(\alpha - l)$. But $F \in \mathscr{F}(\alpha - l + 1, B)$ would imply $F \cup^* L \in \mathscr{F}(\alpha + 1, G) = \phi$ by (α), which is absurd. Let

B_0 be an $\alpha - l$-crit. subgraph on B. If B_0 is not str. $\alpha - l$-crit., then $\rho(B_0, u) = 0$ for some u, so $u \in \bigcap \mathscr{F}(\alpha - l, B_0) \subset \bigcap \mathscr{F}(\alpha - l, B)$. Now let $F \in \mathscr{F}(\alpha, G - A_0)$ be arbitrary. Then $F \subset L \cup^* B$, $L \subset F$ by the definitions. So $F - L \in \mathscr{F}(\alpha - l, B)$. So $u \in F$ proving $u \in \bigcap \mathscr{F}(\alpha, G - A) = L$. This contradicts to $L \cap B = \phi$. So B_0 is str. $\alpha - l$-crit., so Theorem 1 implies $|B_0| = |B| \geqslant 2(\alpha - l)$.

(γ) $yu \in E(G)$ for some $u \in L$ implies $E(G, y) = A \cup \{u\}$. But $u \in L = \bigcap \mathscr{F}(\alpha, G - A_0)$ so $\mathscr{F}(\alpha, G - A_0 - u) = \phi$ and 2.1 implies (γ). (Note $A \cup \{u\} \subset E(G, y)$).

Now we want to prove (a). Suppose therefore $\rho(G, x) = \delta(G) + 1$, y is arbitrary in $G - x$, Γ is an α-crit. subgraph on $G - x$. Put $A = E(\Gamma, y)$. Then we have $\phi = \quad (\alpha, \Gamma - y - A) \supset \quad (\alpha, G - x - y - A)$. In case 1 we have $E(\Gamma, y) = A = E(G, y)$ so (a) is true. In case 2 we prove $E(G, L, \{y\}) \neq \phi$. $E(G, L, \{y\}) = \phi$ implies $E(G, L) \subset I_0 \cup^* A$ by (α). So $x \in L$, $|I| = 2\alpha + \delta(G) - |B| - |L| - 1$ and Theorem 2' gives $\rho(G, x) \leqslant 2\alpha + \delta(G) - |B| - 2l$, so using ($\beta$) we have $\rho(G, x) \leqslant \delta(G)$ a contradiction. So we have some $u \in L$ with $uy \in E(G)$. Then (γ) implies $E(G, y) = A \cup \{u\} = E(\Gamma, y) \cup \{u\}$ proving (a) in this case.

Finally we prove (b). Suppose $z \in V(G)$, $yz \in E(G)$ and put $A = \{z\}$ now. In case 1 we have $E(G, y) = A = \{z\}$, so $\rho(G, y) = \rho(G, z) = 1 < 2$. In case 2 suppose $uy \in E(G)$ for some $u \in L$. Then (γ) gives $\rho(G, y) = 2$. The role of y and z can be changed ($A = \{y\}$) to get $\rho(G, z) = 2$ if $uz \in E(G)$ for some $u \in L$. So finally $E(G, L, \{y, z\}) = \phi$ can be supposed. Then $E(G, L) \subset I_0$ so (β) and $x \in L$ gives by Theorem 2' that $\rho(G, x) \leqslant 2\alpha + \delta(G) - |B| - 2l - 1 \leqslant \delta(G) - 1$ proving (b) in all cases.

3.8. Andrásfai [1, Theorem 3.2]. *Suppose G is str. α-crit., $\delta(G) = 2$ and $S_2(G) \cup S_1(G) = \phi$. Then $G = K_4$.*

This is the main lemma in proving Theorem 3, and was proved by Andrásfai using Petersen's theorem on 3-regular graphs and a result of his own. [1, Theorem 1.3]. We generalize 3.8 in 3.14 for $\delta(G) \geqslant 2$ arbitrary.

Proof. Let G be str. α-crit. with $\delta(G) = 2$, $\min\{\rho(G, x): x \in V(G)\} = n \geqslant 3$. Theorem 2 gives $n = 3$ and $\rho(G, x) = 3$ for all $x \in V(G)$. So there is a str. α-crit. subgraph Γ on $G - x$ by 2.2 Γ has $2\alpha + 1$ vertices, so $V(\Gamma) = C \cup^* R$ with the property, that $C \in \mathfrak{C}_1^1(\Gamma)$, $R \in \mathscr{R}^1(\Gamma)$. (We used 2.7 and 2.8). Suppose $R \neq \phi$, and let $y \in R$, $\{z\} = E(\Gamma, y)$. ($z \in R$). Then clearly $\phi = \mathscr{F}(\alpha, \Gamma - y - z) \supset \mathscr{F}(\alpha, G - x - y - z)$, so using 3.7 (b) we get $\min\{\rho(G, x), \rho(G, y), \delta(G, z)\} \leqslant 2$ a contradiction. So $R = \phi$, i.e. Γ is a cycle without diagonals of length $2\alpha + 1$. Γ cannot have any diagonals in G, since a diagonal of G would decompose the cycle Γ into an odd cycle C' and an even path $R' \neq \phi$. But then the subgraph Γ' defined by $V(\Gamma') = V(G) - \{x\}$, $C' \in \mathfrak{C}_1^1(\Gamma')$, $R' \in \mathscr{R}^1(\Gamma)$, $E(G, C', R') = \phi$ would be a str. α-crit. subgraph on $G - x$, so the same argument as before would give $R' = \phi$ a contradiction. So $G - x$ is an odd cycle without diagonals. So $\rho(G - x, u) = 2$ for all $u \in V(G) - \{x\}$. But for all $u \in V(G)$, $\rho(G, y) \geqslant 3$ holds, so $ux \in E(G)$ for all $u \in V(G) - \{x\}$ and $\rho(G, x) = 3$ gives $|V(G) - \{x\}| = 3$. So $G - x = K_3$ and x is connected to each vertex of $G - x$ i.e. $G = K_4$ as stated.

Now we prove the theorem of Andrásfai which solves the case $\delta(G) = 2$.

3.9. Theorem 3. [1, Theorem 1.2]. *G is connected α-crit. with $\delta(G) = 2$ iff $G \in \mathscr{A}(K_4)$. (So iff $G \in \mathscr{A}_{\alpha-1}(K_4)$).*

Remark. Our proof of the non-trivial only if part differs from Andrásfai's one in the proof of 3.8 and there is also a difference in the way of reducing 3.9 to 3.8: we use 3.6 instead of an an other theorem of Andrásfai (see [1, Theorem 1.3]).

Proof. First we check the if part. K_4 is str. 1-crit., connected. So any $G \in \mathscr{A}_{\alpha-1}(K_4)$ is α-crit. by 3.5, and trivially G is connected α-crit., and $\delta(K_4) = 2$ so $\delta(G) = 2$.

To prove the only if part first we note, that G is connected 1-crit. only if $G = K_4$. Suppose, we have already proved the only if part for some $\alpha \geqslant 1$ and we want to prove it for $\alpha + 1$. So let G be connected

$\alpha + 1$-crit. with $\delta(G) = 2$. Put $\rho = \min\{\rho(G, x) \colon x \in V(G)\}$. $\rho = 1$ is impossible by 2.7, and the connectedness of G. $\alpha + 1 \geqslant 2$, so $\rho \geqslant 3$ is impossible by 3.8. So there is an x in $V(G)$ with $\rho(G, x) = 2$. But then 3.6 implies that $\rho(G, y) \leqslant 2$ for some $y \in E(G, x)$, i.e. $\rho(G, y) = 2$. Put $E(G, x) - \{y\} = \{x'\}$. $x' \notin E(G, y)$. Since $x' \in E(G, y)$ and $\phi = \mathscr{F}(\alpha, G - x - E(G, x) = \mathscr{F}(\alpha, G - x - \{y, x\})$ would give $\{y, x\} = E(G, x')$;, so $xx'y$ would be a connected component of G contradicting to $|V(G)| \geqslant 6$ or to the connectedness of G. So $x' \notin E(G, y)$. So 3.2 (b) gives, that $G(x, y)$ is α-crit. Clearly the connectedness of G implies that $G(x, y)$ is connected α-crit, so by induction $G(x, y) \in \mathscr{A}_{\alpha - 1}(K_4)$. Put $E(G, y) = \{x, y'\}$. Clearly $G(x, y)[x', y', x, y] = G$. So $G \in \mathscr{A}_\alpha(K_4)$ follows by the definition of $\mathscr{A}_\alpha(H)$ if $\rho(G(x, y), x')$, $\rho(G(x, y), y') \geqslant 2$. But clearly $\rho(G(x, y), x') = \rho(G, x') \geqslant \rho = 2$ and $\rho(G(x, y), y') = \rho(G, y') \geqslant \rho = 2$, since $x'y' \notin E(G)$ by the second part of 3.2 (b). So Theorem 3 is proved.

Now we mention some consequences of Theorem 3.

3.10. *Suppose* G *is str.* α*-crit. for some* $\alpha \geqslant 1$ *with* $\delta(G) = 2$. *Then*

(a) $|S_3(G)| = |S'_3(G)| \leqslant 4$;

(b) $|E(G)| \leqslant 2\alpha + 4$.

Proof. Trivial by 2.7, 2.8 and Theorem 3.

3.11. Suppose $n \geqslant 2$, $\mathscr{F}(n, G) = \phi$ for some G with $|V(G)| = 2n$. Then there is a C in $\mathfrak{C}_1^1(G)$ with $|C| \leqslant 2\left[\frac{n}{2}\right] + 1$.

Remark. One can easily check that this result is sharp for all values of n.

Proof. We only have to prove the existence of such a C in $\mathfrak{C}_1^0(G)$, and we do this using induction on n. $|V(G)| = 4$ and $K_3 \in \mathfrak{C}_1^1(K_4)$ impliés $G = K_4$ and so $K_3 \in \mathfrak{C}_1^1(K_4) = \mathfrak{C}_1^1(G)$. So 3.11 is true for $m = 2$. Suppose 3.11 is true for some $n = \alpha \geqslant 2$ and we want to prove it for $n = \alpha + 1$. Suppose $|V(G)| = 2\alpha + 2$, $\mathscr{F}(\alpha + 1, G) = \phi$. Of course we may suppose that G is α-crit. Suppose $x \in V(G)$ is isolated: $\rho(G, x) = 0$. Then clearly $\phi = \mathscr{F}(\alpha, G - x - y)$ for any $y \in V(G) - \{x\}$. So using our inductional hypothesis we have a C in $\mathfrak{C}_1^0(G - x - y) \subset \subset \mathfrak{C}_1^0(G)$ with $|C| \leqslant 2\left[\frac{\alpha}{2}\right] + 1 \leqslant 2\left[\frac{\alpha + 1}{2}\right] + 1$ proving 3.11 for G in this case. So we may assume, that G is str. α-crit. Using 2.7 we get, that 2 cases must be considered.

(a) $V(G) = S_1(G) \cup^* V(G_1) \cup^* V(G_2)$; G_i is connected α_i-crit. for some α_i with $\alpha_1 + \alpha_2 = \alpha - \frac{|S_1(G)|}{2}$. In this case G_i is an odd cycle of length $2\alpha_i + 1$ by 2.8.

So $\alpha_1 + \alpha_2 \leqslant \alpha$ gives $\min\{|V(G_1)|, |V(G_2)|\} \leqslant \alpha + 1 < < 2\left[\frac{\alpha + 1}{2}\right] + 1$ proving 3.11 for G in this case.

(b) $V(G) = S_1(G) \cup^* V(G_1)$, G_1 is connected $|V(G_1)|/2$-crit. But now Theorem 3 gives $G_1 \in \mathscr{A}(K_4)$ because this theorem means the following:

There exist 4 vertices x_1, x_2, x_3, x_4 and $P_{ij} \in \mathscr{P}_0^1(G_1)$ for $1 \leqslant i < j \leqslant 4$ such that $V(G_1) = \bigcup_{1 \leqslant i < j \leqslant 4} P_{ij}$, the endpoints of P_{ij} are x_i and x_j, and $V(G_1) = \{x_1, x_2, x_3, x_4\} \cup^* \bigcup^*_{1 \leqslant i < j \leqslant 4} P'_{ij}$ for $P'_{ij} = = P_i - \{x_i, x_j\}$. Clearly $C_{ijk} = P_{ij} \cup P_{jk} \cup P_{ki} \in \mathfrak{C}_1^1(G_1)$ for $1 \leqslant i < j < < k \leqslant 4$ and there are exactly 2 C_{ijk}-s with $1 \leqslant i < j < k \leqslant 4$ containing x for each $x \in V(G) - \{x_1, x_2, x_3, x_4\}$ and there are exactly 3 C_{ijk}-s containing x_l for each $1 \leqslant l \leqslant 4$. So $\sum_{1 \leqslant i < j < k \leqslant 4} |C_{ijk}| = 2|V(G_1)| + + 4 \leqslant 4\alpha + 8 \leqslant 4(\alpha + 2)$ proving $|C_{ijk}| \leqslant \alpha + 2$ for some $1 \leqslant i < j < k \leqslant 4$. But $2\left[\frac{\alpha + 1}{2}\right] + 1$ is the largest odd integer $\leqslant \alpha + 2$ so $|C_{ijk}| \leqslant \leqslant 2\left[\frac{\alpha + 1}{2}\right] + 1$; so 3.11 holds for in this case, too.

3.12. Now we mention some open problems arising from 3.11.

Problem 1. Does there exist a sequence of constants $c_p \geqslant 0$ with $\lim_{p \to \infty} c_p = 0$ for which the following is true: Whenever $\mathscr{F}(\alpha + 1, G) = \phi$

and $|V(G)| = 2\alpha + p$, we always have a C in $\mathfrak{C}_1^1(G)$ with $|C| \leqslant$ $\leqslant c_p \cdot \alpha + o(\alpha)$.

3.11 proves that $c_1 = 1$. The following conjecture of Erdős (personal communication) is related to this problem.

Conjecture of Erdős. Do there exist 2 functions f, g with $f(m) =$ $= o(m)$, $g(m) = o(m)$ for which the following is true: Suppose $\mathscr{F}(m - f(m), G) = \phi$, $|V(G)| = 2m$. Then there is a C in $\mathfrak{C}_1^1(G)$ with $|C| \leqslant g(m)$.

An other generalization of 3.11 would be the much more difficult

Problem 2. $k \geqslant 2$, $m \geqslant 1$. Determine the minimal function $f(k, m)$ for which the following is true: If $\mathscr{F}(m, G) = \phi$ and $|V(G)| = km$, then G has a $k+1$-chromatical subgraph H with $|H| \leqslant f(k, m)$.

3.11 gives $f(2, m) = m + O(1)$, and clearly $f(k, m) \leqslant m - k + 1$, but we don't know, whether $\lim\limits_{k \text{ fixed}, m \to \infty} f(k, m) = \infty$.

Now we deduce our affirmative answer to a conjecture of Gallai.

3.13. Theorem 4. *If* G *is str.* α*-crit.,* $\delta(G) \geqslant 2$, *then* $|S'_{\delta(G)+1}(G)| = |S_{\delta(G)+1}(G)| \leqslant \delta(G) + 2$. So $|E(G)| \leqslant (\alpha - 1)\delta(G) +$ $+ \binom{\delta(G) + 2}{2}$.

Remark.

(1) This is clearly a strengthening of Theorem 2. $|S_{\delta(G)+1}(G)| =$ $= \delta(G) + 2 = \delta + 2$ holds for all members of $\mathscr{A}(K_{\delta+2})$ proving that Theorem 4 is best possible in this sense. I do not know however, whether $|S_{\delta(G)+1}(G)| = \delta(G) + 2$ can hold for any G which is not contained in $\mathscr{A}(K_{\delta(G)+2})$.

(2) The upper bound of $|E(G)|$ given in this theorem is to be compared to that of Conjecture 1′.

Proof. Goes by induction on $\delta(G)$. The case $\delta(G) = 2$ is stated in 3.10. Suppose we know Theorem 4 for all str. α-crit. G's with $\delta(G) =$ $= \delta - 1$, and G is str. α-crit. graph with $\delta(G) = \delta$. We have to prove

$|S_{\delta+1}(G)| \leq \delta + 2$. ($S'_{\delta+1}(G) = S_{\delta+1}(G)$ by Theorem 2.) $\delta \geq 3$ so we are done if $S_{\delta+1}(G) = \phi$. So let x be a member of $S_{\delta+1}(G)$: $\rho(G, x) = = \delta + 1$. By 2.2 we have a str. α-crit. subgraph Γ on $G - x$. Using 3.7 (a) we get $\rho(G, u) \leq \rho(\Gamma, u) + 1$ for all $u \in V(\Gamma)$. Note, that Γ is str. α-crit. with $\delta(\Gamma) = \delta - 1$, so we may use our inductional hypothesis. We get $|S_\delta(\Gamma)| = \delta + 1$. Suppose $u \in S_{\delta+2}(G) - \{x\}$. Then $\delta + 1 \leq \rho(G, u) \leq \leq \rho(\Gamma, u) + 1$, so $u \in S'_\delta(\Gamma) = S_\delta(\Gamma)$. Thus we get $S_{\delta+1}(G) = S_\delta(\Gamma) \cup^* \cup^* \{x\}$, so $|S_{\delta+1}(G)| \leq \delta + 2$ proving what we wanted.

3.13 clearly proves that

3.14. *If* G *is str.* α*-crit.,* $\delta(G) \geq 2$, *and* $\rho(G, x) = \delta(G) + 1$ *for all* $x \in V(G)$, *then* $G = K_{\delta(G)+2}$.

Remark. This, too, was originally conjectured by Gallai (personal communication) and is a generalization of 3.8.

§4. THE CASE $\delta(G) = 3$

Clearly 2.8 can be restated as follows:

G is connected α-crit. with $\delta(G) = 1$ iff $G \in \mathscr{A}(K_3)$.

This form of 2.8 and Theorem 3 together would suggest, that in general G is connected α-crit. iff $G \in \mathscr{A}(K_{\delta(G)+2})$. But the complement of the graph in [15] proving $(9, 3, 3) \twoheadrightarrow 4$ gives a counterexample to this in case $\delta(G) = 3$. Other counterexamples in any case $\delta(G) \geq 3$ can be easily constructed using 3.5. Still some other examples can be found in Krieger [11] called the $L^{i,k}$ graphs, in Wessel's papers [18, Satz 91, 92 and 93] and [19, Theorem 1] and in my forthcoming paper [14, Theorem 3]. However we prove in this chapter the following

4.1. Theorem 5. *G is connected α-crit. with $\delta(G) = 3$ iff $G \in \in \bigcup_{H \in \mathscr{H}} \mathscr{A}(H)$. Here $|\mathscr{H}| < \infty$, namely $\mathscr{H} = \{H$: H is connected α-crit. with $\delta(H) = 3$ and $\alpha \leq 27\}$.*

Before turning to the proof of this we formulate some conjectures arising here

Conjecture 1. There is a constant $c_0(\delta)$ (depending only on $\delta \geqslant 1$) such that G is connected α-crit. with $\delta(G) = \delta$ iff $G \in \bigcup_{H \in \mathscr{H}_\delta} \mathscr{A}(H)$ where $\mathscr{H}_\delta = \{H\colon H$ is connected β-crit. for some $\beta \leqslant c_0(\delta)$ and $\delta(H) = \delta\}$.

Theorems 1, 3 and 5 give $c_0(1) = c_0(2) = 1$ and $c_0(3) \leqslant 27$. The case $\delta \geqslant 4$ is unsolved. Whenever this conjecture is true for some δ, then it clearly implies the following 2 consequences (see 3.10).

Conjecture 1′. There is a constant $c_1(\delta)$ for each $\delta \geqslant 1$ such that whenever G is connected α-crit. with $\delta(G) = \delta$ then always $|E(G)| \leqslant \leqslant 2\alpha + c_1(\delta)$ is true.

Conjecture 1″. There is a constant $0 \leqslant b(\delta)$ for each $\delta \geqslant 1$ such that whenever $\alpha > b(\delta)$ and G is connected α-crit. with $\delta(G) = \delta$ then there is an x in $V(G)$ with $\rho(G, x) = 2$.

However now we prove that Conjecture 1″ implies Conjectures 1′ and 1, so either all these 3 conjectures are true for some δ, or they all fail for this δ.

4.2. *Suppose Conjecture* 1″ *is true for some* $\delta \geqslant 2$ *with a* $b(\delta)$. *Then we have*

(a) $|E(G)| \leqslant 2\alpha + b(\delta)(\delta - 2) + \binom{\delta + 1}{2} + 1$ *for any connected* α-*crit.* G *with* $\delta(G) = \delta$;

(b) G *is connected* α-*crit. iff* $G \in \bigcup_{H \in \mathscr{H}_\delta} \mathscr{A}(H)$. *Here* $\mathscr{H}_\delta = \{H\colon$ H *is connected* β-*crit. with* $\delta(G) = \delta,\ \beta \leqslant 2b(\delta)(\delta - 2) + (\delta - 1)^2 + 1\}$; *i.e. Conjectures* 1′ *and* 1 *are true with* $c_1(\delta) = b(\delta)(\delta - 2) + \binom{\delta + 1}{2} + 1$, *resp.* $c_0(\delta) = 2b(\delta)(\delta - 2) + (\delta - 1)^2 + 1$.

Proof.

(a) If G is connected α-crit., $\delta(G) = \delta$, $\alpha \leqslant b(\delta)$ then 3.13 gives $|E(G)| \leqslant (\alpha - 1)\delta + \binom{\delta + 2}{2}$. Now $2\alpha + b(\delta)(\delta - 2) + \binom{\delta + 1}{2} + 1 = 2(\alpha - b(\delta)) + (b(\delta) - 1)\delta + \binom{\delta + 2}{2}$, so in this case we only have to prove, that $(\alpha-1)\delta \leqslant 2\alpha - 2b(\delta) + (b(\delta)-1)\delta$, i.e. $0 \leqslant (b(\delta)-\alpha)(\delta-2)$. But $\delta \geqslant 2$, $b(\delta) \geqslant \alpha$, so this is true. Now we want to prove by induction on α: we suppose, that (a) holds for some $\alpha \geqslant b(\delta)$ and we want to prove (a) for $\alpha+1$. Let G be connected $\alpha+1$-crit., $\delta(G)=\delta$. We have an $x \in V(G)$ with $\rho(G,x)=2$. Put $E(G,x)=\{y,x'\}$. $x'y \in E(G)$ would mean $\phi = \mathscr{F}(\alpha, G-x-E(G,x)) = \mathscr{F}(\alpha, G-x'-\{y,x\}) = \mathscr{F}(\alpha, G-y-\{x,x'\})$ so 2.1 (a) gives, that $xx'y$ would be a connected component (which is a triangle) of G, but this is absurd. So $x'y \notin E(G)$ and we may use 3.2 (b). So we get, that $G(x,y)$ is connected α-crit. If we prove, that $|E(G)| = |E(G(x,y)| + 2$ then using the induction hypothesis on $G(x,y)$ we get $|E(G)| \leqslant \leqslant 2 + 2\alpha + b(\delta)(\delta-2) + \binom{\delta+1}{2} + 1$, so (a) is true for G. Now we prove $|E(G)| = |E(G(x,y))| + 2$: 3.2 (b) also gives $E(G,\{x'\},E(G,y)-\{x\}) = \phi$, so $E(G(x,y)) = E(G-x-y) \cup^* \{x'z\colon z \in E(G,y)-\{x\}\}$. On the other hand $E(G) = E(G-x-y) \cup^* \{xy, xx'\} \cup^* \{yz\colon z \in E(G,y)-\{x\}\}$ is trivial, so we clearly have $|E(G)| = |E(G,y))| + 2$. This completes the proof of (a).

(b) First note, that $\frac{3}{2}|S_3'(G)| + |S_2(G)| \leqslant |E(G)|$ for a connected α-crit. G, so $|S_3'(G)| + |S_2(G)| = 2\alpha + \delta(G) = 2\alpha + \delta$ gives $\frac{3}{2}(2\alpha+\delta) - \frac{1}{2}|S_2(G)| \leqslant |E(G)|$. Comparing this with (a) we have $\frac{3}{2}(2\alpha+\delta) - 2\alpha - b(\delta)(\delta-2) - \binom{\delta+1}{2} - 1 \leqslant \frac{1}{2}|S_2(G)|$, so $|S_2(G)| \geqslant \geqslant 2\alpha - 2b(\delta)(\delta-2) - ((\delta-1)^2+1) \geqslant \alpha+1$ if $\alpha \geqslant 2b(\delta)(\delta-2) + + (\delta-1)^2 + 2$. Now we prove (b) by induction on α. (b) is trivial for $\alpha \leqslant 2b(\delta)(\delta-2) + (\delta-1)^2 + 1$. Now let (b) hold for some $\alpha \geqslant 2b(\delta)(\delta-2) + (\delta-1)^2 + 1$ and we prove (b) for any connected $\alpha+1$-crit. G with $\delta(G)=\delta$. Put $\alpha+1=\alpha'$. G is connected α'-crit., $\delta(G)=\delta$ so the above result gives $|S_2(G)| \geqslant \alpha'+1$ since $\alpha' \geqslant 2b(\delta)(\delta-2) + (\delta-1)^2 + 2$. This and $\mathscr{F}(\alpha'+1, G) = \phi$ gives 2 vertices x and y in $S_2(G)$ with $xy \in E(G)$. Put $E(G,x)=\{y,x'\}$, $E(G,y)=\{x,y'\}$, $x' \neq y'$, since $x'=y'$ would give, that xyx' is a connected component of G contradicting to the connectedness of G (see the proof of (a)). So we may use 3.2 (b) to get that $G(x,y)$ is connected $\alpha' = \alpha + 1$-crit. The induction hypothesis gives $G(x, y) \in \in \bigcup_{H \in \mathscr{H}_\delta} \mathscr{A}(H)$, so $G = G(x, y)[x', y', x, y]$, together with $\rho(G(x, y), x') = = \rho(G, x') \geq 2$, $\rho(G(x, y), y') = \rho(G, y') \geq 2$ gives $G \in \bigcup_{H \in \mathscr{H}_\delta} \mathscr{A}(H)$ proving the only if part of (b). The if part is a simple consequence of 3.5.

Now we want to mention that Conjecture 1 is true for all $0 \leq \delta \leq \delta_0$ iff the following conjecture is true for all $0 \leq \delta \leq \delta_0$.

Conjecture 1‴. There is a $b'(\delta) \geq 0$ for each $\delta \geq 1$ such that whenever $\alpha > b'(\delta)$ and G is connected α-critical, with $\delta(G) = \delta$, then G is not 3-fold (vertex)-connected (i.e. G is exactly 2-fold connected.)

This equivalency is a simple consequence of Wessel's theorem in [18, Satz 81]. (see also [1, Theorem of Gallai 2.5]). We omit the details.

The proof of Theorem 5. 4.2 (b) shows that we only have to prove, that Conjecture 1″ is true with $b(3) = 11$. But 3.7 shows that to do this we only need to show the following lemma.

4.3. *Suppose* $\alpha > 11$, G *is str.* α*-crit.*, $\delta(G) = 3$. *Then there are* 3 *vertices* $x, y, z \in V(G)$ *with* $\mathscr{F}(\alpha, G - x - y - z) = \phi$ *and* $yz \in E(G)$.

Proof. Let G be str. α-crit. $\delta(G) = 3$. We are done if we find a $y \in V(G)$ with $\rho(G, y) = 2$. (See 2.1 (a)). So $\rho = \min\{\rho(G, x): x \in V(G)\} \geq 3$ may be assumed. $\rho(G, x) \leq 4$ for all $x \in V(G)$ by Theorem 2. Finally $\rho(G, x) = 3$ for all $x \in V(G)$ cannot hold, since $|V(G)|$ is odd. So we find an $x_1 \in V(G)$ with $\rho(G, x_1) = 4$. Let $x \in E(G, x_1)$. We find a str. α-crit. subgraph Γ on $G - x$ using 2.2.

Clearly we may assume, that whenever $\mathscr{F}\left(\frac{|H|}{2}, H\right) = \phi$, $R \in \in \mathscr{R}^0(G - x)$ and $H \cup^* R = V(G) - \{x\}$ then $R = \phi$. (If namely $y \in R$, then we choose a z in R such that $R - y - z \in \mathscr{R}^0(G - x)$ (we have such a z) and then $\mathscr{F}(\alpha, G - x - y - z) = \phi$, proving 4.3). We will call this the assumption (A). So e.g. we may assume $S_1(\Gamma) = \phi$. On the other hand 2.3 gives $3 = \rho(G, x_1) - 1 = \rho(\Gamma, x_1)$. So using 2.7 and 2.8 we get, that Γ is connected α-crit. so by Theorem 3 $\Gamma \in \mathscr{A}(K_4)$. Now $\Gamma \in \in \mathscr{A}(K_4)$ means the following: $V(\Gamma) = \bigcup_{1 \leq i < j < 4} P_{ij}$ for some $P_{ij} \in \mathscr{P}_0^1(\Gamma)$

with endpoints x_i and x_j and $V(\Gamma) = \{x_1, x_2, x_3, x_4\} \cup^* \bigcup^*_{1 \leq i < j \leq 4} P'_{ij}$ $(P'_{ij} = P_{ij} - \{x_i, x_j\})$.

Now put

$$P_{12} \cup P_{13} \cup P_{14} = A, \; A \cup \{x\} = A', \; P_{23} \cup P_{34} \cup P_{42} = C;$$

$$P_{ij} = P_{ji} \quad \text{for} \quad 1 \leq j < i \leq 4\,;$$

$$A_1 = Q^1_{12} \cup Q^1_{13} \cup Q^1_{14},$$

$$A_0 = Q^0_{12} \cup^* Q^0_{13} \cup^* Q^0_{14} - \{x_2, x_3, x_4\}; \; A'_0 = A_0 \cup^* \{x\}.$$

Here $Q^l_{ij} = \{z \in P_{ij}\colon |P| \equiv l(2)$ where P is the subpath of P_{ij} connecting x_i and $z\}$ for $1 \leq i, j \leq 4$; $(Q^l_{ij} = Q^{1-l}_{ji})$, finally we denote by y_j the endpoint of the path $P_{1j} - \{x_j\}$ for which $x_j y_j \in E(\Gamma)$. $(y_j \in Q^1_{1j})$.

As a matter of fact, $C = C(\Gamma)$, $A = A(\Gamma)$ etc. depends on Γ, but we won't write this out usually. We may assume $|C(\Gamma)| \leq |C(\Gamma')|$ for all str. α-crit. subgraphs Γ' on $G - x$. But $C(\Gamma)$ can also be defined, as the only member of $\mathfrak{C}^1_1$ $(\Gamma - x_1)$. So we may assume, that whenever $C_0 \subsetneqq C$ and $C_0 \in \mathfrak{C}^0_1(G)$, then there is no str. α-crit. subgraph Γ' on $G - x$ with $C_0 \in \mathfrak{C}^1_1(\Gamma')$. Using this we shall be able to prove that

(α) $C \in \mathfrak{C}^1_1(G)$.

On the other hand we shall be able to prove, that the assumption (A) implies that

(β) $E(G, A_1, A') = E(\Gamma, A_1, A') \cup \{xx_1\} \cup E$ with either $E = \phi$ or $E = \{xy\}$ for some $y \in A_1$; and

(γ) $E(G, A_1 - \{y_2, y_3, y_4\}, C) = \phi$.

Now $|E(G, \{y_2, y_3, y_4\}, C)| \leq 9$ is trivial from $\rho(G, y_i) \leq 4$. However one can also check using (A) that $|E(G, \{y_2, y_3, y_4\}, C)| \leq 6$. Put $|E| = e$. (β) and (γ) imply that $2 = \rho(\Gamma, y) = \rho(G, y)$ for all $y \in A'_1 =$ $= A_1 - \{y_2, y_3, y_4\} - E(G, x)$. So $A'_1 = \phi$, and $|A_1| \leq 4 + e$. So we also have $|A_0| \leqslant 3 + e$. (α) gives $\rho(G, y) = 2$ for all $y \in C$ not connected to A', i.e. $\rho(G, y) = 2$ for all $y \in C - E(G, x) - E(G, \{y_2, y_3, y_4\}) - E(G, A_0)$ so this latter must be empty. But $|E(G, x) \cap C| \leqslant 3 - e$, $|C \cap E(G, A_0)| \leqslant 2 \cdot |A_0| \leqslant 6 + 2e$ since any member of A_0 is connected to $\geqslant 2$ members of A' and has the valency $\leqslant 4$ in G. Finally $|E(G, \{y_2, y_3, y_4\}, C)| \leqslant 6$ gives $|C \cap E(G, \{y_2, y_3, y_4\})| \leqslant 6$. So

$$C = (C \cap E(G, x)) \cup (E(G, A_0) \cap C) \cup (C \cap E(G, \{y_2, y_3, y_4\}))$$

has at most $15 + e$ elements. So $|C|$ – being odd – has $\leqslant 15$ elements. We have now $2\alpha + 3 = |V(G)| = |A_1| + |A_0| + 1 + |C| \leqslant 8 + 2e + 15 \leqslant \leqslant 25$, i.e. $\alpha \leqslant 11$, proving 4.3. So we have now to prove (α)-(γ).

We start by (β). Suppose first $y, y' \in A_1$. Then one can easily find a subgraph Γ' of $\Gamma \cup \{yy'\}$ with the property $V(\Gamma') = C \cup^* C_1 \cup^* R$; $C, C_1 \in \mathfrak{C}_1^1(\Gamma \cup \{yy'\})$; $R \in \mathscr{R}^1(\Gamma \cup \{yy'\})$; $E(\Gamma, C, C_1) = E(\Gamma, C \cup^* C_1, R) = \phi$. But then clearly Γ' is str. α-crit. by 2.7 and 2.8, and $\delta(\Gamma') = 2$. Finally $\max\{\rho(\Gamma', x) \colon x \in V(G)\} = 2$, so Γ' cannot be a subgraph of $G - x$ by $\rho(G, x_1) = 4$ and 2.3. So $yy' \notin E(G)$ follows, since Γ is a subgraph of $G - x$.

So we have $E(G, A_1, A_1) = E(\Gamma, A_1, A_1) = \phi$. Next suppose $yz \in E(G, A_2, A_1) - E(\Gamma, A_2, A_1)$ with $A_2 = A_0 - E(\Gamma, x_1)$. One can now easily find a subgraph Γ' of $\Gamma \cup \{yz\}$ with the properties $V(\Gamma') = V(\Gamma_1) \cup^* R$, $E(\Gamma_1, R) = \phi$, $\phi \neq R \in \mathscr{R}^1(\Gamma \cup \{yz\})$, Γ_1 is str. $\dfrac{|V(G_1)|}{2} - 1$-crit. ($\Gamma_1 \in \mathscr{A}(K_4)$). (It is easy to define this Γ' in any case, so we leave it to the reader.) So this contradicts the assumption (A) proving $E(G, A_1, A_2) = E(\Gamma, A_1, A_2)$. In case of $E(G - \Gamma, E(\Gamma, x_1), A_1) \neq \phi$ we easily find a subgraph Γ' on $G - x$ with $\Gamma' \in A(K_4)$ and $\rho(\Gamma', x_1) = 2$, a contradiction. If $|E(G - xx_1, x, A_1)| \geqslant 2$, Then one of the previous contradictions can be easily deduced, this we leave to the reader.

Next we prove (γ). We prove that $E(G, A_1 - \{y_j\}, Q_{jk}^1 \cup Q_{jl}^1) = \phi$ for all permutation $\langle j, k, l\rangle$ of $\langle 2, 3, 4\rangle$. Let $y \in A_1 - \{y_j\}$, $y' \in Q_{jk}^1 \cup Q_{jl}^1$, and define the subgraph Γ' on $G - x$ as follows: $\Gamma' = \Gamma \cup \{yy'\} - \{zz'\}$, where $z'y_j, zz' \in E(\Gamma)$, $z \in Q_{1j}^1$, $z' \in Q_{1j}^0$. There is such a zz' in P_{1j} if

$|P_{1j}| \neq 1$. In this case $\mathscr{F}(\alpha, \Gamma' - z' - y_j) = \mathscr{F}(\alpha, (\Gamma - z' - y_j) \cup \{yy'\}) = \phi$. Namely let $F \in \mathscr{F}(\alpha, \Gamma' - z' - y_j)$. Then $|F \cap C| \leqslant \frac{|C| - 1}{2}$ so $|F \cap A| \geqslant \alpha - \frac{|C| - 1}{2} = \frac{|A| - 1}{2}$. But $F \cap A = F \cap (A - z' - y_j)$ and $\mathscr{F}\left(\frac{|A| - 1}{2}, A - z' - y_j \cup \{yy'\}\right) = A_1 - \{y_j\}$. So $F \cap A = A_1 - \{y_j\}$. Now $\{y_k, y_l, y\} \subset A_1 - \{y_j\} \subset F$ implies $\{x_k, x_l, y'\} \cap F = \phi$. But $C - \{x_k, x_l, y'\}$ consists of 3 even paths in Γ' so $F \cap C \subset C - \{x_k, x_l, y'\}$ implies $|F \cap C| \leqslant \frac{|C| - 3}{2}$. So we have $\alpha = |F| = |F \cap C| + |F \cap A| \leqslant \frac{|C| - 3}{2} + \frac{|A| - 1}{2} = \frac{|C| + |A|}{2} - 2 = \alpha - 1$, a contradiction. So we have proved $\mathscr{F}(\alpha, \Gamma' - z' - y_j) = \mathscr{F}(\alpha, (\Gamma - z' - y_j) \cup \{yy'\}) = \phi$. If $yy' \in E(G)$ held, then $\Gamma - z' - y_j = H$, $\phi \neq R = \{z', y_j\}$ would be a counterexample to (A). So we have $yy' \notin E(G)$ in case $|P_{1j}| \neq 1$. If $|P_{1j}| = 1$ then $y_j = x_1$, $x_1 x_j \in E(G)$. So $\mathscr{F}(\alpha + 1, G - x_1 x_j) \neq \phi$. Let $F \in \mathscr{F}(\alpha + 1, G - x_1 x_j)$, $x_1, x_j \in F$. Clearly $x \notin F$, so $F \in \mathscr{F}(\alpha + 1, \Gamma - x_1 x_j)$. But this means $|F \cap C| \leqslant \frac{|C| - 1}{2}$, so $|F \cap A| \geqslant \frac{|A| + 1}{2}$. $F \cap A$ is independent in Γ so $\mathscr{F}\left(\frac{|A| + 1}{2}, A\right) = \{A_1\}$ gives $F \cap A = A_1$, and $|F \cap C| = \frac{|C| - 1}{2}$. $A_1 \subset F$ means $y_k, y_l \in F$, so $x_k, x_l \notin F$. So $F \cap C = F \cap (C - \{x_k, x_l\})$. But $C - \{x_k, x_l\} = (P_{kl} - \{x_k, x_l\}) \cup^* P^0$ with $P^0 = (P_{kj} - \{x_k\}) \cup (P_{ij} - \{x_l\})$. $F \cap C$ is independent in Γ, so $|F \cap P_{kl} - \{x_k, x_l\}| \leqslant \frac{|P_{kl}|}{2} - 1$ so $|F \cap P^0| \geqslant \frac{|P^0| + 1}{2}$. But this means $F \cap P^0 = Q^1_{jk} \cup Q^1_{jl}$. So we have $Q^1_{jk} \cup Q^1_{jl} \cup A_1 \subset F \in \mathscr{F}(\alpha + 1, G - x_1 x_j)$, and so the only edge connecting $Q^1_{jk} \cup Q^1_{jl}$ and A_1 is $x_1 x_j = y_j x_j$. So $yy' \notin E(G)$ follows in this case, too, proving $E(G, A_1 - \{y_j\}, Q^1_k \cup Q^1_{jl}) = \phi$ in any case. So we have $E(G, A_1 - \{y_2, y_3, y_4\}), Q^1_{23} \cup Q^1_{24} \cup Q^1_{32} \cup Q^1_{34} \cup Q^1_{42} \cup Q^1_{43}) = \phi$. Thus $Q^1_{ij} \cup Q^1_{ji} = P_{ji}$ and $P_{23} \cup P_{34} \cup P_{42} = C$ gives (γ).

So to complete our proof we need to prove (α):

(α1) $E(G, Q^1_{ij}, Q^1_{ji}) = E(\Gamma, Q^1_{ij}, Q^1_{ji})$. Let namely $yy' \in E(G)$, $y \in Q^1_{ij}$, $y' \in Q^1_{ji}$, P be the subpath of P_{ij} connecting y and y', $R = P - \{y, y'\}$.

Clearly $H = \Gamma \cup \{yy'\} - R \in \mathscr{A}(K_4)$ so $\mathscr{F}(|H|/2, H) = \phi$. On the other hand $R \in \mathscr{P}_0^0(G)$, so $R \in \mathscr{R}^0(G)$. So (A) implies $R = \phi$ and thus $yy' \in$ $\in E(\Gamma)$.

(α2) $E(G, Q_{ji}^1, Q_{jk}^0) = E(\Gamma, Q_{ji}^1, Q_{jk}^0)$, $\langle i, j, k \rangle$ is an arbitrary permutation of $\langle 2, 3, 4 \rangle$. Suppose namely $yy' \in E(G)$ with $y \in Q_{ji}^1$, $y' \in Q_{jk}^0$ and let P be the subpath of P_{jk} connecting x_j and y', $R = P - \{x_j, y'\}$. Clearly $R \in \mathscr{R}^0(G)$. On the other hand one can easily check, that $H =$ $= \Gamma \cup \{yy'\} - R \in \mathscr{A}(K_4)$. ($x_1, x_k, x_i$ is the same as in Γ, the "new" x_j is y.) So $\mathscr{F}(|H|/2, H) = \phi$ by Theorem 3, so (A) gives. $R = \phi$. Now $yy' \in E(\Gamma)$ by the minimality of C.

(α3) $E(G, Q_{ji}^0, Q_{jk}^0) = \phi$. Let namely $y \in Q_{ji}^0$, $y' \in Q_{jk}^0$ and assume $yy' \in E(G)$. Then $\Gamma \cup \{yy'\} - P'_{ik} \in \mathscr{A}(K_4)$ (note that $P_{ik} - \{x_i, x_k\} =$ $= P'_{ik}$); x_1, x_j is the same as in Γ, the new x_k is y and the new x_l is y', finally $P'_{ik} \in \mathscr{R}^0(G)$, so there is a subgraph R on P'_{ik} which is a 1-factor, so the graph Γ' defined by $E(\Gamma') = E(R) \cup^* E(\Gamma - P'_{ik}) \cup^*$ $\cup^* \{yy'\}$ is a str. α-crit. subgraph on $G - x$. But $C_0 = P \cup P' < C$ and $C_0 \in \mathfrak{C}_1^1(\Gamma')$ for the subpaths P resp. P' of P_{ji} resp. P_{jk} connecting x_j and y resp. y'. However this contradicts to the minimality of $|C|$.

(α4) $E(G, Q_{ij}^1, Q_{ij}^1) = \phi$. Let namely $y, y' \in Q_{ij}^1$, $yy' \in E(G)$. Then we always get an l and m with $2 \leqslant l < m \leqslant 4$, such that $C - P_{lm} -$ $- P \in \mathscr{R}^0(\Gamma) \subset \mathscr{R}^0(G)$ for the subpath P connecting y and y' in P_{ij}, and so putting $P_{1l} \cup P_{lm} \cup P_{1m} = C'$ we have $V(G) - \{x\} = C \cup^* C' \cup^* R$ with a $R \in \mathscr{R}^0(G - x)$. But $P, C' \in \mathfrak{C}_1^0(G)$ so the subgraph Γ' defined by $P, C' \in \mathfrak{C}_1^1(\Gamma')$, $R \in \mathscr{R}^1(\Gamma')$, $E(\Gamma', C, C') = E(\Gamma', C \cup^* C', R) = \phi$ is str. α-crit. subgraph on $G - x$ with $\max\{\rho(\Gamma', z) \colon z \in V(G) - \{x\}) = 2$. But 2.3 gives $\rho(\Gamma', x_1) = 3$ a contradiction.

(α5) $E(G, Q_{ji}^1, Q_{jk}^1) = \phi$. Let now $y \in Q_{ji}^1$, $y' \in Q_{jk}^1$ P resp. P' the subpaths of P_{ji} resp. P_{jk} connecting x_j and y resp. y'. Put $C_0 = P \cup P'$, $C' = P_{1i} \cup P_{1k} \cup P_{ik}$. Then $C - C_0 - P_{ik} \in \mathscr{R}^0(\Gamma) \subset \mathscr{R}^0(G)$, so $(A - P_{1i} - P_{1k} = P'_{1j} \in \mathscr{R}^0(\Gamma) \subset \mathscr{R}^0(G))$,, we have $V(G) - \{x\} =$ $= C_0 \cup^* C' \cup^* R$ with $C_0, C' \in \mathfrak{C}_1^0(G)$ and $R \in \mathscr{R}^0(G)$. This gives the same contradiction as in (α4).

(α1)-(α5) together prove (α) and so 4.3, and Theorem 5.

§5. ON THE ARROW-SYMBOL $(m, n, k, r) \to n + l$

5.1. *The definition of the symbol.* Erdős in [4] proposed the following problem:

Put $(m, n, k, r) \to n + l$ if the following statement is true (and put $(m, n, k, r) \nrightarrow n + l$ in the opposite case).

Suppose $|S| = m$, $\mathcal{F} \subset [S]^n$ and if $|X| \leqslant k$, then $\mathcal{F}$ has a member F with $F \cap X = \phi$. Then there is an $A \in [S]^{n+l}$ such that there exists an $F \in \mathcal{F}$ for each $B \in [A]^r$ with $B \subset F$.

Put $f_r(n, k, l) = \max\{m: (m, n, k, r) \to n + l]$. Determine the function $f_r(n, k, l)$.

We only consider the case $r = 2$ and put $f_r(n, k, l) = f(n, k, l)$.

The following results have been proved on $f(n, k, l)$ so far.

5.2.

(a) $f(n, k, l) = 2n + k - 2$ *if* $\left[\frac{k+1}{2}\right] \leqslant l \leqslant k$.

(b) $2n + 2 \leqslant f(n, 3, 1) \leqslant 2n + 10\sqrt{n}$, *and in general*

$$f(n, 2k + 1, k) \leqslant 2n + l - 1 + (8l + 2)\sqrt{2(n + l - 1)}$$

(c) $f(n, k, 1) \geqslant 2n + k - 2 + \Delta(k)$ *with*

$$\Delta(k) = \max\{q: q(q + 1) \leqslant k\}.$$

(a) for $k = 1$, $l = 1$ in Theorem 1 (see 5.3), and is proved in general in [4]. The lower bound in (b) is due to Szemerédi [16] whose proof was applied by Milner and Sauer in [7, Theorem 4] to prove (c). The upper bounds of (b) are proved in [14]. Some other results on $f(n, k, l)$ are proved in [4], [12], [14], [15]. In this § we want to prove some new results on $f(n, 2k + 1, k)$ (see 5.4) and finally we give some conjectures (5.5). First we formulate an equivalent from of $(m, n, k, 2) \to n + l$.

5.3. *The following statements are equivalent.*

(a) $(m, n, k, 2) \to n + l$,

(b) *If* G *is a graph,* $|V(G)| = m$, $\mathscr{F}(n + l, G) = \phi$ *then* $\mathscr{F}(n, G - x) = \phi$ *for some* $X \in [V(G)]^k$.

(c) *If* G *is an* $n + l - 1$*-critical graph,* $|V(G)| = m$ *then* $\mathscr{F}(n, G - X) = \phi$ *for some* $X \in [V(G)]^k$.

Proof. Trivial (b) $\Leftrightarrow$ (c) is clear. To prove (a) $\Rightarrow$ (b) chose $\mathscr{F} = \mathscr{F}(n, G)$. To prove (b) $\Rightarrow$ (a) choose G defined by $V(G) = S$, $E(G) = \{xy\colon \neg\exists F \in \mathscr{F}$ with $x, y \in F\}$.

5.4. Theorem 6.

(a) $(2n + 3, n, 3, 2) \to n + 1$ *if* $n \geqslant 12$, *so* $f(n, 3, 1) \geqslant 2n + 3$.

(b) $(2n + 2l, n, 2l + 1) \to n + l$ *if* $1 \leqslant l < n$, *so* $f(n, 2l + 1, l) \geqslant 2n + 2l$ *if* $1 \leqslant l \leqslant n$.

(c) $(2n + 2l + 1, n, 2l + 1) \to n + l$ *if* $1 \leqslant l < \dfrac{n + 16}{43}$ *so* $f(n, 2l + 1, l) \geqslant 2n + 2l + 1$ *in this case.*

Proof. (a) is in fact Lemma 4.3. The proof of (b) uses the same ideas as the proof of 3.11 so we leave it to the reader.

We use inductionoon l to prove (c). Clearly $l = 1$ is proved by (a). Suppose (c) is true for all $1 \leqslant l < l_0 < \dfrac{n + 16}{43}$, and we want to prove it for l_0. Let G be any $n + l_0 - 1$ crit. graph. Suppose $\rho(G, x) \leqslant 1$ for some $x \in V(G)$. Then we always have an $y \in V(G) - \{x\}$ with $\mathscr{F}(n + l_0 - 1, G - x - y) = \phi$. Using the induction hypothesis on $l_0 - 1$ we have $(2n + 2l_0 - 1, n, 2l_0 - 1) \to n + l_0 - 1$, so (see 5.3 (b)) we get an $X_0 \in [V(G) - \{x, y\}]^{2l_0 - 1}$ with $\mathscr{F}(n, G - \{x, y\} - X_0) = \phi$, so $X = X_0 \cup \{x, y\}$ proves $(2n + 2l_0 + 1, n, 2l_0 + 1) \to n + l_0$ in this case. So we may assume $\rho(G, x) \geqslant 2$ for all $x \in V(G)$. So G is str. n-crit, and using 2.7 we have 3 cases.

(α) $V(G) = V(G_1) \cup^* V(G_2) \cup^* V(G_3)$ G_i's are str. n_i-crit. with $\delta(G_i) = 1$, $E(G, G_i, G_j) \neq \phi$, $n_1 + n_2 + n_3 = n + l_0 - 1$, $n_1 \geqslant n_2 \geqslant n_3$.

(β) $V(G) = V(G_1) \cup^* V(G_2)$ G_1 is str. n_1-crit. with $\delta(G_1) = 1$ G_2 is str. n_2-crit. with $\delta(G_2) = 2$, $n_1 + n_2 = n + l_0 - 1$, $E(G, G_1, G_2) = = \phi$.

(γ) G is connected $n + l_0 - 1$-crit.

In both cases (α) and (β) we are done if $n_1 \geqslant l_0$ since $G_1 \in \mathfrak{C}_1^1(G)$ by 2.8 and $|V(G_1)| = 2n_1 + 1 \geqslant 2l_0 + 1$ means the existence of an $X \in [V(G_1)]^{2l_0+1}$ with $\mathscr{F}(n_1 - l_0 + 1, G_1 - X) = \phi$. On the other hand $\mathscr{F}(n + l_0 - n_1, G - V(G_1)) = \phi$. So we get $\phi = \mathscr{F}(n + l_0 - n_1 + (n_1 - l_0 + 1) - 1, G - X) = \mathscr{F}(n, G - X)$. proving $(2n + 2l_0 + 1, n, 2l_0 + 1) \to n + l_0$ for this case. So the case (α) is proved if $\dfrac{n + l_0 - 1}{3} \geqslant l_0$ i.e. $l_0 \leqslant \dfrac{n-1}{2}$, which always holds if $l_0 < \dfrac{n+16}{43}$.

In case (β) we are also done if $2l_0 < n_2 + 1$: in this case we use (b). Put $|V(G_2)| = 2(n' + l_0)$, i.e. $n' = n_2 + 1 - l_0$. We have $l_0 < n_2 + 1 - l_0 = n'$ so $(2n' + 2l_0, n', 2l_0 + 1) \to n + l_0$ gives an $X \in \in [V(G_2)]^{2l_0+1}$ with $\mathscr{F}(n', G_2 - X) = \phi$. Now $\mathscr{F}(n_1 + 1, G_1) = \phi$ gives $\phi = \mathscr{F}(n_1 + n', G - X) = \mathscr{F}(n_1 + n_2 + 1 - l_0, G - X) = = \mathscr{F}(n, G - X) = \phi$ proving $(2n + 2l_0 + 1, n, 2l_0 + 1) \to n + l_0$ in this case. So case (β) is finished if $3l_0 \leqslant n + l_0$, i.e. $l_0 \leqslant \dfrac{n-1}{2}$ which always holds if $l_0 \leqslant \dfrac{n+16}{43}$.

So only case (γ) is remaining, and here we use Theorem 5 to get, that $G \in \mathscr{A}(H)$ for some connected 27-crit. H with $\delta(H) = 3$. But this means $V(G) = \bigcup_{\substack{1 \leqslant i < j \leqslant 57 \\ x_i x_j \in E(H)}} P_{ij}$, where $P_{ij} \in \mathscr{P}_0^1(G)$, P_{ij} has the end points x_i, x_j, and

$$V(G) = \{x_1, \dots, x_{57}\} \cup^* \bigcup^*_{\substack{1 \leqslant i < j \leqslant 57 \\ x_i x_j \in E(H)}} P'_{ij}$$

with $P'_{ij} = P_{ij} - \{x_i, x_j\}$, and $V(H) = \{x_1, \dots, x_{59}\}$.

On the other hand suppose $2l_0 - 1 \leqslant |P'_{ij}|$ for some $1 \leqslant i < j \leqslant 57$, $x_ix_j \in E(H)$. Then we have an $P^0 \in \mathscr{P}^1_1(G)$ with $|P_0| = 2l_0 - 1$ and $\rho(G, x) = 2$ for all $x \in P_0$. Let u, v be the endpoints of P_0 and $uu', vv' \in E(G)$, $u', v' \notin P_0$ (there are such u' and v'). Then put $X = P_0 \cup \{u', v'\}$. X is a path of length $2l_0 + 1$ without diagonals, with endpoints u', v'. Put $X_s = \{z$: the subpath of X connecting u', and z is of length $\equiv s \bmod 2\}$. Clearly $X_0 \cup^* X_1 = X$, $E(G, X_0) = X_1$, $|X_0| = l_0$, $|X_1| = l_0 + 1$. We prove $\mathscr{F}(n, G - x) = \phi$: suppose $F \in \mathscr{F}(n, G - x)$ then $F \cup^* X_0 \in \mathscr{F}(n + l_0, G) = \phi$ which is absurd. So we are done, if $2l_0 - 1 \leqslant |P'_{ij}|$ for some $x_ix_j \in E(H)$.

So only the case when $|P'_{ij}| \leqslant 2l_0 - 2$ for all $x_ix_j \in E(H)$ is left. But in this case $2n + 2l_0 + 1 = |V(G)| = |V(H)| + \sum_{x_ix_j \in E(H)} |P'_{ij}| \leqslant$

$\leqslant 57 + 2|E(H)|(l_0 - 1) \leqslant 57 + \left\{26 \cdot 3 + \binom{5}{2}\right\}(l_0 - 1) = 88l_0 - 31$, (we used Theorem 4) so $n \leqslant 43l_0 - 16$ i.e. $l_0 \geqslant \dfrac{n + 16}{43}$ would follow in contradiction to $l_0 < \dfrac{n + 16}{43}$. This completes the proof of (c).

5.4. We close our paper by stating 2 more problems related to Theorem 6. We cannot prove the following:

Conjecture 2.

(a) There is an $f(n)$ with $\lim\limits_{n \to \infty} f(n) = \infty$ such that $(2n + f(n), n, 3, 2) \to n + 1$ i.e. $\lim\limits_{n \to \infty} f(n, 3, 1) - 2n = \infty$.

(b) There is an $f(l, n)$ with $\lim\limits_{\substack{n \to \infty \\ l \text{ fixed}}} f(l, n)$ such that $(2n + f(l, n), n, 2l + 1, 2) \to n + l$, i.e. $\lim\limits_{\substack{l \text{ fixed} \\ n \to \infty}} f(n, 2l + 1, l) - 2n = \infty$.

Conjecture 1″ would imply (a) and the same argument as in 5.3 shows that Conjecture 1 implies (b). Finally 5.2, (b) shows, that $f(n) > 10\sqrt{n}$, $f(k, n) > (8k + 2)\sqrt{2(n + l)}$ cannot occur. We would be interested, whether $\underline{\lim} \dfrac{f(n, 2l + 1, l) - 2n}{\sqrt{n}} > 0$ holds or not.

REFERENCES

[1] B. Andrásfai, On critical graphs. Theory of Graphs. *International symposium held at Rome,* 1966, Dunod, Paris, Gordon and Breach, New York, 1967, 9-19.

[2] L.N. Beineke – F. Harary – M.D. Plummer, On the critical lines of a graph, *Pacific Journal of Math.,* 22 (1967), 205-212.

[3] G.A. Dirac, On the structure of 5- and 6-chromatic abstract graph, *Journal für Reine u. Angew. Math.,* 214-115 (1964), 43-52.

[4] P. Erdős, On a lemma of Hajnal – Folkmann. Combinatorial theory and its applications, *Colloquia Math. Soc. J. Bolyai,* 4 (1970), 311-316.

[5] P. Erdős – T. Gallai, On the minimal number of vertices representing the edges of a graph. *Publications of the Mathematical Institute of the Hung. Acad. of Sci.,* VI, (1961), 181-203.

[6] P. Erdős – A. Hajnal – J.W. Moon, A problem in graph theory, *American Math. Monthly,* 71 (1964), 1107-1110.

[7] J.H. Folkmann, An upper bound for the chromatic number of a graph. Combinatorial Theory and its applications. *Colloquia Math. Soc. J. Bolyai,* 4 (1970), 437-451.

[8] A. Hajnal, On *k*-saturated graphs, *Canad. Journal of Math.,* 17 (1965), 720-724.

[9] G. Hajós, Über eine Konstruktion nicht *n*-färbbarer Graphen, *Wiss. Zeitschrift der Martin Luther Univ.,* Halle – Wittenberg, Math. Nat., X/1 (1961), 116-117.

[10] F. Harary – M.D. Plummer, On indecomposable graphs, *Canad. Journal of Math.,* 19 (1967), 800-809.

[11] M.M. Krieger, Graphs edge-critical with respect to independence number, *Annals of the New York Acad. of Sciences,* 175, Art. 1, (1970), 265-271.

[12] E.C. Milner – N. Sauer, Generalizations of a lemma of Folkmann, *Annals of the New York Acad. of Sciences,* 175, Art. 1, (1970), 295-307.

[13] M.D. Plummer, On linecritical graphs, *Monatshefte für Math.,* 21 (1967), 40-48.

[14] L. Surányi, On linecritical graphs II. (*to appear.*)

[15] L. Surányi, On a problem of P. Erdős and A. Hajnal. Combinatorial Theory and its applications, *Coll. Math. Soc. J. Bolyai,* 4 (1970), 1029-1041.

[16] E. Szemerédi, On a problem of P. Erdős. Combinatorial Theory and its applications, *Coll. Math. Soc. J. Bolyai,* 4 (1970), 1051-1052.

[17] A. Zykov, On some properties of linear complexes, *Math. Sb.,* 24, 66 (1949), 163-188, *Amer. Math. Soc. Translations,* No. 79.

[18] W. Wessel, Kanten-kritische Graphen mit der Zusammenhangzahl 2, *Manuscripta Math.,* 2 (1970), 309-334.

[19] W. Wessel, On the problem of determining whether a given graph is edge-critical or not. Combinatorial theory and its applications. *Coll. Math. Soc. J. Bolyai,* 4 (1970), 1123-1139.

[20] C. Berge, *Graphs et hypergraphs,* Dunod, Paris.

COLLOQUIA MATHEMATICA SOCIETATIS JÁNOS BOLYAI
10. INFINITE AND FINITE SETS, KESZTHELY (HUNGARY), 1973.

ON COLOUR-CRITICAL HYPERGRAPHS

B. TOFT

ABSTRACT

This paper gives a survey of a few results and unsolved problems concerning colour-critical hypergraphs. Both uniform and non-uniform hypergraphs are considered. Methods for constructing critical hypergraphs are presented. Problems concerning the possible number of vertices and edges and the possible degrees are discussed. Not all proofs are included.

1. INTRODUCTION AND DEFINITIONS

A *hypergraph* H is a finite set $V(H)$ of *vertices* together with a set $E(H)$ of *edges* consisting of different subsets of $V(H)$, such that each edge contains at least two elements of $V(H)$ and $V(H) \cap E(H) = \phi$. An edge of H consisting of the subset A of $V(H)$ is denoted (A). If $(A) \in E(H)$ then $H - (A)$ denotes the hypergraph obtained from H by removing the edge (A) as such, but removing no vertices. If $x \in V(H)$ then $E_x(H)$ denotes the set of edges of H containing x, and $H - x$ denotes the hypergraph obtained from H by removing the edges of $E_x(H)$

as such and the vertex x. Following C. Berge [2, p. 411] we denote by $d_H(x)$ the *degree* of x in H, that is the maximal number of edges of $E_x(H)$ having two and two only x in common. If all edges of H contain the same number r of vertices then H is *r-uniform*. A 2-uniform hypergraph is a *graph*.

A *colouring* of a hypergraph H is a mapping K of $V(H)$ into the positive integers such that the restriction of K to any edge of H is non-constant. If K has at most k different values then it is a *k-colouring* and H is *k-colourable*. The least k for which H is k-colourable is the chromatic number of H, denoted $\chi(H)$. H is *k-critical* if it is connected, $\chi(H) = k$, and $\chi(H - (A)) < \chi(H)$ for all edges $(A) \in E(H)$.

The above definition of chromatic number is due to P. Erdős and A. Hajnal [11]. Critical hypergraphs were introduced by L. Lovász [14], who observed that $d_H(x) \geqslant k - 1$ for each vertex x of a k-critical hypergraph H.

A hypergraph H is 1-critical iff H consists of a single vertex and no edges, i.e. $H = K_1$. H is 2-critical iff $|V(H)| \geqslant 2$ and $E(H) = \{(V(H))\}$. The structure of all 3-critical hypergraphs is not known, however D. König proved [13, Satz X. 12] that the 3-critical graphs are the odd circuits. For 4-critical hypergraphs not even the graphs are known.

If H_1 and H_2 are disjoint hypergraphs then $H_1 * H_2$ denotes the hypergraph obtained from H_1 and H_2 by joining each vertex of H_1 to each vertex of H_2 by an edge of size 2. Obviously $\chi(H_1 * H_2) = \chi(H_1) + \chi(H_2)$. Moreover $H_1 * H_2$ is critical iff H_1 and H_2 are critical.

The main-theme of this paper is to describe constructions providing a method of transforming any k-critical hypergraph $(k \geqslant 3)$ into an r-uniform one ($r \geqslant 2$, and if $k = 3$ then $r \geqslant 3$). This shows that in a certain sense the problem of characterizing all k-critical hypergraphs is equivalent of characterizing the r-uniform ones (except for $(k, r) = (3, 2)$).

Much of this paper was inspired by constructions of 3-critical uniform hypergraphs due to H.L. Abbott and D. Hanson [1].

2. TWO EXAMPLES

I. Let $1 \leqslant d \leqslant k-2$ and $2 \leqslant s \leqslant r$. Let the hypergraph H consist of two disjoint hypergraphs H_1 and H_2 together with all edges of size s consisting of exactly one vertex of H_1 and $s-1$ vertices of H_2, where H_1 has n_1 vertices and is d-critical and where H_2 has $(s-2) \cdot d + (r-1)(k-1-d) + 1$ vertices and all subsets of size r of the set of vertices as edges.

Suppose that H has a $(k-1)$-colouring K. In H_1 at least d different colours, say $1, \ldots, d$, are used. Because of the edges of H of size s at most $(s-2) \cdot d$ vertices of H_2 have colours from $\{1, \ldots, d\}$. For each of the remaining $k-1-d$ colours of K at most $r-1$ vertices of H_2 have this colour. This means that at most $n_1 + (s-2) \cdot d + (r-1)(k-1-d)$ vertices of H have got a colour in K, hence there is at least one vertex without a colour. This contradiction shows that $\chi(H) \geqslant k$. Moreover it is easy to check that $\chi(H-(A)) \leqslant k-1$ for all edges (A) of H, hence $\chi(H) = k$ (since the chromatic number decreases by at most 1 when an edge is removed) and H is k-critical.

If $d = 1$ then H has $(r-1)(k-2) + s$ vertices and $E(H)$ consists of all subsets of $V(H)$ of size r not containing a certain vertex $x \in V(H)$ together with all subsets of size s containing x.

If $d = 1$ and $s = r$ then H has $(r-1)(k-1) + 1$ vertices and $E(H)$ consists of all subsets of $V(H)$ of size r. If $r = 2$ then this is the complete k-graph K_k.

II. The second in example is more special. It was inspired by similar examples in [1].

Let r be even $\geqslant 2$ and let $V(H) = \{x, a_1, \ldots, a_r, b_1, \ldots, b_r\}$. Moreover let $E(H)$ consist of edges (A), where *either* $A = \{x, a_i, b_i\}$, $i \in \{1, \ldots, r\}$, *or* $A = \{c_1, \ldots, c_r\}$ where $c_i = a_i$ or $c_i = b_i$ and the number of those i for which $c_i = a_i$ is either an odd number $\leqslant r/2$ or an even number $\geqslant r/2$. The number of edges of H of size 3 is thus r and the number of edges of size r is $2^{r-1} + \frac{1}{2} \cdot \binom{r}{r/2}$.

Suppose that H has a 2-colouring K with colours 1 and 2. We may suppose that the number of a_i's having colour 1 is an odd number $\leqslant r/2$ or an even number $\geqslant r/2$ (if this is not the case then interchange the colours 1 and 2). Since K is a 2-colouring there exists then a j such that a_j and b_j both have the colour 2 in K, hence x has the colour 1 and if a_i has colour 1 then b_i has colour 2. It follows that the number of a_i's having colour 1 is $\neq r/2$. The set $A = \{c_1, \ldots, c_r\}$, where $c_i = a_i$ if a_i has colour 2 and $i \neq j$ and where $c_i = b_i$ otherwise, is an edge of H and all vertices of this edge have the colour 2. This is a contradiction, hence $\chi(H) \geqslant 3$. It is not difficult to check that $\chi(H - (A)) \leqslant 2$ for all edges (A) of H, hence H is 3-critical.

3. REDUCTION OF CRITICAL HYPERGRAPHS TO GRAPHS

For $k \geqslant 4$ there is a method of reducing any k-critical hypergraph to a k-critical graph. This method was developed in [18] and also described in [21]. Here we shall mention only a special case.

Theorem 1. *Let $k \geqslant 4$ and let H' be a hypergraph and (A) an edge of H'. Let G denote the graph $K_{k-4} * G'$, where G' is an odd circuit (if $k = 4$ then K_{k-4} is empty and $G = G'$). Assume that H' and G are disjoint and $|V(G)| \geqslant |A|$. The hypergraph H is obtained from $H' - (A)$ and G by joining each vertex of G by an edge of size 2 to precisely one vertex of A such that each vertex of A is incident with at least one of these edges.*

Then: H is k-critical iff H' is k-critical.

The proof of Theorem 1 is not difficult and we omit it. In the above construction the edge (A) of H' is removed as such and replaced by some new vertices (those of G) and some new edges of size 2. The number of new vertices that it is necessary to introduce is at most $\max\{k-1, |A|+1\}$. If the construction is carried out successively for all edges (A) of size $\geqslant 3$ of a k-critical hypergraph H' then we obtain in the end a k-critical graph.

P. Erdős asked [4; 10] whether there for each $k \geqslant 4$ exists a positive constant c_k such that there are infinitely many examples of k-critical graphs G having $> c_k \cdot |V(G)|^2$ edges. For $k \geq 6$ such graphs were obtained by G. A. Dirac [4] as follows: Let $G = G_1 * G_2$, where G_1 and G_2 are critical and $\chi(G_i) \geq 3$, $i = 1, 2$, and $|V(G_1)| = |V(G_2)|$. Then G is $(\chi(G_1) + \chi(G_2))$-critical and $|E(G)| > (1/4) \cdot |V(G)|^2$. For $k = 4$ and $k = 5$ this method does not work because in this case either G_1 or G_2 would have at most 2 vertices. We may, however, proceed as follows: Let H_1 and H_2 be two disjoint 2-critical hypergraphs of the same odd size $r \geq 3$. $H = H_1 * H_2$ is 4-critical and has 2 edges of size r and $(1/4) \cdot |V(H)|^2$ edges of size 2. By reducing H to a 4-critical graph G by the method of Theorem 1 we need only introduce r new vertices for each of the two edges of H of size r, hence we need only introduce $2r$ new vertices in all, hence G has $> (1/16) \cdot |V(G)|^2$ edges. This proves the existence of c_4 [19]. $K_1 * G$ shows the existence of c_5.

The above indicates that it is not only because of the generalizing aspect that it is of interest to consider critical hypergraphs. Even if one is primarily interested in graphs it might in some cases be useful and maybe even indispensable to consider hypergraphs as well.

4. INCREASING THE SIZE OF EDGES OF CRITICAL HYPERGRAPHS

Theorem 2. *Let $k \geq 3$ and let H_1 and H_2 be two disjoint hypergraphs. Let $E(H_1)$ be the disjoint union of E_1 and E_2, where $E_1 \neq \phi$, and let $x \in V(H_2)$. The hypergraph H is obtained from H_1 and H_2 as follows:*

$$V(H) = V(H_1) \cup V(H_2 - x),$$

$$E(H) = E_2 \cup E(H_2 - x) \cup \{(A) \mid A = A_i \cup B_j,$$

$$\text{where} \quad (A_i) \in E_1 \quad \text{and} \quad (B_j \cup \{x\}) \in E_x(H_2)\}.$$

Then: If any two of the three hypergraphs H_1, H_2 and H are k-critical then they are all three k-critical.

Proof. We shall consider only the most important case and prove that if H_1 and H_2 are both k-critical then H is k-critical. The proofs in the two other cases are similar.

Thus we assume that H_1 and H_2 are k-critical. Suppose that H has a $(k-1)$-colouring K with colours $1, \ldots, k-1$. Since H_1 is not $(k-1)$-colourable there exists $(A_i) \in E_1$ such that the vertices of A_i all have the same colour, say 1, in K. There exists $(B_j \cup \{x\}) \in E_x(H_2)$ such that all vertices of B_j have colour 1 in K, since otherwise x could be given the colour 1 and H_2 would be $(k-1)$-colourable. Hence all vertices of $A_i \cup B_j$ have got the same colour in K. This is a contradiction, hence $\chi(H) \geqslant k$.

For all edges (A) of H $H-(A)$ is $(k-1)$-colourable. The proof of this falls into three cases: $(A) \in E_2$, $(A) \in E(H_2 - x)$ and $(A) = (A_i \cup B_j)$. In the first case a $(k-1)$-colouring of $H_1 - (A)$ and of $H_2 - x$ provide a $(k-1)$-colouring of $H-(A)$. In the second case a $(k-1)$-colouring of $H_1 - (A_i)$, where $(A_i) \in E_1$, giving all vertices of A_i the same colour 1 and a $(k-1)$-colouring of $H_2 - (A)$ giving x the colour 1 provide a $(k-1)$-colouring of $H-(A)$. In the third case a $(k-1)$-colouring of $H_1 - (A_i)$ giving all vertices of A_i the colour 1 and a $(k-1)$-colouring of $H_2 - (B_j \cup \{x\})$ giving all vertices of $B_j \cup \{x\}$ the colour 1 provide a $(k-1)$-colouring of $H-(A)$. This proves the case of Theorem 2 that we consider.

Theorem 2 may be used to increase the size of edges of a k-critical $(k \geqslant 3)$ hypergraph H_1. Let thus E_1 denote a set of edges of H_1 all of size s_1 and suppose that we want to replace the edges of E_1 by some new vertices and new edges all of size s_2, where $s_2 > s_1$. This may be done using the construction of Theorem 2 with H_2 as the k-critical hypergraph of §2.I with $d = 1$, $r = s_2$, $s = s_2 - s_1 + 1$ and x being the vertex of H_2 contained in all edges of size s. The obtained k-critical hypergraph H has $n_1 + (s_2 - 1)(k-1) - (s_1 - 1)$ vertices, where n_1 is the number of vertices of H_1.

We shall consider two further special cases of the construction of Theorem 2 for 3-critical hypergraphs.

If H_1 is 3-critical $(r-2)$-uniform, r even $\geqslant 4$, and $E_1 = E(H_1)$, and H_2 is the 3-critical example of §2.II with x as the vertex of H_2

contained in all edges of size 3, then the 3-critical hypergraph H obtained from H_1 and H_2 by Theorem 2 is r-uniform. We may formulate this as

Theorem 3. *If r is even $\geqslant 4$ and there exists a 3-critical $(r-2)$-uniform hypergraph with n vertices and e edges, then there also exists a 3-critical r-uniform hypergraph with $n + 2r$ vertices and $e \cdot r + 2^{r-1} + \frac{1}{2} \cdot \binom{r}{r/2}$ edges.*

Theorem 3 is a result in the spirit of [1] and improves [1, Theorem 3 (3)]. P. Erdős asked for the least integer $m(r)$ for which there is a 3-critical r-uniform hypergraph with $m(r)$ edges [6; 7; 9; 12; 15]. $m(4)$ is not known, but Theorem 3 with $r = 4$ and $n = e = 3$ shows that $m(4) \leqslant 23$.

For the second special case let H_1 be an odd circuit of length $2m + 1$ and $E_1 = E(H_1)$. Let H_2 be the 3-critical hypergraph $K_1 * H'$, where $V(K_1) = \{x\}$ and H' is a 2-critical hypergraph consisting of an edge of size 3. The 3-critical hypergraph H^* obtained from H_1 and H_2 by Theorem 2 is 3-uniform, moreover the three vertices of $V(H_2 - x)$ are each contained in $2m + 2$ edges of H^* and have each degree $m + 1$ in H^*. This shows that in a 3-critical 3-uniform hypergraph there is no upper limit on the possible degrees.

5. CONSTRUCTIONS OF CRITICAL UNIFORM HYPERGRAPHS

Let H be any k-critical hypergraph $(k \geqslant 4)$ and let $r \geqslant 2$. We want to transform H into an r-uniform k-critical hypergraph. For those edges whose size have to be increased to size r we use the procedure of §4. For those edges whose size have to be decreased we may first decrease to edges of size 2 by the procedure of §3 and then (if $r \geqslant 3$) increase from size 2 to size r by the procedure of §4.

If $k = 3$ then we may increase the size of edges as above, however Theorem 1 does not work and in general there does not exist a construction similar to the one of Theorem 1 decreasing the size of edges of 3-

critical hypergraphs to size 2. However, there is such a construction decreasing the size of edges to size 3. This goes as follows:

Theorem 4. *Let H' be a hypergraph and (A) an edge of H'. Let H^* be the 3-critical 3-uniform hypergraph of the second special case of §4 and let y be one of the three vertices of H^* of degree $m + 1$ contained in $2m + 2$ edges. Assume that H' and H^* are disjoint and $2m + 2 \geqslant |A|$. The hypergraph H is obtained from $H' - (A)$ and $H^* - y$ as follows: For each of the $2m + 2$ edges $(A_i) \in E_y(H^*)$ the remaining two vertices of A_i in $H^* - y$ are combined with precisely one vertex of A to form a new edge of size 3, moreover each vertex of A shall be contained in at least one of these new edges.*

Then: H is 3-critical iff H' is 3-critical.

The proof of Theorem 4 is not difficult and we omit it. The number of vertices of H is $n' + 2m + 3$, where n' is the number of vertices of H' and $2m + 2 \geqslant |A|$, hence only $|A| + 1$ new vertices need to be added in order to obtain H from H' if $|A|$ is even $\geqslant 4$ (and only $|A| + 2$ if $|A|$ is odd).

We recapitulate: If $k \geqslant 3$ and $r \geqslant 2$ and $(k, r) \neq (3, 2)$ then any k-critical hypergraph may be transformed into an r-uniform one by the above procedure.

6. SOME RESULTS

Theorem 5. *There exists a k-critical graph $(k \geqslant 4)$ with n vertices iff $n \geqslant k$ and $n \neq k + 1$. There exists a k-critical r-uniform hypergraph $(k \geqslant 3$ and $r \geqslant 3)$ with n vertices iff $n \geqslant (r - 1)(k - 1) + 1$.*

Proof. The first part is due to G. Dirac [3, Theorem 6; 5, Theorem 2]. The second part with $k = 3$ is due to H. L. Abbott and D. Hanson [1, Corollary 1 and 2]. We shall therefore only consider the second part with $k \geqslant 4$.

If H is k-critical r-uniform then obviously $n \geqslant (r - 1)(k - 1) + 1$, since otherwise H would be $(k - 1)$-colourable.

Suppose conversely that $n \geqslant (r-1)(k-1)+1$. If $n \geqslant (r-1)(k-1)+k+1$ then let H_1 be a k-critical graph with $n_1 = n-(r-1)(k-1)+1$ vertices. Increase the size of the edges of H_1 from size 2 to size r by the procedure described in §4. The result is a 3-critical r-uniform hypergraph with $n_1 + (r-1)(k-1)-1 = n$ vertices.

If $n = (r-1)(k-1)+\delta$, where $1 \leqslant \delta \leqslant k$, then let M denote a set of $(r-1)(k-1)$ vertices. Let $\mathscr{P}$ denote all possible partitions of M into $k-1$ disjoint sets (or parts) of size $r-1$. Let $x \in M$ and let $B_1, \ldots, B_m$ be all sets of size $r-1$ contained in M and containing x. Obviously $m = \binom{(r-1)(k-1)-1}{r-2} \geqslant k \geqslant \delta$. Let $\mathscr{P}_i^*$ denote the subset of $\mathscr{P}$ consisting of those partitions in which x is contained in B_i, $i = 1, \ldots, m$, and divide $\mathscr{P}$ into δ non-empty classes $\mathscr{P}_1, \ldots, \mathscr{P}_\delta$ such that for each $\mathscr{P}_i^*$ there is a $\mathscr{P}_j$ containing all partitions of $\mathscr{P}_i^*$. This is possible because $m \geqslant \delta$.

An r-uniform hypergraph H is defined as follows: $V(H)$ consists of M together with δ new vertices $x_1, \ldots, x_\delta$. $E(H)$ consists of all subsets of size r of M together with all sets of size r of type $\{x_j\} \cup A$, where A is a subset of size $r-1$ of M, A being a part of a partition contained in $\mathscr{P}_j$.

It is not difficult to prove that H is k-critical. We omit the proof. Moreover H is r-uniform and has $(r-1)(k-1)+\delta = n$ vertices. This gives Theorem 5.

Let $F_k^r(n)$ denote the maximal number of edges possible in a k-critical r-uniform hypergraph having n vertices. Theorem 5 shows for which k, r and n the number $F_k^r(n)$ is defined.

Theorem 6.

(a) *For all* $r \geqslant 2$ *there exists a positive constant* c_3^r *such that for infinitely many values of* n

$$c_3^r \cdot n^{r-1} < F_3^r(n) < n^{r-(1/2^{r-1})}.$$

(b) *For all* $k \geqslant 4$ *and all* $r \geqslant 2$ *there exists a positive constant* c_k^r

such that for infinitely many values of n

$$c_k^r \cdot n^r < F_k^r(n) < (1/r!) \cdot n^r .$$

Proof. The lower bounds of Theorem 6 may be obtained by induction over r. For $r = 2$ (a) is true, since $F_3^2(n) = n$ for n odd $\geqslant 3$, and (b) follows by §3.

Assume now that (a) is true for r and let H_1 be a 3-critical r-uniform hypergraph with n_1 vertices and $> c_3^r \cdot n_1^{r-1}$ edges and let $E_1 = E(H_1)$. Let $H_2 = K_1 * H'$, where $V(K_1) = \{x\}$ and H' is a 2-critical hypergraph consisting of an edge of size n_2. We may assume that n_2 is (constant) $\cdot\, n_1$, e.g. $n_2 = n_1$, and that H_1 and H_2 are disjoint. The hypergraph H obtained from H_1 and H_2 by the construction of Theorem 2 is 3-critical and all edges have size $r + 1$ except one edge of size n_2. This edge may be transformed into edges of size $r + 1$ by the procedure of §5 (first decreasing from size n_2 to size 3 and then increasing from size 3 to size $r + 1$) by adding at most $n_2 + 2r$ new vertices. If n is the number of vertices of the obtained 3-critical $(r + 1)$-uniform hypergraph then the number of edges is $> c_3^r \cdot n_1^{r-1} \cdot n_2 > c_3^{r+1} \cdot n^r$. This proves the lower bound of (a). The lower bound of (b) may be proved similarly, however, in this case H' should be a $(k - 1)$-critical $(r + 1)$-uniform hypergraph with n_2 vertices (H is then k-critical $(r + 1)$-uniform).

The upper bound of (b) is trivial, since $F_k^r(n) \leqslant \binom{n}{r}$. I asked whether it is possible to obtain a 3-critical r-uniform hypergraph with n vertices and as much as (constant) $\cdot\, n^r$ edges, and P. Erdős give a negative answer by obtaining the upper bound of (a). The argument of P. Erdős is this: Let a hypergraph H consist of $2r$ vertices $a_1, b_1, a_2, b_2, \ldots$ $\ldots, a_r, b_r$ and the 2^r edges (A) of size r given by $A = \{c_1, \ldots, c_r\}$, where either $c_i = a_i$ or $c_i = b_i$. H is not a sub-hypergraph of any 3-critical r-uniform hypergraph, because if so then $H - (\{a_1, \ldots, a_r\})$ would have a 2-colouring giving $a_1, \ldots, a_r$ all the same colour, which is not the case. However, P. Erdős proved [8, Theorem 1] that any r-uniform hypergraph with n vertices and $\geqslant n^{r-(1/2^{r-1})}$ edges contains

H as a sub-hypergraph. This proves the upper bound of (a), hence Theorem 6 has been proved.

By carrying out a computation based on the above construction (choose $n_2 = n_1/r$) it can be proved that for $k = 4, 5$ the constant $1/4r^r$ may be used as c_k^r, and for $k \geqslant 6$ $1/r^r$ may be used.

Let $F_k(n)$ denote the maximal number of edges possible in a k-critical hypergraph (not necessarily uniform) having n vertices.

Theorem 7. *Let* $k \geqslant 3$. *For infinitely many values of* n

$$(1/2)^{k-3} \binom{n}{[n/2]} \leqslant F_k(n) \leqslant \binom{n}{[n/2]} .$$

Proof. No edge of a k-critical hypergraph is a subset of another edge, hence the upper bound is an immediate consequence of Sperner's Lemma [17; 2, p. 288]. A hypergraph H_3 with $n = 2r - 1$ vertices and all subsets of size r as edges is 3-critical (§2.I) and gives the lower bound for $k = 3$. $H_k = H_3 * K_{k-3}$ gives the lower bound for $k \geqslant 4$. This proves Theorem 7. If $(1/2)^{k-3}$ is replaced by $(1/2)^{k-2}$ then "for infinitely many values of n" may be replaced by "for all values of $n \geqslant k$".

Theorem 8. *For all* $k \geqslant 4$ *and all* $r \geqslant 2$ *and all natural numbers* α *there exists a* k*-critical* r*-uniform hypergraph* H *for which* $d_H(x) \geqslant \alpha$ *for all vertices* x *of* H.

Proof. We shall prove slightly more, namely that H may be chosen such that it moreover has $\geqslant \alpha$ independent (i.e. disjoint) edges. The proof is by induction over r. For $r = 2$ the result was obtained by M. Simonovits and myself [16; 20]. Let H_1 denote a k-critical r-uniform hypergraph containing α independent edges in which all degrees are $\geqslant \alpha$, and let $E_1 = E(H_1)$. Let $H_2 = K_1 * H'$, where $V(K_1) = \{x\}$ and H' is a $(k-1)$-critical $(r+1)$-uniform hypergraph with $\geqslant \alpha$ vertices. We may assume that H_1 and H_2 are disjoint. The hypergraph H obtained from H_1 and H_2 by the construction of Theorem 2 is k-critical $(r+1)$-uniform. Moreover H has $\geqslant \alpha$ independent edges and all vertices of H have degree $\geqslant \alpha$. This proves Theorem 8.

7. THREE UNSOLVED PROBLEMS

There are of course a lot of unsolved problems concerning critical hypergraphs. Here we shall mention only three with a connection to the results of §6.

Problem 1. What is the possible number of vertices in a k-critical r-uniform hypergraph with the property that any two edges have at most one vertex in common (I heard this question first from H. L. Abbott and D. Hanson).

Problem 2. What is the order of magnitude of $F_3^r(n)$?

Problem 3. Does Theorem 8 with $k = 3$ and $r \geq 3$ hold?

REFERENCES

[1] H.L. Abbott – D. Hanson, On a combinatorial problem of Erdős, *Canad. Math. Bull.*, 12 (1969), 823-829.

[2] C. Berge, *Graphes et hypergraphes,* Dunod, Paris 1970.

[3] G.A. Dirac, Some theorems on abstract graphs, *Proc. London Math. Soc.*, 2 (1952), 69-81.

[4] G.A. Dirac, A property of 4-chromatic graphs and some remarks on critical graphs, *J. London Math. Soc.*, 27 (1952), 85-92.

[5] G.A. Dirac, Circuits in critical graphs, *Monatshefte für Math.*, 59 (1955), 178-187.

[6] P. Erdős, On a combinatorial problem, *Nord. Mat. Tid.*, 11 (1963), 5-10.

[7] P. Erdős, On a combinatorial problem II, *Acta Math. Acad. Sci. Hungar.*, 15 (1964), 445-447.

[8] P. Erdős, On extremal problems of graphs and generalized graphs, *Israel J. Math.*, 2 (1964), 183-190.

[9] P. Erdős, On a combinatorial problem III, *Canad. Math. Bull.*, 12 (1969), 413-416.

[10] P. Erdős, *Problems and results in chromatic graph theory,* Proof techniques in graph theory (Ed. Frank Harary), Academic Press 1969, 27-35.

[11] P. Erdős – A. Hajnal, On chromatic number of graphs and set-systems, *Acta. Math. Acad. Sci. Hungar.*, 17 (1966), 61-99.

[12] M. Herzog – J. Schönheim, The B_r property and chromatic numbers of generalized graphs, *J. Combinatorial Theory,* Ser. B 12 (1972), 41-49.

[13] D. König, *Theorie der endlichen und unendlichen Graphen,* Leipzig 1936, reprinted Chelsea Publ. Comp. New York 1950.

[14] L. Lovász, On chromatic number of finite set-systems, *Acta Math. Acad. Sci. Hungar.*, 19 (1968), 59-67.

[15] W.M. Schmidt, Ein kombinatorisches Problem von P. Erdős und A. Hajnal, *Acta Math. Acad. Sci. Hungar.*, 15 (1964), 373-374.

[16] M. Simonovits, On colour critical graphs, *Studia Sci. Math. Hungar.*, 7 (1972).

[17] E. Sperner, Ein Satz über Untermengen einer endlichen Menge, *Math. Zeitschrift,* 27 (1928), 544-548.

[18] B. Toft, *Some contributions to the theory of colour-critical graphs,* Ph. D.-thesis Univ. of London 1970, published as No. 14 in Various Publication Series, Mathematisk Institut, Aarhus Universitet.

[19] B. Toft, On the maximal number of edges of critical k-chromatic graphs, *Studia Sci. Math. Hungar.*, 5 (1970), 461-470.

[20] B. Toft, Two theorems on critical 4-chromatic graphs, *Studia Sci. Math. Hungar.*, 7 (1972), 83-89.

[21] B. Toft, Colour-critical graphs and hypergraphs, submitted to *J. Combinatorial Theory,* Ser. B.

COLLOQUIA MATHEMATICA SOCIETATIS JÁNOS BOLYAI
10. INFINITE AND FINITE SETS, KESZTHELY (HUNGARY), 1973.

DUALITY AND TRINITY

W.T. TUTTE

1. DERIVED MAPS

Consider a graph G drawn on the sphere. We require G to be finite and connected, and to have at least one edge, but it is otherwise unrestricted. It may have loops, isthmuses and multiple joins. It separates its complementary point-set in the sphere into disjoint simply connected domains called *faces*. The combinatorial structure defined by the vertices, edges and faces in terms of the incidence relations between them is a *map* on the sphere. We denote it by M. We construct the *first derived map* of M as follows.

First we subdivide each edge A of M into two new edges by taking some internal point $s(A)$ of A as a new vertex. We thus convert G into a subdivided graph G' with twice as many edges. We observe that G' has no loops and that its only double joins are those resulting from the loops of G.

Next we adopt one internal point $t(F)$ of each face F as a new vertex, and we subdivide F into triangles by means of new edges joining

$t(F)$ across F to the vertices of G' incident with F. The number of new edges drawn to such a vertex v has to be equal to the multiplicity of the incidence of v with F. More precisely exactly one such new edge is drawn to v in each of the angles at v marked off by G' and occupied by F. (See Figure 1.)

In this way we construct from M a loopless map M' whose faces are triangles. Its vertices are

(i) the vertices of G,

(ii) the vertices $s(A)$ corresponding to the edges A of G and

(iii) the vertices $t(F)$ corresponding to the faces F of M. The edges of M' are the edges of G' together with the new edges drawn from the points $t(F)$. It is a basic property of M' that each of its faces is incident with exactly one vertex of each kind. We can describe it as a 3-coloured triangulation of the sphere, the three colours 1, 2 and 3 corresponding to the three kinds of vertex, in the above order. We observe too that M' is an Eulerian triangulation, each of its vertices being incident with an even number of edges. But it is a rather special kind of Eulerian triangulation since each vertex of colour 2 has to be incident with exactly 4 edges.

We call M' the *first derived* map of M. The construction can be generalized as a method for subdividing cellular n-complexes into simplicial ones. In a topological application we might take the first derived map of M', called the second derived map of M, and so on to subdivide the sphere into arbitrarily small triangles.

An edge in a 3-coloured Eulerian triangulation may conveniently be assigned one of the three colours, the one distinct from the colours of the two ends. Thus an edge incident with a face has the colour of its opposite vertex. The faces can be coloured in two colours, black and white, so that no two of the same colour have a common incident edge. We may suppose the vertex-colours 1, 2 and 3 to appear in clockwise order round the boundary of a black triangle, and in anticlockwise order round the boundary of a white one.

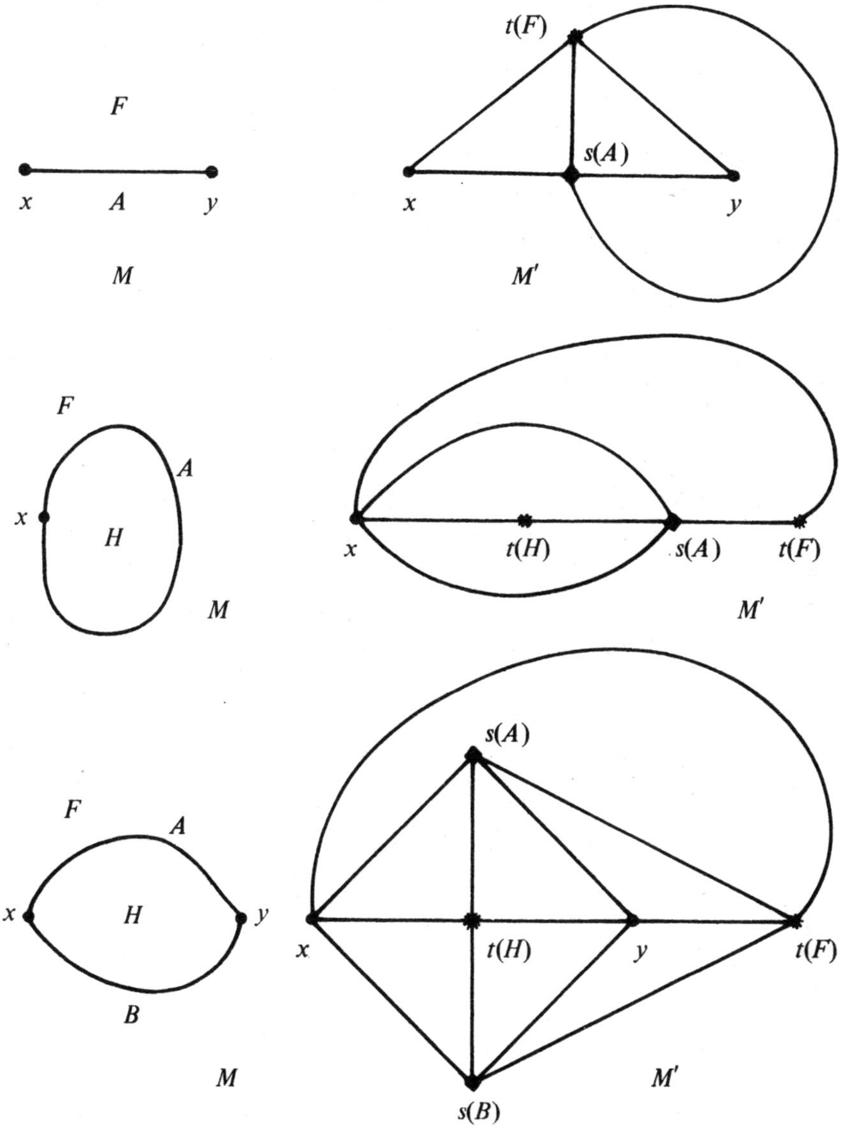

Figure 1

Let us return to the map M' and see how G and M can be reconstructed from it. The vertices of G are distinguished as those of colour 1. The edges are defined by the vertices of colour 2. Each such vertex is incident with exactly two edges of colour 3, and its union with these two edges is the corresponding edge of G and M. Each vertex of colour 3 defines a face of M made up of the vertex itself and the incident open edges and faces of M'. Thus M is completely determined. The same procedure can be applied with the colours 1 and 3 interchanged, and it then yields the *dual map* M^* of M. The graph G^* of M^* is the *dual graph* of G, with respect to M.

The relation between a map, its first derived map and its dual map was found convenient by C.A.B. Smith and W.T. Tutte in a paper on self-dual maps ([5]). It has also been used to apply a known enumerative theory of "rooted" Eulerian triangulations to the problem of enumerating general rooted planar maps of n edges. ([8]).

2. ALTERNATING MAPS

Can we start with an arbitrary 3-coloured Eulerian triangulation T of the sphere and perform an operation analogous to the extraction of the maps M and M^*, even though there be no colour whose vertices all have valency 4? According to the present section the answer is "Yes", but in general we get a triad of directed maps rather than a pair of undirected ones.

We show how to construct a directed graph G_1 having the vertices of T of colour 1 as its vertices. If x is one of these vertices and F_b is a black triangle incident with x we draw a directed edge (or *dart*) of G_1 from x, along a median of F_b, to the mid-point of the opposite side A. We then continue the dart along a median of the adjacent white face F_w to the opposite vertex y (which is of colour 1). We call x the *tail* of the dart and y its *head.* The two may coincide, in which case the dart is a loop taken with a specified sense of description. We do not describe in detail the topological trick for assigning a metric and so justifying our talk of mid-points and medians.

We thus have G_1, a directed graph with exactly one dart bisecting each black triangle, and each white one. At each vertex v of G_1 outgoing and incoming darts alternate in the cyclic order around v. We express this property by saying that G_1 is *alternating*. A loop of G_1 appears twice in this cyclic order if it is incident with v, once as an outgoing dart and once as incoming. We observe that G_1 defines a map M_1 with faces corresponding to the vertices of T of colours 2 and 3. For example let w be a vertex of colour 2. Then the corresponding face of M_1, also said to be of colour 2, is made up of w itself, the incident open edges of colour 3, and the incident open half-edges of colour 1, together with the incident open half-triangles. The expressions "half-edge" and "half-triangle" relate to bisection by a dart of G_1. So G_1 defines a map M_1 with directed edges. It is an Eulerian map and its faces are 2-coloured in the colours 2 and 3. We call it an *alternating map* for the same reason that G_1 is an *alternating graph.*

Evidently the above construction can be repeated with an arbitrary permutation of the three colours. We thus obtain three alternating maps M_1, M_2 and M_3 with vertices of colours 1, 2 and 3 respectively in T. We call them three *trine* alternating maps, just as M and M^* were two "dual" maps. Likewise we speak of the property of *trinity* for alternating maps in analogy with the property of "duality" for undirected maps. There is no reason why the construction should not be carried out on other orientable surfaces, but we do not discuss that generalization here.

Our construction is illustrated in Figure 2. For simplicity this shows only part of the triangulation concerned. Black triangles are shaded.

It is claimed here that trinity is a true generalization of duality. To justify this claim consider a first derived map M', that is a 3-coloured Eulerian triangulation in which every vertex of colour 2 has valency 4. When we construct M_1 from this T we find that each vertex v of colour 2 gives rise to a 2-sided face. This face is bounded by two oppositely directed darts of G_1. (Figure 3). Now in many applications two oppositely directed darts are considered to be equivalent to a single undirected edge having the same ends. So G_1 can be counted as the same graph as G, and G_3 can

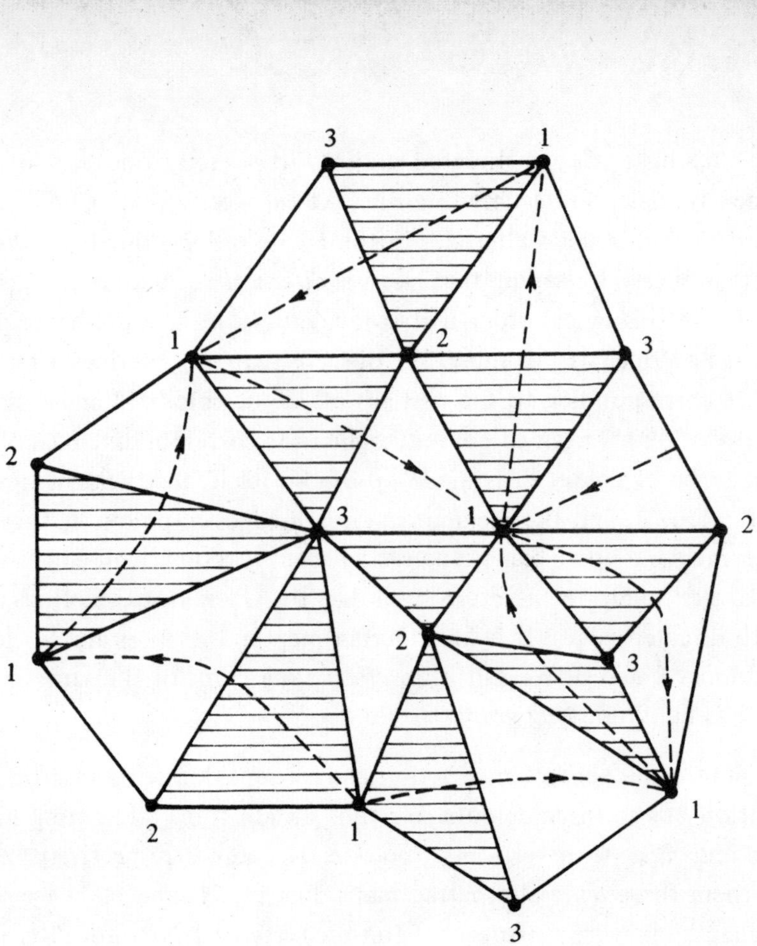

Figure 2

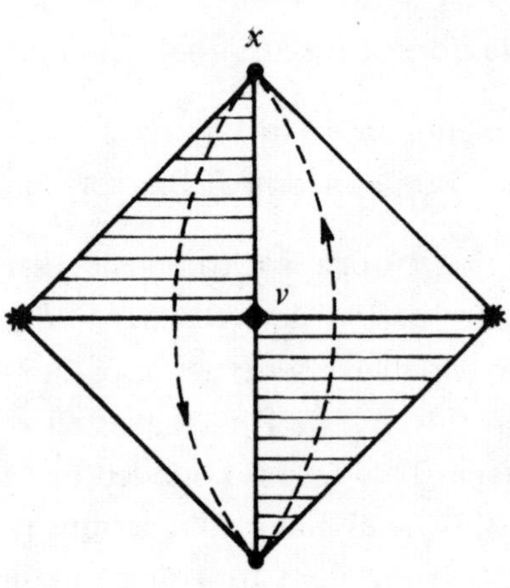

Figure 3

likewise be identified with G^*. The map M_1 can be identified with M, or at least it can be derived from M by a simple process of thickening edges into digons. The same remark applies to M_3 and M^*.

The third trine map M_2 has vertices corresponding to the edges of M (and M^*). In the terminology of [4] it is an oriented form of the *medial* map of M.

3. SPANNING ARBORESCENCES

A *spanning* subgraph of a graph G is a subgraph that includes every vertex. A spanning tree of G is a spanning subgraph that is a tree. The tree-number $T(G)$ of G is the number of spanning trees of G. There is a well-known theorem expressing $T(G)$ as a determinant. It was published in 1940 by R.L. Brooks, C.A.B. Smith, A.H. Stone and W.T. Tutte, who noted that basically it is part of the electrical theory of G. Kirchhoff. In some recent literature it is called the "Kirchhoff – Trent Theorem" in honour of an independent discovery published in 1954. ([3], [6]).

Another theorem on spanning trees asserts that $T(G) = T(G^*)$. It is proved quite easily by setting up a $1-1$ correspondence between the spanning trees of G and those of G^*. If S is a tree of G the edges of the corresponding tree of G^* are those not meeting edges of S (in vertices of M' of colour 2). This is the Tree-Duality Theorem, a corollary of Whitney's Duality Theorem.

In this section we generalize the Tree-Duality Theorem to a theorem about trine graphs, thereby further justifying our claim that trinity is a true generalization of duality. To do this we need the idea of a directed spanning tree.

Let S be a tree whose edges are directed, and let r be one of its vertices. We call S an *arborescence* with *root* r if exactly one dart is outgoing from each of its other vertices. It is then found that at r all the incident darts are incoming. Moreover if we construct a path in the arborescence, following the darts, it always leads to r. Some people emphasize this property by referring to the root as "Rome". A *spanning arborescence*, or spanning directed tree, in a directed graph G, is a spanning tree of G which is an arborescence with respect to the directions of edges induced by the darts of G. Let us write $T_v(G)$ for the number of spanning arborescences of G with root v.

There is a theorem expressing $T_v(G)$ as a determinant. It arose out of the work of Brooks, Smith, Stone and Tutte, and it was incorporated by the present writer in a paper on dissections of triangles. ([7], 1948). It was used by T. van Aardenne – Ehrenfest and N.G. de Bruijn in their classic paper on "Circuits and trees in oriented linear graphs" of 1951. In modern text-books it is called the "Bott – Mayberry Theorem", again in honour of a discovery published in 1954. ([2]).

Let us describe a directed graph as *balanced* if at each vertex the number of incoming darts is equal to the number of outgoing ones. It is a consequence of the above-mentioned theorem, pointed out in [7], that in a balanced directed graph G the number $T_v(G)$ is independent of v. We therefore refer to it simply as the *tree-number* of G. It is found to be unchanged by the operation of reversing every dart of G.

We now go back to the triangulation T and the three corresponding directed graphs G_1, G_2 and G_3. These are alternating and therefore balanced. We endeavour to prove the Tree Trinity Theorem $T(G_1) = T(G_2) = T(G_3)$. Actually this theorem was proved in [7]. Here it is expounded in a somewhat different way.

To prove the theorem we first choose arbitrarily a black triangle F_0 of T. We say that F_0 is *inactive* and that the other black triangles are *active*. We say also that the vertices of T incident with F_0 are *inactive* and that the other vertices of T are *active*.

Let f be a mapping of the set of active black triangles into the colour-set $\{1, 2, 3\}$. Let us suppose that each active vertex v of T is incident with exactly one active black triangle F such that $f(F)$ is the colour of v. Then we call f a *tree-colouring* of T *based on* F_0. We say that $f(F)$ is the *referred colour* of the black triangle F in the tree-colouring.

The tree-colouring f determines a spanning subgraph S_1 of G_1. Its darts are those darts of G_1 that bisect active black triangles with referred colour 1. Similarly we define S_2 and S_3. We say that S_1, S_2 and S_3 are the *constituents* of f.

3.1. *Let f be a tree-colouring of T based on F_0. Then its constituents S_1, S_2 and S_3 are spanning arborescences of G_1, G_2 and G_3 respectively. In each case the root is the inactive vertex of the appropriate colour.*

Proof. Assume that for some colour $j(1)$ the constituent $S_{j(1)}$ contains a directed circuit J_1, a cyclic sequence of darts head to tail. Let the residual domain of J_1 containing the inactive black triangle be D_1, and let the other residual domain be E_1.

Let X be any dart of J_1. It proceeds from its tail x through an active black triangle $F_b(X)$ and then through a white triangle $F_w(X)$ to its head y. The common incident edge of $F_b(X)$ and $F_w(X)$ crossed by the dart, the *cross-bar* of X, has one end $b(X)$ in D_1 and the other end $a(X)$ in E_1. It follows from the fact that black and white triangles alternate around each vertex of T that $a(X)$ has the same colour, $j(2)$ say, for every dart X of J. (See Figure 4).

Now consider a dart Y of $S_{j(2)}$ with tail inside E_1. It proceeds through a black triangle $F_b(Y)$ of referred colour $j(2)$. This is not one of the black triangles bisected by darts of J_1; those triangles have referred colour $j(1) \neq j(2)$. Then Y enters a white triangle $F_w(Y)$ by crossing its side of colour $j(2)$ at its mid-point. But $F_w(Y)$ cannot be one of the white triangles bisected by darts of J_1; those triangles have their sides of colour $j(2)$ in D_1, except for one end. We deduce that Y, its head and its tail are all contained in E_1.

Since each vertex in E_1 must be active it follows that $S_{j(2)}$ has as many darts as vertices lying entirely within E_1. Hence $S_{j(2)}$ has a directed circuit J_2 lying entirely within E_1. Its residual domain E_2 not

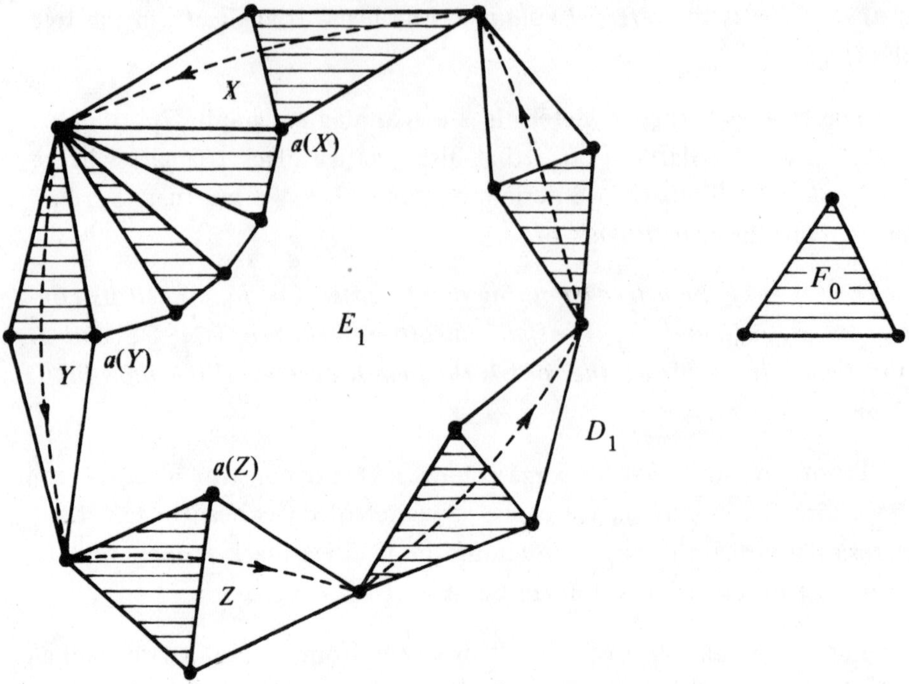

Figure 4

containing F_0 is a proper subset of E_1. By the same reasoning we deduce that E_2 contains a directed circuit J_3 of some $S_{j(3)}$ such that $j(3) \neq j(2)$, and so on. But this is impossible, by the finiteness of T.

We deduce that S_i has no directed circuit, $(i = 1, 2, 3)$. Hence if we start at one of its active vertices and follow the darts we must arrive eventually at the inactive vertex of S_i (incident with F_0). Accordingly S_i is connected. It is a tree because the number of its edges is one less than the number of its vertices, and it must therefore be an arborescence, with its inactive vertex as root.

Let us define a *partial tree-colouring* of T *based on* F_0 as a mapping g of some subset of the set of active black triangles onto $\{1, 2, 3\}$, subject to the condition that for each active vertex v there is at most one active black triangle F such that $g(F)$ is defined and is the colour of v. If g_1 and g_2 are two partial tree-colourings of T based on F_0 such that $g_2(F)$ is defined and equal to $g_1(F)$ whenever $g_1(F)$ is defined, then we say that g_2 *induces* g_1. Evidently a tree-colouring of T based on F_0 is to be counted as one of the partial tree-colourings based on F_0.

3.2. *Let K_i be a spanning arborescence of G_i with an inactive root, $(i = 1, 2,$ or $3)$. Then there is exactly one tree-colouring f of T such that K_i is a constituent of f.*

Proof. Let g_0 be the partial tree-colouring of T which assigns the colour i to each active black triangle bisected by an edge of K_i, and which has no other black triangle withing its domain of definition. It is based on F_0.

Consider an arbitrary partial tree-colouring g of T based on F_0. Suppose that there is an active vertex v of T satisfying the following conditions.

(i) *Just one black triangle, F_1 say, incident with v lies outside the domain of definition of g.*

(ii) *If F_2 is a black triangle distinct from F_1 and incident with v, then $g(F_2)$ is distinct from the colour of v.*

In this case we can define a partial tree-colouring g' of T, based on F_0, whose domain of definition is that of g increased by F_1. When $g(F)$ is defined we write $g'(F) = g(F)$, and we write also that $g(F_2)$ is the colour of v. We refer to this operation as *augmenting* g at v. It is clear that if a tree-colouring t of T, based on F_0, induces g then it induces also g'.

Starting with g_0 we augment it if possible to obtain a partial tree-colouring g_1. Then, if possible, we augment g_1 to obtain g_2, and so on. This process must terminate with a partial tree-colouring $f = g_m$, based of course on F_0. By repeated application of the rule just stated we have the following

Lemma 3.2.1. *If a tree-colouring t of T, based on F_0, induces g_0 then it induces f.*

Let g_k be a member of our sequence of partial tree-colourings. We refer to a black triangle F, active or inactive, as *k-used* or *k-unused* according as it is within or outside the domain of definition of g_k. A vertex v of T of colour j, active or inactive, is *k-used* if it is incident with a k-used black triangle F such that $g_k(F) = j$, and *k-unused* otherwise. Thus F_0 is k-unused for every k. We need another

Lemma 3.2.2. *Let F be an m-unused black triangle. Then its incident vertices of colours other than i are m-unused.*

Proof. Let w be one of these two vertices, with colour j say. Assume it to be m-used. Clearly w is 0-unused. Hence there is a non-negative integer $n < m$ such that w is n-unused but $(n+1)$-used. This means that g_{n+1} was formed from g_n by augmenting it at w. In this process a colour was assigned to an n-unused black triangle F_i incident with w. But this is contradictory; it implies that w is incident with two distinct n-unused black triangles F and F_1, and that therefore g_n is not augmentable at w.

Let us construct the subgraph H of T defined by the sides of colour i of the m-unused black triangles, and the ends of these sides. Then each vertex of H is m-unused, by 3.2.2. We note that H is not null since it includes the edge A_i of F_0 of colour i. Any active vertex x of T appearing in H must be at least divalent in H, for otherwise g_m could be augmented at x.

Suppose that H includes a circuit J. Then K_i must have a vertex in each residual domain of J. Since K_i is connected and J has no vertex of colour i it follows that some dart of K_i crosses an edge of J. This is impossible since the black triangles incident with the edges of J are m-unused.

We deduce that each component of H is a tree, and so has at least two monovalent vertices. But we have seen that every vertex of H not incident with A_i is at least divalent in H. We deduce that H consists solely of A_i and its two incident vertices.

It follows that the domain of definition of f includes all the active black triangles. Thus f is a tree-colouring of T, based on F_0. By 3.2.1 it is the only tree-colouring of T, based on F_0, that induces g_0, i.e., that has K_i as a constituent. This completes the proof of the theorem.

By 3.1 and 3.2 there is a $1-1-1$ correspondence between the spanning arborescences of G_1, G_2 and G_3, the arborescences having an inactive root in each case. Corresponding arborescences are constituents of the same tree-colouring based on F_0. This result establishes the Tree-Trinity Theorem: $T(G_1) = T(G_2) = T(G_3)$.

The present author acknowledges his indebtedness to the other authors of [3] for some of the basic ideas of the foregoing theory. In particular it was A.H. Stone who first isolated three trine alternating graphs from a dissection of equilateral triangles into equilateral triangles, and it was C.A.B. Smith who united them by means of a bicubic map (the dual of an Eulerian triangulation). Smith's reaction to the Tree-Trinity Theorem, as proved in [7], was that the tree-number concerned should be expressed directly as a function of the bicubic map. The author hopes that this requirement is adequately met by the present theory; tree-number is the number of tree-colourings of T, based on an arbitrarily chosen triangle F_0.

REFERENCES

[1] T. van Aardenne-Ehrenfest – N.G. de Bruijn, *Circuits and trees in oriented linear graphs,* Simon Stevin 28 (1951), 203-217.

[2] R. Bott – J.P. Mayberry, *Matrices and trees, Economic activity analysis,* Wiley, New York 1954, 391-400.

[3] R.L. Brooks – C.A.B. Smith – A.H. Stone – W.T. Tutte, The dissection of rectangles into squares, *Duke Math. J.,* 7 (1940), 312-340.

[4] O. Ore, *The Four-Color Problem,* Academic Press, New York, 1967.

[5] C.A.B. Smith – W.T. Tutte, A class of self-dual maps, *Can. J. Math.,* 2 (1950), 179-196.

[6] H. Trent, A note on the enumeration and listing of all possible trees in a connected linear graph. *Proc. Nat. Acad. Sci.,* U.S.A. 40 (1954), 1004-1007.

[7] W.T. Tutte, The dissection of equilateral triangles into equilateral triangles, *Proc. Cambridge Phil. Soc.,* 44 (1948), 463-482.

[8] W.T. Tutte, A census of planar maps, *Can. J. Math.,* 15 (1963), 249-271.

COLLOQUIA MATHEMATICA SOCIETATIS JÁNOS BOLYAI
10. INFINITE AND FINITE SETS, KESZTHELY (HUNGARY), 1973.

DEGREE-CONSTRAINED SUBGRAPHS AND MATROID THEORY

M. LAS VERGNAS

Let G be a graph with vertex-set X: by a well-known theorem of J. Edmonds – D.R. Fulkerson [6] the subsets of X saturated by some matching of G are the independent sets of a matroid. A matching of G may be considered as a subgraph H such that $d_H(x) \leqslant 1$ for all $x \in X$: a natural extension of the preceding result to subgraphs H such that $d_H(x) \leqslant f(x)$ for all $x \in X$, where f is any given integer-valued function defined on X, can be easily derived from a construction due to W.T. Tutte [14] which establishes a $1-1$ correspondence between these subgraphs and certain matchings of an auxiliary graph. Similarly for any constraint (c) on degrees (or generalized degrees) of subgraphs of G such that subgraphs satisfying (c) may be thus reduced to matchings we are able to define a matroid related to (c). Among constraints for which such a construction exists let us mention $d_H(x) \leqslant f$ $(\geqslant f)$, $d_H(x)$ even (odd), $d_H(x) = d_H(y)$, $d_H(x) \leqslant d_H(y)$, etc. . . .

A matroid thus derived from a constraint (c) is defined by means of its independent sets; for constraints such that the corresponding subgraphs have interesting properties we complete this definition by an expression of the rank function. The obtained expression of the rank function always implicitly contains a necessary and sufficient condition for a subgraph of G satisfying (c) to exist, and yields thus a matroidal interpretation of the known conditions of the literature ($d_H(x) = 1$ for all $x \in X$ [12], $d_H(x) = f(x)$ for all $x \in X$ [13], $g(x) \leq d_H(x) \leq f(x)$ for all $x \in X$ [8]).

Definitions and notations concerning graphs are those of [1]. Given a set X an (undirected) *multigraph* $G = (X, U)$ is a family $(e_u)_{u \in U}$ of subsets of X with cardinalities 1 or 2. An element of X is a *vertex* of G, an element of U is an *edge*; an edge $u \in U$ such that $|e_u| = 1$ is called a *loop*. If G has no loops and no multiple edges, that is, if $e_u = e_v$ for $u, v \in U$ implies $u = v$, G is a graph; U is then identified with a set of 2-element subsets of X.

A *subgraph* $H = (Y, V)$ of the multigraph G, $Y \subseteq X$, $V \subseteq U$, is a subfamily $(e_u)_{u \in V}$ such that $e_u \subseteq Y$ for all $u \in V$. If $Y = X$, H is characterized by the subset $V \subseteq U$; we write then $H \subseteq G$. Given $Y \subseteq X$ the *subgraph of* G *induced by* Y is the multigraph $G_Y = (Y, \{u \mid u \in U$ such that $e_u \subseteq Y\})$.

Let $S, T \subseteq X$: and *(S, T)-edge* $u \in U$ is an edge such that $e_u = \{x, y\}$, $x \in S$, $y \in T$. We denote by $m_G(S, T)$ the number of (S, T)-edges of G. Given $x \in X$ we set $d_G(x)$ the *degree of* x *in* $G = m_G(x, X) + m_G(x, x)$ ($m_G(x, X)$ stands for $m_G(\{x\}, X)$), loops are thus counted twice, and $\Gamma_G(x) = \{y \mid y \in X$ such that there exists an edge $\{x, y\}$ in $U\}$.

A *matching* of G is a subset $W \subseteq U$ such that the maximal degree in (X, W) is 1. Vertices of degree 1 in (X, W) are said to be *saturated* by W. Given $Y \subseteq X$ we denote by $s(W, Y)$ the number of vertices in Y saturated by W. Let $H = (Y, V)$ be a subgraph of G: then W_H denote the matching $V \cap W$.

Definitions concerning matroids may be found in [3]. Given a matroid M on a set X we denote by $\mathscr{L}_M$ the set of subsets of X which are independent sets of M, and by r_M the rank function of M. Let $Y \subseteq X$: the *submatroid induced by* Y is denoted by $M(Y)$, the *quotient matroid of* M *by* $X - Y$ by $M/(X - Y)$. We have

$$\mathscr{L}_{M(Y)} = \{S \mid S \subseteq Y \text{ such that } S \in \mathscr{L}_M\}$$

$$\mathscr{L}_{M/(X-Y)} = \{S \cap Y \mid S \in \mathscr{L}_M \text{ such that } S \cap (X - Y) \text{ is a base of } X - Y\}$$

and, for $Z \subseteq Y$,

$$r_{M(Y)}(Z) = r_M(Z)$$

$$r_{M/(X-Y)}(Z) = r_M(Z + (X - Y)) - r_M(X - Y)\ ^*$$

All sets considered in this paper are finite (except $N = \{0, 1, 2, \ldots\}$ and $Z = \{\ldots, -2, -1, 0, 1, 2, \ldots\}$).

§1.

Let $G = (X, U)$ be a graph

(1) (J. Edmonds – D.R. Fulkerson [6]) the subsets $Y \subseteq X$ such that there exists a matching of G saturating all vertices in Y are the independent sets of a matroid $E(G)$ on X,

(2) (C. Berge [2]) the minimum number of vertices of G unsaturated by a matching is $\operatorname*{Max}_{S \subseteq X} (p_i(S) - |X - S|)$ where $p_i(S)$ is the number of connected components of G_S having an odd number of vertices.

This later result gives in fact the rank of X in the matroid $E(G)$: $r_{E(G)}(X) = |X| - \operatorname*{Max}_{S \subseteq X} (p_i(S) - |X - S|)$.

Proposition 1. *For* $Y \subseteq X$

$$r_{E(G)}(Y) = |Y| - \operatorname*{Max}_{S \subseteq X} (p_{Y,i}(S) - |X - S|)$$

where $p_{Y,i}(S)$ *is the number of connected components of* G_S *contained in* Y *which have an odd number of vertices.*

*As usual, we write $A + B$ instead of $A \cup B$ if $A \cap B = \phi$.

Proof. Let H be the union of G and a complete graph on a new set A of vertices, disjoint from X, such that $|A| = |X - Y|$ and all the $(X - Y, A)$-edges. Let H' be a disjoint copy of H and $\bar{G}$ be the union of H and H' plus all the (A, A')-edges. We set $\bar{X} = X + X' + A + A'$.

Lemma 1.1. $|\bar{X}| - r_{E(G)}(\bar{X}) = 2(|Y| - r_{E(G)}(Y))$.

We have $\geqslant$ because the edges of G having at least one vertex in Y (resp. in Y') are the only edges of G (resp. G'). As for let us consider a matching W of G such that $s(W, Y) = r_{E(G)}(Y)$: we can clearly enlarge W into a matching $\bar{W}$ of $\bar{G}$ such that $|\bar{X}| - s(\bar{W}, \bar{X}) = 2(|Y| - s(W, Y))$.

From Lemma 1.1 and Berge's theorem (1) we get: $r_{E(G)}(Y) = 2|X| - \frac{1}{2} \operatorname{Max}_{\bar{S} \subseteq \bar{X}} (p_i(\bar{S}) + |\bar{S}|)$ where $\bar{p}_i(\bar{S})$ is the number of connected components of $\bar{G}_{\bar{S}}$ having an odd number of vertices.

Lemma 1.2. *Given any* $\bar{S} \subseteq \bar{X}$ *there exists* $\bar{T} \subseteq \bar{X}$ *such that* $A + A' \subseteq \bar{T}$ *and* $\bar{p}_i(\bar{S}) + |\bar{S}| \leqslant \bar{p}_i(\bar{T}) + |\bar{T}|$.

We distinguish 3 cases

(1) $\bar{S} \cap A \neq \phi$ and $\bar{S} \cap A' \neq \phi$: take $\bar{T} = \bar{S} \cup A \cup A'$ (if $\bar{S} \cap A \neq A$ or $\bar{S} \cap A' \neq A'$, $\bar{p}_i(\bar{T}) \geqslant \bar{p}_i(\bar{S}) - 1$, $|\bar{T}| \geqslant |\bar{S}| + 1$).

(2) $\bar{S} \cap A = \bar{S} \cap A' = \phi$: take $\bar{T} = \bar{S} \cup A \cup A'$, $(\bar{p}_i(\bar{T}) \geqslant \bar{p}_i(S) - 2|X - Y|$ since there exist at most $|X - Y|$ odd components of $\bar{G}_{\bar{S}}$ meeting $X - Y$ (resp. $X' - Y'$), $|\bar{T}| = |\bar{S}| + 2|X - Y|)$.

(3) $\bar{S} \cap A = \phi$ and $\bar{S} \cap A' \neq \phi$. $\bar{p}_i(\bar{S}) = \bar{p}_i(\bar{S} \cap (X + A)) + \bar{p}_i(\bar{S} \cap (X' + A'))$.

If $\bar{p}_i(\bar{S} \cap (X + A)) + |\bar{S} \cap (X + A)| < \bar{p}_i(\bar{S} \cap (X' + A') + |\bar{S} \cap (X' + A')|$, $\bar{S}'$ equal to 2 copies of $\bar{S} \cap (X' + A')$ is such that $\bar{p}_i(\bar{S}') + |\bar{S}'| \geqslant \bar{p}_i(\bar{S}) + |\bar{S}|$ (since $\bar{p}_i(\bar{S}') \geqslant 2\bar{p}_i(\bar{S} \cap (X' + A')| - 2$ and $\bar{p}_i(\bar{S} \cap (X + A)) + |\bar{S} \cap (X + A)| \leqslant \bar{p}_i(\bar{S} \cap (X' + A') + |\bar{S} \cap (X' + A')| - 2)$; apply Case 1;

If $\bar{p}_i(\bar{S} \cap (X + A)) + |\bar{S} \cap (X + A)| \geqslant \bar{p}_i(\bar{S} \cap (X' + A') + |\bar{S} \cap (X' + A')|$, $\bar{S}'$ equal to 2 copies of $\bar{S} \cap (X + A)$ is such that $\bar{p}_i(\bar{S}') + |\bar{S}'| \geqslant \bar{p}_i(\bar{S}) + |\bar{S}|$ $(\bar{p}_i(\bar{S}') = 2\bar{p}_i(\bar{S} \cap (X + A))$: apply Case 2.

Lemma 1.3. *Given any* $\bar{S} \subseteq \bar{X}$ *such that* $A + A' \subseteq \bar{S}$ *there exists* $\bar{T} \subseteq \bar{X}$ *such that* $A + A' \subseteq \bar{T}$, $\bar{T} \cap X'$ *is the copy of* $\bar{T} \cap X$ *and* $\bar{p}_i(\bar{S}) + |\bar{S}| \leqslant \bar{p}_i(\bar{T}) + |\bar{T}|$.

Let $S = \bar{S} \cap X$, $S' = \bar{S} \cap X'$. We distinguish 2 cases;

(1) the connected component of $\bar{G}_{\bar{S}}$ which contains $A + A'$ has an even number of vertices. $p_i(S) = p_{Y,i}(S) + p_{Y',i}(S')$: take $\bar{T}$ equal to 2 copies of $S + A$ or $S' + A'$ according as $p_{Y,i}(S) + |S| \geqslant p_{Y',i}(S') + |S'|$ or $\leqslant$ respectively

(2) the connected component of $\bar{G}_{\bar{S}}$ which contains $A + A'$ has an odd number of vertices. $p_i(S) = p_{Y,i}(S) + p_{Y',i}(S') + 1$:

if $p_{Y,i}(S) + |S| \geqslant p_{Y',i}(S') + |S'| + 1$, take $\bar{T}$ as 2 copies of $S + A(\bar{p}_i(\bar{T}) = 2p_{Y,i}(S))$

if $p_{Y,i}(S) + |S| < p_{Y',i}(S') + |S'| + 1$, as $\bar{p}_i(\bar{S}) + |\bar{S}|$ is even $p_{Y,i}(S) + |S|$ and $p_{Y',i}(S') + |S'| + 1$ are both odd or both even, then $p_{Y,i}(S) + |S| \leqslant p_{Y',i}(S') + |S'| - 1$: take $\bar{T}$ as 2 copies of $S' + A'$.

By Lemma 1.3 we have $\underset{\bar{S} \subseteq \bar{X}}{\text{Max}} (\bar{p}_i(\bar{S}) + |\bar{S}|) = \underset{S \subseteq X}{\text{Max}} (p_i(S + S' + A + A') + |S + S' + A + A'|)$ where S' is the copy of S, but then $p_i(S + S' + A + A') = 2p_{Y,i}(S)$, from which we get the required expression of $r_{E(G)}(Y)$.

Corollary 1.

$$r_{E(G)}(Y) = |Y| - \underset{S \subseteq Y \cup \Gamma_G(Y)}{\text{Max}} (p_{Y,i}(S) - |(Y \cup \Gamma_G(Y)) - S|) .$$

Proof. Given $S \subseteq Y \cup \Gamma_G(Y)$, set $S' = S \cup (X - (Y \cup \Gamma_G(Y))$: the connected components of G_S and $G_{S'}$ contained in Y are identical, $p_{Y,i}(S') - |X - S'| = p_{Y,i}(S) - |(Y \cup \Gamma_G(Y)) - S|$, from Proposition 1 we get $\geqslant$.

For the reverse inequality given $S \subseteq X$, set $S' = S \cap (Y \cup \Gamma_G(Y))$: the connected components of G_S and $G_{S'}$ contained in Y are again identical, $p_{Y,i}(S) - |X - S| \leqslant p_{Y,i}(S') - |(Y \cup \Gamma_G(Y)) - S')|$.

Corollary 2.

$$r_{E(G)}(Y) = |Y| - \operatorname*{Max}_{S \subseteq Y} (p_i(S) - |\Gamma_G(S) - S|) .$$

Proof. Given $S \subseteq Y$, set $S' = S \cup (X - (S \cup \Gamma_G(S))) = X - (\Gamma_G(S) - S)$: $p_{Y,i}(S') \geqslant p_i(S)$, $|X - S'| = |\Gamma_G(S) - S|$, $p_{Y,i}(S') - |X - S'| \geqslant p_i(S) - |\Gamma_G(S) - S|$, from Proposition 1 we get $\geqslant$.

For the reverse inequality given $S \subseteq X$ let S' be the set of vertices of the connected components of G_S contained in Y: $p_{Y,i}(S) = p_i(S')$, $|X - S| \geqslant |\Gamma_G(S') - S'|$ (since $S \cap (\Gamma_G(S') - S') = \phi$), then $p_{Y,i}(S) - |X - S| \leqslant p_i(S') - |\Gamma_G(S') - S'|$.

A necessary and sufficient condition for a matching saturating all vertices of $Y \subseteq X$ to exist to that $Y \subset \mathscr{L}_{E(G)}$; this is equivalent to $r_{E(G)}(Y) \geqslant |Y|$. Then $p_{Y,i}(S) \leqslant |X - S|$ for all $S \subseteq X$ or $p_i(S) \leqslant |\Gamma_G(S) - S|$ for all $S \subseteq Y$ are, by Proposition 1 and by Corollary 2 respectively, necessary and sufficient conditions of L. Lovász' Theorem 8.2 of [8]. Conversely Proposition 1 can be derived from Theorem 8.2: k new vertices are added to G, joined to all vertices of Y, then the minimal k such that there exists a matching of the new graph saturating all vertices of Y is equal to $|Y| - r_{E(G)}(Y)$.

Let G be the bipartite graph associated with a family $(A_i)_{i \in I}$ of subsets of a set E: for $X \subseteq E$ we have, by Corollary 2, $r_{E(G)}(X) = |X| - \operatorname*{Max}_{Y \subseteq X} (p_i(Y) - |A^{-1}(Y) - Y|) = \operatorname*{Min}_{Y \subseteq X} (|A^{-1}(Y)| + |X - Y|)$ where $A^{-1}(Y) = \{i \mid i \in I \text{ such that } A_i \cap Y \neq \phi\}$ $(p_i(Y) = |Y|)$, which is the classical expression of the rank function of a transversal matroid.

§2.

Let X be a set and f, g be integer-valued functions defined on X. $g(x) \leqslant f(x)$ for all $x \in X$ will be written as $g \leqslant f$. $\operatorname{Min}(f, g)$ denotes the function $x \rightsquigarrow \operatorname{Min}(f(x), g(x))$ defined on X. Given $Y \subseteq X$ we set $f(Y) = \sum_{x \in Y} f(x)$; for $Y = X$ the special notation $\|f\| = f(X)$ will be used. For fixed X and $Y \subseteq X$ we denote by 1_Y the characteristic function of Y with respect to X: $1_Y(x) = 1$ for $x \in Y$, $= 0$ for $x \in X - Y$.

The notion of *integer polymatroid,* or more briefly here ***Z*-*matroid,*** has been introduced by J. Edmonds [5]: given a (finite) set X, a ***Z***-matroid on X is defined by a set $\mathscr{L}_M$ of integer-valued functions defined on X – the *independent functions* of M – satisfying the following axioms:

(1) $0 \in \mathscr{L}_M$

(2) $\xi \in \mathscr{L}_M$ and $\eta \leqslant \xi$ (where $\eta: X \to N$) imply $\eta \in \mathscr{L}_M$

(3) given $\alpha: X \mapsto N$ there exists an integer $r_M(\alpha)$ such that for any $\xi \in \mathscr{L}_M$, $\xi \leqslant \alpha$, maximal for inclusion with these two properties we have $\|\xi\| = r_M(\alpha)$. The function $\alpha \leadsto r_M(\alpha)$ is called the *rank function* of M.

A matroid can clearly be identified with a ***Z***-matroid (take for independent functions the characteristic functions of the independent sets).

Let us say that a ***Z***-matroid M is *bounded* if there exists a constant c such that $\|\xi\| \leqslant c$ for all $\xi \in \mathscr{L}_M$. Let $\bar{M}$ be a matroid on a set $\bar{X}$ and $\bar{X} = \sum_{x \in X} \bar{X}(x)$ be a partition of $\bar{X}$: we say that $\bar{M}$ is *complete* with respect to this partition if $\bar{Y} \in \mathscr{L}_{\bar{M}}$ and $\bar{Y}' \subseteq \bar{X}$ such that $|\bar{Y} \cap \bar{X}(x)| = |\bar{Y}' \cap \bar{X}(x)|$ for all $x \in X$ imply $\bar{Y}' \in \mathscr{L}_{\bar{M}}$. Then we have:

Lemma 1. *Let $\bar{M}$ be a matroid on a (finite) set $\bar{X}$ complete with respect to a partition $\bar{X} = \sum_{x \in X} \bar{X}(x)$: then the functions defined on X by $x \leadsto |\bar{Y} \cap \bar{X}(x)|$ for $\bar{Y} \in \mathscr{L}_{\bar{M}}$ are the independent functions of a (bounded) **Z**-matroid M on X. The rank function of M is given for $\alpha: X \mapsto N$ by $r_M(\alpha) = r_{\bar{M}}(\sum_{x \in X} \bar{A}(x))$ where $\bar{A}(x) \subseteq \bar{X}(x)$ is such that* $|\bar{A}(x)| = \text{Min}\,(\alpha(x), |\bar{X}(x)|)$.

Lemma 1 has the following converse: let M be a bounded $\mathbf{Z}$-matroid on a set X, for every $x \in X$ choose a set $\bar{X}(x)$ such that $|\bar{X}(x)| = \text{Max}\{\xi(x) \mid \xi \in \mathscr{L}_M\}$, the $\bar{X}(x)$ being pairwise disjoint, set $\bar{X} = \sum_{x \in X} \bar{X}(x)$ and $\bar{\mathscr{L}} = \{\bar{Y} \mid \bar{Y} \subseteq \bar{X}$ such that there exists $\xi \in \mathscr{L}_M$ satisfying $|\bar{Y} \cap \bar{X}(x)| = \xi(x)$ for all $x \in X\}$, then

(1) $\bar{\mathscr{L}}$ is the set of independent subsets of a (finite) matroid $\bar{M}$ on $\bar{X}$, complete with respect to the partition $\bar{X} = \sum_{x \in X} \bar{X}(x)$

(2) M is the $\mathbf{Z}$-matroid associated with $\bar{M}$ and the partition $\bar{X} = \sum_{x \in X} \bar{X}(x)$, as in Lemma 1.

In the sequel Lemma 1 will always be used when $\bar{M}$ is complete with respect to the partition $\bar{X} = \sum_{x \in X} \bar{X}(x)$; in the general case the functions $x \rightsquigarrow |\bar{Y} \cap \bar{X}(x)|$ still define a $\mathbf{Z}$-matroid, however the expression of the rank function is less simple. We have the following proposition:

Let $\bar{M}$ be a matroid on a set $\bar{X}$ and $\bar{X} = \sum_{x \in X} \bar{X}(x)$ be a partition of $\bar{X}$. Set $\hat{\bar{\mathscr{L}}} = \{\bar{Y} \mid \bar{Y} \subseteq \bar{X}$ such that there exists $\bar{Z} \in \mathscr{L}_{\bar{M}}$ satisfying $|\bar{Z} \cap \bar{X}(x)| = |\bar{Y} \cap \bar{X}(x)|$ for all $x \in X$, then $\hat{\bar{\mathscr{L}}}$ is the set of independent subsets of a matroid. $\hat{\bar{M}}$ on $\bar{X}$.

(Proof: consider a bipartite graph on $\bar{X}$ and a disjoint copy $\hat{\bar{X}}$ of $\bar{X}$, having for edges the 2-element sets $\{\bar{x}, \bar{y}\}$ $\bar{x} \in \bar{X}(x)$, $\bar{y} \in \bar{X}(x)$, $x \in X$; then $\hat{\bar{\mathscr{L}}}$ is exactly the set of $\bar{Y} \subseteq \hat{\bar{X}}$ such that there exists a matching mapping $\hat{\bar{Y}}$ on a subset $\bar{Y} \in \mathscr{L}_{\bar{M}}$: by a theorem of C. St. J. A. Nash-Williams – H. Perfect [9], [11] $\hat{\bar{\mathscr{L}}}$ is the set of independent sets of a matroid.

By construction $\hat{\bar{M}}$ is complete with respect to the partition $\bar{X} \equiv \sum_{x \in X} \bar{X}(x)$, and the $\mathbf{Z}$-matroids induced from $\bar{M}$ and $\hat{\bar{M}}$ by this partition are identical.

Proposition 2. *Let* $G = (X, U)$ *be a multigraph and* f *an integer-*

valued function defined on X *such that* $f \leqslant d_G$. *Set* $\mathscr{L} = \{\xi \mid \xi\colon X \mapsto N$ *such that there exists* $H \subseteq G$ *satisfying* $\xi \leqslant d_H \leqslant f\}$:

$\mathscr{L}$ *is the set of independent functions of a Z-matroid* $\bar{E}(G;f)$ *on* X. *The rank function of* $\bar{E}(G;f)$ *is given for* $\alpha\colon X \mapsto N$ *such that* $\alpha \leqslant f$ *by*

$$r_{E(G;f)}(\alpha) = \|\alpha\| - \operatorname*{Max}_{T \subseteq S \subseteq X} (-d_G(S-T) + \alpha(S-T) -$$

$$-f(X-S) + m_G(S-T, X-S) + q_G(S,T))$$

where $q_G(S,T) = q_G(f, \alpha, S, T)$ *is the number of connected components* C *of* G_T *such that* $\alpha(x) = f(x)$ *for all* $x \in C$ *and* $f(C) + m_G(C, S-T)$ *is odd.*

Proof. Let $\bar{G}$ be the following graph:

vertex-set: $\bar{X}$, $|\bar{X}| = 2\|d_G\| + \|f\|$, with partitions $\bar{X} = \sum_{x \in X} \bar{X}(x)$, $\bar{X}(x) = \bar{D}(x) + \bar{I}(x) + \bar{A}(x)$ such that $|\bar{D}(x)| = |\bar{I}(x)| = d_G(x)$, and $|\bar{A}(x)| = f(x)$. We set $\bar{D} = \sum_{x \in X} \bar{D}(x)$, $\bar{A} = \sum_{x \in X} \bar{A}(x)$, $\bar{X}_0 = \bar{X} - \bar{A}$. To fix a correspondence between G and $\bar{G}_{\bar{D}} = \bar{D}(G)$ we may suppose that $\bar{D}(x) \subseteq \{x\} \times U \times \{1, 2\}$ and $(x, u, 1) \in \bar{D}(x)$ iff u is incident to x, $(x, u, 2) \in \bar{D}(x)$ iff u is a loop incident to x. For $u \in U$ we set then $\bar{u} = \{(x, u, 1), (y, u, 1)\}$ if u has 2 vertices, x and y, $x \neq y$, while $\bar{u} = \{(x, u, 1), (x, u, 2)\}$ if u is a loop, incident to x. We set $\bar{U} = \{\bar{u} \mid u \in U\}$.

edge-set of G: $\bar{U} + \sum_{x \in X} \{(\bar{D}(x), \bar{I}(x))\text{-edges}\} + \sum_{x \in X} \{(\bar{I}(x), \bar{A}(x))\text{-edges}\}$.

$\bar{U}$ is a perfect matching of $\bar{D}(G)$. Let $H = (X, V)$ $V \subseteq U$ be a subgraph of G; for $x \in X$ we have $d_H(x) = s(\bar{V}, \bar{D}(x))$, the number of vertices of $\bar{D}(x)$ saturated by $\bar{V} = \{\bar{u} \mid u \in V\}$.

Lemma 2.1. $\xi\colon X \mapsto N$ *is such that there exists* $H \subseteq G$ *so that* $\xi \leqslant d_H \leqslant f$ *if and only if there exists a matching of* $\bar{G}$ *saturating* $\bar{X}_0$ *and at least* $\xi(x)$ *vertices of* $\bar{A}(x)$ *for every* $x \in X$.

Let $H = (X, V)$, $V \subseteq U$ be such that $\xi \leqslant d_H \leqslant f$; $\bar{V}$ can clearly be enlarged into a matching $\bar{V}'$ of $\bar{G}$ saturating $\bar{D}$ and $d_G(x) - d_H(x)$ vertices of $\bar{I}(x)$ for every $x \in X$; then $\bar{V}'$ can be enlarged into a matching $\bar{W}$ saturating $\bar{X}_0$ and $d_H(x) \geqslant \xi(x)$ vertices of $\bar{A}(x)$ for every $x \in X$ (this is possible since $d_H(x) \leqslant f(x)$).

Conversely let $\bar{W}$ be a matching of $\bar{G}$ saturating $\bar{X}_0$ and at least $\xi(x)$ vertices of $\bar{A}(x)$: we check easily that $s(\bar{W}, \bar{A}(x)) = s(\bar{W}_{\bar{D}(G)}, \bar{D}(x))$ for every $x \in X$. Let $H = (X, V) \subseteq G$ such that $\bar{V} = \bar{W}_{\bar{D}(G)}$: $\xi(x) \leqslant$ $\leqslant s(\bar{W}, \bar{A}(x)) = s(\bar{W}_{\bar{D}(G)}, \bar{D}(x)) = d_H(x) \leqslant |\bar{A}(x)| = f(x)$.

Now subsets $Y \subseteq A$ wich are saturated by matchings of $\bar{G}$ saturating $\bar{X}_0$ are (by Edmonds – Fulkerson's theorem (1) and the definition of a quotient matroid) the independent subsets of the matroid $E(G)/\bar{X}_0$ on A; Lemma 1 on $\boldsymbol{Z}$-matroids and Lemma 2.1 give the first part of the proposition.

Let $\alpha: X \mapsto N$ be such that $\alpha \leqslant f$: we fix $\bar{Y} \subseteq \bar{A}$ such that $|\bar{Y}(x)| = \alpha(x)$ for all $x \in X$, where $\bar{Y}(x) = \bar{Y} \cap \bar{A}(x)$. By Lemma 1

$$r_{E(G;f)}(\alpha) = \begin{cases} r_{E(\bar{G})/\bar{X}_0}(\bar{Y}), \\ r_{E(\bar{G})}(\bar{X}_0 + \bar{Y}) - r_{E(\bar{G})}(\bar{X}_0) = r_{E(\bar{G})}(\bar{X}_0 + \bar{Y}) - |\bar{X}_0| \end{cases}$$

($\bar{X}_0$ being clearly an independent set of $\boldsymbol{E}(\bar{G})$), by Proposition 1

$$= |\bar{Y}| - \operatorname*{Max}_{\bar{S} \subseteq \bar{X}} (p_{\bar{X}_0 + \bar{Y}, i}(\bar{S}) - |\bar{X} - \bar{S}|)$$

where $p_{\bar{X}_0 + \bar{Y}, i}(\bar{S})$ is the number of connected components of $\bar{G}_{\bar{S}}$ contained in $\bar{X}_0 + \bar{Y}$ which have an odd number of vertices.

Lemma 2.2. *Let $\bar{S} \subseteq \bar{S}' \subseteq \bar{X}$: if $\bar{S}' - \bar{S}$ is adjacent to at most $|\bar{S}' - \bar{S}|$ connected components of $\bar{G}_{\bar{S}}$, or if $\bar{S}' - \bar{S}$ is contained in $\bar{X}_0$ and is adjacent to at most $|\bar{S}' - \bar{S}| + 1$ connected components of $\bar{G}_{\bar{S}}$ then $p_{\bar{X}_0 + \bar{Y}, i}(\bar{S}') + |\bar{S}'| \geqslant p_{\bar{X}_0 + \bar{Y}, i}(\bar{S}) + |\bar{S}|$.*

$p_{\bar{X}_0 + \bar{Y}, i}(\bar{S}) - p_{\bar{X}_0 + \bar{Y}, i}(\bar{S}')$ is at most equal to the number of odd components of $\bar{G}_{\bar{S}}$ contained in $\bar{X}_0 + \bar{Y}$ which are adjacent to $\bar{S}' - \bar{S}$:

the first case is clear. Let us suppose this number is $|\bar{S}' - \bar{S}| + 1$ and let $\bar{C}_1, \bar{C}_2, \ldots, \bar{C}_{|\bar{S}'-\bar{S}|+1}$ be the corresponding components. Set $\bar{D} = = \bar{C}_1 + \bar{C}_2 + \ldots + \bar{C}_{|\bar{S}'-\bar{S}|+1} + (\bar{S}' - \bar{S})$; as $\bar{D} \subseteq \bar{X}_0 + \bar{Y}$ (since $\bar{S}' - \bar{S} \subseteq \bar{X}_0$) and $|\bar{D}| \equiv 1 \pmod 2$, $\bar{G}_{\bar{D}}$ (which is a union of components of $\bar{G}_{\bar{S}'}$) participates at least for one unit in $\bar{p}_{\bar{X}_0+\bar{Y},i}$: therefore also in this case $\bar{p}_{\bar{X}_0+\bar{Y},i}(\bar{S}) - \bar{p}_{\bar{X}_0+\bar{Y},i}(\bar{S}') \leqslant |\bar{S}' - \bar{S}|$.

Lemma 2.3. *Given any* $\bar{S} \subseteq \bar{X}$ *there exists* $\bar{S}' \subseteq \bar{X}$ *such that it is the sum of subsets of types* $\bar{D}(x), \bar{I}(x), \bar{A}(x)$ *and* $\bar{p}_{\bar{X}_0+\bar{Y},i}(\bar{S}') + |\bar{S}'| \geqslant \geqslant \bar{p}_{\bar{X}_0+\bar{Y},i}(\bar{S}) + |\bar{S}|$.

Let $x \in X$. If $\bar{S} \cap \bar{D}(x) \neq \phi$ then $\bar{D}(x) - \bar{S}$ is adjacent to at most $|\bar{D}(x) - \bar{S}| + 1$ connected components of $\bar{G}_{\bar{S}}$: by Lemma 2.2 $\bar{p}_{\bar{X}_0+\bar{Y},i}(\bar{S} \cup \bar{D}(x)) + |\bar{S} \cup \bar{D}(x)| \geqslant \bar{p}_{\bar{X}_0+\bar{Y},i}(\bar{S}) + |\bar{S}|$. If $\bar{S} \cap \bar{I}(x) \neq \phi$ then $\bar{I}(x) - \bar{S}$ is adjacent to at most 2 connected components of $\bar{G}_{\bar{S}}$: by Lemma 2.2 $\bar{p}_{\bar{X}_0+\bar{Y},i}(\bar{S} \cup \bar{I}(x)) + |\bar{S} \cup \bar{I}(x)| \geqslant \bar{p}_{\bar{X}_0+\bar{Y},i}(\bar{S}) + |\bar{S}|$. If $\bar{S} \cap \bar{A}(x) \neq \phi$ then $\bar{A}(x) - \bar{S}$ is adjacent to at most 1 connected component of $\bar{G}_{\bar{S}}$: by Lemma 2.2 $\bar{p}_{\bar{X}_0+\bar{Y},i}(\bar{S} \cup \bar{A}(x)) + |\bar{S} \cup \bar{A}(x)| \geqslant \geqslant \bar{p}_{\bar{X}_0+\bar{Y},i}(\bar{S}) + |\bar{S}|$.

Lemma 2.4. *Given* $\bar{S} \subseteq \bar{X}$ *there exist* $T \subseteq S \subseteq X$ *such that* $\bar{p}_{\bar{X}_0+\bar{Y},i}(\bar{D}(S) + \bar{I}(T + (X - S)) + \bar{A}(S)) + |\bar{D}(S) + \bar{I}(T + (X - S)) + \bar{A}(S)| \geqslant \geqslant \bar{p}_{\bar{X}_0+\bar{Y},i}(\bar{S}) + |\bar{S}|$, *where* $\bar{D}(S) = \sum_{x \in S} \bar{D}(x)$ *etc.*

By Lemma 2.3 we may suppose that $\bar{S}$ is a sum of subsets of types $\bar{D}(x), \bar{I}(x), \bar{A}(x)$: let us set $S = \{x \mid x \in X \;\; \bar{S} \cap \bar{D}(x) \neq \phi\}$, $T = \{x \mid x \in S \;\; \bar{S} \cap \bar{I}(x) \neq \phi\}$. Then

if $x \in S$; $\bar{A}(x) - \bar{S}$ is adjacent to at most 1 connected component of $\bar{G}_{\bar{S}}$, by Lemma 2.2 $\bar{p}_{\bar{X}_0+\bar{Y},i}(\bar{S} \cup \bar{A}(x)) + |\bar{S} \cup \bar{A}(x)| \geqslant \bar{p}_{\bar{X}_0+\bar{Y},i}(\bar{S}) + |\bar{S}|$

if $x \in X - S$; $\bar{p}_{\bar{X}_0+\bar{Y},i}(\bar{S}) + |\bar{S}| = \bar{p}_{\bar{X}_0+\bar{Y},i}(\bar{S} - \bar{X}(x)) + |\bar{S} - \bar{X}(x)| + + \bar{p}_{\bar{X}_0+\bar{Y},i}(\bar{S} \cap \bar{X}(x)) + |\bar{S} \cap \bar{X}(x)|$. Now $\bar{S} \cap \bar{X}(x)$ equales to ϕ, $\bar{I}(x)$,

$\bar{A}(x)$ or $\bar{I}(x) + \bar{A}(x)$; by an immediate calculation $\bar{p}_{\bar{X}_0+\bar{Y},i}(\bar{S} \cap \bar{X}(x)) + |\bar{S} \cap \bar{X}(x)| \leq \bar{p}_{\bar{X}_0+\bar{Y},i}(\bar{I}(x)) + |\bar{I}(x)|$.

By Lemma 2.4 we have

$$\operatorname*{Max}_{\bar{S} \subseteq \bar{X}} (\bar{p}_{\bar{X}_0+\bar{Y},i}(\bar{S}) - |\bar{X} - \bar{S}|) =$$

$$= \operatorname*{Max}_{T \subseteq S \subseteq X} (\bar{p}_{\bar{X}_0+\bar{Y},i}(\bar{D}(S) + \bar{I}(T + (X - S)) + \bar{A}(S)) +$$

$$+ |\bar{D}(S) + \bar{I}(T + (X - S)) + \bar{A}(S)|) .$$

We calculate $\bar{p}_{\bar{X}_0+\bar{Y},i}(\bar{D}(S) + \bar{I}(T + (X - S)) + \bar{A}(S))$: since $\bar{X}(T) + \bar{D}(S - T)$, $\bar{I}(X - S)$ and $\bar{A}(S - T)$ are pairwise nonadjacent, we have

$$\bar{p}_{\bar{X}_0+\bar{Y},i}(\bar{D}(S) + \bar{I}(T + (X - S)) + \bar{A}(S)) =$$

$$= \bar{p}_{\bar{X}_0+\bar{Y},i}(\bar{X}(T) + \bar{D}(S - T)) +$$

$$+ \bar{p}_{\bar{X}_0+\bar{Y},i}(\bar{I}(X - S)) + \bar{p}_{\bar{X}_0+\bar{Y},i}(\bar{A}(S - T)) ;$$

$$\bar{p}_{\bar{X}_0+\bar{Y},i}(\bar{I}(X - S)) = d_G(X - S) ;$$

$$\bar{p}_{\bar{X}_0+\bar{Y},i}(\bar{A}(S - T)) = \alpha(S - T) .$$

Finally, let us calculate the value of $\bar{p}_{\bar{X}_0+\bar{Y},i}(\bar{X}(T) + \bar{D}(S - T))$. $\bar{X}(x)$ being connected and not contained in $\bar{D}$ are connected components of $\bar{G}_{\bar{X}(T)+\bar{D}(S-T)}$ contained in $\bar{D}$ are those components of $\bar{G}_{\bar{D}(S-T)}$ which are nonadjacent to $\bar{X}(T)$: $m_G(S - T, X - S)$ of them consist of one vertex, the others having 2 vertices. Since a component of $\bar{G}_{\bar{D}(S-T)}$ adjacent to $\bar{X}(T)$ is adjacent to exactly one $\bar{X}(x)$ the connected components of $\bar{G}_{\bar{X}(T)+\bar{D}(S-T)}$ not contained in $\bar{D}$ are in $1-1$ correspondance with the connected components of $\bar{G}_{\bar{X}(T)}$, then in $1-1$ correspondence with the connected components of G_T (we may suppose G without isolated vertices). Let C be the vertex-set of a connected component of G_T: the cardinality of the corresponding component of $\bar{G}_{\bar{X}(T)+\bar{D}(S-T)}$ is $|\bar{X}(C)| + m_G(C, S - T) \equiv f(C) + m_G(C, S - T) \pmod 2$, this component being contained in $\bar{X}_0 + \bar{Y}$ if and only if $\alpha(x) = f(x)$

for all $x \in C$. With the notations of the statement of Proposition 2 we have thus $\bar{p}_{\bar{X}_0 + \bar{Y}, i}(\bar{X}(T) + \bar{D}(S - T)) = m_G(S - T, X - S) + q_G(f, \alpha, S, T)$.

From here $r_{\bar{E}(G;f)}(\alpha)$ is obtained by an immediate calculation.

Proposition 2 contains Lovász' theorem 8.3 of [8] (which gives a necessary and sufficient condition for the existence of a subgraph $H \subseteq G$ such that $g \leqslant d_H \leqslant f$ Lovász' condition is $r_{\bar{E}(G;f)}(g) \geqslant \|g\|$). In fact the given expression of $r_{\bar{E}(G;f)}(\alpha)$ is equivalent to Lovász' theorem 7.3 which gives the value of $\underset{H \subseteq G}{\text{Min}} (\| (g - d_H)^+ \| + \| (d_H - f)^+ \|)$ $(= \delta(f, d_G - g)$ in Lovász' notations; $\xi^+ = \text{Max}(\xi, 0)$ for $\xi \colon X \mapsto Z$), by the following equality (which of course may be deduced from Proposition 2 and Lovász' theorem 7.3, but there is a simpler proof!):

Proposition 3.

$$r_{\bar{E}(G;f)}(g) = \|g\| - \underset{H \subseteq G}{\text{Min}} (\| (g - d_H)^+ \| + \| (d_H - f)^+ \| .$$

Proof. Let us consider the graph $\bar{G}$ of the proof of Proposition 2 with $\alpha = g$: we have $r_{\bar{E}(G;f)}(g) = r_{E(\bar{G})}(\bar{X}_0 + \bar{Y}) - 2\|d_G\|$.

Let $H = (X, V) \subseteq G$: we can clearly enlarge $\bar{V}$ into a matching $\bar{V}'$ of $\bar{G}$ saturating $\bar{D}$ and $d_G(x) - d_H(x)$ vertices of $\bar{I}(x)$ for every $x \in X$, then $\bar{V}'$ into a matching $\bar{W}$ saturating $\bar{D}$ and,

if $d_H(x) \leqslant g(x)$, $\bar{I}(x)$ and $d_H(x)$ vertices of $\bar{Y}(x)$

if $g(x) \leqslant d_H(x) \leqslant f(x)$, $\bar{I}(x)$ and $d_H(x)$ vertices of $\bar{A}(x)$ including all vertices of $\bar{Y}(x)$

if $f(x) \leqslant d_H(x)$, $\bar{A}(x)$ and $d_G(x) - d_H(x) + f(x)$ vertices of $\bar{I}(x)$ for every $x \in X$. For this matching $\bar{W}$ we have

$$s(\bar{W}, \bar{X}_0(x) + \bar{Y}(x)) =$$
$$= 2d_G(x) + g(x) - (g(x) - d_H(x))^+ - (d_H(x) - f(x))^+$$

for every $x \in X$; thus for any $H \subseteq G$

$$r_{\bar{E}(G;f)}(g) \geqslant \|g\| - (\| (g - d_H)^+ \| + \| (d_H - f)^+ \|) .$$

For the reverse inequality let $\bar{W}$ be any matching of $\bar{G}$, let $H = (X, V)$ denote the subgraph of G such that $\bar{V} = \bar{W}_{\bar{D}(G)}$: it suffices to see that $\bar{W}$ saturates at most $2d_G(x) + g(x) - (g(x) - d_H(x))^+ - (d_H(x) - f(x))^+$ vertices of $\bar{X}_0(x) + \bar{Y}(x)$ for every $x \in X$ or, equivalently, that $r_{\bar{E}(\bar{G}_{\bar{D}(x) - \bar{D}_H(x) + \bar{I}(x) + \bar{A}(x)})}(\bar{D}(x) - \bar{D}_H(x) + \bar{I}(x) + \bar{Y}(x)) \leqslant$ $\leqslant 2d_G(x) + g(x) - (g(x) - d_H(x))^+ - (d_H(x) - f(x))^+$ where $\bar{D}_H(x)$ is the subset of $\bar{D}(x)$ saturated by $\bar{V}$. A look at $\bar{G}$ may be sufficiently convincing; anyway by the easy part of Proposition 1 the left term is at most equal to

$$|\bar{D}(x) - \bar{D}_H(x) + \bar{I}(x) + \bar{Y}(x)| -$$
$$- (p_{\bar{D}(x) - \bar{D}_H(x) + \bar{I}(x) + \bar{A}(x), i}(\bar{S}) -$$
$$- |(\bar{D}(x) - \bar{D}_H(x) + \bar{I}(x) + \bar{A}(x)) - \bar{S}|)$$

for any $\bar{S} \subseteq \bar{D}(x) - \bar{D}_H(x) + \bar{I}(x) + \bar{A}(x)$: the required inequality is obtained if $d_H(x) \leqslant g(x)$ for $\bar{S} = \bar{D}(x) - \bar{D}_H(x) + \bar{A}(x)$, if $g(x) \leqslant d_H(x) \leqslant$ $\leqslant f(x)$ for $\bar{S} = \bar{D}(x) - \bar{D}_H(x) + \bar{I}(x) + \bar{A}(x)$, if $f(x) \leqslant d_H(x)$ for $\bar{S} = \bar{I}(x)$.

Proposition 3 and Lovász' theorem 7.3 yield an alternative proof for the second part of Proposition 2; the above one, however, which uses a variant of Tutte's construction of [14], is an archetype for proofs of subsequent propositions (§§3, 4).

By definition of $r_{\bar{E}(G;f)}(g)$(= Max $\|\xi\| \mid \xi \leqslant g$ such that there exists $H \subseteq G$ satisfying $\xi \leqslant d_H \leqslant f$) we have

$$r_{\bar{E}(G;f)}(g) = \underset{H \subseteq G,\, d_H \leqslant f}{\text{Max}} (\|\text{Min}\,(d_H, g)\|) = \|g\| - \underset{H \subseteq G,\, d_H \leqslant f}{\text{Max}} (\|(g - d_H)^+\|).$$

In particular $r_{\bar{E}(G;f)}(f) = \underset{H \subseteq G,\, d_H \leqslant f}{\text{Max}} \|d_H\| = 2m(G;f)$ where $m(G;f)$ denotes the maximum number of edges of a subgraph $H \subseteq G$ such that $d_H \leqslant f$. Denoting by $m'(G;f)$ the minimum number of edges of a subgraph $H \subseteq G$ such that $d_H \geqslant f$ we have clearly $2m'(G;f) = \|d_G\| - 2m(G; d_G - f)$. By Proposition 3 and the identity $\|(g - d_H)^+\| + \|(d_H - f)^+\| = \|(d_{G-H} - (d_G - g))^+\| + \|((d_G - f) - d_{G-H})^+\|$ we get

$r_{\bar{E}(G;d_G-f)}(d_G - f) = \|d_G\| - \|f\| - \|g\| + r_{\bar{E}(G;f)}(f)$, and thus for $f = g$ $m(G;f) + m'(G;f) = \|f\|$. For $f \equiv 1$ this equlity is Norman – Robin's theorem. The directed proof of the general case is straightforward, similar to the case $f \equiv 1$.

We give some variants of Proposition 2:

1-ST VARIANT

Lemma 2. *Let M be a Z-matroid on a set X and α an integer-valued function defined on X: then the functions $\xi\colon X \mapsto N$ such that there exists $\eta\colon X \mapsto N$ with the properties $\eta \leqslant \alpha$, $\xi + \eta$ is independent for M and $\|\eta\| = r_M(\alpha)$ are the independent functions of a Z-matroid on X, the Z-matroid quotient of M by α, denoted by M/α. The rank function of M/α is given by $r_{M/\alpha}$ is given by $r_{M/\alpha}(\beta) = r_M(\alpha + \beta) - - r_M(\alpha)$ for $\beta\colon X \mapsto N$.*

With notations of Proposition 2 let us suppose that there exists a $H \subseteq G$ such that $g \leqslant d_H \leqslant f$ i.e. that g is an independent function of $\bar{E}(G;f)$. Then by Lemma 2 *the functions $\xi\colon X \mapsto N$ such that there exists a $H \subseteq G$ satisfying $g + \xi \leqslant d_H \leqslant f$ are the independent functions of a Z-matroid on X with rank function* $\mathrm{rk}(\alpha) = r_{\bar{E}(G;f)}(g + \alpha) - \|g\|$ *for $\alpha\colon X \mapsto N$, $\alpha \leqslant f - g$.*

2-ND VARIANT

Lemma 3. *Let M be a Z-matroid on a set X: the subsets $Y \subseteq X$ such that 1_Y is an independent function of M are the independent sets of a matroid M_S on X. The rank function of M_S is given by $r_{M_S}(Y) = = r_M(1_Y)$ for $Y \subseteq X$.*

With notations of Proposition 2 let us suppose that there exists a $H \subseteq G$ such that $g \leqslant d_H \leqslant f$: by variant 1 and Lemma 3 *the subsets $Y \subseteq X$ such that there exists a $H \subseteq G$ with the properties $g(x) \leqslant d_H(x) \leqslant \leqslant f(x)$ for all $x \in X$ and $g(x) < d_H(x)$ for all $x \in Y$; are the independent sets of a matroid $E(G;f,g)$ on X; the rank function is given by $r_{E(G;f,g)}(Y) = r_{\bar{E}(G;f)}(g + 1_Y) - \|g\|$ for $Y \subseteq X$ i.e.*

$$r_{E(G;f,g)}(Y) = |Y| - \max_{T \subseteq S \subseteq X} (-d_G(S-T) + g(S-T) - $$
$$- f(X-S) + m_G(S-T, X-S) + q_G(f,g,Y,S,T) +$$
$$+ |Y \cap (S-T)|)$$

where $q_G(f,g,Y,S,T)$ *is the number of connected components* C *of* G_T *such that* $f = g+1$ *on* $C \cap Y$, *and* $f = g$ *on* $C - Y$ *and* $f(C) + m_G(S-T, X-S)$ *is odd.*

n.b. for $g \equiv 0$ and $f \equiv 1$ it can easily be seen that given $T \subseteq S \subseteq X$ and $x \in S - T$ the above expression to be maximized does not decrease if $T + x$ is substituted for T; the maximum is thus achieved for $S = T$, in which case it reduces to $p_{Y,i}(S) - |X - S|$: we get thus the expression of $r_{E(G)}(Y)$ given by Proposition 1.

3-RD VARIANT

Let us modify $\bar{G}$ by substituting for $\bar{G}_{\bar{X}(x)}$ a graph having vertex-set $\bar{D}(x) + \bar{A}(x)$, $|\bar{A}(x)| = d_G(x) - f(x)$, edge-set $\{(\bar{D}(x), \bar{A}(x))\text{-edges}\}$, this for all x belonging to a subset $X'' \subseteq X$, and set $\bar{X}_0 = \bar{D} + \bar{I}(X')$ where $X' = X - X''$; then by the proof of Proposition 2, except for some details left to the reader, we obtain the following symmetric form of Proposition 2: *suppose there exists a subgraph* $H \subseteq G$ *such that* $f(x) \geqslant d_H(x)$ *for all* $x \in X'$ *and* $f(x) \leqslant d_H(x)$ *for all* $x \in X''$, *then the integer-valued functions* ξ *defined on* X *such that there exists a subgraph* $H \subseteq G$ *satisfying* $\xi(x) \leqslant d_H(x) \leqslant f(x)$ *for all* $x \in X'$ *and* $f(x) \leqslant d_H(x) \leqslant d_G(x) - \xi(x)$ *for all* $x \in X''$ *are the independent functions of a* **Z**-*matroid on* X; *the rank function is given by*

$$\mathrm{rk}(\alpha) = \|\alpha\| - \max_{T \subseteq S \subseteq X} (-d_G(S-T) - d_G(X'' - S) +$$
$$+ \alpha((S-T) \cap X') + \alpha(X'' - S) - f(X' - S) +$$
$$+ f((S-T) \cap X'') + m_G(S-T, X-S) +$$
$$+ q_G(X', f, \alpha, S, T))$$

for $\alpha \leqslant f$ *on* X', $\alpha \leqslant d_G - f$ on X'', *where* $q_G(X', f, \alpha, S, T)$ *is the* *number of connected components C of G_T such that $\alpha = f$ on $C \cap X'$, $\alpha = d_G - f$ on $C \cap X''$ and $f(C) + m_G(C, S - T)$ is odd.*

§3.

The central idea in [14] is to reduce the considered constraint – $d_H(x) = f(x)$ for all $x \in X$ – to a property of matchings – existence of a perfect matching – for which a necessary and sufficient condition is known [12]. A slight modification of Tutte's construction and a stronger result on matchings has given in §2 an analogous result for constraint $g(x) \leqslant d_H(x) \leqslant f(x)$ for all $x \in X$. At this point a natural question arises: is there other "usual" constraints on subgraphs which can similarly be reduced to manageable properties of matchings (essentially to the existence of matchings saturating a given set)? The core of this question is in fact the following problem:

Given a system $\mathscr{C}$ of subsets of a set $\bar{D}$. Does there exist a graph $\bar{L}$ containing $\bar{D}$ as a stable set of vertices and a subset $\bar{S}_0$ of the vertex-set such that $\mathscr{C}$ is the system of sets of unsaturated vertices of $\bar{D}$ in matchings of $\bar{L}$ saturating $\bar{S}_0$?

Let us take an example: suppose we are looking for a necessary and sufficient condition for the existence of a subgraph $H \subseteq G$ such that $g(x) \leqslant d_H(x) \leqslant f(x)$ and $d_H(x)$ even for all $x \in X$ (such a condition will be given in §4), and we think it possible to reduce this problem to the problem of the existence of matchings saturating a given set. We use $\bar{D}(G)$ to represent subgraphs of $\bar{G}$ by matchings; let $H = (X, V)$ be a subgraph of $\bar{G}$ satisfying the constraints, denote by $\bar{D}_H$ the subset of $\bar{D}$ saturated by $\bar{V}$: $\bar{D}_H$ has the properties $g(x) \leqslant |\bar{D}_H \cap \bar{D}(x)| \leqslant f(x)$ and $|\bar{D}_H \cap \bar{D}(x)|$ even for all $x \in X$. Set $\mathscr{C} = \{\bar{Z} \mid \bar{Z} \subseteq \bar{D}$ such that $g(x) \leqslant |\bar{Z} \cap \bar{D}(x)| \leqslant f(x)$ and $|\bar{Z} \cap \bar{D}(x)|$ even for all $x \in X\}$; if we solve affirmatively the above problem for this $\mathscr{C}$ we are ready: set then $\bar{G} = \bar{L} + \{$edges of $\bar{U}\}$ and $\bar{X}_0 = \bar{S}_0 + \bar{D}$, clearly there exists a subgraph $H \subseteq G$ such that $g(x) \leqslant d_H(x) \leqslant f(x)$, $d_H(x)$ even for all $x \in X$ if and only if there exists a matching of $\bar{G}$ saturating $\bar{X}_0$. Moreover from $E(\bar{G})/\bar{X}_0$ it will be possible to derive a matroid (Z-matroid) associated with the given constraint.

We call a set $\mathscr{C}$ of subsets of $\bar{D}$ such that the above problem for $\mathscr{C}$ has an affirmative solution a *Tutte-class of subsets of* D. Of course, here, we are primarily interested in Tutte-classes which correspond to "usual" constraints on subgraphs.

By the construction of §2 $\{g(x) \leqslant |\bar{Z} \cap \bar{D}(x)| \leqslant f(x)$ for all $x \in X\}$ (we write $\{g(x) \leqslant |\bar{Z} \cap \bar{D}(x)| \leqslant f(x)$ for all $x \in X$ instead of $\{\bar{Z} | \bar{Z} \subseteq \bar{D}$ such that ...$\}$) is a Tutte-class (take $\bar{L} = \sum_{x \in X} \bar{G}_{\bar{X}(x)}$, $\bar{S}_0 = \bar{X}_0 - \bar{D} + \bar{Y}$ with $\alpha = g$). One other basic example is $\{|\bar{Z} \cap \bar{E}|$ even$\}$ (resp. odd) for $\bar{E} \subseteq \bar{D}$ (take $\bar{L}$: vertex-set $\bar{D} + \bar{I}$ such that $|\bar{I}| = |\bar{E}|$ (resp. $|\bar{I}| = |\bar{E}| - 1$), edges $(\bar{I} + \bar{E}, \bar{I})$-edges, $\bar{S}_0 = \bar{I}$).

To give further examples let us consider the following graphs:

$\bar{K}_{(g,f)}(\bar{E})$ for g, f integers $g \leqslant f \leqslant |\bar{E}|$

vertex-set: $\bar{E} + \bar{I} + \bar{A}$ such that $|\bar{I}| = |\bar{A}| = |\bar{E}|$

edges: $(\bar{E}, \bar{J})$- and $(\bar{I}, \bar{B})$-edges where $\bar{J} \subseteq \bar{I}$ $|\bar{J}| = |\bar{E}| - g$, $\bar{B} \subseteq \bar{A}$ $|\bar{B}| = f$

given any matching saturating $\bar{I}$ the number of unsaturated vertices of $\bar{E}$ is $\geqslant g$ and $\leqslant f$, and equal to the number of saturated vertices of $\bar{A}$.

$\bar{K}_{(\text{even})}(\bar{E})$ (resp. $\bar{K}_{(\text{odd})}(\bar{E})$)

vertex-set: $\bar{E} + \bar{I}$ such that $|\bar{I}| = |\bar{E}|$ (resp. $|\bar{I}| = |\bar{E}| - 1$)

edges: $(\bar{E} + \bar{I}, \bar{I})$-edges

given any matching saturating $\bar{I}$ the number of saturated vertices of $\bar{E}$ is even (resp. odd)

$\bar{K}_{(\text{diff})}(\bar{E}, \bar{E}')$ for $\bar{E} \cap \bar{E}' = \phi$

vertex-set: $\bar{E} + \bar{E}' + \bar{I} + \bar{I}' + \bar{A}$ such that $|\bar{I}| = |\bar{E}|$, $|\bar{I}'| = |\bar{A}| = |\bar{E}| = |\bar{E}'|$

edges: $(\bar{E}, \bar{I})$-, $(\bar{E}', \bar{I}')$-, $(\bar{I}, \bar{I}')$- and $(\bar{I}', \bar{A})$-edges

given any matching $\bar{W}$ saturating $\bar{I} + \bar{I}'$ we have $s(\bar{W}, \bar{A}) =$

$= \text{uns}(\bar{W}, \bar{E}') - \text{uns}(\bar{W}, \bar{E}) + |\bar{E}|$ where $\text{uns}(\bar{W}, \bar{E}) = |\bar{E}| - s(\bar{W}, \bar{E})$ is the number of vertices of $\bar{E}$ unsaturated by $\bar{W}$. Furthermore given any subsets of $\bar{E}, \bar{E}'$ there exists a matching saturating $\bar{I} + \bar{I}'$ such that these subsets are the sets of unsaturated vertices of $\bar{E}$ and $\bar{E}'$ respectively.

Given a graph $\bar{L}$ with vertex-set $\bar{X}$ we define

for $\bar{E} \subseteq \bar{X}$ $\bar{L} *_{\bar{E}} \bar{K}$ ($\bar{K}$ being one of $\bar{K}_{(g,f)}$, $\bar{K}_{(\text{even})}$ or $\bar{K}_{(\text{odd})}$): take the union of $\bar{L}$ with a copy of $\bar{K}(\bar{E})$ ($\bar{K}_{(g,f)}(\bar{E})$, etc. . .) having only the vertices of $\bar{E}$ in common with $\bar{L}$,

for $\bar{E}, \bar{E}' \subseteq \bar{X}$ such that $\bar{E} \cap \bar{E}' = \phi$ $\bar{L} *_{\bar{E},\bar{E}'} \bar{K}_{(\text{diff})}$: take the union of $\bar{L}$ with a copy of $\bar{K}_{(\text{diff})}(\bar{E}, \bar{E}')$ having only the vertices of $\bar{E}, \bar{E}'$ in common with $\bar{L}$.

Let $\bar{D}$ denote a set and also the graph having $\bar{D}$ for vertex-set and no edges. Other examples of Tutte classes are:

$\{g \leqslant |\bar{Z} \cap \bar{E}'| - |\bar{Z} \cap \bar{E}| \leqslant f\}$ for $\bar{E}, \bar{E}' \subseteq \bar{D}$ such that $\bar{E} \cap \bar{E}' = \phi$, $-|\bar{E}| \leqslant g \leqslant f \leqslant |\bar{E}'|$ (take $\bar{L} = \bar{D} *_{\bar{E},\bar{E}'} \bar{K}^1_{(\text{diff})} *_{\bar{A}_1} \bar{K}^2_{(g+|\bar{E}|, f+|\bar{E}|)}$, $\bar{S}_0 = \bar{I}_1 + \bar{I}'_1 + \bar{A}_1 + \bar{I}_2$ (we omit brackets, composition of $*$-operations is from left to right; convention: in $\bar{K}^1_{(\text{diff})}$ $\bar{I}, \bar{I}', \bar{A}$ are labelled $\bar{I}_1, \bar{I}'_1, \bar{A}_1$ etc. . .))

$\{|\bar{Z} \cap \bar{E}'| - |\bar{Z} \cap \bar{E}|$ even$\}$ (resp. odd) for $\bar{E}, \bar{E}' \subseteq \bar{D}$ such that $\bar{E} \cap \bar{E}' \neq \phi$ (take $\bar{L} = \bar{D} *_{\bar{E},\bar{E}'} \bar{K}^1_{(\text{diff})} *_{\bar{A}_1} \bar{K}^2_{(\text{parity of } |\bar{E}|)}$ (resp. $\bar{K}^2_{(\text{parity of } |\bar{E}|+1)}$) $\bar{S}_0 = \bar{I}_1 + \bar{I}'_1 + \bar{A}_1 + \bar{I}_2$).

Let us say that the above 3 Tutte-classes are elementary ones: intersections of elementary Tutte-classes may give further Tutte-classes. Let us consider a sequence of elementary Tutte-classes $\mathscr{C}_1, \mathscr{C}_2, \ldots, \mathscr{C}_n$ satisfying the following absorption rules:

if $\mathscr{C}_k$ is $\{g_k \leqslant |\bar{Z} \cap \bar{E}'_k| - |\bar{Z} \cap \bar{E}_k| \leqslant f_k\}$ then a subsequent $\mathscr{C}_h$ containing $|\bar{Z} \cap \{\bar{x}\}|$ for a $\bar{x} \in \bar{E}_k + \bar{E}'_k$ always contains $|\bar{Z} \cap \bar{E}'_k| - |\bar{Z} \cap \bar{E}_k|$ i.e. if $\mathscr{C}_h$ $k+1 \leqslant h \leqslant n$ is defined by a condition on $|\bar{Z} \cap \bar{E}'_h| - |\bar{Z} \cap \bar{E}_h|$ then $\bar{E}_k \cap \bar{E}_h \neq \phi$ implies $\bar{E}_k \subseteq \bar{E}_h$ and $\bar{E}'_k \subseteq \bar{E}'_h$ (and similar implications if $\bar{E}'_k \cap \bar{E}_h$, $\bar{E}_k \cap \bar{E}'_h$ or $\bar{E}'_k \cap \bar{E}'_h \neq \phi$)

if $\mathscr{C}_k$ is $\{|\bar{Z} \cap \bar{E}'_k| - |\bar{Z} \cap \bar{E}_k| \text{ even}\}$ (resp. odd) then a $|\bar{Z} \cap \{\bar{x}\}|$ for $\bar{x} \in E_k + E'_k$ is contained in no subsequent $\mathscr{C}_h$. Then we assert (omitting the proof which is straightforward but tedious) that there are suitable subscripts such that the natural sequence of $*$-operations does provide a $\bar{L}(\mathscr{C}_1 \cap \mathscr{C}_2 \cap \ldots \cap \mathscr{C}_n)$ and a $\bar{S}_0$ i.e. that $\mathscr{C}_1 \cap \mathscr{C}_2 \cap \ldots$ $\ldots \cap \mathscr{C}_n$ is a Tutte-class.

The absorption rules are in particular obviously satisfied (for any ordering) by sequences $\mathscr{C}_1, \mathscr{C}_2, \ldots, \mathscr{C}_n$ such that the $\bar{E}_k + \bar{E}'_k$ are pairwise disjoint.

Corresponding to any given Tutte-class other graphs may be used for $*$-operations – or example $\bar{K}_{(=)}(\bar{E}, \bar{E}')$ for $\bar{E} \cap \bar{E}' = \phi$ (vertex-set $\bar{E} + \bar{E}' + \bar{I} + \bar{I}'$ such that $|\bar{I}'| = |\bar{E}|$, $|\bar{I}'| = |\bar{E}'|$, $(\bar{E}, \bar{I})$-, $(\bar{E}', \bar{I}')$- and $(\bar{I}, \bar{I}')$-edges) corresponding to $\{|\bar{Z} \cap \bar{E}| = |\bar{Z} \cap \bar{E}'|\}$ $(\bar{S}_0 = \bar{I} + \bar{I}')$ – but if they are to be used, suitable absorption rules may have to be formulated (here: $|\bar{Z} \cap \{\bar{x}\}|$ for $\bar{x} \in \bar{E} + \bar{E}'$ must not occur anymore is subsequent Tutte-classes). Note that $\bar{L} = \bar{D} *_{\bar{E}, \bar{E}'} \bar{K}_{(=)}$, $\bar{S}_0 = \bar{I} + \bar{I}'$ is a simpler solution for $\{|\bar{Z} \cap \bar{E}| = |\bar{Z} \cap \bar{E}'|\}$ than $\bar{D} *_{\bar{E}, \bar{E}'} \bar{K}^1_{(\text{diff})} *_{\bar{A}_1} \bar{K}^2_{(0,0)}$, $\bar{I}_1 + \bar{I}'_1 +$ $+ \bar{A}_1 + \bar{I}_2$ given by the above construction.

A necessary (not sufficient) condition for a set $\mathscr{C}$ of subsets of $\bar{D}$ to be a Tutte-class is that $\text{Min}(\mathscr{C})$, the subset of $\mathscr{C}$ constituted by its elements, minimal for inclusion, be the set of bases of a matroid.

$\{|\bar{Z} \cap \bar{E}| = 0$ or $2\}$ is a Tutte-class (cf. §4.1); we show that $\{|\bar{Z} \cap \bar{E}| = 0$ or $3\}$ is not. Suppose it were; let G be the graph of Fig. 1, then the subsets of a, b, c contained in some subgraph H of G satisfying the constraints would be the independent sets of a matroid: $\{a\}$ and $\{b, c\}$ are such independent sets but $\{a, b\}$ and $\{a, c\}$ are not, a contradiction.

A similar argument shows that $\{|\bar{Z} \cap \bar{E}| = |\bar{Z} \cap \bar{E}'| = 0$ or $2\}$ $\bar{E}, \bar{E}' \subseteq \bar{D}$, $\bar{E} \cap \bar{E}' = \phi$, or $\{|\bar{Z} \cap \bar{E}| = |\bar{Z} \cap \bar{E}'| = |\bar{Z} \cap \bar{E}''|\}$, $\bar{E}, \bar{E}', \bar{E}'' \subseteq \bar{D}$ are pairwise disjoint, are not Tutte-classes of subsets.

Let us point out that in the definition of a Tutte-class the convenience hypotheses, $\bar{D}$ stable set of $\bar{L}$ and $\bar{S}_0 \subseteq \bar{X} - \bar{D}$, result in no loss of

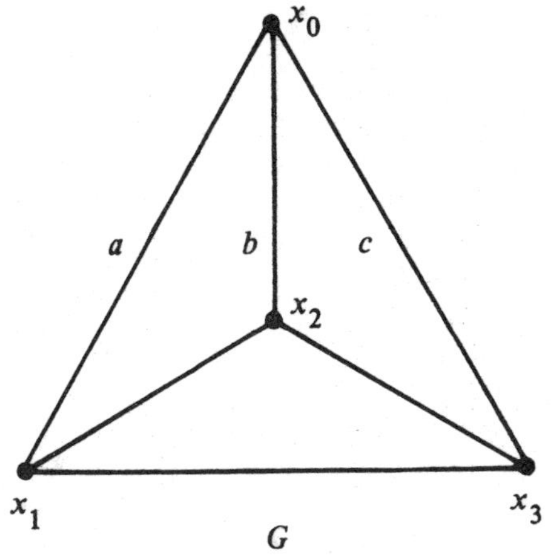

$d_H(x_1) \leqslant 2$

$d_H(x_2) = 0$ or 3

$d_H(x_3) = 0$ or 3

Fig. 1

generality. Suppose that given a set $\mathscr{C}$ of subsets of $\bar{D}$ there exists a graph $\bar{L}$ and a subset $\bar{S}_0$ of its vertex-set such that $\mathscr{C}$ is the set of subsets $\bar{Z} \subseteq \bar{D}$ for which there exists a matching of $\bar{L}$ saturating $\bar{S}_0$ and in $\bar{D}$ exactly those vertices belonging to $\bar{Z}$. Let $\bar{L}''$ be a copy of $\bar{L}$ (superscript $''$ giving the $1-1$ correspondence) and $\bar{D}'$ be a copy of $\bar{D}$; we denote by $\bar{L}^*$ the graph obtained from $\bar{L}''$ by adding $\bar{D} + \bar{D}'$ to its vertex-set and all edges $\{\bar{x}, \bar{x}'\}$ for $\bar{x} \in \bar{D}$, $\{\bar{x}', \bar{x}''\}$ for $\bar{x} \in \bar{D} - \bar{S}_0$, $\{\bar{x}, \bar{x}''\}$ for $\bar{x} \in \bar{D} \cap \bar{S}_0$. Then $\bar{L}$ and $\bar{S}_0 = \bar{S}_0'' \cup \bar{D}'' \cup \bar{D}'$ is a solution of the original problem.

§4. APPLICATIONS

1. Let $\bar{D} = \sum_{x \in X} \bar{D}(x)$: then by §3 $\{g(x) \leqslant |\bar{Z} \cap \bar{D}(x)| \leqslant f(x),$ $|\bar{Z} \cap \bar{D}(x)|$ even for all $x \in X\}$ is a Tutte-class of subsets of $\bar{D}$ (for $\{g(x) \leqslant |\bar{Z} \cap \bar{D}(x)| \leqslant f(x),\ |\bar{Z} \cap \bar{D}(x)|$ even$\}$ take $\bar{L} = \bar{D} *_{\bar{D}(x)} \bar{K}^1_{(g(x), f(x))} *_{\bar{A}_1} \bar{K}^2_{(\text{even})}$, $\bar{S}_0 = \bar{I}_1 + \bar{A}_1 + \bar{I}_2$).

Proposition 4. *Let* $G = (X, U)$ *be a multigraph and* f, g *be two functions defined on* X *with values in the set of even integers, such that*

$g \leqslant f \leqslant d_G$. *Then a necessary and sufficient condition for a subgraph* $H \subseteq G$ *such that* $g(x) \leqslant d_H(x) \leqslant f(x)$ *and* $d_H(x)$ *is even for all* $x \in X$ *to exist is that*

$$q'_G(S, T) + m_G(S - T, X - S) \leqslant$$
$$\leqslant d_G(S - T) - g(S - T) + f(X - S) \quad \textit{for all} \quad T \subseteq S \subseteq X$$

where $q'_G(S, T)$ *is the number of connected components* C *of* G_T *such that* $m_G(C, S - T)$ *is odd.*

Proof. Instead of the auxiliary graph $\bar{G}$ given by the above construction we rather take the simpler $\bar{G} = \bar{D}(G) \cup \left(\sum_{x \in X} \bar{G}_{\bar{X}(x)} \right)$ where $\bar{G}_{\bar{X}(x)}$ is a graph with vertex-set $\bar{X}(x) = \bar{D}(x) + \bar{I}_1(x) + \bar{I}_2(x)$ such that $|\bar{I}_1(x)| = d_G(x) - g(x)$, $|\bar{I}_2(x)| = f(x) - g(x)$, and $(\bar{D}(x), \bar{I}_1(x))$- and $(\bar{I}_1(x) + \bar{I}_2(x), \bar{I}_2(x))$-edges. The proof is similar to the proof of Proposition 2: there exists a subgraph $H \subseteq G$ such that $g(x) \leqslant d_H(x) \leqslant f(x)$ and $d_H(x)$ even for all $x \in X$ if and only if $\bar{G}$ has a perfect matching, then if and only if $\bar{p}_i(\bar{S}) \leqslant |\bar{X} - \bar{S}|$ for all $\bar{S} \subseteq \bar{X}$, but quantification may be restricted to $\bar{S} = \bar{X}(T) + \bar{D}(S - T) + \bar{I}_2(S - T) + \bar{I}_1(X - S)$ for $S \subseteq X$ and $T \subseteq S^{g<d}G$ ($S^{g<d}G = \{x \mid x \in S$ such that $g(x) < d_G(x)$). Now for such a set $\bar{S}$

$$\bar{p}_i(\bar{S}) = \bar{p}_i(\bar{X}(T) + \bar{D}(S - T)) + \bar{p}_i(\bar{I}_2(S - T)) + \bar{p}_i(\bar{I}_1(X - S))$$

$\bar{p}_i(\bar{I}_2(S - T)) = 0$ (for $x \in S - T$ $\bar{G}_{\bar{I}_2(x)}$ is a connected component of $\bar{G}_{\bar{S}}$ but has an even number of vertices)

$\bar{p}_i(\bar{I}_1(X - S)) = d_G(X - S) - g(X - S)$

$\bar{p}_i(\bar{X}(T) + \bar{D}(S - T)) = m_G(S - T, X - S) + q'_G(S, T)$ where $q'_G(S, T)$ is the number of connected components C of G_T such that the corresponding component of G_S has an odd number of vertices i.e. such that $d_G(C) + d_G(C) - g(C)) + (f(C) - g(C)) + m_G(C, S - T) \equiv m_G(C, S - T)$ (mod 2) is odd.

We find a quantification "for all $S \subseteq X$ and $T \subseteq S^{g<d}G$" owing to the fact that if $g(x) = d_G(x)$ $\bar{G}_{\bar{X}(x)}$ is not connected; to get a quanti-

fication "for all $T \subseteq S \subseteq X$" we can modify the construction of $\bar{G}$ so that $\bar{G}_{\bar{X}(x)}$ be connected for all $x \in X$ or else show directly that the condition is necessary for all $T \subseteq S \subseteq X$.

Let H be a subgraph of G such that $g(x) \leqslant d_G(x) \leqslant f(x)$ and $d_G(x)$ even for all $x \in X$, and let $T \subseteq S \subseteq X$. Clearly

$$m_G(S-T, X-S) + m_H(T, X-S) + m_{G-H}(T, S-T) =$$

$$= m_H(X-S, S) + m_{G-H}(S-T, X-(S-T)) .$$

Let C be a connected component of G_T. We have $2m_H(C, X) = = d_G(C) + m_H(C, X-S) + m_H(C, S-T)$; $d_G(C)$ being even $m_H(C, X-S) \equiv m_H(C, S-T) \pmod 2$, thus $m_H(C, X-S) + + m_{G-H}(C, S-T) \equiv m_H(C, S-T) + m_{G-H}(C, S-T) = m_G(C, S-T) \pmod 2$ so that $m_H(T, X-S) + m_{G-H}(T, S-T) \geqslant q'_G(S, T)$. On the other hand $m_H(X-S, S) \leqslant d_H(X-S) \leqslant f(X-S)$, $m_{G-H}(S-T, X-(S-T)) \leqslant d_{G-H}(S-T) \leqslant d_G(S-T) - g(S-T)$.

Corollary. *Let $G = (X, U)$ be a multigraph and A be a subset of X: a necessary and sufficient condition for the existence of pairwise disjoint (elementary) circuits of G covering A is that*

$$q'_G(S, T) + m_G(S-T, X-S) \leqslant$$

$$\leqslant d_G(S-T) - 2|(S-T) \cap A| + 2|X-S|$$

for all $T \subseteq S \subseteq X$.

Proof. Pairwise disjoint elementary circuits of G covering A constitute a subgraph H of G such that $d_H = 0$ or 2 on $X-A$, $= 2$ on A and conversely: we apply Proposition 4 with $g = 2.1_A$, $f = 2.1_X$.

n.b. In the condition of the corollary the quantification may be restricted to $T \subseteq S \subseteq X$ such that $T - A = S - A$. More generally, in Proposition 4 we may restrict the quantification to $T \subseteq S \subseteq X$ such that $S^{g=0} \subseteq T$ (since for $x \in S^{g=0} - T$ if $\bar{S} \cap \bar{X}(x) = \bar{D}(x) + \bar{I}_2(x)$ $\bar{I}_1(x)$ being adjacent to at most $d_G(x) + 1 = |\bar{I}_1(x)| + 1$ components of $\bar{G}_{\bar{S}}$, by Lemma 2.2 $\bar{p}_i(\bar{S} + \bar{I}_1(x)) + |\bar{S} + \bar{I}_1(x)| \geqslant \bar{p}_i(\bar{S}) + |\bar{S}|$).

Proposition 5. *Let* $G = (X, U)$ *be a graph,* $x_0 \in X$ *and* U_0 *be the set of edges of* G *incident to* x_0*: the subsets of* U_0 *contained in unions of (elementary) circuits of* G *containing* x_0 *and pairwise without common vertices other than* x_0*, are the independent sets of a matroid on* U_0*. The rank function of this matroid is given by*

$$rk(V) = |V| - \max_{S \subseteq X - x_0} (p'_{V,x_0}(S) + m_V(x_0, X - S - x_0) - 2|X - S - x_0|)$$

for $V \subseteq U_0$*, where* $p'_{V,x_0}(S)$ *is the number of connected components* C *of* G_S *such that all* (x_0, C)*-edges belong to* V *and* $m_G(x_0, C)$ *is odd* ($m_V(x_0, X - S - x_0)$ *is the number of* $(x_0, X - S - x_0)$ *edges belonging to* V*).*

Proof. *A union of circuits pairwise* without common vertices except perhaps x_0 is a subgraph H of G satisfying the constraint $d_H(x) = 0$ or 2 for all $x \in X - x_0$. Let $\bar{G} = \bar{D}(G) \cup \sum_{x \in \bar{X} - x_0} \bar{G}_{\bar{X}(x)}$ where $\bar{G}_{\bar{X}(x)}$ is the graph defined in the proof of Proposition 5. Then subsets $V \subseteq U_0$ contained in unions of circuits of G, pairwise disjoint on $X - x_0$, are in $1-1$ correspondence with subsets of $\bar{D}(x_0)$ saturated by matchings of $\bar{G}$ saturating $\bar{X}_0 = \bar{X} - \bar{D}(x_0)$: these $V \subseteq U_0$ are thus the independent subsets of a matroid on U_0 isomorphic to $E(\bar{G})/\bar{X}_0$.

Let $V \subseteq U_0$; we denote by $\bar{D}_V(x_0)$ the set of vertices of $\bar{D}(x_0)$ saturated by $\bar{V}$. The rank of V in the above matroid is given by $\text{rk}(V) = r_{E(G)}(\bar{X}_0 + \bar{D}_V(x_0)) - r_{E(\bar{G})}(\bar{X}_0) = r_{E(\bar{G})}(\bar{X}_0 + \bar{D}_V(x_0)) - |\bar{X}_0|$, $|\bar{X}_0|$ being clearly an independent set of $E(\bar{G})$. By Proposition 1 $r_{E(G)}(\bar{X}_0 + \bar{D}_V(x_0)) = |\bar{X}_0 + \bar{D}_V(x_0)| - \max_{\bar{S} \subseteq \bar{X}} (p_{\bar{X}_0 + \bar{D}_V(x_0), i}(\bar{S}) - |\bar{X} - \bar{S}|)$, however we may restrict the quantification in the Max to $\bar{S} \subseteq \bar{X}$ such that $\bar{S} \cap \bar{X}(x) = \bar{D}(x) + \bar{I}_1(x) + \bar{I}_2(x)$, $\bar{D}(x) + \bar{I}_2(x)$ or $\bar{I}_1(x)$ for all $x \in X - x_0$ (cf. the proof of Prop. 2). Furthermore the case $\bar{S} \cap \bar{X}(x) = \bar{D}(x) + \bar{I}_2(x)$ is covered by the case $\bar{S} \cap \bar{X}(x) = \bar{D}(x) + \bar{I}_1(x) + \bar{I}_2(x)$ (since $g = 0$, cf. the note after the corollary of Proposition 4), and by Lemma 2.2 we may suppose $S \supseteq D(x_0)$: finally the quantification may be restricted to $\bar{S} = \bar{X}(S) + \bar{I}_1(X - S - x_0) + \bar{D}(x_0)$ for $S \subseteq X - x_0$. Then

$$\bar{p}_{\bar{X}_0+\bar{D}_V(x_0),i}(\bar{S}) = \bar{p}_{\bar{X}_0+\bar{D}_V(\bar{x}_0),i}(\bar{X}(S) + \bar{D}(x_0)) +$$
$$+ \bar{p}_{\bar{X}_0+\bar{D}_V(x_0),i}(\bar{I}_1(X - S - x_0))$$

$$\bar{p}_{\bar{X}_0+\bar{D}_V(x_0),i}(\bar{I}_1(X - S - x_0) = d_G(X - S - x_0)\,.$$

Among the connected components of $\bar{G}$ induced by $\bar{X}(S) + \bar{D}(x_0)$ $m_V(x_0, X - S - x_0)$ there are one-vertex components induced by $\bar{D}(x_0)$, the others being in $1 - 1$ correspondence with connected components of G_S. In order that for a component C of G_S the corresponding component of $\bar{G}_{\bar{S}}$ has an odd cardinality and is contained in $\bar{X}_0 + \bar{D}_V(x_0)$ we must have $2d_G(C) + 2|C| + m_G(C, x_0) \equiv m_G(C, x_0)$ (mod 2) odd and all (C, x_0)-edges contained in V: $\bar{p}_{\bar{X}_0+\bar{D}_V(x_0),i}(\bar{X}(S) + \bar{D}(x_0)) = = m_V(x_0, X - S - x_0) + p'_{V,x_0}(S)$.

Let us remove the vertex x_0 from G: U_0 being in $1 - 1$ correspondence with $A = \Gamma_G(x_0)$ Proposition 5 has the following equivalent formulation:

Proposition 5′. *Let* $G = (X, U)$ *be a graph and* A *be a subset of* X: *the subsets* $Y \subseteq A$ *such that there exists a set of pairwise disjoint (elementary) paths of length* $\geqslant 1$ *with end-point set containing* Y *and contained in* A, *are the independent set of a matroid on* A. *The rank function of this matroid is given by*

$$rk(Y) = |Y| - \operatorname*{Max}_{S \subseteq X} (p_{A;Y,i}(S) - |X - S| - |X - (S \cup Y)|)$$

for $Y \subseteq A$, *where* $p_{A;Y,i}(S)$ *is the number of connected components* C *of* G_S *such that* $C \cap A \subseteq Y$ *and* $|C \cap A|$ *is odd.*

The particular case $A = X$ may be formulated

Corollary. *Let* $G = (X, U)$ *be graph: the subsets* $Y \subseteq X$ *contained in end-point sets of pairwise disjoint paths of length* 1 *or* 2 *are the independent sets of a matroid on* X. *The rank function of this matroid is given by*

$$rk(Y) = |Y| - \operatorname*{Max}_{S \subseteq X} (p_{Y,i}(S) - |X - S| - |X - (S \cup Y)|)\,.$$

2. Let $\bar{D} = \bar{E}_1 + \bar{E}_2 + \ldots + \bar{E}_{n+n'}$, $n, n' \geqslant 1$, and g_k, f_k be integers such that $g_k \leqslant f_k \leqslant |\bar{E}_k|$ $k = 1, 2, \ldots, n + n'$; then

$$\Big\{g_k \leqslant |\bar{Z} \cap \bar{E}_k| \leqslant f_k \quad k = 1, 2, \ldots, n + n', \quad \sum_{k=1}^{k=n} |\bar{Z} \cap \bar{E}_k| = \sum_{k=n+1}^{k=n+n'} |\bar{Z} \cap \bar{E}_k|\Big\}$$

is a Tutte-class of subsets of $\bar{D}$ (take $\bar{L} = \bar{D} *_{\bar{E}_1} \bar{K}^1_{(g_1,f_1)} *_{\bar{E}_2} \bar{K}^2_{(g_2,f_2)} * \ldots$ $\ldots *_{\bar{E}_{n+n'}} \bar{K}^{n+n'}_{(g_{n+n'},f_{n+n'})} *_{\bar{A}_1+\bar{A}_2+\ldots+\bar{A}_n, \bar{A}_{n+1}+\bar{A}_{n+2}+\ldots+\bar{A}_{n+n'}} \bar{K}_{(=)}$ and $\bar{S}_0 = \bar{I}_1 + \bar{I}_2 + \ldots + \bar{I}_{n+n'} + \bar{A}_1 + \bar{A}_2 + \ldots + \bar{A}_{n+n'} + \bar{I} + \bar{I}'$).

Let $N = (X, U)$ be a transportation network – a directed 1-graph $(U \subseteq X \times X - \{(x, x) | x \in X\})$ together with 2 distinguished vertices, the source a (such that $\omega^-(a) = U \cap (X \times \{a\}) = \phi$) and the sink b (such that $\omega^+(b) = U \cap (\{b\} \times X) = \phi\}$, and f, g two integer-valued functions defined on U such that $g \leqslant f$; then the above construction shows that integer flows of N (i.e. functions φ: $U \mapsto N$ such that $\sum_{u \in \omega^-(x)} \varphi(u) = \sum_{u \in \omega^+(x)} \varphi(u)$ for all $x \in X - a - b$ and $g(u) \leqslant \varphi(u) \leqslant f(u)$ for all $u \in U$ – write $\varphi \in \Phi(N)$) can be reduced to matchings.

Explicitly, let c_0 be any integer such that $c_0 \geqslant f(u)$ for all $u \in U$: for $u = (x, y) \in U$ set $\bar{E}^+(u) = \{(x, u, k) | k = 1, 2, \ldots, c_0\}$, $\bar{E}^-(u) = \{(y, x, k) | k = 1, 2, \ldots, c_0\}$, for $x \in X$ set $\bar{D}^+(x) = \sum_{u \in \omega^+(x)} \bar{E}^+(u)$, $\bar{D}^-(x) = \sum_{u \in \omega^-(x)} \bar{E}^-(u)$, then set $\bar{D} = \sum_{x \in X} \bar{D}^-(x) + \sum_{x \in X} \bar{D}^+(x)$; let $\bar{D}(N)$ be the (undirected) graph on vertex-set $\bar{D}$ with an edge $\bar{u}_k = \{(x, u, k), (y, u, k)\}$ $1 \leqslant k \leqslant c_0$ if and only if $u = (x, y) \in U$; with a function φ: $U \mapsto N$ associate the matching $\bar{\varphi} = \{u_k | u \in U, 1 \leqslant k \leqslant \varphi(u)\}$ of $\bar{D}(N)$. By the above construction the intersection for $x \in X - a - b$ of $\{g(u) \leqslant |\bar{Z} \cap \bar{E}^-(u)| \leqslant f(u)$ for all $u \in \omega^-(x)$, $g(u) \leqslant |\bar{Z} \cap \bar{E}^+(u)| \leqslant f(u)$ for all $u \in \omega^+(x)$, $\sum_{u \in \omega^-(x)} |\bar{Z} \cap \bar{E}^-(u)| = \sum_{u \in \omega^+(x)} |\bar{Z} \cap \bar{E}^+(u)|\}$ is a Tutte-class of subsets of $\bar{D}$; let $\bar{L}$ be the corresponding graph and $\bar{S}_0$ (note that $\bar{S}_0$ is the vertex-set of $\bar{L}$), set $\bar{G} = \bar{D}(N) \cup \bar{L}$, $\bar{X}_0 = (\bar{D} - \bar{D}^+(a) - \bar{D}^-(b)) \cup \bar{S}_0$: then φ: $U \mapsto N$ belongs to $\Phi(N)$ if and only if there is a matching of $\bar{G}$ saturating $\bar{X}_0$ and containing $\bar{\varphi}$.

$E(\bar{G})/\bar{X}_0$ is a matroid on $\bar{D}^+(a) + \bar{D}^-(b)$; $\bar{G}$ being bipartite (easy verification) with $\bar{D}^+(a)$ in one colour-set, $\bar{D}^-(b)$ in the other $E(\bar{G})/\bar{X}_0 = (E(\bar{G})/\bar{X}_0)(\bar{D}^+(a)) \oplus (E(\bar{G})/\bar{X}_0)(\bar{D}^-(b))$ (direct sum) by Ore's theorem 7.4.1 of [10]. Let us denote by $F(N)$ the Z-matroid on $\omega^+(a)$ associated with $(E(\bar{G})/\bar{X}_0)(\bar{D}^+(a))$ and the partition $\bar{D}^+(a) = = \sum_{u \in \omega^+(a)} \bar{E}^+(u)$: $\xi\colon \omega^+(a) \mapsto N$ is an independent function of $F(N)$ if and only if there exists $\varphi \in \Phi(N)$ such that $\xi(u) \leqslant \varphi(u)$ for all $u \in \in \omega^+(a)$ (the fact that functions $\xi \leqslant \varphi \upharpoonright \omega^+(a)$ $\varphi \in \Phi(N)$ define a Z-matroid seems to be known, at least in certain particular cases (cf [4] p. 79 ex. 7), however we have not found any explicit reference to a proof).

Proposition 6. *Let N be a transportation network: (with the above notations) if $\Phi(N) \neq \phi$ then the functions $\xi\colon \omega^+(a) \mapsto N$ such that there exists $\varphi \in \Phi(N)$ such that $\xi(u) \leqslant \varphi(u)$ for all $u \in \omega^+(a)$ are the independent functions of Z-matroid $F(N)$ on $\omega^+(a)$. The rank function is given by*

$$r_{F(N)}(\alpha) = \operatorname*{Min}_{a \in S \subseteq X - b} \Big(\sum_{u \in \omega^+(S)} f_\alpha(u) - \sum_{u \in \omega^-(S)} g(u) \Big)$$

for $\alpha\colon \omega^+(a) \mapsto N$ such that $g(u) \leqslant \alpha(u) \leqslant f(u)$ for all $u \in \omega^+(a)$, where $f_\alpha(u) = \alpha(u)$ for $u \in \omega^+(a)$, $= f(u)$ for $u \in U - \omega^+(a)$ (and $\omega^+(S) = U \cap (S \times (X - S))$, $\omega^-(S) = U \cap ((X - S) \times S)$).

Proof. Let $\alpha\colon \omega^+(a) \mapsto N$; we set $\omega^+(a) = \{u_1, u_2, \ldots, u_n\}$ and $\alpha(u_k) = \alpha_k$ $k = 1, 2, \ldots, n$. Let N' be constructed from N in the following way: let $u_k = (a, a_k)$ $k = 1, 2, \ldots, n$, we add n new vertices $a'_1, a'_2, \ldots, a'_n$; let u_k denote in N' the directed edge (a'_k, a_k) and add $2n$ new directed edges $u'_k = (a, a'_k)$, $v_k = (b, a'_k)$ $k = 1, 2, \ldots, n$. We set $U' = U + \{u'_1, u'_2, \ldots, u'_n\} + \{v_1, v_2, \ldots, v_n\}$; let $f', g'\colon U' \mapsto N$ having f, g for restrictions on U and such that $f'(u'_k) = \alpha_k$, $f'(v_k) = = f(u_k) - \alpha_k$, $g'(u'_k) = g'(v_k) = 0$ $k = 1, 2, \ldots, n$.

By definition $r_{F(N)}(\alpha) = \operatorname*{Max}_{\varphi \in \Phi(N)} \Big(\sum_{k=1}^{k=n} \operatorname{Min}(\alpha_k, \varphi(u_k)) \Big)$; we assert that $r_{F(N)}(\alpha) = \operatorname*{Max}_{\varphi' \in \Phi(N')} \mathscr{V}(\varphi')$ where $\mathscr{V}(\varphi') = \sum_{k=1}^{k=n} \varphi'(u'_k)$ (N' is not exactly a transportation network, since $\omega^-(b) = \{a'_1, a'_2, \ldots, a'_n\}$; by definition here $\Phi(N')$ is the set of $\varphi'\colon U' \mapsto N$ satisfying the constraints of flows for all $u \in U'$ and all $x \in X - a - b + \{a'_1, a'_2, \ldots, a'_n\}$). Let $\varphi \in \Phi(N)$, we define $\varphi'\colon U' \mapsto N$ by $\varphi' = \varphi$ on U and $\varphi'(u'_k) = \operatorname{Min}(\alpha_k, \varphi(u_k))$, $\varphi'(v_k) = (\varphi(u_k) - \alpha_k)^+$ for $k = 1, 2, \ldots, n$: then clearly $\varphi' \in \Phi(N')$ and is such that $\mathscr{V}(\varphi') = \sum \operatorname{Min}(\alpha_k, \varphi(u_k))$. It follows that $\mathscr{V}(N') \geqslant r_{F(N)}(\alpha)$ (where $\mathscr{V}(N') = \operatorname*{Max}_{\varphi' \in \Phi(N')} \mathscr{V}(\varphi')$). For the reverse inequality let $\varphi' \in \Phi(N')$ (not empty by the preceding construction) such that $\mathscr{V}(N') = \mathscr{V}(\varphi')$. Then $\varphi'(v_k) > 0$ implies $\varphi'(u'_k) = \alpha_k$: if not i.e. $\varphi'(u'_{k_0}) < \alpha_{k_0}$, φ'_1 defined by $\varphi'_1 = \varphi'$ on $U' - u'_{k_0} - v_{k_0}$ and $\varphi'_1(u'_{k_0}) = \varphi(u'_{k_0}) + 1$, $\varphi'_1(v_{k_0}) = \varphi'(v_{k_0}) - 1$ belongs to $\Phi(N')$ and is such that $\mathscr{V}(\varphi'_1) = \mathscr{V}(\varphi') + 1$, a contradiction. We have thus $\varphi'(u'_k) = \operatorname{Min}(\alpha_k, \varphi'(u_k))$: then since the restriction of φ' to U belongs to $\Phi(N)$, $\mathscr{V}(N') \leqslant r_{F(N)}(\alpha)$.

For $\alpha \in S' \subseteq X - B + \{a'_1, a'_2, \ldots, a'_n\} = X' - b$ set $c'(S') = \sum_{u \in \omega'^+(S')} f'(S') - \sum_{u \in \omega'^-(S')} g'(S')$. By Ford – Fulkerson's theorem ([7], cf. [1] Chap. 5 Th. 1) $\mathscr{V}(N') = \operatorname*{Min}_{a \in S' \subseteq X' - b} c'(S')$; but

(1) if $a_k \in S'$ and $a'_k \notin S'$ for $a \in S' \subseteq X' - b$ then $c'(S' + a'_k) = c'(S') - \alpha_k + g(u_k) \leqslant c'(S')$

(2) if $a_k \notin S'$ and $a_k \in S'$ for $a \in S' \subseteq X' - b$ then $c'(S' - a'_k) = c'(S') - f(u_k) + \alpha_k \leqslant c'(S')$

so that the quantification may be restricted to S' $a \in S' \subseteq X' - b$ such that $a'_k \in S'$ if and only if $a_k \in S'$ for $k = 1, 2, \ldots, n$. Such a quantification is equivalent to a quantification on S such that $a \in S \subseteq X - b$ and we find for $\mathscr{V}(N')$ the expression given in the statement of the proposition.

The calculation of $r_{F(N)}(\alpha)$ for any $\alpha\colon \omega^+(a) \mapsto N$ is easily reduced to the case of α satisfying $g(u) \leqslant \alpha(u) \leqslant f(u)$ for all $u \in \omega^+(a)$. On the other hand if $g = 0$ the given expression for $r_{F(N)}(\alpha)$ is just Ford – Fulkerson's theorem since in this case given any $\varphi \in \Phi(N)$ and α: $\omega^+(a) \mapsto N$ such that $\alpha(u) \leqslant f(u)$ for all $u \in \omega^+(a)$; there always exist

$\psi \in \Phi(N)$ such that $\psi(u) = \text{Min}(\alpha(u), \varphi(u))$ for all $u \in \omega^+(a)$; this is no more necessarily true if $g \neq 0$.

N.B. All constructions which can be performed on $\bar{D}(G)$ to express constraints on subgraphs of a multigraph G have counterparts directly on G (the corresponding constructions on G amounting to identify in $\bar{D}(G) \cup \bar{L}$ all vertices of $\bar{D}(x)$ to x for every $x \in X$). To obtain an expression for the rank function we have then to apply Proposition 2; however, proofs are perhaps simpler by using $\bar{G}$, reducing all constraints to the two basic ones $0 \leqslant d_H(x) \leqslant 1$, $d_H(x) = 1$. In the case of transportation networks it can be seen that all constraints may be added to the classical ones by using $\bar{D}(N)$ and §3, can also be expressed by the device of a suitable auxiliary network.

REFERENCES

[1] C. Berge, *Graphes et hypergraphes,* Dunod, Paris 1970.

[2] C. Berge, Sur le couplage maximum d'un graphe, C. R. *Acad. Sci. Paris,* 247 (1958), 258-259.

[3] H.H. Crapo – G.-C. Rota, *Combinatorial Geometries,* preliminary ed., the M.I.T. Press, 1970.

[4] F.D.J. Dunstan – A.W. Ingleton – D.J.A. Welsh, Supermatroids, *Proceedings of the* 1972 *Oxford Conference on Combinatorics,* 72-122.

[5] J. Edmonds, Submodular functions, matroids and certain polyhedra, *Combinatorial Structures and their Applications,* (Proc. Calgary International Conference, 1969), 69-87, Gordon – Breach, New-York.

[6] J. Edmonds – D.R. Fulkerson, Transversals and matroid partition, *J. Res. Nat. Bur. Stand.,* 69B (1965), 147-153.

[7] L.R. Ford – D.R. Fulkerson, Maximal flow through a network, *Canad. J. Math.,* 8 (1956), 399-

[8] L. Lovász, Subgraphs with prescribed valencies, *J. Comb. Theory*, 8 (1970), 391-416.

[9] C.St.J.A. Nash-Williams, An applications of matroids to graph theory, *Théorie des Graphes* (Proc. Symp. Rome 1966), 263-265, Dunod, Paris, 1967.

[10] O. Oré, *Theory of Graphs,* Amer. Math. Soc. Colloq. Publ. 38, Providence 1962.

[11] H. Perfect, Independence spaces and combinatorial problems, *Proc. London. Math. Soc.*, 19 (1969), 17-30.

[12] W. T. Tutte. The factorization of linear graphs, *J. London. Math. Soc.*, 22 (1947), 107-111.

[13] W.T. Tutte, The factors of graphs, *Canad. J. Math.*, 4 (1952), 314-328.

[14] W.T. Tutte, A short proof of the factor theorem for finite graphs *Canad. J. Math.*, 6 (1954), 347-352.

COLLOQUIA MATHEMATICA SOCIETATIS JÁNOS BOLYAI
10. INFINITE AND FINITE SETS, KESZTHELY (HUNGARY), 1973.

SOME REMARKS ON SUPPLEMENTARY DIFFERENCE SETS

JENNIFER SEBERRY WALLIS

1. INTRODUCTION AND DEFINITIONS

Let $S_1, S_2, \ldots, S_n$ be subsets of V, a finite abelian group of order v written in additive notation, containing $k_1, k_2, \ldots, k_n$ elements respectively. Write T_i for the totality of all differences between elements of S_i (with repetitions), and T for the totality of elements of all the T_i. If T contains each non-zero element of V a fixed number of times, λ say, then the sets $S_1, S_2, \ldots, S_n$ will be called $n - \{v; k_1, k_2, \ldots, k_n; \lambda\}$ *supplementary difference sets.* Throughout this paper this will be abbreviated as sds.

The parameters of $n - \{v; k_1, k_2, \ldots, k_n; \lambda\}$ supplementary difference sets satisfy

$$\text{(1)} \qquad \lambda(v-1) = \sum_{i=1}^{n} k_i(k_i - 1) .$$

If $k_1 = k_2 = \ldots = k_n = k$ we will write $n - \{v; k; \lambda\}$ to denote the n supplementary difference sets and (1) becomes

(2) $\lambda(v-1) = nk(k-1)$.

We shall be concerned with collections, (denoted by square brackets [ ]) defined on a fixed group V of order v, in which repeated elements are counted multiply, rather than with sets (denoted by braces { }). If T_1 and T_2 are two collections then T_1 and T_2 will denote the result of adjoining the elements of T_1 to T_2 with total multiplicities retained. For example: $x_1, x_2, x_3 \in V$ and $T_1 = [x_1, x_2, x_3, x_3]$, $T_2 = [x_1, x_2, x_4]$ then

(3) $T_1 + T_2 = [x_1, x_1, x_2, x_2, x_2, x_3, x_4]$.

Suppose $x_1, x_2, \ldots, x_v$ are the elements of V ordered in some fixed way. Let X be a subset of V. Further let φ and ψ be two maps from V into a commutative ring with unity (1). Then the matrix $M = [m_{ij}]$ of order v defined by

(4) $m_{ij} = \psi(x_j - x_i)$

will be called *type* 1 and the matrix $N = [n_{ij}]$ of order v defined by

(5) $n_{ij} = \varphi(x_j + x_i)$

will be called *type* 2.

If φ and ψ are defined by

(6) $$\varphi(x) = \psi(x) = \begin{cases} 1 & x \in X, \\ 0 & x \notin X, \end{cases}$$

then M and N will be called the *type* 1 *incidence matrix of* X *(in* V*)* and the *type* 2 *incidence matrix of* X *(in* V*)*, respectively. These are discussed further in [10].

Notation. Let $A = \{a_1, a_2, \ldots, a_k\}$ be a k-set then we will use ΔA for the collection of differences between distinct elements of A, i.e.,

$$\Delta A = [a_i - a_j\colon i \neq j,\ 1 \leqslant i, j \leqslant k].$$

Notation. If $k_1 = k_2 = \ldots = k_n = k$ we will write $n - \{v; k; \lambda\}$ to denote $n - \{v; k_1, k_2, \ldots, k_n; \lambda\}$ sds. If $k_1 = k_2 = \ldots = k_i, k_{i+1} =$

$= k_{i+2} = \ldots = k_{i+j}, \ldots, k_l = \ldots = k_n$ then we will sometimes write $n - \{v;\ i\colon k_1,\ j\colon k_{i+1}, \ldots;\ \lambda\}$. A (v, k, λ) difference set repeated n-times will be denoted $n - (v, k, \lambda)$.

A *balanced incomplete block design* or BIBD (v, b, r, k, λ) may be considered to be a $(0, 1)$ matrix B of size $v \times b$, with row sum r and column sum k, such that the inner product of any pair of distinct row vectors is λ. B satisfies

$$BB^T = (r - \lambda)I + \lambda J\,,$$

where I is the identity matrix and J the matrix of all 1's.

An *Hadamard matrix* H of order h has every element $+1$ or -1 and satisfies $HH^T = hI_h$. A *skew-Hadamard matrix* $H = I + R$ is an Hadamard matrix with $R^T = -R$.

2. CYCLOTOMY

We now turn to Storer [2; p. 24-25] for the elementary theory of cyclotomy:

Let x by a primitive root of $F = GF(q)$ where $q = q^\alpha = ef + 1$ is a prime power. Write $G = \langle x \rangle \setminus \{0\}$. The *cyclotomic classes* C_i in F are:

$$C_i = \{x^{es+i}\colon s = 0, 1, \ldots, f-1\} \quad i = 0, 1, \ldots, e-1\,.$$

We note that C_i are pairwise disjoint and their union is G.

For fixed i and j, the *cyclotomic number* (i, j) is defined to be the number of solutions of the equation

$$z_i + 1 = z_j \quad (z_i \in C_i,\ z_j \in C_j)\,,$$

where $1 = x^0$ is the multiplicative unit of F. That is (i, j) is the number of ordered pairs s, t such that

$$x^{es+i} + 1 = x^{et+j} \quad (0 \leqslant s,\ t \leqslant f-1)\,.$$

Note that the number of times

$$x^{es+i} - x^{et+k} \in C_j$$

is the number of solutions of

$$x^{es+i} - x^{et+k} = x^{er+j}$$

$$x^{et+k} + x^{er+j} = x^{es+i}$$

$$x^{e(t-r)+k-j} + 1 = x^{e(s-r)+i-j}$$

which is the cyclotomic number $(k-j, i-j)$. Using [2, p. 25]

$$(k-j, i-j) = (e-(j-k), (i-k)-(j-k)) = (j-k, i-k)\,.$$

Hence

$$\Delta C_i = [x^{es+i} - x^{et+i}\colon\ s \neq t,\ 0 \leqslant s,\ t \leqslant f-1] =$$

$$= [x^{es+i} - x^{et+i}\colon\ 0 \leqslant s,\ t \leqslant f-1]/f\{0\} =$$

$$= (-i, 0)C_0 + (1-i, 0)C_1 + (2-i, 0)C_2 + \ldots$$

$$\ldots = (0, 0)C_i + (1, 0)C_{i+1} + (2, 0)C_{i+2} + \ldots$$

and

$$\Delta(C_i - C_j) = (-i, 0)C_0 + (1-i, 0)C_1 + \ldots$$

$$\ldots + (-j, 0)C_0 + (1-j, 0)C_1 + \ldots$$

$$\ldots + (-j, i-j)C_0 + (1-j, i-j)C_1 + \ldots$$

$$\ldots + (-i, j-i)C_0 + (1-i, j-i)C_1 + \ldots$$

$$= (0, 0)C_j + (1, 0)C_{j+1} + \ldots$$

$$\ldots + (0, 0)C_i + (1, 0)C_{i+1} + \ldots$$

$$\ldots + (0, i-j)C_j + (1, i-j)C_{j+1} + \ldots$$

$$\ldots + (0, j-i)C_i + (1, j-i)C_{i+1} + \ldots$$

3. SOME KNOWN APPLICATIONS OF SDS

Perhaps the most important application of *sds* is in constructing BIBD's. Here the module theorems of Bose are used, viz.,

If there exist $m - \{v; k; \lambda\}$ sds then there exists a BIBD $(v, b = mv, r = mk, k, \lambda)$.

If there exist $(t + s) - \{u = kt/s + k - 1;\ t{:}\ k,\ s{:}\ (k - 1); \lambda = (k - 1)s\}$ sds then there exists a BIBD

$$(kt/s + k, b = u(s + t), r = us, k, \lambda = (k - 1)s)\,.$$

In [6] it is pointed but that

If there exist $(n + m) - \{v; n{:}\ k, m{:}\ (k + r); \lambda\}$ sds then there exist $\left[n\binom{k+r-2}{r} + m\binom{k+r}{r}\right] - \left\{v; k; \binom{k+r-2}{r}\lambda\right\}$ sds.

Similarly, if there exist $n - \{v; k_1, k_2, \ldots, k_n; \lambda\}$ sds there exist $n' - - \{v; k; \lambda'\}$ sds for some n', λ' and $k \leqslant \min(k_1, k_2, \ldots, k_n)$. Also in [6] and [7] a number of constructions, by various authors, are quoted.

Sds have also proved to be very important in constructing Hadamard matrices. In particular the existence of

(i) $4 - \left\{v; k_1, k_2, k_3, k_4; \sum_{i=1}^{4} k_i - v\right\}$ sds; or

(ii) $4 - \left\{v; k_1, k_2, k_3, k_4; \sum_{i=1}^{4} k_i - v - 1\right\}$ sds (here necessarily $k_i = \frac{1}{2}(v \pm 1)$ for v odd or $k_1 = \frac{1}{2}(v \pm 2)$, $k_2 = k_3 = k_4 = \frac{1}{2}v$ for v even); or

(iii) $2 - \left\{v; k_1, k_2; k_1 + k_2 - \frac{1}{2}v\right\}$ sds (here v is necessarily even);

or

(iv) $2 - \left\{v; k_1, k_2; k_1 + k_2 - \frac{1}{2}(v + 1)\right\}$ sds (here v is necessarily odd)

imply the existence of Hadamard matrices of order

(i) $4v$;

(ii) $4(v+1)$;

(iii) $2v$;

(iv) $2(v+1)$; respectively.

Sds which can be used to form Hadamard matrices of various types can be found, using cylotomy, in the following cases:

(i) $2-\{4f+1; 2f; 2f-1\}$ sds when $4f+1$ is a prime power, [10, p. 282];

(ii) $2-\{4f+1; 2f; 2f-1\}$ sds when $4f+1 \equiv 5 \pmod 8 = s^2+4$ is a prime power, [4], [5];

(iii) $4-\{4f+1; 2f; 4f-2\}$ sds when $4f+1 \equiv 5 \pmod 8 = s^2+36$ is a prime power, [5];

(iv) $2-\{2f+1; f; f-1\}$ sds when $2f+1 \equiv 5 \pmod 8$ is a prime power, [10, p. 321];

(v) $2-\{8f+1; 4f; 4f-1\}$ sds when $8f+1 = p^t$ is a prime power, $p \equiv 5 \pmod 8$, $t \equiv 2 \pmod 4$ [10, p. 323];

(vi) $4-\{8f+1; 4f; 2(2f-1)\}$ sds when $8f+1 \equiv 9 \pmod{16}$ is a prime power, [10, p. 334].

4. CONSTRUCTIONS FOR SDS

The case $e = 2$.

The cyclotomic matrices and constraints are:

	0	1	
0	A	B	$2A = f-1$
1	A	A	$A+B = f$

f odd

	0	1	
0	A	B	$2B = f$
1	B	B	$A+B = f-1$

f even

Using these cyclotomic matrices, we have

Lemma 1. *If* $q = 2f + 1$ *(f even), then* C_0, C_1 *are* $2 - \{2f + 1; 2 - \{2f + 1; f; f - 1\}$ sds.

Proof.

$$\Delta C_0 + \Delta C_1 = AC_0 + BC_1 + BC_0 + AC_1 = (f - 1)G \setminus \{0\} .$$

Lemma 2. *If* $q = 2f + 1$ *(f even), then* C_0, $\{0\} + C_0$ *are* $2 - \{2f + 1; f, f + 1; f\}$ sds.

Proof.

$$\Delta C_0 + \Delta(\{0\} + C_0) = AC_0 + BC_1 + (A + 2)C_0 + BC_1 =$$
$$= fG \setminus \{0\} .$$

The case $e = 4$:

The cyclotomic matrices and constraints are

	0	1	2	3
0	A	B	C	D
1	E	E	D	B
2	A	E	A	E
3	E	D	B	E

$$2A + 2E = f - 1$$
$$B + D + 2E = f$$
$$A + B + C + D = f$$

f odd

	0	1	2	3
0	A	B	C	D
1	B	D	E	E
2	C	E	C	E
3	D	E	E	B

$$A + B + C + D = f - 1$$
$$B + D + 2E = f$$
$$2C + 2E = f$$

f even

where for f odd, we have

$$16A = q - 7 + 2s$$

$$16B = q + 1 + 2s - 8t$$

$$16C = q + 1 - 6s$$

$$16D = q + 1 + 2s + 8t$$

$$16E = q - 3 - 2s$$

and for f even

$$16A = q - 11 - 6s$$

$$16B = q - 3 + 2s + 8t$$

$$16C = q - 3 + 2s$$

$$16D = q - 3 + 2s - 8t$$

$$16E = q + 1 - 2s$$

where $q = s^2 + 4t^2$, $s \equiv 1 \pmod 4$ is the proper representation of q if $p \equiv 1 \pmod 4$; the sign of t is ambiguously determined.

Lemma 3. *If* $q = 4f + 1$ *(f odd), then* C_0, C_1 *or* C_0, C_3 *are* $2 - \left\{4f + 1; f; \frac{1}{2}(f - 1)\right\}$ sds.

Proof.

$$\begin{aligned} \Delta C_0 + \Delta C_1 = \Delta C_0 + \Delta C_3 &= \\ &= AC_0 + EC_1 + AC_2 + EC_3 + \\ &+ EC_0 + AC_1 + EC_2 + AC_3 = \\ &= \frac{1}{2}(f - 1)G \setminus \{0\}. \end{aligned}$$

Lemma 4. *If* $q = 4f + 1$ *(f odd), then* $\{0\} + C_0$, $\{0\} + C_1$ *are* $2 - \left\{4f + 1; f + 1; \frac{1}{2}(f + 1)\right\}$ sds.

Proof.

$$\Delta(\{0\} + C_0) + \Delta(\{0\} + C_1) = (A + 1)C_0 + EC_1 + (A + 1)C_2 + EC_3 + EC_0 + (A + 1)C_1 + EC_2 + (A + 1)C_3$$

Lemma 5. *If* $q = 4f + 1$ *(f odd), then* $A_1 = C_0 + C_1$ *and* $A_2 = C_0 + C_3$ *are* $2 - \{4f + 1; 2f; 2f - 1\}$ sds *with the property that* $a \in A_i \Rightarrow -a \notin A_i$, $i = 1, 2$.

Proof.

$$\begin{aligned}\Delta(C_0 + C_1) + \Delta(C_0 + C_3) = {} & AC_0 + EC_1 + AC_2 + EC_3 + \\ & + EC_0 + AC_1 + EC_2 + AC_3 + \\ & + BC_0 + EC_1 + EC_2 + DC_3 + \\ & + EC_0 + DC_1 + BC_2 + EC_3 + \\ & + AC_0 + EC_1 + AC_2 + EC_3 + \\ & + EC_0 + AC_1 + EC_2 + AC_3 + \\ & + DC_0 + BC_1 + EC_2 + EC_3 + \\ & + EC_0 + EC_1 + DC_2 + BC_3 = \\ & = (2f - 1)G \setminus \{0\} .\end{aligned}$$

Since f is odd, $-1 \in C_2$, and we have the result.

Lemma 6. *If* $q = s^2 + 4t^2 = 4f + 1$ *(f odd), then* $A = C_0 + C_1$ *and* $|t|$ *copies of* $B = C_0 + C_2$ *are* $(|t| + 1) - \left\{4f + 1; 2f; \frac{1}{2}(|t| + 1)(2f - 1)\right\}$ sds *with the properties* $a \in A \Rightarrow -a \notin A$, $b \in B \Rightarrow -b \in B$.

Proof. $a \in A \Rightarrow -a \notin A$, $b \in B \Rightarrow -b \in B$ follows because f is odd and $-1 \in C_2$. We choose our primitive root so that t is negative, i.e., $|t| = -t$.

$$
\begin{aligned}
\Delta(C_0 + C_1) + |t|\Delta(C_0 + C_2) = {} & AC_0 + EC_1 + AC_2 + EC_3 + \\
& + EC_0 + AC_1 + EC_2 + AC_3 + \\
& + BC_0 + EC_1 + AC_2 + DC_3 + \\
& + EC_0 + DC_1 + BC_2 + EC_3 + \\
& + |t|\{2AC_0 + 2EC_1 + 2AC_2 + 2EC_3 + \\
& + \quad CC_0 + DC_1 + AC_2 + BC_3 + \\
& + \quad AC_0 + BC_1 + CC_2 + DC_3\} = \\
& = \frac{1}{16}[(4q - 12 - 8t)(C_0 + C_2) + \\
& + (4q - 12 + 8t)(C_1 + C_3) + \\
& + |t|(4q - 20)(C_0 + C_2) + \\
& + |t|(4q - 4)(C_1 + C_3)] = \\
& = \frac{1}{16}[(4q(1 - t) - 12 + 12t)G \setminus \{0\}] = \\
& = \frac{1}{2}(2f - 1)(|t| + 1)G \setminus \{0\} .
\end{aligned}
$$

Lemma 7. *If* $q = 25 + 4t^2 = 4f + 1$ *(f odd), then* $C_0 + C_2$, C_0 *are* $2 - \left\{4f + 1; 2f, f; \frac{1}{4}(5f - 3)\right\}$ sds.

Proof.

$$
\begin{aligned}
\Delta(C_0 + C_2) + \Delta C_0 = {} & \frac{1}{16}((4q - 20)(C_0 + C_2) + (4q - 4)(C_1 + C_3) + \\
& + A(C_0 + C_2) + E(C_1 + C_3)) = \\
& = \frac{1}{16}((5q - 27 + 2s)(C_0 + C_2) + \\
& + (5q - 7 - 2s)(C_1 + C_3)) =
\end{aligned}
$$

$$= \frac{1}{16}((5q - 17)G \setminus \{0\}) =$$

$$= \frac{1}{4}(5f - 3)G \setminus \{0\}, \quad \text{where} \quad s = 5 .$$

Lemma 8. *If* $q = 9 + 4t^2 = 4f + 1$ *(f odd) then* $C_0 + C_2$, C_1 *are* $2 - \{4f + 1; 2f, f; \frac{1}{4}(5f - 3)\}$ sds.

Proof.

$$\Delta(C_0 + C_2) + \Delta C_1 = (5q - 23 - 2s)(C_0 + C_2) +$$

$$= (5q - 11 + 2s)(C_1 + C_3) =$$

$$= \frac{1}{16}((5q - 17)G \setminus \{0\}) =$$

$$= \frac{1}{4}(5f - 3)G \setminus \{0\}, \quad \text{where} \quad s = -3 .$$

Lemma 9. *If* $q = s^2 + 4t^2 = 4f + 1$ *(f odd), then* $|s - 1|$ *copies of* $(C_0 + C_1)$ *and*

(i) for $s > 0$ $4|t|$ *copies of* C_0,

(ii) *for* $s < 0$ $4|t|$ *copies of* C_1,

are $(|s - 1| + 4|t|) - \{4f + 1; |s - 1|: 2f, |t|: f; \frac{1}{2}((2f - 1)(|s - 1| + |t|)\}$ sds.

Proof. We choose the primitive root so that $t = |t|$. Now

$$|s - 1|\Delta(C_0 + C_1) = \frac{1}{16}(|s - 1|(4q - 12 - 8t)(C_0 + C_2) +$$

$$+ |s - 1|(4q - 12 + 8t)(C_1 + C_3))$$

$$4t\Delta C_0 = \frac{1}{16}(4t(q - 7 + 2s)(C_0 + C_2) +$$

$$+ 4t(q - 3 - 2s)(C_1 + C_3))$$

$$4t\Delta C_1 = \frac{1}{16}(4t(q-3-2s)(C_0+C_2)+$$

$$+4t(q-7+2s)(C_1+C_3)),$$

so for $s > 0$, $|s-1| = s-1$ and

$$|s-1|\Delta(C_0+C_1)+4t\Delta C_0 = \frac{1}{4}((q-3)(s-1+t)-2t)G/\{0\};$$

while for $s < 0$, $|s-1| = 1-s$ and

$$|s-1|\Delta(C_0+C_1)+4t\Delta C_1 = \frac{1}{4}((q-3)(1-s+t)-2t)G/\{0\},$$

which gives the result.

Lemma 10. *If* $q = 4f+1$ *(f even), then* C_0, C_1, C_2, C_3 *are* $4-\{4f+1; f; f-1\}$ sds *and* $\{0\}+C_0$, $\{0\}+C_1$, $\{0\}+C_2$, $\{0\}+C_3$ *are* $4-\{4f+1; f+1; f+1\}$ sds.

Proof.

$$\Delta C_0 + \Delta C_1 + \Delta C_2 + \Delta C_3 = (A+B+C+D)G \setminus \{0\} =$$

$$= (f-1)G \setminus \{0\}.$$

Lemma 11. *If* $q = 1+4t^2 = 4f+1$ *(f even), then* C_0+C_2, C_1, C_3 *are* $3-\left\{4f+1; 2f, f; \frac{1}{2}(3f-2)\right.$ sds *and* $\{0\}+C_0+C_2$, $\{0\}+C_1$, $\{0\}+C_3$ *are* $3-\left. 4f+1; 2f+1, f+1, f+1; \frac{1}{2}(3f+2)\right\}$ sds.

Proof.

$$\Delta(C_0+C_2)+\Delta C_1+\Delta C_3 = (A+B+C+D+2C)(C_0+C_2)+$$

$$+(A+B+C+D+2E)(C_1+C_3) =$$

$$= \left(f-1+\frac{1}{4}(2f-1+s)\right)(C_0+C_2)+$$

$$+\left(f-1+\frac{1}{4}(2f+1-s)\right)(C_1+C_3)+$$

$$= \frac{1}{2}(3f-2)G \setminus \{0\}, \quad \textbf{with} \quad s = 1.$$

Lemma 12. *If* $q = 25 + 4t^2 = 4f + 1$ *(f even), then* $C_0 + C_2$, $\{0\} + C_1$, $\{0\} + C_3$ *are* $3 - \{4f + 1; 2f, f + 1; 3f/2\}$ sds *(here* $s = 5$*).*

Lemma 13. *If* $q = 9 + 4t^2 = 4f + 1$ *(f even), then* $\{0\} + C_0 + C_2$, $\{0\} + C_1$, $\{0\} + C_3$ *are* $3 - \{4f + 1; 2f + 1, f, f; 3f/2\}$ sds *(here* $s = - 3$*).*

Lemma 14. $q = s^2 = 4f + 1$ *(f even), then* $C_0 + C_1$, $C_2 + C_3$ *are* $2 - \{4f + 1; 2f; 2f - 1\}$ sds *and* $\{0\} + C_0 + C_1$, $\{0\} + C_2 + C_3$ *are* $2 - \{4f + 1; 2f + 1; 2f + 1\}$ sds.

Proof.

$$\begin{aligned}\Delta(C_0 + C_1) + \Delta(C_2 + C_3) = {} & \Delta C_0 + \Delta C_1 + \Delta C_2 + \Delta C_3 + \\ & + BC_0 + EC_1 + EC_2 + DC_3 + \\ & + EC_0 + DC_1 + BC_2 + EC_3 + \\ & + EC_0 + DC_1 + BC_2 + EC_3 + \\ & + BC_0 + EC_1 + EC_2 + DC_3 + \\ & = (2f - 1)G \setminus \{0\}, \quad \text{where} \quad t = 0 .\end{aligned}$$

Lemma 15. *If* $q = 4f + 1$ *(f even), then* $C_0 + C_1$, $C_1 + C_2$, $C_2 + C_3$, $C_3 + C_0$ *are* $4 - \{4f + 1; 2f; 2(2f - 1)\}$ sds.

The case $e = 6$:

The cyclotomic matrix for $e = 6$ (f odd) is

	0	1	2	3	4	5
0	A	B	C	D	E	F
1	G	H	I	E	C	I
2	H	J	G	F	I	B
3	A	G	H	A	G	H
4	G	F	I	B	H	J
5	H	I	E	C	I	G

$$2A + 2G + 2H = f - 1$$
$$B + F + G + H + I + J = f$$
$$C + E + G + H + 2I = f$$
$$A + B + C + D + E + F = f,$$

and in this case, the cyclotomic numbers are given by the relations

$$72A = 2q - 22 - 8x - 2a + 2c$$
$$72B = 2q + 2 - 3a - c - 9b + 9d$$
$$72C = 2q + 2 - 8x + a - c + 24y - 3b - 9d$$
$$72D = 2q + 2 + 24x + 6a + 2c$$
$$72E = 2q + 2 - 8x + a - c - 24y + 3b + 9d$$
$$72F = 2q + 2 - 3a - c + 9b - 9d$$
$$72G = 2q - 10 + 4x + a - c + 12y + 3b + 9d$$
$$72H = 2q - 10 + 4x + a - c - 12y - 3b - 9d$$
$$72I = 2q + 2 + 4x - 2a + 2c$$
$$72J = 2q + 2 - 12x + 6a + 2c$$

where $q = x^2 + 3y^2$, $4q = a^2 + 3b^2 = c^2 + 27d^2$.

Lemma 16. *If* $q = 6f + 1$ *(f odd),* C_0, C_1, C_2 *are* $3 - \left\{6f + 1; f; \frac{1}{2}(f - 1)\right\}$ sds *and* $\{0\} + C_0$, $\{0\} + C_1$, $\{0\} + C_2$ *are* $3 - \left\{6f + 1; f + 1; \frac{1}{2}(f + 1)\right\}$ sds.

Lemma 17. *If* $q = 6f + 1$ *(f odd),* $C_0 + C_i$, $C_1 + C_{i+1}$, $C_2 + C_{i+2}$, $i = 1, 2,$ *or* 3, *are* $3 - \{6f + 1; 2f; 2f - 1\}$ sds *and* $\{0\} + C_0 + C_i$, $\{0\} + C_1 + C_{i+1}$, $\{0\} + C_2 + C_{i+2}$, $i = 1, 2,$ *or* 3 *are* $3 - \{6f + 1; 2f + 1; 2f + 1\}$ sds.

The case $e = 8$:

The cyclotomic matrix for $e = 8$ (f odd) is

	0	1	2	3	4	5	6	7
0	A	B	C	D	E	F	G	H
1	I	J	K	L	F	D	L	M
2	N	O	N	M	G	L	C	K
3	J	O	O	I	H	M	K	B
4	A	I	N	J	A	I	N	J
5	I	H	M	K	B	J	O	O
6	N	M	G	L	C	K	N	O
7	J	K	L	F	D	L	M	I

$$2A + 2I + 2J + 2N = f - 1$$

$$B + H + I + J + K + M + 2O = f$$

$$C + G + K + L + M + 2N + O = f$$

$$D + F + I + J + K + 2L + M = f$$

$$A + B + C + D + E + F + G + H = f$$

and the cyclotomic numbers are given by the relations:

I. If 2 is a fourth power in G	II. If 2 is not a fourth power in G
$64A = q - 15 - 2x$	$64A = q - 15 - 10x - 8a$
$64B = q + 1 + 2x - 4a + 16y$	$64B = q + 1 + 2x - 4a - 16b$
$64C = q + 1 + 6x + 8a - 16y$	$64C = q + 1 - 2x + 16y$
$64D = q + 1 + 2x - 4a - 16y$	$64D = q + 1 + 2x - 4a - 16b$
$64E = q + 1 - 18x$	$64E = q + 1 + 6x + 24a$
$64F = q + 1 + 2x - 4a + 16y$	$64F = q + 1 + 2x - 4a + 16b$
$64G = q + 1 + 6x + 8a + 16y$	$64G = q + 1 + 2x - 16y$
$64H = q + 1 + 2x - 4a - 16y$	$64H = q + 1 + 2x - 4a + 16b$
$64I = q - 7 + 2x + 4a$	$64I = q - 7 + 2x + 4a + 16y$
$64J = q - 7 + 2x + 4a$	$64J = q - 7 + 2x + 4a - 16y$
$64K = q + 1 - 6x + 4a + 16b$	$64K = q + 1 + 2x - 4a$
$64L = q + 1 + 2x - 4a$	$64L = q + 1 - 6x + 4a$
$64M = q + 1 - 6x + 4a - 16b$	$64M = q + 1 + 2x - 4a$
$64N = q - 7 - 2x - 8a$	$64N = q - 7 + 6x$
$640 = q + 1 + 2x - 4a$	$640 = q + 1 - 6x + 4a$

where x, y, a and b are specified by:

I. $q = x^2 + 4y^2$, $x \equiv 1 \pmod 4$ is the unique proper representation of $q = p^\alpha$ if $p \equiv 1 \pmod 4$; otherwise,

$$q = (\pm p^{\alpha/2})^2 + 4.0^2; \quad \text{i.e.,} \quad x = \pm p^{\alpha/2}, \quad y = 0.$$

II. $q = a^2 + 2b^2$, $a \equiv 1 \pmod 4$ is the unique proper representation of $q = p^\alpha$ if $p \equiv 1$ or $3 \pmod 8$; otherwise,

$$q = (\pm p^{\alpha/2})^2 + 2.0^2; \quad \text{i.e.,} \quad a = \pm p^{\alpha/2}, \quad b = 0.$$

The signs of y and b are ambiguously determined.

Lemma 18. *If* $q = 8f + 1$ *(f odd),* C_0, C_1, C_2, C_3 *are* $4 - 4 - \left\{8f + 1; f; \frac{1}{2}(f - 1)\right\}$ sds *and* $\{0\} + C_0$, $\{0\} + C_1$, $\{0\} + C_2$ $\{0\} + C_3$ *are* $4 - \left\{8f + 1; f + 1; \frac{1}{2}(f + 1)\right\}$ sds.

Lemma 19. *If* $q = 8f + 1 = 1 + 2b^2$ *($f \equiv 1 \pmod 4$) and* 2 *is a 4-th power, then* C_0, C_1 *are* $2 - \left\{8f + 1; f; \frac{1}{4}(f - 1)\right\}$ sds *and* $\{0\} + C_0$, $\{0\} + C_1$ *are* $2 - \left\{8f + 1; f + 1; \frac{1}{4}(f + 1)\right\}$ sds.

Proof.

$$\begin{aligned}\Delta C_0 + \Delta C_1 &= (A + J)(C_0 + C_4) + (I + A)(C_1 + C_5) + \\ &+ (N + I)(C_2 + C_6) + (J + N)(C_3 + C_7) = \\ &= \frac{1}{4}(f - 1)G \setminus \{0\}, \quad \text{when} \quad a = 1 .\end{aligned}$$

Lemma 20. *If* $q = 8f + 1 = 49 + 2b^2$ *($f \equiv -1 \pmod 4$) and* 2 *is a 4-th power, then* $\{0\} + C_0$, $\{0\} + C_1$ *are* $2 - \left\{8f + 1; f + 1; \frac{1}{4}(f + 1)\right\}$ sds.

Proof.

$$\begin{aligned}\Delta(\{0\} + C_0) + \Delta(\{0\} + C_1) &= (A + J + 1)(C_0 + C_4) + \\ &+ (A + I + 1)(C_1 + C_5) + \\ &+ (N + I)(C_2 + C_6) + \\ &+ (J + N)(C_3 + C_7) + \\ &= \frac{1}{4}(f + 1)G \setminus \{0\}, \quad \text{when} \quad a = -7.\end{aligned}$$

Lemma 21. *If* $q = 8f + 1 = 81 + 4y^2 = 9 + 2b^2$ *(f odd) is a prime power and* 2 *is a fourth power in* G, C_0 *and* $\{0\} + C_0$ *are each repeated* 4 *times, are* $8 - \{8f + 1; 4{:}f, 4{:}(f + 1); f\}$ sds.

Proof.

$$\Delta C_0 + \Delta(\{0\} + C_0) = (2A + 1)(C_0 + C_4) + 2I(C_1 + C_5) +$$
$$+ 2N(C_2 + C_6) + 2J(C_3 + C_7) =$$
$$= [(2q + 34 - 4x)(C_0 + C_4) +$$
$$+ (2q - 14 + 4x + 8a)(C_1 + C_3 + C_5 + C_7) +$$
$$+ (2q - 14 - 4x - 16a)(C_2 + C_6)]/64 =$$
$$= \frac{(2q - 2)}{64} G \setminus \{0\} \quad \text{with} \quad a = -3, \; x = 9 .$$

Lemma 22. *If* $q = 8f + 1 = (-3)^2 + 2b^2$ *(i.e.,* $a = -3$*) then* C_0, C_1, $\{0\} + C_0$, $\{0\} + C_1$ *each repeated twice are* $8 - \{8f + 1; 4{:}f, 4{:}(f + 1); f\}$ sds *when* 2 *is a fourth power in* G.

Proof.

$$\Delta C_0 + \Delta C_1 + \Delta(\{0\} + C_0) + \Delta(\{0\} + C_1) =$$
$$= [(4q + 20 + 8a)(C_0 + C_1 + C_4 + C_5) +$$
$$+ (4q - 28 - 8a)(C_2 + C_3 + C_6 + C_7)]/64 =$$
$$= \frac{4q - 4}{64} G \setminus \{0\} \quad \text{with} \quad a = -3 .$$

Lemma 24. *Suppose* $q = 8f + 1$ *(*f *odd) is a prime power, then if in the unique proper representation of* q

(i) $y = 0$, $C_0 + C_1$, $C_3 + C_6$, $C_2 + C_3$, $C_0 + C_5$ *are* $4 - \{8f + 1; 2f; 2f - 1\}$ sds; *and if*

(ii) $b = 0$, $C_0 + C_1$, $C_3 + C_6$, $C_1 + C_2$, $C_4 + C_7$ *are* $4 - \{8f + 1; 2f; 2f - 1\}$ sds.

Proof. (i) If 2 is not a fourth power

$$\Delta(C_0 + C_1) + \Delta(C_3 + C_6) = (A + J + B + I)(C_0 + C_4) +$$
$$+ (I + A + J + H)(C_1 + C_5) +$$

$$+ (N + I + O + M)(C_2 + C_6) +$$

$$+ (J + N + O + K)(C_3 + C_7) +$$

$$+ (I + N + K + L)(C_0 + C_4) +$$

$$+ (N + J + L + M)(C_1 + C_5) +$$

$$+ (J + A + F + I)(C_2 + C_6) +$$

$$+ (A + I + D + J)(C_3 + C_7) =$$

$$= \frac{1}{8} [(q - 5 + 2y - 2b)(C_0 + C_4) +$$

$$+ (q - 5 - 2y + 2b)(C_1 + C_5) +$$

$$+ (q - 5 + 2y + 2b)(C_2 + C_6) +$$

$$+ (q - 5 - 2y - 2b)(C_3 + C_7)] .$$

So

$$\Delta(C_0 + C_1) + \Delta(C_3 + C_6) + \Delta(C_2 + C_3) + \Delta(C_0 + C_5) =$$

$$= \frac{1}{8} [(2q - 10 + 4y)(C_0 + C_4 + C_2 + C_6) +$$

$$+ (2q - 10 - 4y)(C_1 + C_3 + C_5 + C_7)] =$$

$$= (2f - 1)G \setminus \{0\} \quad \text{when} \quad y = 0 .$$

Also

$$\Delta(C_0 + C_1) + \Delta(C_3 + C_6) + \Delta(C_1 + C_2) + \Delta(C_4 + C_7) =$$

$$= \frac{1}{8} [(2q - 10 - 4b)(C_0 + C_4) +$$

$$+ (2q - 10)(C_1 + C_5) +$$

$$+ (2q - 10 + 4b)(C_2 + C_6) +$$

$$+ (2q - 10)(C_3 + C_7)] =$$

$$= (2f - 1)G \setminus \{0\} \quad \text{when} \quad b = 0 .$$

(ii) If 2 is a fourth power

$$\Delta(C_0 + C_1) + \Delta(C_3 + C_6) = \frac{1}{8}[(q-5+2y+2b)(C_0+C_4) +$$
$$+ (q-5-2y-2b)(C_1+C_5) +$$
$$+ (q-5+2y-2b)(C_2+C_6) +$$
$$+ (q-5-2y+2b)(C_3+C_7)]$$

and proceeding as before we have the result.

5. AN APPLICATION OF SUPPLEMENTARY DIFFERENCE SETS TO THE CONSTRUCTION OF BALANCED INCOMPLETE BLOCK DESIGNS

The following theorem is a generalization of a result shown us by S. Lin.

Theorem 23. *Suppose* $B_1, B_2, \ldots, B_n$ *are* $n - \{v; k; \lambda\}$ *supplementary difference sets from an abelian group* G *of order* v. *Let* $A_1, A_2, \ldots, A_k$ *be obtained from* B_1 *by successively removing one element.*

Write Ⓐ$_i$ *for the incidence matrix of* A_i *and* Ⓑ$_j$ *for the incidence matrix of* B_j. *Then*

$$C = \begin{matrix} \overbrace{Ⓐ_1\ Ⓐ_2 \cdots Ⓐ_k}^{\text{each repeated } \alpha \text{ times}} & \overbrace{Ⓑ_1}^{\beta \text{ times}} & \overbrace{Ⓑ_2 \cdots Ⓑ_n}^{\text{each repeated } \gamma \text{ times}} \\ 11 \ldots\ldots 11 & 00 \ldots & \ldots\ldots 0 \end{matrix}$$

is the incidence matrix of a BIBD $(v+1, \alpha v(v+1), \alpha v k, k, \alpha k(k-1))$

$$r = \alpha v k = (\beta + (n-1)\gamma)k + k(k-1)\alpha$$

$$\alpha = k(k-1)\alpha = \lambda\gamma$$

$$\gamma = \alpha(k-2) + \beta\,.$$

Proof. Each matrix $\textcircled{A}_i$ has $k-1$ elements per row and column, so C has k elements per column and

$$r = \alpha k v = \alpha k(k-1) + (\beta + (n-1)\gamma)k$$

elements per row.

Let $\Delta\beta_1$ be differences between elements of B_1, then

$$\sum_{i=1}^{k} \Delta A_i = (k-2)\Delta B_1 \, .$$

Now since the sets $B_1, \ldots, B_n$ are supplementary difference sets,

$$\lambda(v-1) = nk(k-1)$$

and

$$\sum_{j=1}^{n} \Delta B_j = \Delta B_1 + \sum_{j=2}^{n} \Delta B_j = \lambda G/\{0\} \, .$$

So we want

$$\alpha \sum_{i=1}^{k} \Delta A_i + \beta \Delta B_1 + \gamma \sum_{j=2}^{n} \Delta B_j = \mu G/\{0\} \, ,$$

that is

$$(\alpha(k-2) + \beta)\Delta B_1 + \gamma(\lambda G/\{0\} - \Delta B_1) = \mu G/\{0\}$$

and

$$\mu = \lambda\gamma, \quad \alpha(k-2) + \beta = \gamma$$

(unless B_1 is a difference set).

Also we need the inner product of the last row of C with the other rows to be μ, so

$$\mu = \alpha k(k-1) \, .$$

Now we wish to simplify

(7) $$\mu = \alpha k(k-1) = \lambda\gamma$$

(8) $\quad r = \alpha v k = (\beta + (n-1)\gamma)k + k(k-1)\alpha$

(9) $\quad \lambda(v-1) = nk(k-1)$

(10) $\quad \mu v = r(k-1)$

(11) $\quad \alpha(k-2) + \beta = \gamma$

If $\alpha = 1$, then (11) and (8) give

$$n\gamma = v - 1$$

$$\beta = \gamma - k + 2\,,$$

and a BIBD $(v+1, b, vk, k, k(k-1))$ = BIBD $(v+1, v(v+1), vk, k, k(k-1))$.

Corollary. *If* p *is an odd prime power, there exist* $2-\{p; \frac{p-1}{2}; \frac{p-3}{2}\}$ *and* $2-\{p; \frac{p+1}{2}; \frac{p+1}{2}\}$ sds *so there exist*

$$\text{BIBD}\left(p+1, p(p+1), p\left(\frac{p-1}{2}\right), \left(\frac{p-1}{2}\right), \left(\frac{p-1}{2}\right)\left(\frac{p-3}{2}\right)\right)$$

and

$$\text{BIBD}\left(p+1, p(p+1), p\left(\frac{p+1}{2}\right), \left(\frac{p+1}{2}\right), \left(\frac{p-1}{2}\right)\left(\frac{p+1}{2}\right)\right)$$

Corollary. *If there exists a* (v, k, λ) *and a* $(v, v-k, v-2+\lambda)$ *difference set, then there exist*

$$\text{BIBD}\,(v+1, v(v+1), vk, k, k(k-1))$$

and

$$\text{BIBD}\,(v+1, v(v+1), v(v-k), v-k, (v-k)(v-k-1))$$

Corollary. *If* $p = ef + 1$ *is a prime power, there exist*

$$e - \{ef+1; f; f-1\} \quad \textit{and} \quad e - \{ef+1; f+1; f+1\}$$

supplementary difference sets and so there exist

$$\text{BIBD}\,(ef+2,(ef+2)(ef+1),(ef+1)f,f,f(f-1))$$

and

$$\text{BIBD}\,(ef+2,(ef+1)(ef+2),(ef+1)(f+1),f+1,f(f+1))\,.$$

Comment. It was pointed out to us by Mr. Dean Hoffman that in the above corollaries the BIBD's obtained from complementary sds are not themselves complementary.

Clearly the method of this section also has application to $n-\{v;k_1,k_2,\ldots,k_n;\lambda\}$ sds.

6. AN APPLICATION OF SDS IN CONSTRUCTING SKEW-HADAMARD MATRICES

Lemma 24. *Let* $p=4f+1=s^2+4t^2\equiv 5\ (\text{mod}\ 8)$ *be a prime power. Suppose there exists a skew-Hadamard matrix* $S+I$ *of order* $|t|+1$. *Then there exists a skew-Hadamard matrix of order* $(|t|+1)(p+1)$.

Proof. From lemma 6

$$C_0+C_1 \quad \text{and} \quad |t| \quad \text{copies of} \quad C_0+C_2$$

are $(|t|+1)-\ 4f+1;2f;\frac{1}{2}(|t|+1)(2f-1)\}$ supplementary difference sets with the property that

$$x\in C_0+C_1\Rightarrow -x\notin C_0+C_1\,,$$

and

$$y\in C_0+C_2\Rightarrow -y\in C_0+C_2\,.$$

Then the type 1 $(1,-1)$ incidence matrix A and the type 2 $(1,-1)$ incidence matrix B of C_0+C_1 and C_0+C_2 satisfy.

$$AJ=BJ=-J,\ \ (A+I)^T=-(A+I),\ \ B^T=B\,,$$

$$AA^T+|t|BB^T=(|t|+1)(p+1)I-(|t|+1)J\,.$$

Now $SS^T=|t|I,\ S^T=-S$ and we consider

$$H = \left[\begin{array}{c|c} S+I & (S+I) \times e \\ \hline (S-I) \times e^T & \bar{I} \times -A + S \times B \end{array} \right],$$

where e is the $1 \times p$ matrix of ones. It is easily verified that H is a skew-Hadamard matrix of order $(|t|+1)(p+1)$.

REFERENCES

[1] Đ.C. Hunt – J. Wallis, Cyclotomy, Hadamard arrays and supplementary difference sets, *Proceedings Second Manitoba Conference on Numerical Mathematics,* (1972), 351-381.

[2] T. Storer, *Cyclotomy and Difference Sets,* Markham Publishing Company, Chicago, 1967.

[3] G. Szekeres, Cyclotomy and complementary difference sets, *Acta Arithmetica,* 18 (1971), 349-353.

[4] J. Wallis, Amicable Hadamard matrices, *J. Combinatorial Th. Ser. A*, 11 (1971), 296-298.

[5] J. Wallis, A note on Amicable Hadamard matrices, *Utilitas Math.*, 3 (1973), 119-125.

[6] J. Wallis, On supplementary difference sets, *Aeq. Math.*, 8 (1973), 242-257.

[7] J. Wallis, A note on supplementary difference sets, *Aeq. Math.* 10 (1974), 46-49.

[8] J.S. Wallis, Some matrices of Williamson type, *Utilitas Math.* 4 (1973), 147-154.

[9] J.S. Wallis, Construction of Williamson type matrices, (to appear).

[10] J. Wallis – A.L. Whiteman, Some classes of Hadamard matrices with constant diagonal, *Bull. Austral. Math. Soc.*, 7 (1972), 233-249.

COLLOQUIA MATHEMATICA SOCIETATIS JÁNOS BOLYAI
10. INFINITE AND FINITE SETS, KESZTHELY (HUNGARY), 1973.

SOME REMARKS ON A PAPER OF GRÜNBAUM

H. WALTHER

In this paper $V(G)$, $E(G)$ denotes the vertex set resp. edge set of the finite graph G, $\nu(G)$ is the cardinality of $V(G)$. $c(G), p(G)$ denotes the maximum length of a circuit resp. path in G. We want to denote by C_k^j resp. P_k^j ($j \geqslant 1$, $k \geqslant 1$ are integers) the least integer ν that has the following property: There is a k-connected graph G with $\nu(G) = \nu$ such that for any j vertices of G there exists a circuit resp. path of length $c(G)$ resp. $p(G)$ that contains none of these j vertices. $\bar{C}_k^j$, $\bar{P}_k^j$ are the corresponding numbers if G runs only over planar graphs. The problem whether $P_k^1 = \infty$ holds for all $k \geqslant 1$ was asked by T. Gallai, (see Problem 4 in [8], p. 362) and an analogous question whether $\bar{C}_k^2 = \infty$ holds for $k \geqslant 1$ was asked by H. Sachs (see (Problem 50 in [8] p. 368, where the restriction that G is planar is missing although he stated his problem explicitly restricting it to planar graphs). I decided the problem of Gallai in the negative direction. The following results are known so far:

$C_3^1 \leq 10$	(Petersen-graph)	$P_1^1 \leq 12$	[6]
$C_2^2 \leq 220$	[5]	$P_1^2 \leq 108$	[6]
$C_3^2 \leq 90$	[2], [6]	$P_3^2 \leq 324$	[2]
$\bar{C}_2^1 \leq 105$	[4]	$P_3^1 \leq 90$	[6]
$\bar{C}_3^1 \leq 124$	[2]	$\bar{P}_1^1 \leq 25$	[4]
$\bar{C}_4^1 = \infty$	[3]	$\bar{P}_2^1 \leq 82$	[7]
		$\bar{P}_3^1 \leq 484$	[2]
		$\bar{P}_4^1 = \infty$	[3]

The aim of this paper is to give a negative solution of Sachs' problem by proving the following:

Theorem 1. $\bar{C}_3^2 \leq 15004$.

We mention that the analogous result for $\bar{P}_3^2$ can be proved similarly to the proof of Theorem 1 (see Theorem 2).

The proof of Theorem 1. First we construct a graph T with $v(T) = 15004$ and then we prove that T has the desired properties. We start with the graph of Figure 1, that is denoted by G_1. (We mention that G_1 was in effect constructed by Grinberg in [1] and was used by Grünbaum in [2] to prove that $\bar{C}_3^1 \leq 124$.)

Now we substitute each vertex u of G_1 by a copy G^u of G_2 which can be seen in Figure 2. We can do this since any vertex of G_1 has degree 3. (Note, that the role of o, o', o'' is symmetrical in G_2.) T has $121 \times 124 = 15004$ vertices, T is planar and 3-connected. All we have to prove is, that

(+) Given any 2 vertices x_0, x_1 of T there is a circuit C of T containing neither x_0 nor x_1 and having cardinality $c(T)$.

To prove (+) first we make some remarks.

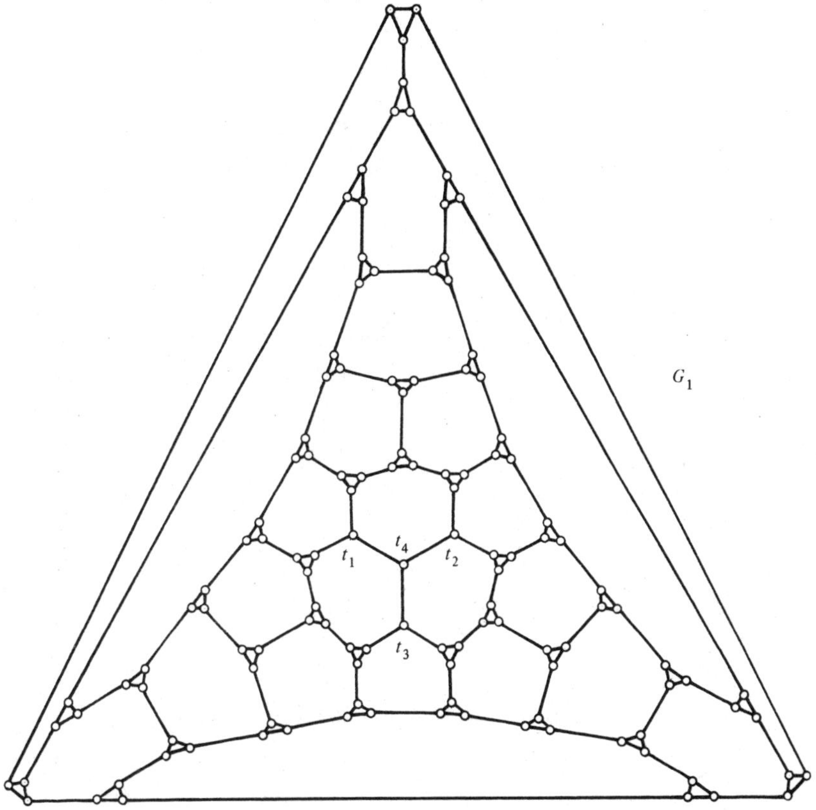

Figure 1

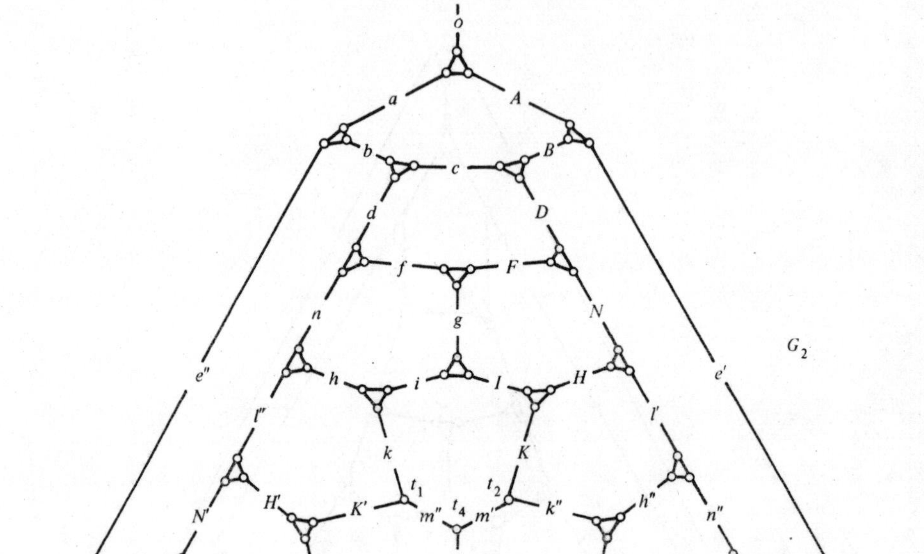
o
a
A
b
B
c
d
D
f
F
g
n
N
G_2
e″
e′
h
i
I
H
l″
l′
k
K
t_1
t_2
t_4
m″
m′
K′
k″
H′
h″
N′
n″
m
I′
i″
D′
F′
g′
i′
k′
t_3
K″
I″
g″
f″
d″
B′
b″
A′
c′
f′
h′
H″
F″
c″
a″
o′
d′
n′
l
N‴
D″
o″
a′
b′
B″
A″
e

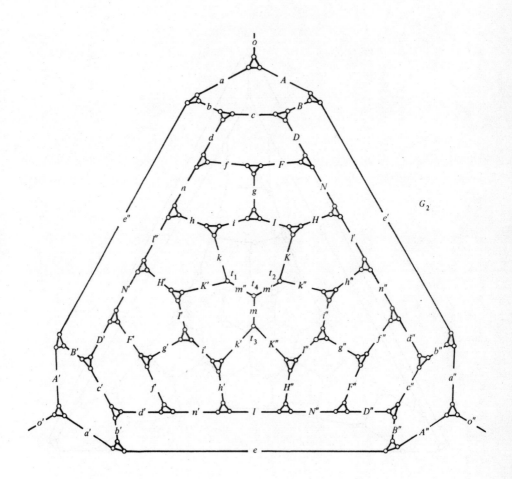

Figure 2

(i) If we omit from G_1 the "top"-triangel we get G_2.

When proving $\bar{C}_3^1 \leqslant 124$ in [2] Grünbaum in fact proved somewhat more, namely he proved:

(ii) If C is a longest circuit of G_1 then it contains 121 vertices, i.e. C omits exactly 3 vertices of G_1. If the vertices x_1, x_2, x_3 span a triangle Δ in G_1, then there is a longest circuit C_Δ of G_1 (i.e. a circuit C_Δ of cardinality $c(G_1)$) which avoids x_1, x_2, x_3 (and contains all other vertices.) Finally for any vertex t_i $(i = 1, 2, 3, 4)$ there is a longest circuit C_i of G_1 which avoids t_i and 2 other t_j's (and contains all other vertices of G_1). (Note, that t_1, t_2, t_3, t_4 are the only 4 vertices which are not contained in some triangle of G_1.)

To state our next remark we need the expression "P is a path through G_2", that means, that P is a path in G_2, and 2 of the 3 "half-edges" o, o', o'' lie in P (in fact P connects these 2 half-edges through G_2). We will also use the expression "P is a path through G^u" for any copy G^u of G_2, finally we say, that P is a longest path through G_2 if P is a path through G_2 having maximal cardinality.

(iii) C is a longest circuit of T iff $C = \bigcup_{v \in V(C_0)} P^v$ where C_0 is a longest circuit of G_1, P^v is a longest path through G^v that connects those 2 half-edges of G^v which correspond to the 2 edges $v', v'' \in E(C_0)$ incident to v.

Now the following remark is an easy consequence of (ii).

(iv) Any longest path P through G_2 avoids exactly 3 vertices. For any vertex u of G_2 there exists a longest path R^u through G_2 avoiding u; and $G_2 - \{o, o', o''\}$ has a Hamiltonian circuit.

Throughout this proof we fix an R^u for any u with the above property.

We are going to prove the following

Lemma. *Given any 2 edges* y, y' *of* G_1. G_1 *has a circuit* $C_{yy'}$

of cardinality $c(G_1)$ *containing neither* y *nor* y'.

Before proving this lemma we show how to deduce (+) from it. Suppose, x_0 and x_1 are 2 vertices of T. We have to find a circuit C in T that contains neither x_0 nor x_1 and has cardinality $c(T)$. We distinguish 2 cases:

(α) x_0 and x_1 lie in the same G^u. Let C_u be a circuit of G_1 not containing u and having cardinality $c(G_1)$. (We have such a C_u according to (ii).) For any vertex v of C_u let v', v'' be the 2 edges of C_u incident to v, and let v'_0, v''_0 be the 2 corresponding half-edges in G^v, finally let P^v be any longest path through G^v connecting v'_0 and v''_0. (Such a P^v exists.) Clearly $C = \bigcup_{v \in V(C_u)} P^v$ avoids the whole G^u and so it also avoids x_0, x_1.

C is a longest path according to (iii) proving (+) in this case.

(β) x_i lies in G^{u_i}, $u_0 \neq u_1$. According to (iv) there is a longest path R^{x_i} through G^{u_i} that avoids x_i. Clearly, it always avoids one of the 3 half-edges of G^{u_i}. We denote this half-edge by e'_i and we denote the edge of G_1 by e_i corresponding to e'_i. (Note, that $e_0 = e_1$ may occur.) According to the lemma there is a longest circuit $C_{e_0 e_1}$ of G_1 containing neither e_0 nor e_1. For any $v \in V(C_{e_0 e_1})$ let v', v'' be the 2 edges of $C_{e_0 e_1}$ incident to v, and v'_0, v''_0 the two half-edges of G^v corresponding to v', v''. Let P^v be any longest path through G^v connecting v'_0, v''_0 if $v \neq u_0, u_1$ and let $P^v = R^{x_i}$ if $v = u_i$. (Note, that P^v connects the 2 half-edges v'_0, v''_0 in this case, too.) Thus $C = \bigcup_{v \in V(C_{e_0 e_1})} P^v$ avoids both x_0 and x_1. By (iii) C is a longest circuit of T. This proves (+) in the second case, too.

So to prove our theorem we only have to prove the lemma. To do this we make some simple observations:

(v) To prove the lemma we may assume $y \neq y'$.

Now let r_1, r_2, r_3 be any 3 vertices of G_1 which span a triangle, we denote the edges incident to these three vertices as indicated on Figure 3.

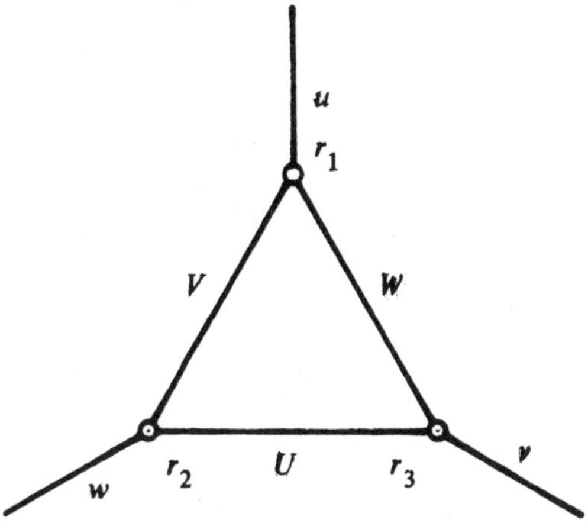

Figure 3

It is obvious that any longest cirucit of G_1 avoiding u (v, resp. w) must also avoid U (V, resp. W). Thus

(vi) We may assume that y and y' are edges indicated by some letters in Figure 2. To prove the lemma for such a y and y' one only has to prove that there is a longest path $P_{yy'}$ through G_2 avoiding both y and y'.

Suppose y and y' are incident to some vertices of the same triangle Δ. Then any longest circuit C_Δ avoiding Δ must also avoid y and y'. Since such a C_Δ exists by (iii), we get the following

(vii) We may assume, that y and y' are not incident in G_3, see Figure 4, and $\{y, y'\} \not\subset \{o, o', o''\}$.

Here G_3 is the graph which arises if each triangle of G_2 is contracted to a single vertex (t_i remains unchanged). We define the expression "P is a longest path through G_3" as we defined it for G_2. Clearly if

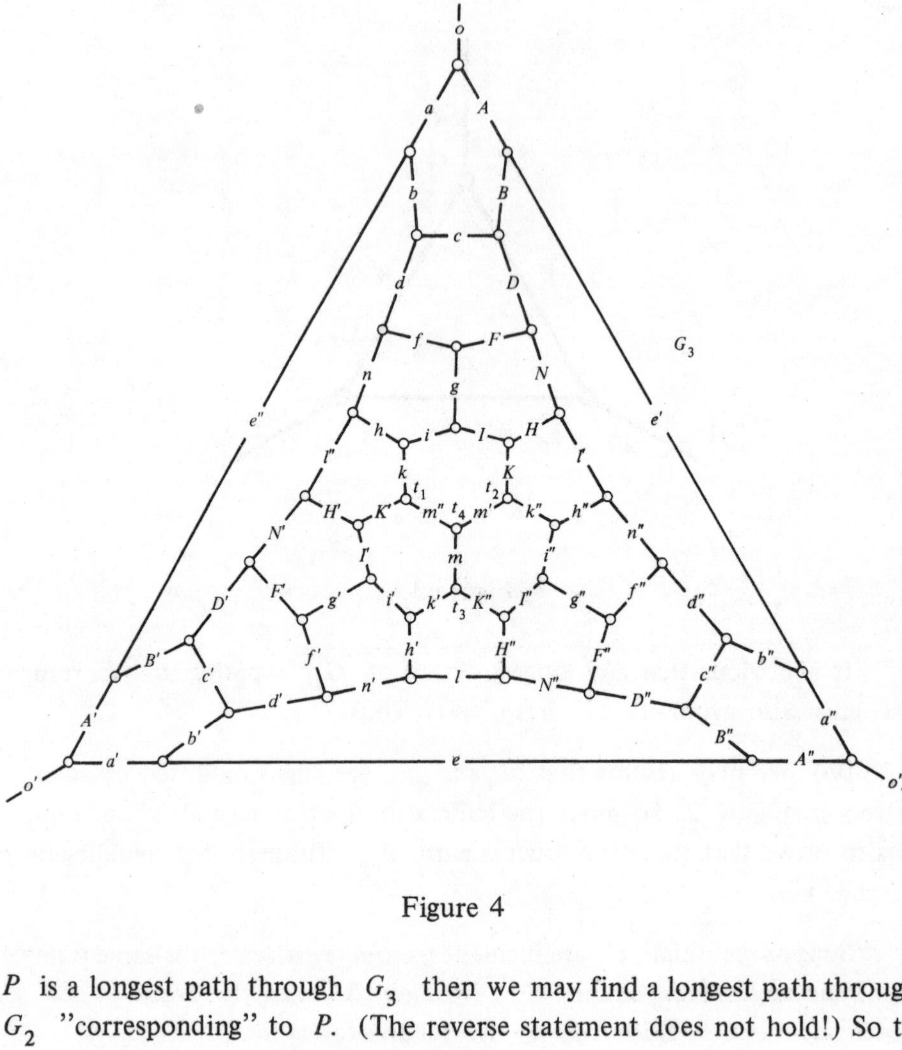

Figure 4

P is a longest path through G_3 then we may find a longest path through G_2 "corresponding" to P. (The reverse statement does not hold!) So to prove the lemma we only have to prove that

(++) Given any 2 edges y_0 and y_1 of G_3 such that $\{y_0, y_1\} \not\subset \not\subset \{o, o', o''\}$ and y_0 and y_1 are not adjacent in G_3, there always exists a longest path $P_{y_0 y_1}$, through G_3 avoiding both y_0 and y_1.

Now let W_i $(i = 1, \ldots, 6)$ be the path through G_3 containing exactly those edges of G_3 which are *not* mentioned in the i-th row below:

W_1: $AaA'a''cc'D''d''eFf'g''Hh''ii'K'K''lm'nN'o$

W_2: $AA''BbB'b'c''e'FF'f''H'Ii'i''kK''l'm'nn'N''o'$

W_3: $A''a'b''Bc'D''de''f''f'fh''I''i'iK'KlmN'nNo$

W_4: $aA'cc'c''e'ef'F''f''gg''h'H''h''I'KkmNnN'o''$

W_5: $Aa'B''b''cc'e''Ff'F''h''II'I''kk'l'lm'nN'n''o''$

W_6: $a'A''bB'cdd'd''D''e'Fg'g''hh'H''Ik''K'l'mN'o$

W_i are longest paths through G_3.

On the other hand note the following: G_3 has 3 symmetries $(0°, +120°, -120°)$ and also 3 axial symmetries. (++) will be established if we prove that given any 2 edges y_0, y_1 of G_3 not adjacent in G_3 such that $\{y_0, y_1\} \not\subset \{o, o', o''\}$ holds we can always find an i and a symmetrical image $\{\bar{y}_0, \bar{y}_1\}$ of the pair $\{y_0, y_1\}$ for which $\bar{y}_0$ and $\bar{y}_1$ lie in the i-th row above (i.e. $\bar{y}_0$ and $\bar{y}_1$ do not lie in W_i.)

Since this can be checked mechanically, we omit the details. So (++) and theorem 1 are proved.

Finally we mention that we can prove the following

Theorem 2. $\bar{P}_3^2 \leqslant 58564$.

The graph T proving this assertion can be constructed on the following way: We start with the tetrahedron-graph and replace each of its vertices by a copy of G_2. We get the graph T'. If we again replace each of T''s vertices by a copy of G_2, we get the graph T which has the desired properties.

REFERENCES

[1] E. Grinberg, Plane homogeneous graphs of degree three without Hamiltonian circuits. *Latvian Math. Yearbook,* 4 (1968), 51-58 (Russian).

[2] B. Grünbaum, Vertices missed by longest paths or circuits. *J. Combin. Theory,* (to appear).

[3] W.T. Tutte, A theorem on planar graphs. *Trans. Amer. Math. Soc.,* 82 (1956), 99-116.

[4] H. Walther, Über die Nichtexistenz eines Knotenpunktes, durch den alle längsten Wege eines Graphen gehen. *J. Combin. Theory,* 6 (1969), 1-6.

[5] H. Walther, Über die Nichtexistenz zweier Knotenpunkte eines Graphen, die alle längsten Kreise fassen. *J. Combin. Theory,* 8 (1970), 330-333.

[6] H. Walther, Extremale Kreise und Wege in Graphen. *Habilitationsschrift TH Ilmenau,* 1969.

[7] T. Zamfirescu, A two-connected planar graph without concurrent longest paths. *J. Combin. Theory,* 13 (1972), 116-121.

[8] P. Erdős – G. Katona, *Theory of Graphs. Proceedings of the Coll. held at Tihany,* Hungary, 1966.

COLLOQUIA MATHEMATICA SOCIETATIS JÁNOS BOLYAI
10. INFINITE AND FINITE SETS, KESZTHELY (HUNGARY), 1973.

INEQUALITIES AND ASYMPTOTIC BOUNDS FOR RAMSEY NUMBERS II

J. YACKEL*

0. INTRODUCTION

Professor Erdős has stimulated a great deal of interest in extremal problems in graph theory, [2], [3]. In particular, his work on Ramsey's theorem which dates back to 1935 in his joint paper with Szekeres [4] has influenced a large amount of research. To this date no one has improved on Erdős' lower bound for Ramsey numbers [1] which was established by the ingenious probabilistic methods which he invented.

This paper is devoted to obtaining upper bounds for Ramsey numbers. Our methods are an extension of the methods of Graver and Yackel [5]. The main result which we obtain is that the Ramsey number $R(n_1, n_2)$ satisfies the inequality

$$R(n_1, n_2) \leq C\left(\frac{\log\log n_2}{\log n_2}\right)^{n_1 - 2} n_2^{n_1 - 1}$$

*This work was partially supported by the Office of Naval Research Contract N00014-67-A-0226-00014 at Purdue University.

where C is bounded for all n_1. This inequality is only of value when n_1 is fixed and n_2 is large. In particular, it does not cover the case of $R(n, n)$ which was treated in Yackel, [6].

1. DEFINITIONS

We consider Ramsey's theorem as it pertains to partitions of pairs of elements of a finite set S into two disjoint classes. Our presentation will use the graph theoretic representation in which the pairs of elements of S in one class determine the edges of the graph.

Definition 1. $I(G)$, the independence number of the graph G, is the maximum number of points of G that can be chosen so that no two are joined by an edge.

Definition 2. $C(G)$, the clique number of the graph G, is the maximum number of points in any complete subgraph of G.

Definition 3. G is a Ramsey (n_1, n_2)-graph if $n_1 > C(G)$ and $n_2 > I(G)$.

Definition 4. $R(n_1, n_2)$ is the largest integer such that there is a Ramsey (n_1, n_2)-graph on $R(n_1, n_2)$ points.

Our primary concern in this paper is to study the local connectedness of a Ramsey (n_1, n_2)-graph and to use this connectedness in obtaining bounds for $R(n_1, n_2)$. To facilitate this study we define several symbols for notation to be used throughout this paper.

Definition 5. With respect to a given independent set H of G the *support* of a point is that subset of H adjacent (joined by an edge) to that point. A *i-point* is a point of G for which the support contains exactly i members.

2. BASIC INEQUALITIES

Our main inequality results from the application of Proposition 6, p. 140 of Graver and Yackel [5]. We will first obtain extensions of Lemma 9, p. 155 of [5], for which it will be necessary to study the connectedness to estimate the intersection of the support of i-points.

We now prove several lemmas for the purpose of estimating the intersection of support of i-points. To that end we let $p(i;\ n_1, n_2)$ be the maximum number of i-points with respect to an independent $(n_2 - 1)$ set for a Ramsey (n_1, n_2)-graph. We also denote by $e(i, j;\ n_1, n_2)$ the maximum number of edges which join two i-points, for which the intersection of the support of the two points is j, all with respect to an independent $(n_2 - 1)$ set for a Ramsey (n_1, n_2)-graph.

Throughout we will assume that n_1 is much smaller than n_2.

Lemma 1. $e(i, 0;\ n_1, n_2) \leqslant p(i;\ n_1, n_2)R(n_1 - 1, n_2)/2$.

Proof. This is an upper bound for the number of edges joining two i-points.

Lemma 2. $e(i, 1;\ n_1, n_2) \leqslant ip(i;\ n_1, n_2)R(n_1 - 2, n_2)/2$.

Proof. Each i-point is adjacent to i elements of the independent set. Among the i-points adjacent to any one element of the independent set the maximum valence is $R(n_1 - 2, n_2)$ since the set of all points adjacent to any point in a Ramsey (n_1, n_2)-graph is a Ramsey $(n_1 - 1, n_2)$-graph and the valence of any point of a Ramsey $(n_1 - 1, n_2)$-graph is at most $R(n_1 - 2, n_2)$. Thus $ip(i;\ n_1, n_2)R(n_1 - 2, n_2)/2$ gives an upper bound for the number of edges between two i-points both of which have common support of at least one point. This completes the proof of the lemma.

Lemma 3.

$$e(i, 2;\ n_1, n_2) \leqslant \frac{p(i;\ n_1, n_2)}{2}\binom{i}{2} R(n_1 - 2, n_2)\frac{a(n_1 - 1, n_2)}{n_2 - 1}$$

where $a(n_1 - 1, n_2)$ *is the average support of the points in a Ramsey* $(n_1 - 1, n_2)$*-graph with respect to an independent* $(n_2 - 1)$ *set.*

Proof. Let an arbitrary i-point, z, be given. Choose two of the support points, say x and y. There are at most $R(n_1 - 2, n_2)$ edges with z as one end point and for which the other endpoint has support containing x. Next we find an independent $n_2 - 1$ set among the points adjacent to z and including x, y.

The points adjacent to z form a Ramsey $(n_1 - 1, n_2)$-graph and the average support of those points with respect to the independent set choosen is $a(n_1 - 1, n_2)$. As we consider all points adjacent to z we thus find that

$$\frac{\binom{n_2 - 3}{a(n_1 - 1, n_2) - 2}}{\binom{n_2 - 1}{a(n_1 - 1, n_2) - 1}} R(n_1 - 2, n_2)$$

of those points will also have support containing x and y.

When we take account of all i-points, all pairs x, y and the fact that each i-point is counted twice we find that

$$e(i, 2;\ n_1, n_2) \leqslant \frac{p(i;\ n_1, n_2)}{2} \binom{i}{2} R(n_1 - 2, n_2) \frac{a(n_1 - 1, n_2)}{n_2 - 1}$$

as stated.

Lemma 4.

$$e(i, j;\ n_1, n_2) \leqslant$$
$$\leqslant \frac{p(i;\ n_1, n_2)}{2} \binom{i}{j} R(n_1 - 2, n_2) \left(\frac{a(n_1 - 1, n_2)}{n_2 - 1}\right)^{j-1} .$$

Proof. The argument is the same as in Lemma 3. We must consider all adjacencies and so we obtain an average but with less freedom when j points of support must be common.

For a fixed value of n_1 and as n_2 is taken to be large it is convenient to write

$$R(n_1, n_2) \leqslant f(n_1) n_2^{n_1 - 1}$$

see Graver and Yackel [5], or Yackel [6].

Theorem 1. *For* n_1 *a fixed integer and* n_2 *sufficiently large we have*

$$p(i;\ n_1, n_2) \leqslant (f(n_1-1))^{\frac{i-1}{i}}\ Cn_2^{\,(n_1-1)-\frac{n_1-2}{i}}$$

where C *is bounded for all* n_1, n_2 *and* $i \leqslant \log n_2$.

Proof. As a direct application of proposition 6 in [5] we determine that

$$(1)\qquad \begin{aligned} k\binom{n_2-1}{k} &\geqslant p(i;\ n_1,n_2)\binom{n_2-1-i}{k-i} - \\ &- \sum_{j=0}^{i} e(i,j;\ n_1,n_2)\binom{n_2-1-2i+j}{k-2i+j}. \end{aligned}$$

We leave it as an exercise for the reader to verify that

$$\frac{\sum_{j=0}^{i} e(i,j;\ n_1,n_2)\binom{n_2-1-2i+j}{k-2i+j}}{\binom{n_2-1-i}{k-i}} =$$

$$= e(i,0;\ n_1,n_2)\frac{\binom{n_2-1-2i}{k-2i}}{\binom{n_2-1-i}{k-i}}(1+o(1))$$

as n_2 approaches ∞ *for* n_1 fixed and $i \leqslant \log n_2$. Lemma 1, Lemma 2, and Lemma 3, together with the fact that $a(n_1,n_2) = o(\log n_2)$ for fixed n_1 as n_2 approaches ∞, suffice for that assertion. If $a(n_1,n_2)$ were not $o(\log n_2)$ then our principal result, Theorem 2, would follow with no more additional work.

To complete the proof for the theorem we need only estimate $e(i,0;\ n_1,n_2)$ by $\frac{p(i;\ n_1,n_2)}{2} f(n_1-1)n_2^{n_1-2}$ using Lemma 1 and the remark preceeding the statement of this theorem. Then we make standard estimates of the quantities in (1) using (2) as well to complete the upper bound. In obtaining the final result we must choose the value of k. Thus we choose

$$k = \text{int}\left[\frac{n_2^{1-\frac{n_1-2}{i}}}{f(n_1-1)^{\frac{1}{i}}}\right]$$

in making our final estimates. This completes the theorem.

3. ASYMPTOTIC BOUNDS

In this section the bounds obtained for $p(i;\, n_1, n_2)$ are used to determine bounds on $R(n_1, n_2)$. Since the results in Section 2 are obtained piecemeal for each $p(i;\, n_1, n_2)$ there is some work yet to be done in order to find the best bounds available from the results stated in Theorem 1.

Theorem 2. $R(n_1, n_2) \leqslant C\left(\frac{\log\log n_2}{\log n_2}\right)^{n_1-2} n_2^{n_1-1}$ *for large values of* n_2, *where* C *is an absolute constant for all* n_1.

Proof. For any Ramsey (n_1, n_2)-graph, with respect to an independent $(n_2 - 1)$ set we find

$$\text{(3)}\qquad \sum_{i=1}^{n_2-1} ip(i;\, n_1, n_2) \leqslant n_2 R(n_1 - 1, n_2)$$

by counting edges.

Since $R(n_1, n_2) = n_2 + \sum_{i=1}^{n_2-1} p(i;\, n_1, n_2)$ we are interested in finding an upperbound for $\sum_{i=1}^{n_2-1} p(i;\, n_1, n_2)$. The upperbound can most easily be established by stating the linear programming problem

Find the maximum of $\sum_{i=1}^{n_2-1} p(i;\, n_1, n_2)$ where

$$p(i;\ n_1, n_2) \leqslant (f(n_1 - 1))^{\frac{i-1}{i}}\ Cn_2^{(n_1 - 1) - \frac{n_1 - 2}{i}}$$

for $\quad i = 1, 2, \ldots, \log n_2$

$$p(i;\ n_1, n_2) \leqslant n_2 R(n_1 - 1, n_2)/i \quad \text{for} \quad n_2 > i > \log n_2$$

and

$$\sum_{i=1}^{n_2 - 1} ip(i;\ n_1, n_2) \leqslant n_2 R(n_1 - 1, n_2)\,.$$

Now, we do no intend to solve this problem but will instead use the dual problem to find an easy and useful bound on the maximum. We accomplish this by finding a feasible point for the dual problem and estimating the value of the objective function for the dual problem at that point. This procedure gives a reasonable bound on the maximum of the original problem when it is carefully done.

The dual problem states:

Find the minimum of

$$\sum_{i=1}^{\log n_2} z_i (f(n_1 - 1))^{\frac{i-1}{i}}\ Cn_2^{(n_1 - 1) - \frac{n_1 - 2}{i}} +$$

$$+ \sum_{i = \log n_2 + 1}^{n_2 - 1} z_i n_2 R(n_1 - 1, n_2)/i + z_{n_2} n_2 R(n_1 - 1, n_2)$$

where $z_i + iz_{n_2} \geqslant 1$ for $i = 1, 2, \ldots, n_2 - 1$.

To establish the theorem we propose the choice

$$z_i = \begin{cases} 1 - \dfrac{1}{M} & \text{if} \quad i \leqslant M \\ 0 & \text{if} \quad i > M \end{cases} \qquad \text{for} \quad i = 1, 2, \ldots, n_2 - 1$$

and $z_{n_2} = \dfrac{1}{M}$. Next we propose $M = \dfrac{n_1 - 2}{3} \dfrac{\log n_2}{\log \log n_2}$ to complete the

the description of the feasible solution for the minimum problem. Finally, with this choice we find

$$\sum_{i=1}^{M} z_i (f(n_1 - 1))^{\frac{i-1}{i}} C n_2^{(n_1 - 1) - \frac{n_1 - 2}{i}} + z_{n_2} n_2 R(n_1 - 1, n_2) \leqslant$$

$$\leqslant \frac{n_2}{M} R(n_1 - 1, n_2)(1 + o(1)) \quad \text{as} \quad n_2 \to \infty$$

and the theorem is established.

4. CONCLUDING REMARKS

It is surprising that the local connectedness properties we studied here in Section 2 should give global results. This seems to suggest that a deeper study of the structure of Ramsey graphs would significantly improve these results.

There is clearly no point in attempting to evaluate constants nor in improving the statement of Theorem 2 by more careful optimization of the linear programming problem.

A study of the constructions of Erdős [1] would be of interest to compare the connectedness of his graphs with the results of this paper. This study has not yet been done to my knowledge.

REFERENCES

[1] P. Erdős, Graph Theory and Probability II. *Canad. J. Math.*, 13 (1961), 346-352.

[2] P. Erdős, Some recent results on extremal problems in graph theory. *Theory of Graphs, Proc. Coll.*, Tihany, Hungary 1966, Academic Press, (1968), 83-98.

[3] P. Erdős, Unsolved Problems in Graph Theory and Combinatorial Analysis. *Combinatorial Mathematics and its applications,* Academic Press, (1971), 97-109.

[4] P. Erdős – G. Szekeres, A combinatorial problem in geometry. *Comp. Math.*, 2(1935), 463-470.

[5] J. Graver – J. Yackel, Some graph theoretic results associated with Ramsey's theorem. *J. Combinatorial Theory*, 4 (1968), 125-175.

[6] J. Yackel, Inequalities and Asymptotic Bounds for Ramsey Numbers. *J. Combinatorial Theory*, 13 (1972), 56-68.

J.E. Baumgartner

An uncountable order type φ is a *Specker type* provided that ω_1, $\omega_1^* \nleq \varphi$ and no uncountable subtype of φ is embeddable in the reals.

Problem 1. Is it true (i.e., is it provable in ZFC) that there exist Specker types φ, ψ such that $\varphi \nleq \psi$?

A partial order $(T, <_T)$ is a *tree* if $\{s\colon s <_T t\}$ is well-ordered by $<_T$ for every $t \in T$. Let $l(t)$ be the order type of $\{s\colon s <_T t\}$. An *Aronszajn tree* is a tree $(T, <_T)$ such that (1) for all $\alpha < \omega_1$, $\{t\colon l(t) = \alpha\}$ is countable, (2) for every $t \in T$ and every $\alpha < \omega_1$, if $\alpha > l(t)$ then there is $s >_T t$ such that $l(s) = \alpha$, and (3) $(T, <)$ has no uncountable chains. Two Aronszajn trees $(S, <_S)$ and $(T, <_T)$ are *almost-isomorphic* if there is a closed unbounded set $C \subseteq \omega_1$ such that $\{s \in S\colon l(s) \in C\}$ is isomorphic to $\{t \in T\colon l(t) \in C\}$ (under the appropriate reduced orderings).

Problem 2. Is it consistent with ZFC that all Aronszajn trees are almost-isomorphic?

D.E. Daykin

1. Let J be a family of distinct subsets X of cardinality c of the interval $S = \{1, 2, 3, \ldots, n\}$. Let δJ denote the family of sets of cardinality $c - 1$ which are subsets of members X of J. The Kruskal – Katona theorem says that the minimal size of δJ is attained when the sets X of J are, roughly speaking, all crowded up at one end of S. In many cases we want the corresponding result when the sets of J are uniformly distributed in S. So we ask for the minimum value of the size of δJ when:

a) Each element of S is in the same number of sets X of J.

b) The family J is invariant under a group of permutations of S.

* * *

2. Given integers $n \geqslant r \geqslant 1$ let X, Y be sets of n elements each and $B(n, r)$ be the set of all bipartite graphs $\Gamma: X \to Y$ of regular degree r. Further let $\mu(n, r)$ be the least integer μ with the property that for each graph Γ of $B(n, r)$ there is a subset W of μ members of X with $\Gamma W = Y$. What can be said about $\mu(n, r)$ and the extremal graphs?

I. Gutman

Let T_t $t = 0, 1, \ldots$ be trees with the same number N of vertices and let T_0 be the chain. The spectrum of T_t is $\{x_i^t \mid i = 1, \ldots, N\}$ and E_t and π_t are defined as:

$$E_t = \sum_{i=1}^{N} |x_i^t| \qquad \pi_t = \prod_{i=1}^{N} |x_i^t| .$$

Conjecture 1.

$$E_0 \geqslant E_t$$

For $N =$ even it can be shown that $\pi_t = 0$ or 1.

Conjecture 2.

$$\pi_t = 1 \quad \text{and} \quad \pi_s = 0 \Rightarrow E_t > E_s .$$

D.J. Kleitman

1. Erdős, Ko and Rado showed that if one considers a collection of N k-element sets that are subsets of an n set, with $k \leqslant \frac{n}{2}$; if each pair have non vanishing intersection, then

$$N \leqslant \binom{n-1}{k-1}.$$

Suppose one has a collection of k dimensional subspaces of a vector space of dimension n over a finite field. Hsieh has proven that analogue of the above result for $k < \frac{n}{2}$: if each pair of spaces intersect non trivially then $N \leqslant \begin{bmatrix} n-1 \\ k-1 \end{bmatrix}$ (Gaussian coefficient). Does this result hold for $2k = n$?

* * *

2. A directed graph is k strongly connected if upon removal of any $k-1$ vertices or edges it remains strongly connected. What conditions must a degree sequence satisfy to have a k strongly connected realization? Results for k-connectedness in undirected graphs are known.

* * *

3. Can one find an algorithm for finding a maximum matching in a graph in fewer than cN^3 steps? Hopcroft and Karp have found a method in finding such a matching in a partite graph in $cN^{5/2}$ steps.

J. Pelikán

Suppose $n \geqslant 5$, and $a_1, \ldots, a_n$ is a sequence with each $a_i = 0$ or 1. Let

$$A_k = \sum_{i=1}^{n-k} a_i a_{i+k} \qquad (k = 0, \ldots, n-1).$$

Conjecture. Not all A_k's are odd.

M.D. Plummer

By a well-known theorem of Tutte, every 4-connected planar graph has a Hamilton cycle.

Conjecture 1. Every 4-connected planar graph is Hamilton-connected (i.e. every pair of points are joined by a Hamilton path).

Donald Nelson has proved: Every 4-connected planar graph G is 1-Hamiltonian (i.e. $G - v$ has a Hamilton cycle for each point v in G).

Conjecture 2. Every 4-connected planar graph is 2-Hamiltonian (i.e. $G - u - v$ has a Hamilton cycle for every pair of points u and v).

A graph G is called a Halin graph if G may be constructed as follows: Let T be any tree in which the inner points (i.e. non-endpoints) all have degree at least 3. Let C be a cycle drawn through the set of endpoints of T in such a way that $G = T \cup C$ is planar.

Example:

G:

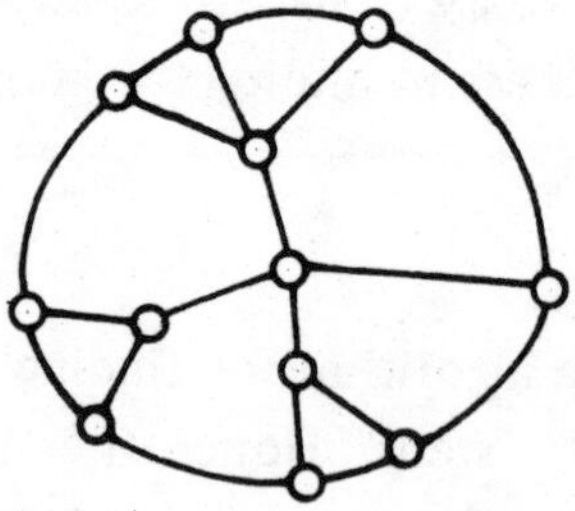

Conjecture 3. Every 4-connected planar triangulation contains a spanning Halin subgraph.

If this were true, then Whitney's theorem that says that such graphs must contain Hamiltonian cycles would follow immediately.

R. Radó

Let I be a set and $f(M)$ be a cardinal for $\phi \subset M \subseteq I$. Required are conditions on f in order that there should exist a family $(A_i\colon i \in I)$ of sets such that

Problem P_1: $\quad |\bigcap_{i \in M} A_i| = f(M)$ for every M,

P_2: $\quad |\bigcup_{i \in M} A_i| = f(M)$ for every M, $\phi \subset M \subseteq I$.

Both P_1 and P_2 are solved if either (i) I is finite or (ii) $f(M)$ is finite for every M.

I. Ruzsa

Can the (open or closed) circle be split into three congruent sets? It is well-known, that an open or closed interval is not the union of $1 < n < \aleph_0$ disjoint congruent sets, and I can prove it for $n = 2$ for any open or closed, bounded, convex region in R^n; the above question is the simplest unsolved case.

Conjecture. If an open or closed, bounded, one-connected region is the union of disjoint congruent sets, then their number is 1 or continuum.

A. Schinzel

1. Let $N(n)$ be the maximal length of a sequence formed of n letters without immediate repetitions and without any subsequence of the form $a \dots b \dots a \dots b \dots a$. Decide whether or not $N(n) = O(n)$.

References. H. Davenport and A. Schinzel, On a combinatorial problem connected with linear differential equations, *Amer. J. Math.*, 87 (1965).

H. Davenport, On a combinatorial problem connected with linear differential equations II, *Acta Arith.*, 17 (1970).

D. Stanton and D. Roselle, On the combinatorial problem of Davenport and Schinzel, *Acta Arith.*, 17 (1970).

E. Szemerédi, On the combinatorial problem of Davenport and Schinzel, to appear in *Acta Arith.*, 25.

* * *

2. Let G be a primitive solvable permutation group and H a subgroup of G. If every $h \in H$ has a fixed point, does it follow that H a fixed point?

References. A. Schinzel, On a theorem of Bauer and some of its applications II, *Acta Arith.*, 22 (1973).

G.J. Simmons

1. Given a distance d and two homothetic continuous and closed loci in E^n construct a graph G_d by connecting each point on either locus to the points at distance d from it on the other locus.

a. Show that if the loci are circles and if G_d is 2-regular and has every point in either circle as a vertex, then G_d has a finite component if and only if all components are finite.

b. Is it true that if G_d is 2-regular and has every point in either loci as a vertex and all components are finite, then the loci must be circles.

Note. Continuity of the loci is essential to part b. of the problem as the following construction by J.A. Davis shows. Let the loci be two collinear closed unit intervals spaced two units apart, and let $d = \sqrt{5}$. All components of G_d are rhombuses of side $\sqrt{5}$:

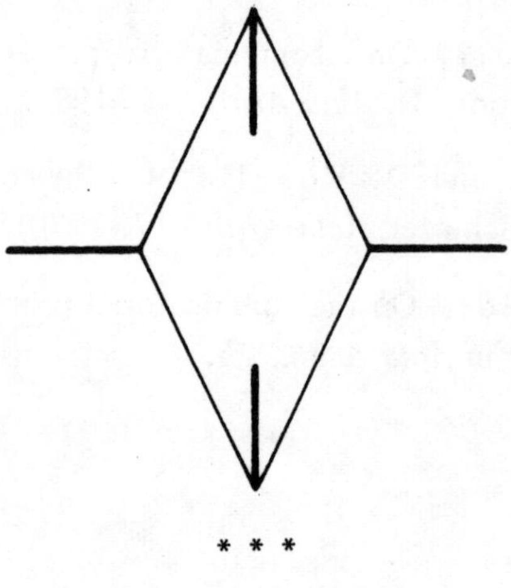

* * *

2. Construct a graph G by connecting each point on the surface of a unit radius sphere to all other points on the surface of the sphere at distance $\sqrt{3}$ from it. What is the chromatic number, $\chi(G)$, of G.

Note. G. Simmons has shown [1] that $3 < \chi(G)$ and E. Straus that $\chi(G) \leqslant 5$.

[1] G. Simmons, On a problem of Erdős concerning a 3-coloring of the unit sphere, to appear in *Discrete Mathematics.*

* * *

3. The well-known problem of tiling the chess-board, which has two diagonally opposite corners removed, using 1×2 dominoes demonstrates the necessity for a tileable configuration to have equally many squares of each color. The following construction shows the insufficiency of this condition

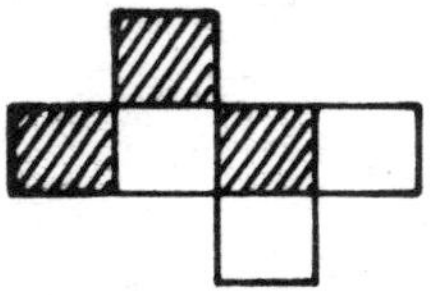

Our problem concerns the number of squares in an edge connected square configuration which must be examined to determine whether the configuration is tileable with 1×2 dominoes. This number can be as few as 3, irrespective of the number of squares in the configuration

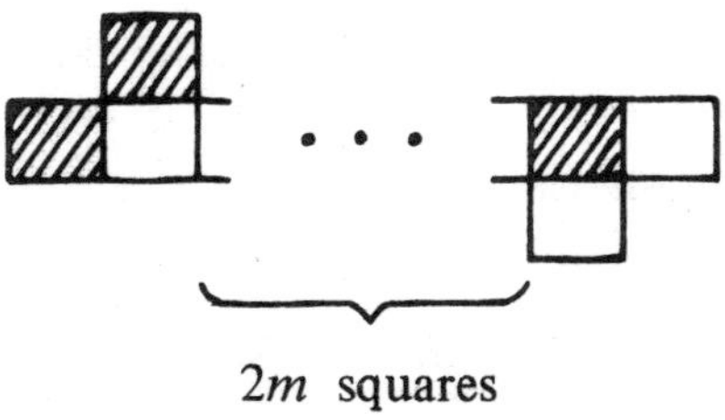

but may require the examination of asymptotically as many as $n(1 - \epsilon)$ squares when the configuration contains $2n$ squares, as the following construction shows:

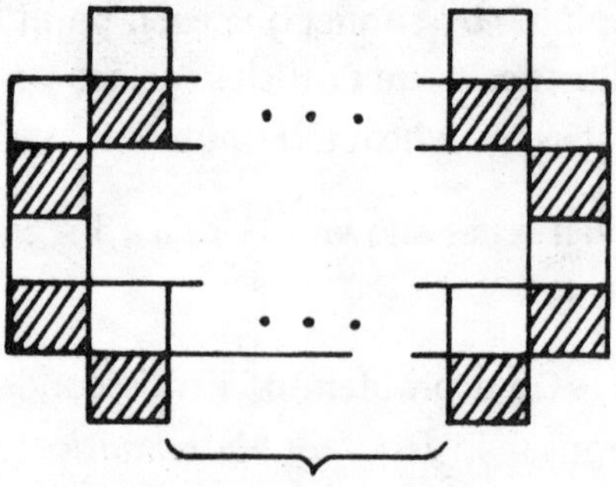

2m − 1 squares

We conjecture that the maximum number of squares which must be considered to decide whether a particular configuration on $2n$ squares is tileable is strictly less than n. Can this be proven?

If the conjecture is false, are there configurations requiring the examination of asymptotically all of the squares to determine tileability – in other words is tileability essentially a local or a global property?

J.M.S. Simões-Pereira

Given a set of arbitrary numbers (we may require integers, real, positive) $a_1 \leqslant a_2 \leqslant \ldots \leqslant a_m$, where $m = n^2$; are there any algorithms to find an $n \times n$ matrix A such that $\det(A)$ is maximum among all $n \times n$ matrices formed with the numbers a_i, each one appearing exactly once?

For $n = 2$, the answer is trivial. Taking positive real numbers $a_1 \leqslant a_2 \leqslant a_3 \leqslant a_4$ the matrix is $A = \begin{pmatrix} a_1 & a_3 \\ a_2 & a_4 \end{pmatrix}$.

(See Hardy, Littlewood, Pólya, *Inequalities.*)

J. Spencer

Define $g = g(n)$ as the least integer so that if $S_1, \ldots, S_n \subseteq X$ and $|S_i| \geqslant \frac{1}{2}|X|$ for $1 \leqslant i \leqslant n$, then there exists A, $A \cap S_i \neq \phi$ for $1 \leqslant i \leqslant n$, $|A| \leqslant g$.

One may easily show $g(2x) \leqslant 1 + g(x)$. Since $g(4) = 2$, $g(2^n) \leqslant n$ for $n \geqslant 2$. By nonconstructive means I can show

$$g(2^n) \geqslant n - c \log n .$$

(?) Is $\lim_{n \to \infty} n - g(2^n) = \infty$?